Biogenic Nanomaterials
Synthesis, Characterization, Applications, and Future Remarks

by

Fatih ŞEN

Sen Research Group, Department of Biochemistry, Kutahya Dumlupinar University, Turkey

Published by **Materials Research Forum LLC**
Millersville, PA 17551, USA

Published as part of the book series
Materials Research Foundations
Volume 180 (2025)
ISSN 2471-8890 (Print)
ISSN 2471-8904 (Online)

Print ISBN 978-1-64490-374-2
eBook ISBN 978-1-64490-375-9

This book contains information obtained from authentic and highly regarded sources. Reasonable efforts have been made to publish reliable data and information, but the author and publisher cannot assume responsibility for the validity of all materials or the consequences of their use. The authors and publishers have attempted to trace the copyright holders of all material reproduced in this publication and apologize to copyright holders if permission to publish in this form has not been obtained. If any copyright material has not been acknowledged please write and let us know so we may rectify this in any future reprints.

Distributed worldwide by

Materials Research Forum LLC
105 Springdale Lane
Millersville, PA 17551
USA
https://mrforum.com

Manufactured in the United States of America
10 9 8 7 6 5 4 3 2 1

Table of Contents

Preface

It is a distinct honor to introduce this book, which consolidates state-of-the-art research and emerging perspectives in the dynamic and transformative field of nanomaterials. In recent years, nanotechnology has become a cornerstone of modern science, influencing disciplines as diverse as chemistry, materials science, biology, medicine, and environmental engineering. The development of novel nanomaterials—whether for energy conversion, environmental remediation, biomedical applications, or advanced sensing technologies—continues to reshape our understanding of what is scientifically and technologically possible.

This volume brings together the collective expertise of researchers from around the world, offering a comprehensive overview of synthesis methods, characterization techniques, and cutting-edge applications. The chapters not only provide a foundation for understanding the fundamental principles of nanomaterials but also highlight innovative approaches and breakthroughs that pave the way for future discoveries.

One of the core aims of this book is to bridge the gap between fundamental research and real-world applications, inspiring readers to envision how nanomaterials can address some of today's most pressing global challenges. It is my hope that this collection will serve not only as a valuable reference for established scholars and industry professionals but also as an inspiration for young researchers and students who are entering this exciting field.

I would like to express my heartfelt gratitude to all contributing authors for their exceptional efforts, expertise, and dedication. Their valuable contributions have made it possible to produce a resource that reflects the diversity and depth of current research in nanoscience. I am equally grateful to the editorial team for their invaluable support and guidance throughout this process.

May this book spark curiosity, foster collaboration, and catalyze the next wave of innovations in the fascinating world of nanomaterials.

Editor

Prof. Dr. Fatih ŞEN

Biogenic Nanomaterials: Synthesis, Characterization, Applications, and Future Remarks   Materials Research Forum LLC
Materials Research Foundations 180 (2025)   https://doi.org/21741/9781644903759

Chapter 1

History Introduction and Physicochemical Properties of Biogenic Nanomaterials

H. Elik[1*], A. R. Kul[1], S. Kaptanoğlu[2], M. S. Erdoğan[3], F. Sen[4*]

[1]Department of Chemistry, Faculty of Science, Yuzuncu Yıl University, Van, Turkiye

[2] Vocational School of Health Services, Yuzuncu Yıl University, Van, Turkiye

[3]Altıntaş Vocational School, Kutahya Dumlupınar University, Kutahya, Turkiye

[4]Sen Research Group, Department of Biochemistry, Kutahya Dumlupinar University, Kutahya, Turkiye

elikhasan866@gmail.com; fatih.sen@dpu.edu.tr

Abstract

Biogenic nanoparticles have emerged as a sustainable option in nanotechnology that utilizes biological systems for synthesis. This chapter explores the historical development of biogenic nanoparticles, emphasizing biosynthesis by microorganisms via extracellular and intracellular pathways. Primary synthesis parameters and the chemical-physical properties of biogenic nanomaterials are discussed. The classification of nanoparticles based on chemical composition, dimensions, and material types is presented. Key application areas such as health, electronics, agriculture, textiles, energy, and environmental sectors, are reviewed. Additionally, natural polymer nanoparticles and future prospects of nanotechnology are mentioned. The chapter concludes with insights into challenges and potential advancements in this field.

Keywords

Biogenic Nanoparticles, Biosynthesis, Physical-Chemical Properties of Nanoparticles, Nanoparticle Classification, Nanotechnology Applications, Future Nanomaterials

1.   Introduction

1.1   History of Biogenic Nanoparticles

The Rise of Nanotechnology In 1959, After Dr. Feynman's speech stating that the tools and equipment used at that time could be produced at the nanoscale and molecular level, the development of nanotechnology accelerated worldwide. Although Doctor Feynman is an important name in the emergence of nanotechnology, other scientists around the world have also contributed to the development of nanotechnology. For this reason, in recent years, advanced research has been carried out on atomic level nanoparticles and biogenic nanoparticles. One of these is hygienic nanoparticles for cleaning purposes. Thanks to the extremely valuable nanotechnology, which attracts great attention today and is researched by many advanced scientists, a new era will begin in improving the quality of life of future generations [1]. Another reason for the increasing interest in the field of nanotechnology is the formation of diverse and complete molecular structures suitable for the purpose of use by structuring atoms, anions, and cations with unusual properties and functions. Unlike the volumetric structures of substances at the nanometer scale, molecular arrangements are formed at the nanometer scale. It is thought that this new technology, which has measurable dimensions, can be developed much more thanks to its technological advantages. Quantum chemistry and quantum effects are expressed when nanometer-sized particles are examined in relation to the known physical and chemical properties of substances. New concepts and properties that differ depending on the size of the material are obtained. The field of nanotechnology, which includes the synthesis of nanostructured materials, is the production of nanoparticles [2].

2.   Synthesis of Biogenic Nanoparticles

Synthesis of nanostructured materials, which is the backbone of nanotechnology; It involves synthesizing nanoscale materials of different shapes and sizes for applicability in different fields including chemistry, electronics, medicine and energy. Nanoparticles are synthesized by various methods, including bottom-up and top-down. The bottom-up or constructive method is the deposition of material from atoms to clusters and nanoparticles. Spinning, sol-gel, chemical vapor deposition (CVD), biosynthesis and pyrolysis are the most commonly used bottom-up methods for nanoparticle production [3].

Top-down or destructive method is the reduction of a bulk material into nanometric scale particles. Mechanical milling, laser ablation, exfoliation, nanolithography, sputtering and thermal decomposition are some of the commonly used nanoparticle synthesis methods [3].

Biosynthesis of nanomaterials is a tramendous and increasingly intensive research area due to its applications in various fields, especially biomedical and pharmaceutical fields. Green nanotechnology, that biologically mediated synthesis of nanoparticles is an arising field to design new NPs. Biological methods of NP synthesis provide a new opportunity to synthesize NPs using natural stabilizing and reducing agents. It is an environmentally friendly and economical alternative to physical and chemical methods that does not require the use of energy and toxic chemicals [4]. Biosynthesized nanomaterials have been successfully applied in big variety of fields, such as nanomaterials for biosensors, antibacterial nanomaterials, drug delivery, and cancer therapy [5]. Biological synthesis of nanomaterials varies depending on the type of biosystems and biomaterials utilized in their synthesis. The threat that bacterial infection poses to human health is a worldwide hazard due to the economic and social burden it brings. Furthermore, the urgency of developing treatments for clinically important bacteria that are becoming increasingly resistant to antibiotics means that new bactericidal agents will need to be developed in the near future [6].

2.1   Biosynthesis of Nanoparticles From Microorganisms

To meet the exponentially growing need for nanostructured materials, biological methods for the synthesis of nanomaterials need to be explored; these methods will offer advantages over chemical and physical methods as a more cost-effective, simpler and environmentally friendly method, and will not require high pressures and temperatures or toxic chemicals. While nanomaterials are produced with conventional methods, organic solvents such as toluene, benzene and acetone, and toxic chemicals such as mercury metal salts that create a threat to the environment and human health can be used. The new trend of nanotechnology attempts to synthesize improved, nontoxic, and biocompatible nanomaterials with sustainable benefits using eco-friendly materials. Plants, bacteria and microorganisms that exist today are used to the greatest extent all over the world for the cost and eco friendly biosynthesis of nanoparticles [7]. Biological synthesis of NPs by microbial biomasses is more efficient than by plants because microbes produce more proteins, resulting in higher NP production and providing higher stability [5]. The fabrication of healthy nanoparticles has been selected as important new research worldwide due to the rapid growth and easy multiplication of microorganisms such as bacteria and fungi, stress factor and regulation of pH. Favorable conditions at the scale of bacteria and fungi microorganisms have the structural ability to synthesize nanoparticles through intracellular and extracellular pathways due to their ability to adapt to metal-containing environments. Microorganisms convert these metal ions into the main form with enzymatic activity (Li). Researchers have demonstrated the ability of some bacteria, which can reduce metal ions, to precipitate metals at the nanometer level. The main advantage of enzymatic activity is that bacteria-based synthesis of nanoparticles enables sustainable, large-scale production with minimal use of toxic chemicals, and fungi also have well-defined geometry and size. It has various intracellular and extracellular enzymes to produce monodisperse nanoparticles. The efficiency of nanoparticles in fungi is higher than in bacteria because they have higher biomass than others. Collitotrichum sp., Fusarium oxysporum, etc. Different fungal species have been reported to produce nanoparticles of different sizes and shapes [8].

Biosynthesis of metal sulfide nanoparticles from microorganisms generally requires metal and sulfur precursors in the form of soluble salts. During intracellular synthesis, these soluble ions tend to enter the cytoplasm of the cell via the manganese or magnesium transport chain and are converted into nanoparticles by the action of intracellular enzymes located in the cytoplasm. Nanoparticles obtained in this way can adhere to the cell membrane or remain in the environment. The formation of extracellular nanoparticles in microorganisms can take place under stress-free conditions, which may be caused by the presence of metal ions in the environment. It also makes the synthesis of nanoparticles easier, cheaper and on a larger scale [9].

Metal oxides have important physical and chemical properties and are multiple structural forms with metallic, insulating or semiconducting properties. Only a few studies have demonstrated the microbial synthesis of metal oxide nanoparticles [8]. Intracellular synthesis of nanoparticles by microorganisms. Additional processing stages such as sonication or reaction with appropriate detergents are required to release the synthesized nanoparticles into the cell [10].

2.1.1 *Extracellular Synthesis of Nanoparticles by Microorganisms*

Synthesis of metal nanoparticles by microbial means depends on the localization of reducing components into cells. When the cell wall secretes reductases or enzymes in soluble form outside the cell, these enzymes participate in the reduction of extracellular metal ions, resulting in the synthesis of extracellular metal nanoparticles [11]. The production of extracellular nanoparticles has broader applications in many different fields such as optoelectronics, electronics, bioimaging and intracellular accumulation detection techniques. [12].

2.1.2   Intracellular Biosynthesis of Bacterial Nanoparticles

It is known that some metals have a toxic effect on some bacteria. However, there are also bacteria that are capable of growing in the presence of toxic metals. These bacteria, which can grow in the presence of metals, have been reported to provide toxicity resistance to the cell by reducing the metals in question to less toxic nanoparticles within the cell. Synthesis occurs in the cytoplasm or cytosol [13].

2.2   Critical Parameters for the Biological Synthesis of Nanoparticles

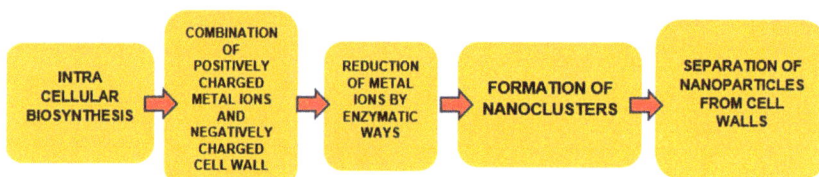

Figure 1. *Flow chart of intracellular synthesis of nanoparticles*

Nanoparticles formed inside microorganisms can be smaller compared to the size of nanoparticles synthesized in reduced form outside the cell. The reason for this may be related to the particles that form the nuclei within organisms. For example, Phoma PT35 is a type of fungus, this microorganism can only select silver atoms and accumulate these silver atoms. Therefore, Phoma PT35 is actually a suitable biosorbent for the preparation of silver nanoparticles [14].

Also, *Verticillium sp.* It has been reported that when fungal biomass is placed in a silver nitrate solution, it accumulates a significant amount of silver nanoparticles within its cells [15].

In many studies, the morphology and size of metal nanoparticles have been controlled by restricting their size in the medium or modifying their functional molecules [16][17]. For example, Ganoderma, the production of biocompatible 20 nm gold nanoparticles using this strain has been accomplished by optimizing various reaction conditions such as pH, temperature, incubation time, aeration, salt concentration, mixing ratio and aeration [11]. For the synthesis of nanoparticles using microorganisms, the synthesis is desirable at optimal temperatures because the enzymes in charge of nanoparticle synthesis are more active at higher temperatures. Another factor that influences culture is pH, and different nanoparticles can be formed at different pH values [16].

Among fungi, alkaline pH (*Isaria Futurarosea*), acidic pH (*Fusarium acuminatum*) and pH 6.0 (*Penicillium Fallutanum*) are ideal for the production of nanoparticles [18].

In addition, as reported in a study, *Aeromonas sp* found that dried cells of strain SH10 produced silver nanoparticles with a diameter of 6.4 nm, which was $[Ag(NH_3)_2]^+$ reduced to Ag^0 in 4 h. These particles are uniformly distributed and of uniform size and remain stationary for more than 6 months after accumulation and precipitation [19]. In another study of upstream culture of Enterobacter (*Klebsiella pneumonia*), the upstream culture of non-pathogenic *B. licheniformis* strain was used to produce extracellular silver nanoparticles with a size of 50 nm [20].

In recent years, researchers have shown considerable interest in the synthesis of bio-based nanoparticles from plant extracts. The reason is that the trees are removed; It is relatively inexpensive and allows mass production, does not require special storage conditions, poses no risk of contamination, and plant extracts are highly resistant to adverse conditions (such as high temperatures, high pH values) and salinity [21].

3.   Physical-Chemical Properties of Biogenic Nanomaterials

Its size, stability, geometry, etc. Many factors affect the potential of nanomaterials in biomedicine. Small nanoparticles easily enter the plasma membrane through absorption mechanisms, allowing their use in the biomedical field. Environmental concentration, pH, temperature, time, pressure, etc. The stability and geometry of nanoparticles can be controlled by arranging the physicochemical parameters involved in the synthesis procedures [22].

4.   Classification of Nanoparticles

Classification of nanoparticles depends on the application or research context in which they are used. There are a wide variety of criteria taken into consideration by researchers, such as origin, morphology, agglomeration, chemical components, dimensionality, toxicity and geometry. Before settling on just one of them, we briefly mention some important criteria that are frequently used.

4.1   Dimensional Classification of Nanoparticles

Nanoparticles are materials with general dimensions on the nanoscale, that is, below 100 nm. Pokropivny and Skorokhod created a new classification scheme for nanomaterials in 2007 that included newly developed 0D, 1D, 2D, and 3D composite nanostructured materials. Accordingly, the dimensional classification of nanoparticles is based on the fact that dimensions smaller than 100 nm are not considered as dimensions. In this scheme, different nanostructures are analyzed and classified considering the dimensions of the nanostructure itself and its components [23].

4.1.1   Zero-dimension (0D) Nanostructures

Zero-dimensional (0D) nanostructures include atomic clusters, filaments, clusters, and particles. All properties or dimensions of these materials are less than 100 nanometers, and their length is equal to their width. 0D spherical nanostructures also include cubes, nanorods, polygons, hollow spheres, and quantum dots (QDs) [24].

4.1.2   One-dimensional (1D) Nanostructures

One-dimensional (1D) nanostructures have two dimensions on the nanometer scale and their length is greater than their width. One-dimensional nanostructures consist of nanotubes, nanowires, nanorods, and nanofibers [24].

4.1.3   Two-dimensional (2D) Nanostructures

Two-dimensional (2D) nanostructures have only one nanoscale dimension. These structures include monolayers and multilayers, amorphous or crystalline thin films, nanosheets, and nanocoatings. 2D nanostructures have potential for new applications including sensors, electronics/optoelectronics and biomedicine [24].

4.1.4   Three-dimensional (3D) Nanostructures

Three-dimensional (3D) nanostructures have different sizes exceeding 100 nanometers. 3D nanostructures are formed by combining multiple nanocrystals in different directions. Multilayered, polycrystalline and fibrous materials, as well as some powders, fall into this category. Examples include carbon nanobuttons, nanotubes, fullerenes (also called as carbon 60), dendrimers, fibers, pillars, honeycombs, polycrystals, and layered skeletons [25].

4.2   Shape classification of Nanoparticles

In applications with nanoparticles, it may be desired that the synthesized nanoparticles be within a certain range in terms of size or shape. Nanoparticles; They can be in various shapes such as spiral, spherical, rod-like, triangular, cubic, hexagonal, prism-shaped, oval-shaped, tube-like [26].

4.3   Original Classification of Nanoparticles

In terms of origin, nanoparticles can be classified under three main headings.

4.3.1   Natural Nanoparticles

As their names suggest, these nanoparticles exist naturally in nature in ultrafine form without being subjected to any artificial processing. Their general origins in nature are volcanoes, sea waves, oil, forest fires, sand storms, crystalline structures, space, and living organisms [26].

4.3.2   Incidental Nanoparticles

Incidental nanoparticles are emitted as ultrafine particles as undesirable byproducts from processes controlled by humans such as food processing, cigarette smoke, engine exhaust, energy production, material sourcing, building demolition, etc. [26].

4.3.3   Engineered Nanoparticles

Taking advantage of modern equipment, a wide range of chemical catalysts and new facile synthesis methods, nanoparticles are now being artificially engineered to enable the development of targeted properties suitable for specific functions [25].

4.4   Chemical Composition Classification of Nanoparticles

According to chemical composition, nanoparticles can be classified into three categories basicly as organic, inorganic and carbon-based NPs.

4.4.1   Organic Nanoparticles

Liposomes, alginates, dextran, ferritin, micelles and dendrimers etc. are commonly known as organic nanoparticles or polymers. Organic nanoparticles are most commonly used in the biomedical field, for example in drug delivery systems, because they are non-toxic, effective and can be injected into specific areas of the body (also known as targeted drug delivery) [3].

4.4.2   Inorganic Nanoparticles

Metal and metal oxide-based NPs and NPs that are not made up of carbon atoms are commonly considered as inorganic NPs.

NPs which are synthesised from metals to nanoscales either by top-down or bottom-up techniques are metal based nanoparticles. Metals commonly used in nanoparticle synthesis are platinum (Pt), gold (Au), silver (Ag), cobalt (Co), Nickel (Ni), aluminum (Al), lead (Pb), cadmium (Cd), palladium (Pd), ruthenium (Ru), copper (Cu), iron (Fe), and zinc (Zn) [3].

Metal oxide nanoparticles are synthesized fundamentally thanks to their improved efficiency and reactivity. For example, Iron (Fe) nanoparticles are instantly oxidized to iron oxide (Fe_2O_3) in the presence of oxygen at room temperature, which increases its reactivity compared to iron nanoparticles [3].

4.4.3   Carbon based Nanoparticles

These carbon based NPs can be classified as graphene, fullerenes, carbon nanotubes (CNTs), carbon nanofibers and carbon black and nanosized activated carbon [3].

5.   Application Areas of Nanoparticles

Today, nanotechnology can be used in all fields where human beings exist. So much so that nanotechnology can be used in many areas to create products or applications that will increase the quality of life of living beings, facilitate work and provide economic advantages through the use of different and advantageous properties of nanoscale material.

5.1   Nanotechnology in Health

Nanotechnology is used in various applications in the health sector. It is based on the use of nanoparticles for applications such as biosensors for detecting biomolecules with high sensitivity and reliability in human serum, drug and gene delivery and active targeting, which play a role in delivering drugs to diseased tissues and organs. Nanoparticles allow them to travel through blood capillaries, entering diseased organs and tissues thanks to their nanoscale size [27]. The small size of nanoparticles is not the main reason for their functionality. The important and unique properties of nanoparticles used for medical purposes, such as their large surface area, their quantum properties and their ability to adsorb and transport other compounds, have caused them to become the focus of attention in medical applications [28].

Biosensors in medicine based on nanotechnology can detect biological molecules with biocatalysts that improve catalytic activity mostly by enhancing the electron carrier capacity and increasing the electroactive surface area of the sensor.

Another area where the use of nanotechnology in medicine is recently and rapidly becoming widespread is drug delivery. Various nanoscale materials are currently being investigated for drug delivery and, more specifically, cancer treatment. In researching nano-biotechnologies in drug delivery; More specific drug targeting and delivery, reduction of toxicity while achieving therapeutic effects, greater safety and biocompatibility and rapid development of new safe drugs are the main goals [28].

NPs can be utilized in many medical imaging applications such as optical imaging, magnetic resonance imaging (MRI) and contrast enhancement for various purposes, biological imaging, tissue repair and immunological testing. In addition to these applications, the unique properties of nanometer size are also used to achieve effective contrast in photothermal therapeutic applications [29].

Conventional optical imaging of cell and tissue parts is carried out by staining the sample with organic dyes. Insufficient fluorescence intensity and photobleaching are two problems frequently encountered in this optical imaging mode. Quantum dots (QDs) are nanoparticles made up of semiconductor inorganic molecules. Under ultraviolet (UV) illumination, these nanoparticles emit strong fluorescent light, and the wavelength (color) of the emitted fluorescent light depends precisely on the particle size. This size dependence is a unique feature of these nanomaterials. Compared to conventional organic dyes, QDs can emit light that is much more intense and significantly more stable against photobleaching [30].

5.2   Nanotechnology in Electric/electronics

Optimizing nanoelectronics through advanced storage and interconnection technologies helps increase electronic performance. By using low-energy materials such as graphene, topological insulators and superconductors, we can talk about energy-efficient nanoelectronics. Different combinations of nanomaterials and nanoelectronics have led to advances in many applications, such as medicine, biotechnology and energy [25]. Molecular and semiconductor electronics, nanotubes and nanowires are accepted as components of nanoelectronics. The miniaturization of electronics thanks to nanodimensional processing plays a significant role in increasing the performance and speed of information exchange. Another benefit of nanoelectronics is the manufacturing of more advanced memory devices and computer processors [31].

In addition, NPs is utilized as additives in organic electronics for enhancing device performance, modifying dielectric properties, improving semiconductor coverage in the charge transport channel, improving the charge-carrier mobility of organic thin film transistors (OTFTs), organic transistor fabrication, enhancing the electrical property of semiconductor polymers etc. [32].

5.3   Nanotechnology in Food and Agriculture

Nanofood is the use of nanotechnology in food processing, production and packaging. Furthermore, nanofoods allow development of new applications and methods that have positive effects on the structure, texture and quality of foods through the use of nanotechnology techniques [33]. Nanotechnology is used in many ways in the food industry. Nanotechnology allows us to design and control foods with desired properties at the molecular level. Nanotechnology has ensured a platform to figure out how the physiognomy of nanomaterials can dramatically alter the appearance, taste, smell and feel of foods while increasing their nutritional value. There are also applications for the creation of new functional products by adding nanostructures of different colors, flavors and nutritional contents. Nanoprocessed foods are generally less susceptible to premature decay than fresh foods and are therefore better suited to long-distance transportation from source to consumer. [34].

Nanotechnology is also used in areas such as smart and healthy packaging of food products with nanostructures used to keep them safe and extend their shelf life and ensure that food reaches the consumer in a healthy manner through nanosensors developed to detect various pathogens, toxic molecules and harmful levels of additives in food. In summary, nanotechnology in food packaging provides antimicrobial functioning, desired thermal, mechanical and optical properties, aroma, gas and vapour barrier, environmentally friendly packaging materials [34][35].

Despite the desired properties and significant advantages of nanoparticles in the dairy and food industries, there is intense public concern about environmental impacts and the toxicity of these nanomaterials. Additionally, while the potential effects of nanoparticles in the gastrointestinal tract are largely unknown, scientific evidence suggests that free engineered nanoparticles can cross cellular barriers and lead to increased production of oxyradicals that cause oxidative damage to cells [36].

In agriculture, nanotechnology is used to increase food production with safety, quality and equivalent or higher nutritional value. Efficient use of pesticides, herbicides, fertilizers and plant growth regulators are the most important ways to increase crop production. Certain nanomaterials can act as pesticides on their own with reduced toxicity and improved sensitivity. Metal oxide nanomaterials are widely studied to protect plant from pathogen infections owing to their intrinsic toxicity [37].

5.4   Nanotechnology in the Textile Sector

The nanotechnology applications in the textile materials can be called nanotextiles. Nanotechnology makes it possible to modify the surface properties of the fabric by adding nanostructures (nanoparticles, nanocomposites, etc.) to the existing properties of the fibers, threads or fabrics used in textile products. Through nanotextile applications, textile products can be produced with characteristics such as hydrophobicity, water resistance, resistance to dirt, absence of fabric wrinkles, antimicrobial effect, flammability, antistatic behavior, air permeability and retention of the applied dye [38] [39].

5.5   Nanotechnology in the Energy Sector

Energy conversion and transport in nanomaterials is very different from that in bulk materials due to classical size and quantum size effects on various energy carriers, such as electrons, phonons, and photons. Nanotechnology for energy applications focuses on adapting nanoscale effects to clean and efficient energy technologies, such as photovoltaics, photochemical solar cells, thermoelectrics, fuel cells, and solar cells [30].

Also, in battery related devices like lithium ion cells, hydrogen storages and fuel cells efficiency gains can be achieved both at the anode and cathode side by utilizing the nanoscale materials [40].

5.6   Nanotechnology for Environment

Environmental pollution caused by anthropogenic activities is becoming a serious problem nowadays. The photocatalytic oxidation method, which uses semiconductors as catalysts for environmental applications such as air purification, water cleaning and soil improvement to eliminate environmental pollution, emerged as an environmentally friendly candidate. Photocatalysis has been used to degrade several toxic organic compounds, pesticides, dyes, harmful gases and bacteria in different environments [41].

5.7   Nanotechnology in the Materials and Manufacturing Sector

Nanotechnology offers many advantages in terms of building and construction materials development. Some of these advantages are the following. • Nanoparticles, carbon nanotubes and carbon nanofibers contribute to the wear and durability of structures [42], • Production of steel that does not corrode due to oxidation or other chemical effects with economical methods [43], • Production of thermal insulation materials with greater efficiency [44], • Ability to detect and self-heal using nanosensors and other materials [45]. Nanotechnology is used in applications such as materials production.

When materials are operated at the nanoscale, their characteristics change at the nanoscale and exhibit enormous electrical, optical and magnetic characteristics. These unique nanoproperties are used in batteries, solar cells, electronics, fuel additives, catalysts, electrochemical industries, defense, cosmetics, pharmaceuticals, food additives and packaging, diagnostic imaging, agriculture, biosensors, chemotherapy, antimicrobials and vaccines [29].

The fields of application of biological nanotechnologies are quite broad and continue to develop. Nanotechnology, which is a multidisciplinary science, has begun to gain importance in the field of medicine and is being used in fields such as biology, physics, chemistry, computer science, electronics and materials science. Furthermore, it also has applications in cosmetics, agricultural and medical fields.

6.   Natural Polymer Nanoparticles

Polymers commonly are derived into two classes as synthetic and natural polymers. Synthetic polymers are widely used in biomedical implants and devices because they can be manufactured in a variety of ways. Natural polymers are basically polysaccharides, so they are better biocompatible than synthetic ones and do not have any side effects in drug delivery systems [46].

6.1   Dextran

Dextran is a complex and branched polysaccharide composed of numerous α-glucose units, made up of chains of different lengths. These types of polysaccharides are made up of glucose, a simple saccharide monomer, and are stored as an energy source in bacteria and yeast. Besides being a natural product, dextrans are used in many fields due to their superior wetting properties, excellent biocompatibility and clinical safety [47]. It has a wide usage area, particularly in the food industry, as a thickener, stabilizer and emulsifier in food applications. It is easily resolved in formamide, methyl sulfoxide, ethylene glycol, glycerol and water. However, some fractions of dextran can dissolve only under a strong thermal effect. Thanks to its most important feature being a neutral polysaccharide, dextran has a wide scope of applications [47][48].

6.2   Alginates

Alginate, a brown alga originated polysaccharide, is also a natural copolymer of mannuronic and glucuronic acid. Like chitosan, it is commonly used in food products, pharmaceutical applications and wound dressings due to its properties such as being easily processed in water, having very little toxicity and inflammation, being biodegradable, controlling its porosity and binding to biologically active molecules. [47][49][48]. In drug delivery applications, the simple formation of an aqueous sodium alginate gel in the presence of divalent cations such as calcium (Ca^{2+}) is preferred. In these applications, a drug trapping capacity of up to 70-90% has been achieved. Since they can be obtained in sizes less than 100 nm, they are also intended for use in gene delivery studies. [47][50][49][48].

6.3   Dendrimers

Research in polymer chemistry and technology have traditionally focused on widely used linear (straight) polymers. Straight macromolecules can contain shorter or longer branches. Recently, it has been discovered that the properties of hyperbranched polymers are different from those of traditional polymers. Dendrimers are discovered firstly in the early 1980s by Donald Tomalia and colleagues, these hyperbranched molecules are called dendrimers. The word comes from the Greek word dendron, which means tree. Although it has had different names in the past, the most commonly used is "dendrimer" [51]. Dendrimers are well-defined, nanometer-sized, uniform and monodisperse structures, composed of woody branches. These hyperbranched molecules include branched units built around a flat polymeric core or small molecule symmetrically. The dendrimer is not a composite material, but simply an engineered architectural model. Its size, shape and flexibility evolve over generations. Thanks to the functionalized external groups, their physicochemical or biological properties change. Therefore, dendrimers have unique properties that make them key materials for very specific applications. The advantages of these well-defined materials make them a new class of nanoscale polymer delivery devices. The diameter of these dendritic macromolecules tends to increase linearly and become more spherical as dendrimer production increases. Therefore, its use in diagnostic imaging and cancer treatment has attracted attention [51].

6.4   Nanoparticles of the Future

Due to their fascinating properties, many nanoparticles have been widely studied. Metallic nanoparticles have attracted great attention in biomedical applications due to their numerous properties and potential [52]. Its binding performance can be increased with different types of biopolymers. Until now, as mentioned above, many attempts have been made to synthesize metallic nanoparticles coated with numerous biopolymers [53]. However, this issue is still not fully explained in the field of nanotechnology. More work is needed, such as identifying nanoparticles, modifying surfaces and identifying new production methods. This area is still in the early phases of its development and looks promising.

7.   Discussion and Conclusion

The nanoscale particles produced have many uses. There are different methods to obtain these particles. Methods using biological systems as environmentally friendly reducing agents attract great attention [54]. Being biocompatible and low cytotoxic is essential for nanoparticles involved in biomedical processes and food sector. When nanoparticles produced by biogenic techniques are compared to physicochemically produced nanoparticles, the toxic structures of unwanted formations that adhere to the nanostructures during physicochemical formation constitute a limitation for biomedical processes. The fact that they have practical and environmentally friendly production techniques, and the economical and biocompatible content of the particles obtained offer many advantages in the synthesis of nanoparticles with green nanotechnologies. Apart from this, nanoparticles produced by biological methods do not require different stabilizing agents due to their self-coating of living systems such as green plants and prokaryotes and their mobility system in the form of stabilization [55].

Nanoparticulate substances are becoming part of more products every day due to their industrial benefits. An adequate legal framework does not yet exist to regulate the use of these substances, their release into the environment or their effects on health. However, the number and diversity of studies examining the effects of nanoparticles on ecosystems and living things have not yet reached a sufficient level to give us a clear idea. Studies generally aim to examine the effects of nanoparticles on aerobic organisms. This study provides relevant information on the anaerobic degradation of sewage sludge as well as the properties and environmental fate of nanoparticulate substances. Furthermore, a very limited number of studies in literature examining the effects of nanoparticles on the anaerobic degradation process of sewage sludge were collected and the results achieved under diverse conditions were compared. Additionally, the acute and chronic effects of the Ag, Al, Ce, Cu, and Fe NPs specifically examined in this study were compared over a wide range of concentrations. Nanoparticles, whose use has increased enormously today, are a potentially toxic substance for the anaerobic degradation process, which has undeniable environmental and economic benefits. Further research is needed to determine the effects of nanoparticles, whose transport to treatment facilities will increase in the future, on the anaerobic process and the precautions that can be taken in this regard. In particular, focusing on determining the amount of nanoparticles arriving at wastewater treatment facilities and what form they are in when they arrive, and focusing on studies on the removal of nanoparticles from wastewater treatment systems will provide very important information in terms of environmental engineering. In recent decades, nanoparticles have attracted attention in recent years in terms of environmental engineering, and scientific studies have begun to be carried out on the characteristics and effects of these substances in various environments [56].

References

[1]   F. Gulbagca, A. Aygün, M. Gülcan, S. Ozdemir, S. Gonca, F. Şen, Green Synthesis of Palladium Nanoparticles: Preparation, Characterization, and Investigation of Antioxidant, Antimicrobial, Anticancer, and DNA Cleavage Activities, Appl. Organomet. Chem. 35 (2021). https://doi.org/10.1002/aoc.6272

[2]   A. Aygun, F. Gulbagca, E.E. Altuner, M. Bekmezci, T. Gur, H. Karimi-Maleh, F. Karimi, Y. Vasseghian, F. Sen, Highly active PdPt bimetallic nanoparticles synthesized by one-step bioreduction method: Characterizations, anticancer, antibacterial activities and evaluation of their catalytic effect for hydrogen generation, Int. J. Hydrogen Energy (2022). https://doi.org/10.1016/J.IJHYDENE.2021.12.144

[3]   S. Anu Mary Ealia, M.P. Saravanakumar, A review on the classification, characterisation, synthesis of nanoparticles and their application, IOP Conf. Ser. Mater. Sci. Eng. 263 (2017) 032019. https://doi.org/10.1088/1757-899X/263/3/032019

[4]   S. Patil, R. Chandrasekaran, Biogenic nanoparticles: a comprehensive perspective in synthesis, characterization, application and its challenges, J. Genet. Eng. Biotechnol. 2020 181 18 (2020) 1–23. https://doi.org/10.1186/S43141-020-00081-3

[5]   R.M. Tripathi, S.J. Chung, Biogenic nanomaterials: Synthesis, characterization, growth mechanism, and biomedical applications, J. Microbiol. Methods 157 (2019) 65–80. https://doi.org/10.1016/J.MIMET.2018.12.008

[6]   P. Singh, A. Garg, S. Pandit, V.R.S.S. Mokkapati, I. Mijakovic, Antimicrobial Effects of Biogenic Nanoparticles, Nanomater. (Basel, Switzerland) 8 (2018). https://doi.org/10.3390/NANO8121009

[7]   A. Fariq, T. Khan, A. Yasmin, Microbial synthesis of nanoparticles and their potential applications in biomedicine, J. Appl. Biomed. 15 (2017) 241–248.

[8]   J. Jeevanandam, Y.S. Chan, M.K. Danquah, Biosynthesis of metal and metal oxide nanoparticles, ChemBioEng Rev. 3 (2016) 55–67.

[9]   M.R. Hosseini, M.N. Sarvi, Recent achievements in the microbial synthesis of semiconductor metal sulfide nanoparticles, Mater. Sci. Semicond. Process. 40 (2015) 293–301.

[10]   F. Okafor, A. Janen, T. Kukhtareva, V. Edwards, M. Curley, Green synthesis of silver nanoparticles, their characterization, application and antibacterial activity, Int. J. Environ. Res. Public Health 10 (2013) 5221–5238.

[11]   K. Chitra, G. Annadurai, Antibacterial activity of pH-dependent biosynthesized silver nanoparticles against clinical pathogen, Biomed Res. Int. 2014 (2014).

[12]   B. Nair, T. Pradeep, Coalescence of nanoclusters and formation of submicron crystallites assisted by Lactobacillus strains, Cryst. Growth Des. 2 (2002) 293–298.

[13]   A.L. Campaña, A. Saragliadis, P. Mikheenko, D. Linke, Insights into the bacterial synthesis of

metal nanoparticles, Front. Nanotechnol. 5 (2023) 1216921.
https://doi.org/10.3389/FNANO.2023.1216921/PDF

[14]   K.B. Narayanan, N. Sakthivel, Biological synthesis of metal nanoparticles by microbes, Adv. Colloid Interface Sci. 156 (2010) 1–13. https://doi.org/10.1016/J.CIS.2010.02.001

[15]   M.A. Yassin, A.M. Elgorban, A.E.-R.M.A. El-Samawaty, B.M.A. Almunqedhi, Biosynthesis of silver nanoparticles using Penicillium verrucosum and analysis of their antifungal activity, Saudi J. Biol. Sci. 28 (2021) 2123–2127.

[16]   P. Singh, Y.J. Kim, C. Wang, R. Mathiyalagan, D.C. Yang, Microbial synthesis of flower-shaped gold nanoparticles, Artif. Cells, Nanomedicine, Biotechnol. 44 (2016) 1469–1474.

[17]   K. Kathiresan, S. Manivannan, M.A. Nabeel, B. Dhivya, Studies on silver nanoparticles synthesized by a marine fungus, Penicillium fellutanum isolated from coastal mangrove sediment, Colloids Surfaces B Biointerfaces 71 (2009) 133–137.

[18]   A.N. Banu, C. Balasubramanian, Optimization and synthesis of silver nanoparticles using Isaria fumosorosea against human vector mosquitoes, Parasitol. Res. 113 (2014) 3843–3851.
https://doi.org/10.1007/s00436-014-4052-0

[19]   H. Zhang, Q. Li, H. Wang, D. Sun, Y. Lu, N. He, Accumulation of Silver(I) Ion and Diamine Silver Complex by Aeromonas SH10 biomass, Appl. Biochem. Biotechnol. 143 (2007) 54–62.
https://doi.org/10.1007/s12010-007-8006-1

[20]   M.H. Siddique, B. Aslam, M. Imran, A. Ashraf, H. Nadeem, S. Hayat, M. Khurshid, M. Afzal, I.R. Malik, M. Shahzad, U. Qureshi, Z.U.H. Khan, S. Muzammil, Effect of Silver Nanoparticles on Biofilm Formation and EPS Production of Multidrug-Resistant Klebsiella pneumoniae, Biomed Res. Int. 2020 (2020) 1–9. https://doi.org/10.1155/2020/6398165

[21]   F. Duman, I. Ocsoy, F.O. Kup, Chamomile flower extract-directed CuO nanoparticle formation for its antioxidant and DNA cleavage properties, Mater. Sci. Eng. C 60 (2016) 333–338.
https://doi.org/10.1016/j.msec.2015.11.052

[22]   H. Barabadi, S. Honary, P. Ebrahimi, M.A. Mohammadi, A. Alizadeh, F. Naghibi, Microbial mediated preparation, characterization and optimization of gold nanoparticles, Brazilian J. Microbiol. 45 (2014) 1493–1501. https://doi.org/10.1590/S1517-83822014000400046

[23]   V.V. Pokropivny, V.V. Skorokhod, Classification of nanostructures by dimensionality and concept of surface forms engineering in nanomaterial science, Mater. Sci. Eng. C 27 (2007) 990–993.
https://doi.org/10.1016/j.msec.2006.09.023.

[24]   P.N. Sudha, K. Sangeetha, K. Vijayalakshmi, A. Barhoum, Nanomaterials history, classification, unique properties, production and market, in: Emerg. Appl. Nanoparticles Archit. Nanostructures, Elsevier, 2018: pp. 341–384. https://doi.org/10.1016/B978-0-323-51254-1.00012-9

[25]   T.A. Saleh, Nanomaterials: Classification, properties, and environmental toxicities, Environ. Technol. Innov. 20 (2020) 101067. https://doi.org/10.1016/J.ETI.2020.101067

[26]   S. Khan, M.K. Hossain, Classification and properties of nanoparticles, in: Nanoparticle-Based Polym. Compos., Elsevier, 2022: pp. 15–54. https://doi.org/10.1016/B978-0-12-824272-8.00009-9

[27]   A. Saharan, P. Mittal, K. Wilson, I. Verma, Introduction and Basics of Nanotechnology, in: Nanotechnology, Jenny Stanford Publishing, 2021: pp. 1–40.

[28]   de Jong, Drug delivery and nanoparticles: Applications and hazards, Int. J. Nanomedicine (2008) 133. https://doi.org/10.2147/IJN.S596

[29]   I. Khan, K. Saeed, I. Khan, Nanoparticles: Properties, applications and toxicities, Arab. J. Chem. 12 (2019) 908–931. https://doi.org/10.1016/J.ARABJC.2017.05.011

[30]   S.K. Murthy, Nanoparticles in modern medicine: state of the art and future challenges, Int. J. Nanomedicine 2 (2007) 129–141.

[31]   S. Kargozar, M. Mozafari, Nanotechnology and Nanomedicine: Start small, think big, Mater. Today Proc. 5 (2018) 15492–15500.

[32]   Z. He, Z. Zhang, S. Bi, Nanoparticles for organic electronics applications, Mater. Res. Express 7 (2020) 012004. https://doi.org/10.1088/2053-1591/ab636f

[33]   H. Bai, X. Liu, Food nanotechnology and nano food safety, in: 2015 IEEE Nanotechnol. Mater. Devices Conf., IEEE, 2015: pp. 1–4.

[34]   D. Kalita, S. Baruah, The Impact of Nanotechnology on Food, in: Nanomater. Appl. Environ. Matrices, Elsevier, 2019: pp. 369–379. https://doi.org/10.1016/B978-0-12-814829-7.00011-2

[35]   C. Catal, A. Aygun, R. Nour, E. Tiri, F. Sen, Green synthesis of nanoparticles the importance of use in food packaging : an overview, (2024) 13–25.

[36]   A. Patel, F. Patra, N. Shah, C. Khedkar, Application of Nanotechnology in the Food Industry: Present Status and Future Prospects, Impact Nanosci. Food Ind. (2018) 1–27. https://doi.org/10.1016/B978-0-12-811441-4.00001-7

[37]   X. He, H. Deng, H. Hwang, The current application of nanotechnology in food and agriculture, J. Food Drug Anal. 27 (2019) 1–21. https://doi.org/10.1016/j.jfda.2018.12.002

[38]   S. Chakrabarty, K. Jasuja, Applications of Nanomaterials in the Textile Industry, Nanoscale Eng. Biomater. Prop. Appl. (2022) 567–587. https://doi.org/10.1007/978-981-16-3667-7_20

[39]   A.K. Yetisen, H. Qu, A. Manbachi, H. Butt, M.R. Dokmeci, J.P. Hinestroza, M. Skorobogatiy, A. Khademhosseini, S.H. Yun, Nanotechnology in Textiles, ACS Nano 10 (2016) 3042–3068. https://doi.org/10.1021/acsnano.5b08176

[40]   I. Matsui, Nanoparticles for electronic device applications: a brief review, J. Chem. Eng. Japan 38 (2005) 535–546

[41]   C. Han, J. Andersen, S.C. Pillai, R. Fagan, P. Falaras, J.A. Byrne, P.S.M. Dunlop, H. Choi, W. Jiang, K. O'Shea, Chapter green nanotechnology: development of nanomaterials for environmental and energy applications, in: Sustain. Nanotechnol. Environ. Adv. Achiev., ACS Publications, 2013: pp. 201–229.

[42]   Z. V Pisarenko, L.A. Ivanov, Q. Wang, Nanotechnology in construction: State of the art and future trends, Nanotekhnologii v Stroit. 12 (2020) 223–231.

[43]   C.I. Idumah, C.M. Obele, E.O. Emmanuel, A. Hassan, Recently emerging nanotechnological advancements in polymer nanocomposite coatings for anti-corrosion, anti-fouling and self-healing, Surfaces and Interfaces 21 (2020) 100734.

[44]   B. Wicklein, A. Kocjan, G. Salazar-Alvarez, F. Carosio, G. Camino, M. Antonietti, L. Bergström, Thermally insulating and fire-retardant lightweight anisotropic foams based on nanocellulose and graphene oxide, Nat. Nanotechnol. 10 (2015) 277–283.

[45]   A. Thakur, S. Kaya, A. Kumar, Recent innovations in nano container-based self-healing coatings in the construction industry, Curr. Nanosci. 18 (2022) 203–216.

[46]   S. Bhatia, Natural Polymers vs Synthetic Polymer, in: Nat. Polym. Drug Deliv. Syst., Springer International Publishing, Cham, 2016: pp. 95–118. https://doi.org/10.1007/978-3-319-41129-3_3.

[47]   A. Martínez, A. Fernández, E. Pérez, M. Benito, J.M. Teijón, M.D. Blanco, Polysaccharide-based nanoparticles for controlled release formulations, Deliv. Nanoparticles (2012) 185–222.

[48]   J. Yang, S. Han, H. Zheng, H. Dong, J. Liu, Preparation and application of micro/nanoparticles based on natural polysaccharides, Carbohydr. Polym. 123 (2015) 53–66.

[49]   Y.D. Reddy, A brief review on polymeric nanoparticles for drug delivery and targeting, J. Med. Pharm. Innov. 2 (2015).

[50]   D. Hudson, A. Margaritis, Biopolymer nanoparticle production for controlled release of biopharmaceuticals, Crit. Rev. Biotechnol. 34 (2014) 161–179.

[51]   B. Klajnert, M. Bryszewska, Dendrimers: properties and applications., Acta Biochim. Pol. 48 (2001) 199–208.

[52]   K. Nesrin, C. Yusuf, K. Ahmet, S.B. Ali, N.A. Muhammad, S. Suna, Ş. Fatih, Biogenic silver nanoparticles synthesized from Rhododendron ponticum and their antibacterial, antibiofilm and cytotoxic activities, J. Pharm. Biomed. Anal. 179 (2020) 112993. https://doi.org/10.1016/J.JPBA.2019.112993

[53]   I. Meydan, A. Aygun, R.N.E. Tiri, T. Gur, Y. Kocak, H. Seckin, F. Sen, Chitosan/PVA-supported silver nanoparticles for azo dyes removal: fabrication, characterization, and assessment of antioxidant activity, Environ. Sci. Adv. 3 (2024) 28–35. https://doi.org/10.1039/D3VA00224A

[54]   A. Baran, Eco-friendly, rapid synthesis of silver nanomaterials and their use for biomedical applications, Dicle Univ. J. Eng. 12 (2021) 329–336.

[55]   A. Hojjati-Najafabadi, A. Aygun, R.N.E. Tiri, F. Gulbagca, M.I. Lounissaa, P. Feng, F. Karimi, F. Sen, Bacillus thuringiensis Based Ruthenium/Nickel Co-Doped Zinc as a Green Nanocatalyst: Enhanced Photocatalytic Activity, Mechanism, and Efficient H2Production from Sodium Borohydride Methanolysis, Ind. Eng. Chem. Res. 62 (2023) 4655–4664. https://doi.org/10.1021/ACS.IECR.2C03833

[56]   E. Kokdemir Ünşar, A. Perendeci, Environmental fate of nanoparticles and their impacts on anaerobic digestion process, Pamukkale Univ. J. Eng. Sci. 22 (2016) 503–512. https://doi.org/10.5505/PAJES.2015.71354

Chapter 2

Basic Concepts of Biogenic Nanomaterials

A. Aygun[1], F. Sen[1]*

[1]Sen Research Group, Biochemistry Department, Faculty of Arts and Science, Kutahya Dumlupinar University, Evliya Celebi Campus, 43100, Kutahya, Turkiye

fatih.sen@dpu.edu.tr

Abstract

Since nanoparticles' physical and chemical synthesis have been used for many years, they are referred to as traditional nanoparticle synthesis methods. However, conventional methods have some limitations regarding sustainability and biotechnological applications. Particularly, the toxicity problem limits the use of nanoparticles and prevents their use in many biotechnological applications, especially clinical applications. Due to all these limitations, the tendency towards biological resources has increased in recent years to obtain nanoparticles with broader applications. This nanotechnological approach, which is in line with the goal of sustainability and is called "green nanotechnology", aims to synthesize nanomaterials with biological organisms or their metabolic products.

Keywords

Biogenic Synthesis, Characterization, Green Nanotechnology, Green Synthesis, Nanomaterials

1. Introduction

In recent years, research into nano-sized materials has made great progress. For this reason, nanotechnology is one of the rapidly developing fields of research [1–3]. Nanotechnology is the branch of science that deals with the production of functional materials, devices, and systems at atomic and molecular sizes (1-100 nm), which have very different optical, electrical, mechanical, and magnetic properties compared to their larger dimensions [4,5]. Nanotechnology deals with fundamental physical and chemical issues such as the existence of new physical properties at the nanoscale, analysis at the atomic and molecular level, the control of substances at the atomic level, and the production of functional complex systems with qualitatively new properties. In summary, this branch of science is essentially concerned with the synthesis, characterization, research, and use of nanostructured materials [6,7]. Nanotechnology enables the creation and design of more precise systems through the miniaturization of existing technologies. Nanotechnological developments will enable the use of more durable, cleaner, safer, and smarter products in all areas of the agriculture, medicine, physics, chemistry, biotechnology, communication, home, transportation, and industry [8].

Some examples of applications of nanotechnology today and in the future are the development of sensors that can detect corrosion and stresses in bridges, drug delivery systems, cancer diagnosis, and treatment, wearable health monitoring devices, cleaning of wastewater, computers with large memory and long battery life, and send signals, food packages that will extend shelf life [9–11]. Although it has received a lot of attention recently, studies in the field of nanotechnology date back to an earlier date. The first mention of the term "nanometer" was by Richard Zsigmondy, who was awarded the Nobel Prize in Chemistry in 1925. Zsigmondy used the term nanometer to characterize particle size and was the first to measure the size of particles such as gold colloids with a microscope [12].

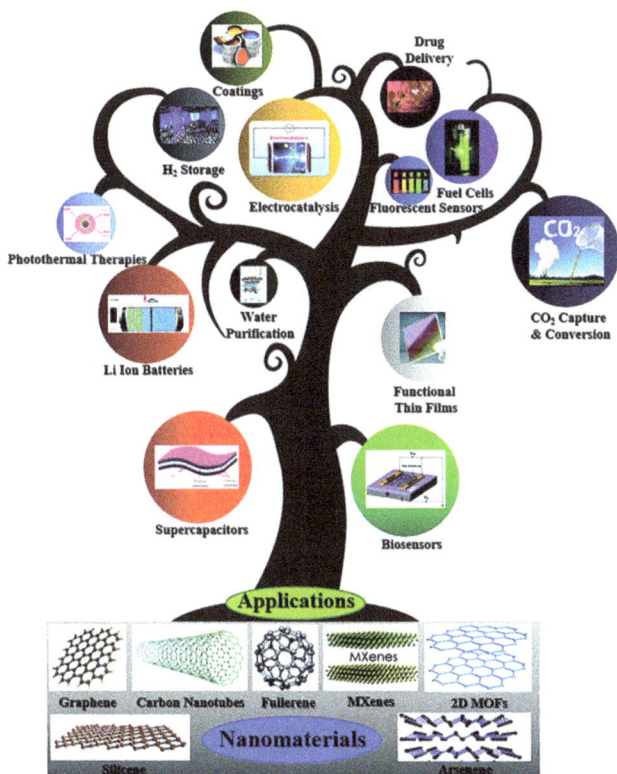

Figure 1. *Usage areas of nanoparticles. (Reprinted with permission from ref. [19]. Copyright: ©2021, Royal Society of Chemistry (RSC)).*

Basically, nanotechnology refers to the controllability of molecules and atomic components and is said to encompass product developments between 1-100 nm. This potentially allows scientists to create specific molecular structures and devices [13,14]. In general, nanotechnology is known as a branch of engineering that deals with structures smaller than 100 nanometers, particularly the structure of molecules. It is a science that reduces the size of matter and gives it new chemical and physical properties [15].

Nanotechnology products (nanoparticles, nanotubes, nanowires, etc.), which are synthesized using various methods, have the potential for interdisciplinary use depending on their physicochemical properties [16–18]. It is opening up new areas of application day by day, offering innovative solutions to many problems across a broad spectrum of both the industrial sector and the health sciences. Biomedical applications, applications in agriculture, food and textile industries, drug delivery systems, energy, sensor, environment, and antimicrobial agents in health sciences are the most common areas where nanotechnology is used (Fig. 1) [19–25].

2.   Nanoparticles and Types

The term "nano" means one billionth of a physical size. A nanometer is a unit of length corresponding to one billionth of a meter [26]. According to the definition by the International Organization for Standardization (ISO), particles with a diameter of less than 100 nm that have geometric, aerodynamic, mobile, or similar properties are referred to as nanoparticles [27,28]. Nanoparticles differ from the larger structure of the same substance in their chemical and physical properties and their material structure [29]. For these reasons, nanoparticles can have different properties compared to the main structure. The reason for these changes is that the periodic boundary conditions are violated when the particle sizes approach or are smaller than the wavelengths of the conduction electrons [30]. Particles have different structures such as spheres, flakes, sheets, tubes, and rods. In addition, the structures can become complex three-dimensional structures such as springs, rollers, and brushes through the manufacturing process [31].

3.   Nanoparticle Synthesis Methods

The production of nanoparticles is based on two different approaches: Bottom-up and top-down. In the top-down approach, the material to be used is excited from the outside by mechanical and/or chemical processes and thus comminuted to nanosize and separated. Examples of the top-down approach include techniques such as mechanical grinding and abrasion. In these processes, the materials are broken down into nanoparticles using much more energy. In the bottom-up approach, the particles are formed by the growth of atom- or molecule-sized structures through chemical reactions (Fig. 2). On the other hand, the methods for synthesizing nanoparticles are based on chemical, physical and biogenic processes, which have become very important in recent years [32–34].

3.1   Chemical Synthesis Methods

Chemical precursors or compounds are used to reduce metal salts for the chemical synthesis of nanomaterials. Organic or inorganic reducing agents such as sodium sodium citrate, N, N-dimethylformamide (DMF), borohydride (NaBH$_4$), ascorbate, Tollens reagent, elemental hydrogen, polyol process, and poly (ethylene glycol) are used in chemical synthesis methods [35]. Chemical synthesis methods of nanomaterials have a bottom-up approach. These methods include chemical reduction, sol-gel, quiescent condensation, solvent evaporation, hydrothermal, solvothermal, thermal decomposition, and microemulsion [36].

The Sol-Gel method is one of the most frequently used methods among chemical synthesis methods [37]. Inorganic compounds such as metal powders or metal alkoxide solutions, nitrates, oxides, and hydroxides are added to a mixture of water and acid and mixed at certain temperatures to form a solution, and a series of sequential chemical reactions and electrochemical interactions of the surface charges of the particles occurs in this solution [38]. A network is formed (gelation) and gradually grows, reaching all points in the system and forming a complete structure (gel). It is a stable suspension of colloidal solid particles in a liquid. These solid particles must be small enough to be responsible for dispersion forces greater than gravity. The smaller these particles are, the more accurate it is to speak of the molecules in the solution. Particles defined as colloids are particles with a size of 500 nm and less, which are too small to be seen with the naked eye. These particles cannot be seen with a normal optical microscope. This is because their maximum size is equal to the wavelength of light. Gel; precipitates formed by precipitation of colloidal particles and containing large amounts of water are called precipitates. The gel is an intermediate phase between the solid and the liquid phase. The main areas of application of the sol-gel method are as follows: Production of wear-resistant coatings, coatings for optical purposes, production of high-strength fibers in fiber optics, and production of electronic and magnetic materials [39,40].

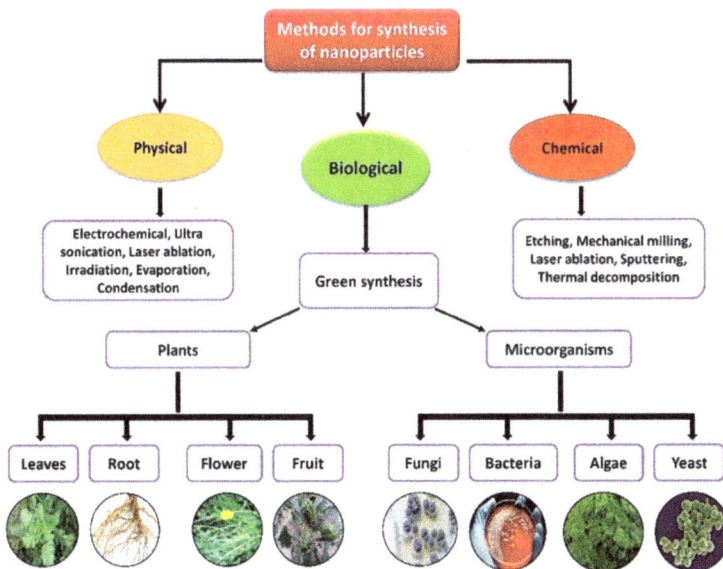

Figure 2. Schematic representation of general nanoparticle production methods with emphasis on the biological synthesis method. (Reprinted with permission from ref. [33]. Copyright: ©2022, MDPI).

3.2   Physical Synthesis Methods

Methods that focus on physical properties, such as the use of mechanical energy, are referred to as physical methods. Methods that use physical factors such as sound, radiation, and radio waves are summarized under physical applications. The aim here is to break down the volumetric material into nanoparticles using physical applications. Physical synthesis methods of nanomaterials generally involve top-down methods. The most commonly used physical methods are ultrasound, radiation, lasers, microwaves, electrospinning, ion sputtering, arc discharge, laser pyrolysis, mechanical ball milling, and electrochemical techniques [41].

3.3   Biogenic (Green) Synthesis Methods

Chemical and physical methods, which can also be called classical synthesis methods, to obtain nanoparticles with the desired size and morphology have some disadvantages. To overcome these undesirable situations, green chemistry practices have been increasingly used in recent years. These green methods, which can be referred to by various terms such as phytoremediation, biogenic synthesis, etc., are more environmentally friendly than the classical methods. In addition, these methods are simple, affordable, and contain no harmful chemicals [42,43].

The basis of biosynthesis is the chemical activities in living organisms and the utilization of energy and substances in living organisms to obtain more complex products. Although the production of metal nanoparticles by classical synthesis methods has disadvantages, such as being complex and very expensive, requiring the use of toxic substances, and not suitable for pharmacological and biomedical applications, these problems can be eliminated by biosynthesis. In other words, a simple and economical systematic production process that does not require toxic substances is suitable for pharmacological and biomedical applications and is suitable for a wide range of commercial production can be achieved [44]. Therefore, it has become a commonly used method today [45]. To produce a nanoparticle biosynthetically, the process is expected to have the following properties: (a) safe, (b) single reaction step, (c) no toxic waste, (d) the raw material used is renewable, (e) the synthesized product can be produced in a simple way ((f) 100% yield. For many reactions, it is tough to fulfill all these conditions [46]. To take advantage of biosynthesis in nanotechnology, it is desirable to synthesize nanoparticles using plant extracts and living organisms or chemicals such as algae, bacteria, yeast, and fungi [47]. Various bioactive molecules found in extracts of different organisms, including microorganisms, plants, algae and fungi, are involved in the synthesis of biogenic nanoparticles (Fig. 3). These bioactive molecules act as stabilizing, reducing and capping agents in the synthesis process. Then, the physical and chemical properties of the synthesized nanomaterials are investigated. The advantages and disadvantages of different synthesis strategies for nanomaterials are listed in relative detail in Table 1.

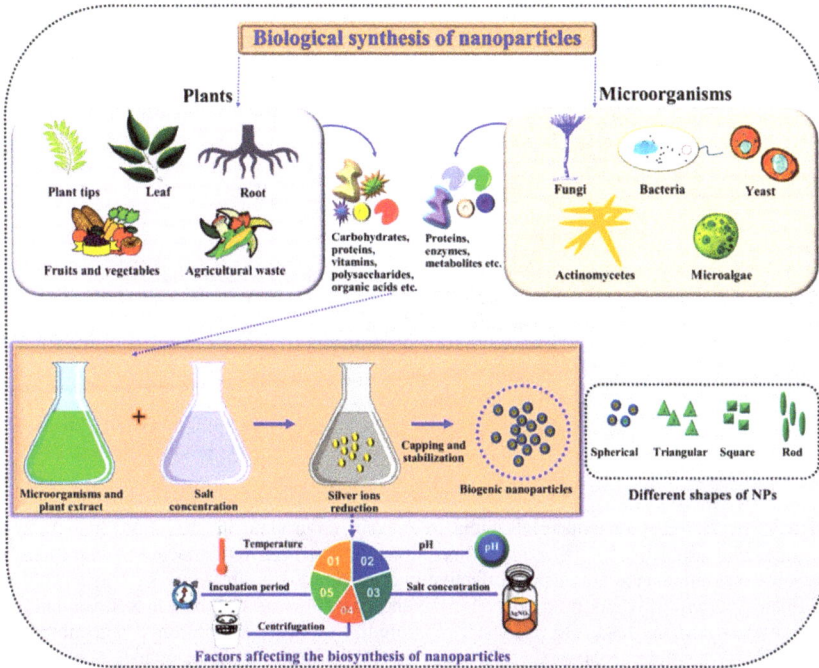

Figure 3. *Schematic illustration for biogenic nanoparticles synthesis (Reprinted with permission from ref. [48]. Copyright: ©2020, MDPI).*

Green, environmentally friendly, and biogenic syntheses take precedence over traditional strategies that use toxic substances that can cause carcinogenicity, environmental toxicity, and cytotoxicity. Using strategies with organic starting materials to synthesize nanoparticles is environmentally friendly and reduces cost. The agricultural crop residues and food industry provide inexpensive, living components for the synthesis of nanomaterials [49]. The nanomaterials that are synthesized using organic strategies are referred to as biogenic nanomaterials [50]. These nanomaterials are characterized by high biocompatibility and sensitivity and are therefore used in biomedicine, agriculture, electronics, and environmental remediation. When evaluating conventional and biogenic nanomaterials, biogenic nanomaterials (BNMs) tend to be larger and chemically more complicated due to the composition of the stabilizers [51].

Table 1. *Comparative advantages and disadvantages of different synthesis methods for nanomaterials [52–54].*

Synthesis method	Advantages	Disadvantages
Chemical method	Large-scale production Controlled sized	Use of harmful chemicals Non-ecofriendly Energy-intensive processes
Physical method	High purity Controlled crystallinity Controlled sized Uniform shape	Hgh cost Difficulty of synthesis conditions (high energy, pressure requirement, etc.)
Biological method	Low cost No toxic Simple (one-pot synthesis) Fast Eco-friendly	Possible ecological imbalance as a result of the use of natural biological resources (microorganisms, plants, etc.) The problem of industrial- scale production

Biogenic synthesis of nanomaterials aims to prevent environmental pollution and eliminate environmental problems by using natural resources. It has been reported in the literature that magnetite biogenic nanomaterials are effective in removing toxic substances such as arsenic and chlorinated organic solvents from water [55]. Nanosensors are used to detect coliform bacteria, heavy metals such as lead, etc. Nanosensors simplify analysis and improve detection limits compared to existing analytical techniques [56,57]. Green nanotechnology helps solve environmental problems in a biological and non-toxic way. Therefore, green nanotechnology approaches environmental problems in a way that respects the principles of green science by using and producing products that do not harm the environment and human health with as little energy as possible [58]. It should also keep pace with the expected trend of chemical reduction and produce as little waste/pollution as possible. For this purpose, nanotechnology methods and approaches, such as the use of solid-state reactions, the absence of harmful organic solvents, energy savings, low toxicity, recyclability, and reusability, are of great importance. With the green synthesis approach that aims to reduce the use of toxic chemicals, green solutions, bio-based converters and materials, molecular studies, alternative energy sciences, the design of new generation catalysts, the production of solar and fuel cells, and the production of new generation batteries that can be used for energy storage, pollution monitoring, prevention, and purification have increased their use in areas [43,59,60].

The nature of biogenic nanomaterials may be very distinct from metallic oxide micelles, metals, and biomolecules. Therefore, the synthesis and characterization of those biogenic nanomaterials may be tough, and it isn't unusual to gain distinct physicochemical information for the equal biogenic nanomaterial. Most literature protocols for the synthesis of biogenic nanomaterial use bottom-up tactics. For example, inside the instruction of biogenic metallic nanoparticles (MNP), distinct synthesis tactics are used: extracellular and intracellular processes (Fig. 4) [61–63]. Other techniques are required for cellular synthesis and isolation of MNPs. The techniques examine the boom of debris beginning from metallic atoms derived from molecular or ionic precursors, much like chemical techniques, and in assessment to bodily techniques primarily based totally on the bulk metallic subdivision. The intracellular instruction approach is extra appropriate for acquiring uniform and small nanoparticles, whilst the maximum crucial step inside the extracellular approach is to govern atom aggregation to ensure the scale and uniformity of MNPs, which is hard to obtain with biogenic artificial techniques. Synthesis technology of biogenic nanomaterials may include an artificial biology technique that allows the intracellular formation of various nanoparticles. For example, ferromagnetic debris is produced with the aid of using magnetotactic bacteria (MTB) in such a manner that they act like a compass needle, allowing their orientation and motion inside a particular geomagnetic field. This synthesis manner has additionally been proven in a few different species, which include yeast and mammalian cells [64,65].

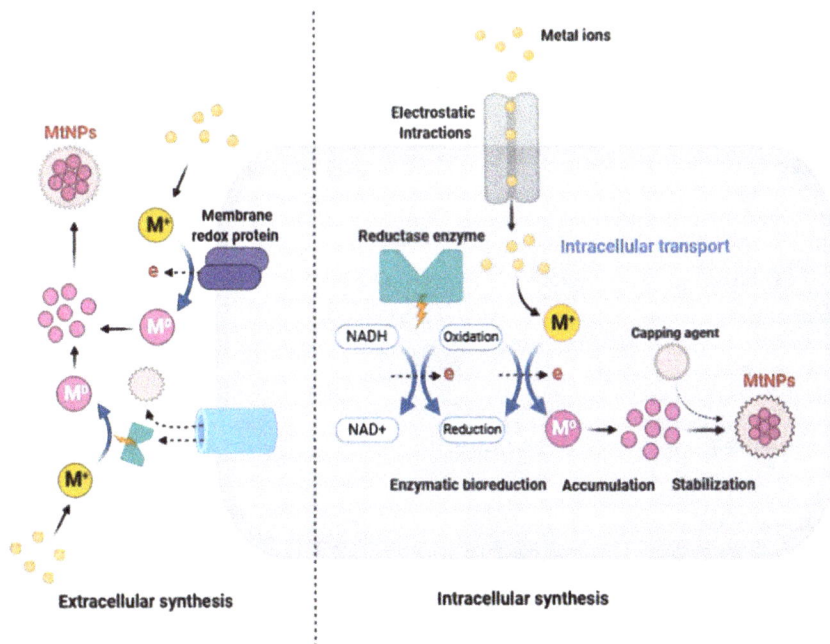

Figure 4. Schematic presentation of intracellular and extracellular methods in the synthesis of biogenic nanomaterials. (Reprinted with permission from ref. [63]. Copyright: ©2021, BMC part of Springer Nature).

Bottom-up processes are a way extra famous for the synthesis of nanoparticles and lots of techniques have been developed. For example, nanoparticles are synthesized via way of means of homogeneous nucleation from drinks or vapors or via way of means of heterogeneous nucleation on chemical, vapor deposition, substrates, electrochemical precipitation, sol-gel methods, aerosol methods, laser pyrolysis, and spray pyrolysis. Nanoparticles or quantum dots also can be produced via way of means of segment separation via way of means of annealing as it should be designed strong substances at improved temperature. Nanoparticles may be synthesized via way of means of confining nucleation, chemical reactions, and increase methods to small spaces, along with micelles [66,67]. Plant-mediated and microbe-mediated synthesis methods of biogenic nanoparticles do not involve using and producing risky chemicals, and the synthesis processes are generally associated with bottom-up processes. Different amounts of plant extract or microbe lysate and metal salts dissolved in deionized water are used in extracellular biobased synthesis. Nanoparticles are synthesized with surfactants and protein from microbial and plant extracts. The potential of microorganisms to behave in a huge variety of pH, pressures, different situations, and temperatures makes them very appealing for this application.

In particular, the pH of the response medium is an essential parameter in biogenic nanoparticle synthesis pushed via way of means of plant extracts or microbes and might result in the manufacturing of nanoparticles with exclusive morphologies [68,69]. The biochemical mechanism for intracellular and extracellular synthesis of biogenic nanomaterials in methods managed via way of means of cells and microorganisms includes simple microbial resistance as a consequence of cell detoxification. The intracellular pathway begins with biosorption, a response observed through reduction and increase, complexation, stabilization, and nucleation with proteins and carbohydrates that form the protective biofilm. For each extracellular and intracellular mechanism, enzymatic discount of strong poisonous materials and precipitation of metals or metallic oxides into non-poisonous insoluble nanoforms are function methods inside the manufacturing of biogenic nanomaterials [70].

4.   Characterization Techniques of Biogenic Nanoparticles

Nanoparticles have unique properties, such as their surface charge, morphology, size, and size distribution in suspension, and various characterization methods are used to study and determine these properties [71]. It is important to accurately determine the physical, chemical, and biological characterization of nanoparticles for the intended application. For example, particle size is one of the most important parameters for the use of nanoparticles in drug delivery systems, which is their main application. Advanced microscopy techniques such as scanning electron microscopy (SEM), atomic force microscopy (AFM), and transmission electron microscopy (TEM) are used to characterize the morphological properties of particles such as size and shape; dynamic light scattering (DLS) is used to measure the average size distribution of particles in suspension; and zeta potential measurement is used to detect the surface charge distribution of particles. UV-Vis spectroscopy is used to determine the wavelength at which particles exhibit maximum absorption according to their surface plasmon resonances. X-ray diffraction analysis (XRD) analysis is performed to determine the crystal structure of nanomaterials, and Fourier Transform Infrared Spectroscopy (FTIR) analysis is performed to determine functional groups and molecular compounds.

4.1   Fourier Transform Infrared Spectrophotometry (FTIR)

FTIR is a powerful analytical method that is fast, non-destructive, and requires minimal sample preparation. Since the intensity of spectral bands obtained in FTIR analysis is proportional to sample concentrations, FTIR is one of the important characterization techniques for quantitative analysis [72]. FTIR analysis is used to identify nanomaterials, inorganic, and organic using infrared light to scan the samples. Changes in the characteristic pattern of absorption bands provide information about material composition. FTIR is used to identify and characterize unknown materials, detect impurities in a material, find additives, and detect its oxidation and decomposition. The FTIR spectrometer generally consists of six components: light source, sample cell, detector, amplifier, converter, and computer. The FTIR working principle can be briefly explained as follows: The radiation from the light source passes through the interferometer and reaches the detector. The signal coming to the detector is amplified by the converter and amplifier and converted into a digital signal. In the final stage, the signal is displayed on a computer screen. Each material has a unique fingerprint region, making FTIR an invaluable analysis for chemical identification [73].

4.2   X-Ray Diffraction (XRD)

The crystal structure of solids is formed when groups of atoms or molecules come together in a geometric arrangement specific to the solid. The atomic structure of a substance can be imaged using various electron microscopes. However, in order to identify unknown structures, diffraction techniques are necessary. XRD is a widely used technique for determining the crystal structures of solids [74]. The structure of the sample is determined by XRD analysis. The interaction between the X-rays and the atomic plane causes the beam to be partially transmitted and the rest to be absorbed, refracted, and scattered by the sample. The X-rays are refracted differently by each element, depending on the type of atoms [75]. X-rays have wavelengths in the range of about 10 to 10^{-3} nm. In XRD analysis, a very small amount of the substance to be analyzed is sufficient for the analysis. The material used is not degraded during the analysis [76].

4.3   Scanning Electron Microscope (SEM)

It is a widely used technique for the characterization of nanoparticles [77]. It provides information about the size and shape of the particles through direct imaging. However, since the results depend on the scanned surface, the data on average particle diameter and size distribution are limited for all synthesized nanoparticles. The average particle diameter obtained after imaging should be compared with the dynamic light scattering results, and the results should be confirmed with more than one technique. In addition to this disadvantage, the particles need to be pulverized for SEM characterization and their surfaces need to be coated with a conductive material. For this reason, it is an expensive and time-consuming characterization technique, even though it is often preferred.

4.4   Transmission Electron Microscopy (TEM)

The TEM works according to a different principle than the SEM, and sample preparation is more demanding and time-consuming. The sample is prepared very thinly so that the electrons can penetrate the material. In contrast to SEM, imaging can also be carried out at dimensions below the nanoscale. TEM is mainly used in materials science to investigate the atomic crystal structure [78]. For the characterization of nanomaterials, SEM is preferred over cost- and time-intensive TEM, as TEM and SEM provide similar data.

4.5   Atomic Force Microscopy (AFM)

Thanks to its probe, it scans the sample surface with or without contact and provides a topographical image of the surface [79]. This microscopic technique provides the most accurate information on the particle size distribution in the submicron range, as it is scanned with high resolution. In addition, the surface does not need to be conductive to scan from it, allowing biological and polymeric structures to be imaged [79].

4.6   Dynamic Light Scattering Technique (DLS)

It is a commonly used and faster method than others for determining the size distributions of submicron and nanometric particles in a colloidal suspension. When the monochromatic beam coming from the light source hits the spherical particles performing Brownian motion in the suspension, the change in the wavelength of the light causes a Doppler shift, and data on the size distribution of the particles in the suspension is generated [80]. The change in wavelength varies depending on the size of the particles. Photon correlation spectroscopy is the most commonly used technique that works according to the DLS principle. Although it is a preferred technique because it is faster and simpler, does not require much equipment and the samples are easy to prepare, particle size analysis with DLS does not provide very reliable results, especially for samples with high polydispersity, and the data obtained must be evaluated and interpreted in detail [81].

4.7   Zeta Potential Measurement

The surface charge and charge density of nanoparticles are crucial for their interaction with the biological environment [82]. Electrostatic forces between nanoparticles and bioactive components are directly related to surface charge, especially when their use in drug delivery systems and cosmetic product formulations is sought [83]. Measurements of zeta potential not only provide information on the charge distribution of nanoparticles in suspension but also on the stability and surface hydrophobicity of the particles during storage. Regardless of whether it is positive or negative, a high value of the potential difference indicates that the stability of the particles is high and their tendency to form aggregates is low. Measuring the zeta potential is a necessary analysis, especially for nanoparticles that are to be used for encapsulation.

4.8   Ultraviolet and Visible Absorption Spectroscopy (UV-Vis)

When describing the UV-Vis absorption spectra of metal nanoparticles, the term surface plasmon is used to describe the electron clouds oscillating at the interface between metal and solution [84]. The particle size influences the absorption spectrum on the nanometer scale. As the particle size decreases, the wavelength of the absorbed light also decreases. When nanoparticles form aggregates, the bandwidth increases, and the surface plasmon resonance shifts into the red range. In addition to the size, the shape of the particles also has an effect on the surface plasmon band [85]. Spherical gold nanoparticles, for example, have a maximum absorption at 520 nm and silver nanoparticles at 400 nm.

5.   Application Areas of Biogenic Nanoparticles

Environmental pollution has become one of the biggest global threats, increasing day by day and causing irreversible damage to societies [86,87]. Continuous urbanization and rapid industrialization have disturbed the balance of the environment by releasing hazardous substances, smoke, and harmful gasses, leading to toxic effects on living beings. In addition, overpopulation, a large number of vehicles running on fossil fuels, factory chimneys, and many other factors cause the destruction and depletion of natural resources [88,89]. Examples of some hazardous substances are pesticides, fertilizers, herbicides, pharmaceutical residues, industrial effluents, toxic gasses, heavy metals, sulfur-containing compounds, particulate matter, oil spills, pathogens, sewage, organic compounds, etc [90–93]. These harmful substances enter the environment and pollute the water, soil, and air, threatening the ecosystem and living health. Under the current circumstances, maintaining a clean and healthy water and air environment is a major challenge. New technologies are being used to clean these waste materials from the air, water, and soil environment. Nanotechnology offers great potential for the development and use of new and low-cost techniques for the detection and monitoring of pollutants, catalytic degradation, adsorption removal, and treatment of environmental pollutants [10,94]. Compared to bulk materials, nanotechnology products exhibit novel physical and chemical properties due to their smaller size (<100 nm), resulting in a higher surface-to-volume ratio and making them efficient catalysts [95]. Moreover, nanomaterials with high surface area/volume ratio and specific functionalization can be made into sensitive nanosensors for the detection of harmful materials and efficient catalysts for their removal.

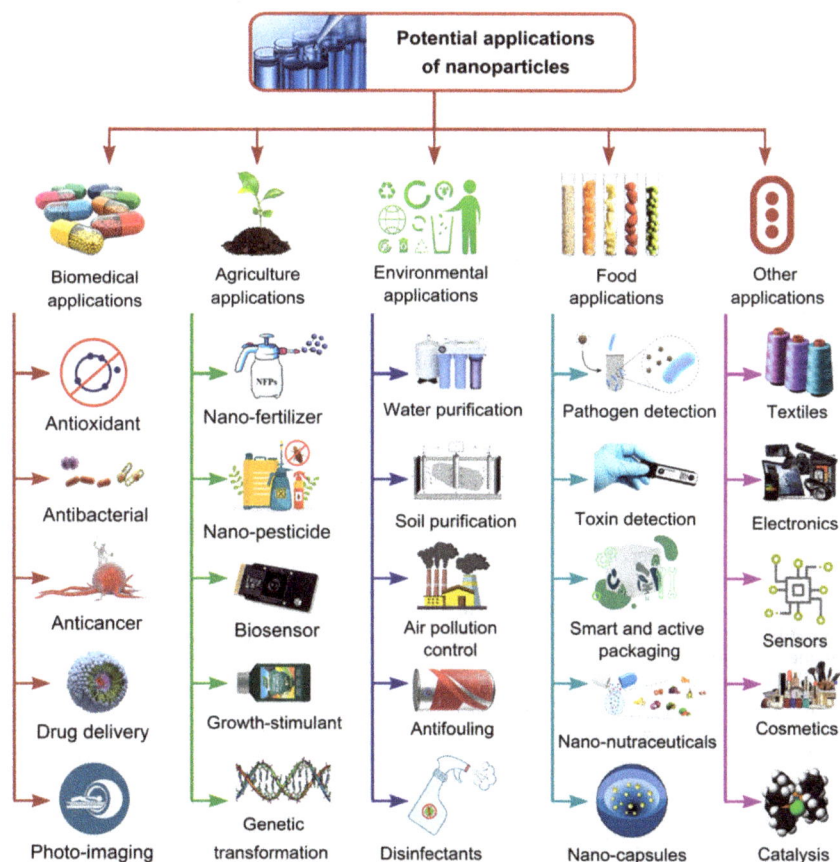

Figure 5. *Applications of green nanomaterials in various industries and sectors. (Reprinted with permission from ref. [43]. Copyright: ©2024, Springer Nature).*

Today, due to the problem of bacterial resistance to antibiotics, studies on the production of new antibacterial agents have accelerated. Metallic and metal oxide nanomaterials such as silver, palladium, and copper have come to the fore in the development of new antibacterial agents [96]. Researchers prefer biogenic nanomaterials in different biomedical applications. Due to their biocompatible properties, biogenic nanomaterials are frequently used in the field of biotechnology. Gold NPs produced by green synthesis were tested for the determination of HCG hormone in urine samples of pregnant women [96,97]. Biogenic nanomaterials have biological activities such as anticancer, antimicrobial, antioxidant, antiviral, and DNA damage prevention. Biogenic nanomaterials have found use in biomedical systems such as medicine, dentistry, surgical applications, malaria treatment, and controlled drug release systems. It is used in sensor/diagnostics, hydrogen/energy storage applications, and as a catalyst in fuel cells. These application areas are discussed in the following sections. An overview of the application areas of biogenic NPs is shown in Fig. 5.

6. Conclusion

The widespread use and commercialization of biogenic nanomaterials are expected to have a greater impact in the near future. However, there is a need to focus on maximizing the stability, reusability, and non-toxicity of existing methods, increasing their efficiency, and making them more cost-effective. On the other hand, due to the capping and stabilization capabilities of the precursors of biogenic nanomaterials, they are easier to recover for reuse than their inorganic counterparts made from inorganic chemicals. In short, biogenic nanomaterials have been found to be an energy-efficient, cost-effective, sustainable, and ecologically sound technique and can serve a critical function in many areas such as energy, medicine, food, agriculture, pharmaceuticals, fuel cells, biomedical applications, and the environment.

References

[1]   S. Malik, K. Muhammad, Y. Waheed, Nanotechnology: A Revolution in Modern Industry, Molecules 28 (2023) 661. https://doi.org/10.3390/MOLECULES28020661

[2]   A. Aygun, R.N.E. Tiri, R. Bayat, F. Sen, Hydrothermal synthesis of BCQD@g-C3N4 nanocomposites supporting environmental sustainability: Organic dye removal and bacterial inactivation, J. Hazard. Mater. Adv. 16 (2024) 100464. https://doi.org/10.1016/J.HAZADV.2024.100464

[3]   K. Nesrin, C. Yusuf, K. Ahmet, S.B. Ali, N.A. Muhammad, S. Suna, Ş. Fatih, Biogenic silver nanoparticles synthesized from Rhododendron ponticum and their antibacterial, antibiofilm and cytotoxic activities, J. Pharm. Biomed. Anal. 179 (2020) 112993. https://doi.org/10.1016/J.JPBA.2019.112993

[4]   R. Darabi, F.E.D. Alown, A. Aygun, Q. Gu, F. Gulbagca, E.E. Altuner, H. Seckin, I. Meydan, G. Kaymak, F. Sen, H. Karimi-Maleh, Biogenic platinum-based bimetallic nanoparticles: Synthesis, characterization, antimicrobial activity and hydrogen evolution, Int. J. Hydrogen Energy 48 (2023) 21270–21284. https://doi.org/10.1016/J.IJHYDENE.2022.12.072

[5]   N. Joudeh, D. Linke, Nanoparticle classification, physicochemical properties, characterization, and applications: a comprehensive review for biologists, J. Nanobiotechnology 2022 201 20 (2022) 1–29. https://doi.org/10.1186/S12951-022-01477-8

[6]   I. Meydan, A. Aygun, R.N.E. Tiri, T. Gur, Y. Kocak, H. Seckin, F. Sen, Chitosan/PVA-supported silver nanoparticles for azo dyes removal: fabrication, characterization, and assessment of antioxidant activity, Environ. Sci. Adv. 3 (2024) 28–35. https://doi.org/10.1039/D3VA00224A

[7]   W. Bouchal, F. Djani, D. Eddine Mazouzi, R.N.E. Tiri, S. Makhloufi, C. Laiadi, A. Martinez-Arias, A. Aygün, F. Sen, Bi-doped BaBiO 3 (x = 0%, 5%, 10%, 15%, and 20%) perovskite oxides by a

sol–gel method: comprehensive biological assessment and RhB photodegradation, RSC Adv. 14 (2024) 7359–7370. https://doi.org/10.1039/D3RA06354B

[8]   B. Elzein, Nano Revolution: "Tiny tech, big impact: How nanotechnology is driving SDGs progress", Heliyon 10 (2024) e31393. https://doi.org/10.1016/J.HELIYON.2024.E31393

[9]   F.J. Tovar-Lopez, Recent Progress in Micro- and Nanotechnology-Enabled Sensors for Biomedical and Environmental Challenges, Sensors 2023, Vol. 23, Page 5406 23 (2023) 5406. https://doi.org/10.3390/S23125406

[10]   N.H.H. Hairom, C.F. Soon, R.M.S.R. Mohamed, M. Morsin, N. Zainal, N. Nayan, C.Z. Zulkifli, N.H. Harun, A review of nanotechnological applications to detect and control surface water pollution, Environ. Technol. Innov. 24 (2021) 102032. https://doi.org/10.1016/J.ETI.2021.102032

[11]   C. Kher, S. Kumar, The Application of Nanotechnology and Nanomaterials in Cancer Diagnosis and Treatment: A Review, Cureus 14 (2022) e29059. https://doi.org/10.7759/CUREUS.29059

[12]   J.E. Hulla, S.C. Sahu, A.W. Hayes, Nanotechnology: History and future, Hum. Exp. Toxicol. 34 (2015) 1318–1321. https://doi.org/10.1177/0960327115603588/FORMAT/EPUB

[13]   S. Bayda, M. Adeel, T. Tuccinardi, M. Cordani, F. Rizzolio, The History of Nanoscience and Nanotechnology: From Chemical–Physical Applications to Nanomedicine, Molecules 25 (2020) 112. https://doi.org/10.3390/MOLECULES25010112

[14]   C. Demir, A. Aygun, M.K. Gunduz, B.Y. Altınok, T. Karahan, I. Meydan, E. Halvaci, R.N.E. Tiri, F. Sen, Production of plant-based ZnO NPs by green synthesis; anticancer activities and photodegradation of methylene red dye under sunlight, Biomass Convers. Biorefinery 2024 (2024) 1–16. https://doi.org/10.1007/S13399-024-06172-2

[15]   K.R. Khedir, G.K. Kannarpady, C. Ryerson, A.S. Biris, An outlook on tunable superhydrophobic nanostructural surfaces and their possible impact on ice mitigation, Prog. Org. Coatings 112 (2017) 304–318. https://doi.org/10.1016/J.PORGCOAT.2017.05.019

[16]   S. Kumari, S. Raturi, S. Kulshrestha, K. Chauhan, S. Dhingra, K. András, K. Thu, R. Khargotra, T. Singh, A comprehensive review on various techniques used for synthesizing nanoparticles, J. Mater. Res. Technol. 27 (2023) 1739–1763. https://doi.org/10.1016/J.JMRT.2023.09.291

[17]   F. Gol, A. Aygun, C. Ture, R.N.E. Tiri, Z.G. Sarıtaş, E. Kaçar, M. Arslan, F. Sen, Environmentally Friendly Synthesis of 2D Cu2O Nanoleaves: Morphological Evaluation, Their Photocatalytic Activity Against Azo Dye And Antibacterial Activity For Ceramic Structures, Bionanoscience (2024) 1–10. https://doi.org/10.1007/S12668-024-01440-X

[18]   B. Şen, B. Demirkan, A. Savk, R. Kartop, M.S. Nas, M.H. Alma, S. Sürdem, F. Şen, High-performance graphite-supported ruthenium nanocatalyst for hydrogen evolution reaction, J. Mol. Liq. 268 (2018) 807–812. https://doi.org/10.1016/J.MOLLIQ.2018.07.117

[19]   N. Baig, I. Kammakakam, W. Falath, I. Kammakakam, Nanomaterials: a review of synthesis methods, properties, recent progress, and challenges, Mater. Adv. 2 (2021) 1821–1871. https://doi.org/10.1039/D0MA00807A

[20]   A. Aygun, E. Ozveren, E. Halvaci, D. Ikballi, R.N.E. Tiri, C. Catal, M. Bekmezci, A. Ozengul, I. Kaynak, F. Sen, The performance of a very sensitive glucose sensor developed with copper nanostructure-supported nitrogen-doped carbon quantum dots, RSC Adv. 14 (2024) 34964–34970. https://doi.org/10.1039/D4RA06566B

[21]   M. Akin, H. Kars, M. Bekmezci, A. Aygun, M. Gul, G. Kaya, F. Sen, Synthesis of MOF-supported Pt catalyst with high electrochemical oxidation activity for methanol oxidation, RSC Adv. 14 (2024) 36370–36377. https://doi.org/10.1039/D4RA06393G

[22]   F. Şen, Ö. Demirbaş, M.H. Çalımlı, A. Aygün, M.H. Alma, M.S. Nas, The dye Removal from Aqueous Solution Using Polymer Composite Films, Appl. Water Sci. 8 (2018) 206. https://doi.org/10.1007/s13201-018-0856-x

[23]   R. Ayranci, G. Başkaya, M. Güzel, S. Bozkurt, F. Şen, M. Ak, Carbon Based Nanomaterials for High Performance Optoelectrochemical Systems, ChemistrySelect 2 (2017) 1548–1555. https://doi.org/10.1002/slct.201601632

[24]   N.H. Khand, A.R. Solangi, S. Ameen, A. Fatima, J.A. Buledi, A. Mallah, S.Q. Memon, F. Sen, F. Karimi, Y. Orooji, A new electrochemical method for the detection of quercetin in onion, honey and green tea using Co3O4 modified GCE, J. Food Meas. Charact. 2021 154 15 (2021) 3720–3730. https://doi.org/10.1007/S11694-021-00956-0

[25]   H. Göksu, Y. Yıldız, B. Çelik, M. Yazıcı, B. Kılbaş, F. Şen, Highly Efficient and Monodisperse Graphene Oxide Furnished Ru/Pd Nanoparticles for the Dehalogenation of Aryl Halides via Ammonia Borane, ChemistrySelect 1 (2016) 953–958. https://doi.org/10.1002/slct.201600207

[26]   A.M. Allahverdiyev, E.S. Abamor, M. Bagirova, S.Y. Baydar, S.C. Ates, F. Kaya, C. Kaya, M. Rafailovich, Investigation of antileishmanial activities of Tio2@Ag nanoparticles on biological properties of L. tropica and L. infantum parasites, in vitro, Exp. Parasitol. 135 (2013) 55–63. https://doi.org/10.1016/J.EXPPARA.2013.06.001

[27]   ISO/TS 27687:2008 - Nanotechnologies — Terminology and definitions for nano-objects — Nanoparticle, nanofibre and nanoplate, (n.d.). https://www.iso.org/standard/44278.html (accessed October 31, 2024).

[28]   R. Gupta, H. Xie, Nanoparticles in Daily Life: Applications, Toxicity and Regulations, J. Environ. Pathol. Toxicol. Oncol. 37 (2018) 230. https://doi.org/10.1615/JENVIRONPATHOLTOXICOLONCOL.2018026009

[29]   H. Seçkin, İ. Meydan, Synthesis and Characterization of Veronica beccabunga Green Synthesized Silver Nanoparticles for The Antioxidant and Antimicrobial Activity, Turkish J. Agric. Res. 8 (2021) 49–55. https://doi.org/10.19159/TUTAD.805463

[30]   K.A. Altammar, A review on nanoparticles: characteristics, synthesis, applications, and challenges, Front. Microbiol. 14 (2023) 1155622. https://doi.org/10.3389/FMICB.2023.1155622

[31]   M.J. Pitkethly, Nanomaterials – the driving force, Mater. Today 7 (2004) 20–29. https://doi.org/10.1016/S1369-7021(04)00627-3

[32]   I. Khan, K. Saeed, I. Khan, Nanoparticles: Properties, applications and toxicities, Arab. J. Chem. 12 (2019) 908–931. https://doi.org/10.1016/J.ARABJC.2017.05.011

[33]   F. Khan, M. Shariq, M. Asif, M.A. Siddiqui, P. Malan, F. Ahmad, Green Nanotechnology: Plant-Mediated Nanoparticle Synthesis and Application, Nanomater. 2022, Vol. 12, Page 673 12 (2022) 673. https://doi.org/10.3390/NANO12040673

[34]   A. Aygun, G. Sahin, R.N.E. Tiri, Y. Tekeli, F. Sen, Colorimetric sensor based on biogenic nanomaterials for high sensitive detection of hydrogen peroxide and multi-metals, Chemosphere 339 (2023) 139702. https://doi.org/10.1016/J.CHEMOSPHERE.2023.139702

[35]   S. Iravani, H. Korbekandi, S. V. Mirmohammadi, B. Zolfaghari, Synthesis of silver nanoparticles: Chemical, physical and biological methods, Res. Pharm. Sci. 9 (2014) 385–406.

[36]   K. Hachem, M.J. Ansari, R.O. Saleh, H.H. Kzar, M.E. Al-Gazally, U.S. Altimari, S.A. Hussein, H.T. Mohammed, A.T. Hammid, E. Kianfar, Methods of Chemical Synthesis in the Synthesis of Nanomaterial and Nanoparticles by the Chemical Deposition Method: A Review, BioNanoScience 2022 123 12 (2022) 1032–1057. https://doi.org/10.1007/S12668-022-00996-W

[37]   D. Bokov, A. Turki Jalil, S. Chupradit, W. Suksatan, M. Javed Ansari, I.H. Shewael, G.H. Valiev, E. Kianfar, Nanomaterial by Sol-Gel Method: Synthesis and Application, Adv. Mater. Sci. Eng. 2021 (2021) 1–21. https://doi.org/10.1155/2021/5102014

[38]   S. Edebali, Y. Oztekin, G. Arslan, Metallic Engineered Nanomaterial for Industrial Use, Handb. Nanomater. Ind. Appl. (2018) 67–73. https://doi.org/10.1016/B978-0-12-813351-4.00004-3

[39]   E. Yilmaz, M. Soylak, Functionalized nanomaterials for sample preparation methods, Handb.

Nanomater. Anal. Chem. Mod. Trends Anal. (2020) 375–413. https://doi.org/10.1016/B978-0-12-816699-4.00015-3

[40]   M.A. Azam, M. Mupit, Carbon nanomaterial-based sensor: Synthesis and characterization, Carbon Nanomater. Sensors Emerg. Res. Trends Devices Appl. (2022) 15–28. https://doi.org/10.1016/B978-0-323-91174-0.00015-9

[41]   N. Al-Harbi, N.K. Abd-Elrahman, Physical methods for preparation of nanomaterials, their characterization and applications: a review, J. Umm Al-Qura Univ. Appl. Sci. (2024) 1–22. https://doi.org/10.1007/S43994-024-00165-7

[42]   U.O. Aigbe, O.A. Osibote, Green synthesis of metal oxide nanoparticles, and their various applications, J. Hazard. Mater. Adv. 13 (2024) 100401. https://doi.org/10.1016/J.HAZADV.2024.100401

[43]   A.I. Osman, Y. Zhang, M. Farghali, A.K. Rashwan, A.S. Eltaweil, E.M. Abd El-Monaem, I.M.A. Mohamed, M.M. Badr, I. Ihara, D.W. Rooney, P.S. Yap, Synthesis of green nanoparticles for energy, biomedical, environmental, agricultural, and food applications: A review, Environ. Chem. Lett. 2024 222 22 (2024) 841–887. https://doi.org/10.1007/S10311-023-01682-3

[44]   H.M. Abuzeid, C.M. Julien, L. Zhu, A.M. Hashem, Green Synthesis of Nanoparticles and Their Energy Storage, Environmental, and Biomedical Applications, Cryst. 2023, Vol. 13, Page 1576 13 (2023) 1576. https://doi.org/10.3390/CRYST13111576

[45]   T. Gur, I. Meydan, H. Seckin, M. Bekmezci, F. Sen, Green synthesis, characterization and bioactivity of biogenic zinc oxide nanoparticles, Environ. Res. 204 (2022) 111897. https://doi.org/10.1016/J.ENVRES.2021.111897

[46]   D. Philip, Green synthesis of gold and silver nanoparticles using Hibiscus rosa sinensis, Phys. E Low-Dimensional Syst. Nanostructures 42 (2010) 1417–1424. https://doi.org/10.1016/J.PHYSE.2009.11.081

[47]   Y. Ju-Nam, J.R. Lead, Manufactured nanoparticles: An overview of their chemistry, interactions and potential environmental implications, Sci. Total Environ. 400 (2008) 396–414. https://doi.org/10.1016/J.SCITOTENV.2008.06.042

[48]   M.A. Ali, T. Ahmed, W. Wu, A. Hossain, R. Hafeez, M.M.I. Masum, Y. Wang, Q. An, G. Sun, B. Li, Advancements in Plant and Microbe-Based Synthesis of Metallic Nanoparticles and Their Antimicrobial Activity against Plant Pathogens, Nanomater. 2020, Vol. 10, Page 1146 10 (2020) 1146. https://doi.org/10.3390/NANO10061146

[49]   Ç. Öter, Solid Phase Extraction for the Determination of Methylene Blue Using Lignocellulosic Biosorbent in Aqueous Solutions, Bull. Environ. Contam. Toxicol. 109 (2022) 352–357. https://doi.org/10.1007/S00128-022-03543-1

[50]   H.R. El-Seedi, R.M. El-Shabasy, S.A.M. Khalifa, A. Saeed, A. Shah, R. Shah, F.J. Iftikhar, M.M. Abdel-Daim, A. Omri, N.H. Hajrahand, J.S.M. Sabir, X. Zou, M.F. Halabi, W. Sarhan, W. Guo, Metal nanoparticles fabricated by green chemistry using natural extracts: Biosynthesis, mechanisms, and applications, RSC Adv. 9 (2019) 24539–24559. https://doi.org/10.1039/c9ra02225b

[51]   D. Kulkarni, R. Sherkar, C. Shirsathe, R. Sonwane, N. Varpe, S. Shelke, M.P. More, S.R. Pardeshi, G. Dhaneshwar, V. Junnuthula, S. Dyawanapelly, Biofabrication of nanoparticles: sources, synthesis, and biomedical applications, Front. Bioeng. Biotechnol. 11 (2023) 1159193. https://doi.org/10.3389/FBIOE.2023.1159193

[52]   S.F. Ahmed, M. Mofijur, N. Rafa, A.T. Chowdhury, S. Chowdhury, M. Nahrin, A.B.M.S. Islam, H.C. Ong, Green approaches in synthesising nanomaterials for environmental nanobioremediation: Technological advancements, applications, benefits and challenges, Environ. Res. 204 (2022) 111967. https://doi.org/10.1016/J.ENVRES.2021.111967

[53]   S.A. Akintelu, A.S. Folorunso, F.A. Folorunso, A.K. Oyebamiji, Green synthesis of copper oxide nanoparticles for biomedical application and environmental remediation, Heliyon 6 (2020) e04508.

https://doi.org/10.1016/J.HELIYON.2020.E04508/ASSET/E239261F-8532-4BE4-BD48-B2BEDCA1907E

[54] S. Mukherjee, C.R. Patra, Biologically synthesized metal nanoparticles: Recent advancement and future perspectives in cancer theranostics, Futur. Sci. OA 3 (2017) 203. https://doi.org/10.4155/FSOA-2017-0035

[55] R. Jain, Recent advances of magnetite nanomaterials to remove arsenic from water, RSC Adv. 12 (2022) 32197–32209. https://doi.org/10.1039/D2RA05832D

[56] H.M. Valenzuela-Amaro, A. Aguayo-Acosta, E.R. Meléndez-Sánchez, O. de la Rosa, P.G. Vázquez-Ortega, M.A. Oyervides-Muñoz, J.E. Sosa-Hernández, R. Parra-Saldívar, Emerging Applications of Nanobiosensors in Pathogen Detection in Water and Food, Biosens. 2023, Vol. 13, Page 922 13 (2023) 922. https://doi.org/10.3390/BIOS13100922

[57] H. Singh, A. Bamrah, S.K. Bhardwaj, A. Deep, M. Khatri, K.H. Kim, N. Bhardwaj, Nanomaterial-based fluorescent sensors for the detection of lead ions, J. Hazard. Mater. 407 (2021) 124379. https://doi.org/10.1016/J.JHAZMAT.2020.124379

[58] A. Verma, S.P. Gautam, K.K. Bansal, N. Prabhakar, J.M. Rosenholm, Green Nanotechnology: Advancement in Phytoformulation Research, Med. 2019, Vol. 6, Page 39 6 (2019) 39. https://doi.org/10.3390/MEDICINES6010039

[59] O. V. Kharissova, B.I. Kharisov, C.M.O. González, Y.P. Méndez, I. López, Greener synthesis of chemical compounds and materials, R. Soc. Open Sci. 6 (2019). https://doi.org/10.1098/RSOS.191378

[60] L. Pokrajac, A. Abbas, W. Chrzanowski, G.M. Dias, B.J. Eggleton, S. Maguire, E. Maine, T. Malloy, J. Nathwani, L. Nazar, A. Sips, J. Sone, A. Van Den Berg, P.S. Weiss, S. Mitra, Nanotechnology for a Sustainable Future: Addressing Global Challenges with the International Network4Sustainable Nanotechnology, ACS Nano 15 (2021) 18608–18623. https://doi.org/10.1021/ACSNANO.1C10919

[61] E.R. Bandala, D. Stanisic, L. Tasic, Biogenic nanomaterials for photocatalytic degradation and water disinfection: a review, Environ. Sci. Water Res. Technol. 6 (2020) 3195–3213. https://doi.org/10.1039/D0EW00705F

[62] M. Jones, M. Goel, A. Sharma, B. Sharma, Recent Advances in Biogenic Silver Nanoparticles for Their Biomedical Applications, Sustain. Chem. 2023, Vol. 4, Pages 61-94 4 (2023) 61–94. https://doi.org/10.3390/SUSCHEM4010007

[63] H. Bahrulolum, S. Nooraei, N. Javanshir, H. Tarrahimofrad, V.S. Mirbagheri, A.J. Easton, G. Ahmadian, Green synthesis of metal nanoparticles using microorganisms and their application in the agrifood sector, J. Nanobiotechnology 2021 191 19 (2021) 1–26. https://doi.org/10.1186/S12951-021-00834-3

[64] K. Nishida, P.A. Silver, Induction of Biogenic Magnetization and Redox Control by a Component of the Target of Rapamycin Complex 1 Signaling Pathway, PLOS Biol. 10 (2012) e1001269. https://doi.org/10.1371/JOURNAL.PBIO.1001269

[65] M.S. Kushwaha, Plasmons and magnetoplasmons in semiconductor heterostructures, Surf. Sci. Rep. 41 (2001) 1–416. https://doi.org/10.1016/S0167-5729(00)00007-8

[66] A.K. Mittal, Y. Chisti, U.C. Banerjee, Synthesis of metallic nanoparticles using plant extracts, Biotechnol. Adv. 31 (2013) 346–356. https://doi.org/10.1016/J.BIOTECHADV.2013.01.003

[67] Y. Kocak, R. Nour, E. Tiri, · Aysenur Aygun, I. Meydan, · Nihed Bennini, T. Karahan, · Fatih Sen, Microwave-Assisted Fabrication of AgRuNi Trimetallic NPs with Their Antibacterial vs Photocatalytic Efficiency for Remediation of Persistent Organic Pollutants, BioNanoScience 2023 1 (2023) 1–9. https://doi.org/10.1007/S12668-023-01237-4

[68] Ç. Öter, Ö.S. Zorer, Synthesis and characterization of a molecularly imprinted polymer adsorbent for selective solid-phase extraction from wastewater of propineb, Polym. Bull. 79 (2022) 8503–8516. https://doi.org/10.1007/S00289-021-03927-Z

[69]   N. Aziz, M. Faraz, R. Pandey, M. Shakir, T. Fatma, A. Varma, I. Barman, R. Prasad, Facile Algae-Derived Route to Biogenic Silver Nanoparticles: Synthesis, Antibacterial, and Photocatalytic Properties, Langmuir 31 (2015) 11605–11612. https://doi.org/10.1021/ACS.LANGMUIR.5B03081

[70]   M. Xu, G. Wei, N. Liu, L. Zhou, C. Fu, M. Chubik, A. Gromov, W. Han, Novel fungus–titanate bio-nanocomposites as high performance adsorbents for the efficient removal of radioactive ions from wastewater, Nanoscale 6 (2013) 722–725. https://doi.org/10.1039/C3NR03467D

[71]   Y.; Khan, H. Sadia, A. Shah, S.Z.; Khan, M.N.; Shah, Z. Chen, Y. Khan, H. Sadia, S. Zeeshan, M.N. Khan, A.A. Shah, N. Ullah, M.F. Ullah, H. Bibi, O.T. Bafakeeh, N. Ben Khedher, S.M. Eldin, B.M. Fadhl, M.I. Khan, Classification, Synthetic, and Characterization Approaches to Nanoparticles, and Their Applications in Various Fields of Nanotechnology: A Review, Catal. 2022, Vol. 12, Page 1386 12 (2022) 1386. https://doi.org/10.3390/CATAL12111386

[72]   Y.R. Herrero, K.L. Camas, A. Ullah, Characterization of biobased materials, Adv. Appl. Biobased Mater. Food, Biomed. Environ. Appl. (2023) 111–143. https://doi.org/10.1016/B978-0-323-91677-6.00005-2

[73]   J.M. Costa-Fernandez, G. Redondo-Fernandez, M.T. Fernandez-Arguelles, A.B. Soldado, Analytical tools for the characterization and quantification of metal nanoclusters, Lumin. Met. Nanoclusters Synth. Charact. Appl. (2022) 57–88. https://doi.org/10.1016/B978-0-323-88657-4.00010-7

[74]   P.B. Raja, K.R. Munusamy, V. Perumal, M.N.M. Ibrahim, Characterization of nanomaterial used in nanobioremediation, Nano-Bioremediation Fundam. Appl. (2022) 57–83. https://doi.org/10.1016/B978-0-12-823962-9.00037-4

[75]   D. Titus, E. James Jebaseelan Samuel, S.M. Roopan, Nanoparticle characterization techniques, Green Synth. Charact. Appl. Nanoparticles (2019) 303–319. https://doi.org/10.1016/B978-0-08-102579-6.00012-5

[76]   Y. Waseda, E. Matsubara, K. Shinoda, X-Ray Diffraction Crystallography, X-Ray Diffr. Crystallogr. (2011). https://doi.org/10.1007/978-3-642-16635-8

[77]   S. Mourdikoudis, R.M. Pallares, N.T.K. Thanh, Characterization techniques for nanoparticles: comparison and complementarity upon studying nanoparticle properties, Nanoscale 10 (2018) 12871–12934. https://doi.org/10.1039/C8NR02278J

[78]   A. Barhoum, M. Luisa García-Betancourt, Physicochemical characterization of nanomaterials: size, morphology, optical, magnetic, and electrical properties, Emerg. Appl. Nanoparticles Archit. Nanostructures Curr. Prospect. Futur. Trends (2018) 279–304. https://doi.org/10.1016/B978-0-323-51254-1.00010-5

[79]   R. Bellotti, G.B. Picotto, L. Ribotta, AFM Measurements and Tip Characterization of Nanoparticles with Different Shapes, Nanomanufacturing Metrol. 5 (2022) 127–138. https://doi.org/10.1007/S41871-022-00125-X

[80]   M.M. Tosi, A.P. Ramos, B.S. Esposto, S.M. Jafari, Dynamic light scattering (DLS) of nanoencapsulated food ingredients, Charact. Nanoencapsulated Food Ingredients (2020) 191–211. https://doi.org/10.1016/B978-0-12-815667-4.00006-7

[81]   S. Falke, C. Betzel, Dynamic Light Scattering (DLS), (2019) 173–193. https://doi.org/10.1007/978-3-030-28247-9_6

[82]   A. Serrano-Lotina, R. Portela, P. Baeza, V. Alcolea-Rodriguez, M. Villarroel, P. Ávila, Zeta potential as a tool for functional materials development, Catal. Today 423 (2023) 113862. https://doi.org/10.1016/J.CATTOD.2022.08.004

[83]   M.J. Mitchell, M.M. Billingsley, R.M. Haley, M.E. Wechsler, N.A. Peppas, R. Langer, Engineering precision nanoparticles for drug delivery, Nat. Rev. Drug Discov. 2020 202 20 (2020) 101–124. https://doi.org/10.1038/s41573-020-0090-8

[84]   J. Jana, M. Ganguly, T. Pal, Enlightening surface plasmon resonance effect of metal nanoparticles

for practical spectroscopic application, RSC Adv. 6 (2016) 86174–86211.
https://doi.org/10.1039/C6RA14173K

[85]   M.A. Garcia, Surface plasmons in metallic nanoparticles: fundamentals and applications, J. Phys.
D. Appl. Phys. 44 (2011) 283001. https://doi.org/10.1088/0022-3727/44/28/283001

[86]   I. Manisalidis, E. Stavropoulou, A. Stavropoulos, E. Bezirtzoglou, Environmental and Health
Impacts of Air Pollution: A Review, Front. Public Heal. (2020).
https://doi.org/10.3389/fpubh.2020.00014

[87]   S.A. Mousa, D.A. Wissa, H.H. Hassan, A.A. Ebnalwaled, S.A. Khairy, Enhanced photocatalytic
activity of green synthesized zinc oxide nanoparticles using low-cost plant extracts, Sci. Reports 2024 141
14 (2024) 1–18. https://doi.org/10.1038/s41598-024-66975-1

[88]   M.B. Tahir, M. Sohaib, M. Sagir, M. Rafique, Role of Nanotechnology in Photocatalysis, Encycl.
Smart Mater. (2022) 578. https://doi.org/10.1016/B978-0-12-815732-9.00006-1

[89]   H. AlMohamadi, S.A. Awad, A.K. Sharma, N. Fayzullaev, A. Távara-Aponte, L. Chiguala-
Contreras, A. Amari, C. Rodriguez-Benites, M.A. Tahoon, H. Esmaeili, Photocatalytic Activity of Metal-
and Non-Metal-Anchored ZnO and TiO2 Nanocatalysts for Advanced Photocatalysis: Comparative
Study, Catal. 2024, Vol. 14, Page 420 14 (2024) 420. https://doi.org/10.3390/CATAL14070420

[90]   P.B. Angon, M.S. Islam, S. KC, A. Das, N. Anjum, A. Poudel, S.A. Suchi, Sources, effects and
present perspectives of heavy metals contamination: Soil, plants and human food chain, Heliyon 10
(2024) e28357. https://doi.org/10.1016/J.HELIYON.2024.E28357

[91]   K.L. Wasewar, S. Singh, S.K. Kansal, Process intensification of treatment of inorganic water
pollutants, Inorg. Pollut. Water (2020) 245–271. https://doi.org/10.1016/B978-0-12-818965-8.00013-5

[92]   M. Kumar, P. Borah, P. Devi, Priority and emerging pollutants in water, Inorg. Pollut. Water
(2020) 33–49. https://doi.org/10.1016/B978-0-12-818965-8.00003-2

[93]   P. Rajak, A. Ganguly, S. Nanda, M. Mandi, S. Ghanty, K. Das, G. Biswas, S. Sarkar, Toxic
contaminants and their impacts on aquatic ecology and habitats, Spat. Model. Environ. Pollut. Ecol. Risk
(2024) 255–273. https://doi.org/10.1016/B978-0-323-95282-8.00040-7

[94]   Nishu, S. Kumar, Smart and innovative nanotechnology applications for water purification,
Hybrid Adv. 3 (2023) 100044. https://doi.org/10.1016/J.HYBADV.2023.100044

[95]   Paras, K. Yadav, P. Kumar, D.R. Teja, S. Chakraborty, M. Chakraborty, S.S. Mohapatra, A.
Sahoo, M.M.C. Chou, C. Te Liang, D.R. Hang, A Review on Low-Dimensional Nanomaterials:
Nanofabrication, Characterization and Applications, Nanomater. 2023, Vol. 13, Page 160 13 (2022) 160.
https://doi.org/10.3390/NANO13010160

[96]   R.M. Tripathi, S.J. Chung, Biogenic nanomaterials: Synthesis, characterization, growth
mechanism, and biomedical applications, J. Microbiol. Methods 157 (2019) 65–80.
https://doi.org/10.1016/J.MIMET.2018.12.008

[97]   P. Kuppusamy, M.Y. Mashitah, G.P. Maniam, N. Govindan, Biosynthesized gold nanoparticle
developed as a tool for detection of HCG hormone in pregnant women urine sample, Asian Pacific J.
Trop. Dis. 4 (2014) 237. https://doi.org/10.1016/S2222-1808(14)60538-7

Biogenic Nanomaterials: Synthesis, Characterization, Applications, and Future Remarks   Materials Research Forum LLC
Materials Research Foundations 180 (2025)   https://doi.org/21741/9781644903759

Chapter 3

Synthesis Methods and Characterization Parameters of Biogenic Nanomaterials

N. Bazancir[1*], I. Meydan[1], S. Celikozlu[2], F. Sen[3*]

[1]Van Vocational School of Health Services, Van Yuzuncu Yil University, Van, Turkey.

[2]Department of Food Processing, Altıntaş Vocational School, Kutahya Dumlupınar University, Kutahya, Turkiye

[3]Sen Research Group, Department of Biochemistry, Kutahya Dumlupinar University, Kutahya, Turkiye.

nuranbazencir@yyu.edu.tr; fatih.sen@dpu.edu.tr

Abstract

Biological nanomaterials are nanoparticles synthesized by living cells, plants or microorganisms. These nanomaterials are environmentally friendly and biocompatible because they consist of naturally occurring components in biological systems. Various methods are used to synthesize biological nanomaterials such as biological reaction, biological mineralization, biological photosynthesis. Characterization of biological nanomaterials is important to determine the size, shape, composition and surface properties of nanoparticles. Various techniques are used to characterize the properties of biological nanomaterials such as XRD, SEM, XRF, AAS, MS, IR, AFM and, XPS. These techniques allow us to understand the properties and performance of nanoparticles.

Keywords

Biological Nanoparticles, Nanoparticle Types, Synthesis, Characterization, Applications

1.   Introduction

The emergence of the concept of nanotechnology was introduced in 1959 by scientist Dr. Feynman's mention of the possibility of configuring tools and equipment used at the molecular level has gained importance [1]. Feynman was a key figure in the emergence of nanotechnology, but other scientists pioneered the field. Nanoparticle research, as an important research area, has made great strides in recent years [2]. It is predicted that a new era will begin for future generations with the realization of possible changes thanks to nanotechnology, which has become an interesting and valuable research area today. The reason for the increased interest in this subject is that, in contrast to the volumetric structure of matter, it exhibits unusual properties and functions in certain size ranges. Nanotechnology is based on working at the molecular level, structuring atom by atom, and combining large structures with essentially new molecular arrangements [3]. Due to their nano size, nanoparticles have some unique properties. The demand for metal nanoparticles is increasing day by day. The main reason for this situation is that these structures can be used in various research fields such as catalysis, magnetic recording devices and electricity [3]. The unique structure of metal nanoparticles enhances their functionality when these particles are used in the electronics and materials industries. It is key to various technologies such as the synthesis of antimicrobial compounds, drug delivery, medical applications such as disease diagnosis and treatment, energy, sensors, and environment [4–11].

Although nanotechnology is a concept that has been mentioned frequently in the last decade, the foundations of worldwide research on the subject date back to the 1950s. It is believed that this technology, which can measure in the nanometer range, can be carried to higher levels thanks to its technical advantages. Current techniques focus on known physical properties of matter. Exploring the nanosize provides new properties, expressed as quantum effects, that vary with the size of the material [4]. The first step in new developments in nanotechnology, including the synthesis of nanostructured materials, is the production of nanoparticles. This review focuses on methods recently used for synthesis and outlines their biomedical applications and their commercial and environmentally hazardous or harmless applications. Synthetic strategies include greener chemical, physical, and biogenic methods, and their role in surface modifiers encompasses a variety of biomedical, commercial, and environmental applications. We will also consider the current situation and future prospects.

2. Nanoparticles

Nanoparticles are typically powders smaller than 100 nm and aggregate into very fine, nanometer-sized particles. Definitions of nanoparticles vary by area, material, and application. For solid materials with very different physical properties, particles smaller than 10-20 nm are considered, while for other applications, particles in the 1 nm1 μm range are considered nanoparticles [12]. Nanoparticles have found applications in many fields from basic materials science to biomedical applications. It is frequently used for analytical purposes in sensor applications, especially due to its optical properties. Nanoparticle-based sensors have been developed for protein, metal, DNA and virus analysis. Various chemical and physical methods are used to synthesize these particles. Many particles obtained by these methods contain toxic chemicals. The resulting particles are therefore toxic. Particle synthesis by biological methods has gained importance in recent years in order to reduce the toxic effects of chemical and physical methods [13].

This new method can kill plants, bacteria, fungi, yeast, algae, viruses and more. Biology is involved in nanoparticle synthesis. Increasing interest in green chemistry and other biological processes is encouraging scientists to develop simple, inexpensive and environmentally friendly nanoparticle synthesis methods.

2.1   Nanoparticle Types

2.1.1   Metallic Nanoparticles

Metal oxides play important roles in materials science such as microelectronics, sensors, piezoelectric devices, fuel cells, surface passivation coatings and corrosion catalysis. Metal oxides are also used as absorbents for environmental pollutants. Various physical and chemical processes synthesize large quantities of metal nanoparticles in a relatively short time. Chemical methods make heavy use of toxic chemicals that can adversely affect medical practices [4]. Currently, plant-mediated biosynthesis of nanoparticles is gaining increasing importance due to its simplicity, environmental friendliness and wide range of biological activities. Nanocrystalline sized (less than 100 nm) nanomaterials may behave more like atoms when dissociated into atomic sizes with large surface area due to valence and conduction bands [14]. For example, in environmental applications of water nanotechnology, iron nanoparticles derived from inexpensive and common natural precursors may be an option for heavy metal processing and disinfection. Systematic studies are also needed to elucidate some reaction mechanisms and to obtain more accurate results. For example, there are various theories about the antiseptic effect of reducing Ag^+ to Ag^0 and silver nanoparticles [15].

The application areas of silver nanoparticles are not only agriculture, textile industry, cosmetics and food industry, but also biosensors, chemical reaction catalysts, photoreceptors, antibacterial agents, antiseptics, water disinfection of medical equipment, etc. It is also widely used in the fields [16]. Copper nanoparticles (Cu-NPs) are of great interest to scientists due to their applications in wound dressings and their biocidal properties in advanced applications (gas sensors, catalytic processes, high temperature superconductors, solar cells, etc.). Metallic nanoparticles are among the most promising and outstanding biomedical agents. Among metal NPs, silver, aluminum, gold, zinc, platinum, titanium, palladium, iron and copper are commonly used. As early as the 16th century, gold nanoparticles were used for processing and recoloring, so the eco-friendly method had to be extended to other organic methods [17]. A production scheme for stable and high-yield biological nanoparticles [18]. ZnO nanoparticles are widely used in many industrial fields such as solar cells, UV light emitting devices, gas sensors, photocatalysts, pharmaceutical and cosmetic industries [19]. Moreover, metal nanoparticles exhibit surface plasmon resonance absorption in the UV-visible region. In biomedical applications, it is used as a sunscreen and non-toxic, self-cleaning antibacterial agent for many skin types [20] and as a blocking agent for dermatology and UV resistance [21].

2.1.2 Polymeric Nanoparticles

Poparticles, nanospheres or nanocapsules are matrix systems made of natural or synthetic polymers, with sizes ranging from 10 to 1000 nm, depending on the production method [22]. Nanoparticles made from biodegradable and biocompatible polymers (polylactic acid (PLA), polyglycolic acid (PGA) or their copolymers, poly(d,l-lactic glycolic acid) (PLGA)) are used as a delivery system [23]. The number of studies and applications of nanotechnology systems in the fields of nanotechnology, medicine and biotechnology has increased significantly [24]. Medical applications include the use of drugs as gene and antigen carriers, in vitro/in vivo diagnostic applications, nutritional supplements, and the production of improved biocompatible materials [24],[25].

Nanoparticles made using natural or synthetic polymers have two main advantages when targeting proteins, peptides, genes and drugs. The first of these properties is the small size of nanoparticles. For this reason, they enter the cells from small capillaries and allow active substances to accumulate in target areas [13].

A second consideration is the use of biodegradable materials in the manufacture of nanoparticles. The use of biodegradable materials allows for controlled release of drugs into target tissues within days or weeks. Also nanoparticles. They increase drug/protein or peptide stability [26].

2.1.3   Quantum Dots

Nanomaterials have various applications in improving human life and the environment [27]. Modern science has begun to explore quantum dots in the 21st century. Quantum dots (QDs) are inorganic semiconductor nanocrystals that emit light in all spectral colors depending on their size [28]. Quantum dots are artificially charged droplets that can contain anything from a single electron to thousands of electrons. Typical sizes range from nanometers to several microns, and their size, shape and interactions can be precisely controlled using advanced nanofabrication techniques. The physics of quantum dots shares many similarities with the behavior of naturally occurring quantum systems in nuclear physics. In fact, quantum dots represent an important trend in condensed matter physics, where people study real atoms and nuclei instead of man-made objects. Similar to atoms, the energy levels of quantum dots are measured from electron compression [29]. The QD size decreases near the blue end of the spectrum and increases near the red end. In addition to visible light, it also has unique properties such as being able to be tuned to the infrared or ultraviolet spectrum. Quantum dots are valuable tools in biotechnology, particularly in cell imaging [30] and labeling. It is considered to be an excellent alternative to conventional fluorescent dyes used for imaging [31,32].

2.2   Application Areas of Nanoparticles

2.2.1   Pharmaceutical Industry

Nanoparticles are small-scale deployable particles that offer great advantages in drug discovery, drug delivery, genes and antigens, diagnostics, and the production of biocompatible materials and pharmaceuticals. One of the applications of nanoparticles is their usefulness as drug delivery systems [33]. As drug delivery systems, nanoparticles can be defined as particles in the particle size range of 10-1000 nm that release encapsulated, encapsulated or surface modified drugs. Nanoparticles have long been used to deliver different molecules to different parts of the human body. The main objectives of nanoparticle delivery systems are particle size, surface properties, and drug or active ingredient release to achieve the highest efficiency [34]. Recently, various applications of metal nanoparticles have been investigated in biomedical, agricultural, environmental and biochemical fields. For example, gold nanoparticles have been used to specifically disperse drugs such as paclitaxel, methotrexate, and doxorubicin [35].

Especially medicinal plants contain many metabolites with pharmacological activity. Many studies have shown that as these metabolites are synthesized, they become stronger by binding to nanoparticles and give them more properties [36]. Particles of different shapes and sizes have different properties depending on the materials used and production methods. Liposomes, solid lipid particles, micelles, dendrimers, hydrogels, conjugates, etc. are the systems examined in this context [37]. The emergence of drug delivery as an effective tool to treat various diseases such as cancer is one of the greatest advances in nanotechnology. Nanoparticles are one of the key components in drug delivery. In various studies, ZnO has been used for drug release in various diseases. One study used ZnO quantum dots as a drug delivery system to detect lung cancer cells with doxorubicin [38]. The reason for encapsulating ZnO nanoparticles with chitosan is to increase the stability of nanomaterials [39]. Their results show that the drug delivery system can be used as an effective construct to identify cancer cells.

2.2.2 Biomedical Applications of Nanoparticles

New applications for nanoparticles and nanomaterials are rapidly developing with new or improved properties based on size, distribution and morphology. Healthcare, Cosmetics, Biomedical, Food and Feed, Pharmaceutical Gene Delivery, Environment, Healthcare, Mechanics, Optics, Chemical Industry, Electronics, Aerospace Industry, Energy Science, Catalysts, Lumophors, Single Electron Transistors, Non-Linear Optical Devices and Optoelectronic Chemical applications. It is becoming more and more effective in many areas. For example: [40]. Among the nanoparticles used in these applications, metal nanoparticles are promising. Metal ions have important antibacterial properties due to their large surface area. Therefore, the increasing resistance of microorganisms to antibiotics and the emergence of these resistant strains have attracted the attention of researchers [41]. These structures are added as an antiseptic to wounds, topical creams, antiseptic sprays, and biomedical agents and exert a broad bactericidal action by disrupting the cell membranes of microorganisms and inhibiting enzymatic activity [42]. Nanoparticles (NPs) can be obtained from natural sources as one of the synthetic products or by-products of chemosynthesis [43]. Its high surface-to-volume ratio and antimicrobial properties make it useful for medical application [44].

The unique properties of nanoparticles make them ideal for the development of electrochemical and biosensors. For example, nanosensors have been developed to detect algal toxins, mycobacteria and mercury in drinking water. Researchers have also developed nanosensors that use nanomaterials to detect pests, viruses, soil nutrient levels and stressors, as well as hormonal regulation. For example, nanosensors have been developed to detect oxygen and its distribution. In addition, copper and palladium nanoparticles have been used in batteries, polymer and plastic plasma waveguides, and optical limiting devices [35].

Microorganisms emerged as tiny nanofactories and the synthesis of nanoparticles by microorganisms brought together biotechnology, microbiology and nanotechnology in a new field, nanobiotechnology. Although metal-microbial interactions, biomineralization and bioextraction are widely used, nanobiotechnology is still in the early stages of development [45]. Microbial synthesis has broad prospects for potential applications in healthcare and nanotechnology [46]. Many factors such as size, stability and shape influence the potential of nanomaterials in biomedicine. Small NPs are useful in biomedical fields as they readily enter cell membranes through absorption mechanisms. Media concentration, pH, temperature, time, pressure, etc. For example, the shape and stability of nanoparticles can be controlled by adjusting the physical and chemical parameters of the synthesis process [47].

Recent advances in nanotechnology have multiplied nanoparticles, wires and tubes that can be used in various fields. Quantum size effects in semiconductors and supermagnetism in nanoparticles in magnetic materials are believed to be the basis of next-generation photoelectric, electronic and various chemical and biochemical sensors [45]. However, their interaction with the body disrupts normal activity and causes illness and disease. An important question is whether the unknown risks of engineered nanoparticles (NPs), particularly the health and environmental effects, outweigh the benefits to society [48]. If nanomaterials are of great interest, it is because of their potential to interact with biological systems [49]. This is due to recent developments in physics and chemistry. The ability of optical systems and electrical and magnetic methods to detect the state of biological systems and organisms has led to potential applications in biology and medicine [50]. Therefore, NPs can be designed to exhibit different properties such as fluorescent and magnetic moments, and these properties can be used as natural nanoparticles in biological and medical applications. In recent years, the combination of NPs and biomolecules has been used successfully in materials science and biology research [51]. Nanoscale materials hold great promise for industrial and biomedical applications. Toxicology studies show that nanoparticles may have adverse health effects, but the primary causal link is still unclear. The interaction of nanoparticles with biological systems, including living cells, has therefore become one of the most pressing areas of joint research in materials science and biology. For example, gold nanoparticles were used as tracers and cell trajectories changed in response to bio-signals applied to the host material, again suggesting that the gold particles themselves were not toxic. In addition, oligonucleotide-modified 13 nm gold particles were used for intracellular gene regulation. In a recent report, rheumatoid arthritis reactions suggest an anti-inflammatory/rheumatoid molecular mechanism where gold(III) salts reduce antigen presentation, thereby reducing the activity of gold drugs in autoimmune arthritis [52]. Due to the different physicochemical properties of nanoparticles from different sources, their cytotoxic potential has been investigated. Water-soluble rosette nanotube structures have been shown to have low pulmonary toxicity due to their biologically inspired design and self-assembly structure. In a review of widely used metal oxide and carbon nanomaterials, he emphasized that he found that the physicochemical characterization of nanomaterials and their interactions with the biological environment are essential for reliable research [53]. There are many reports showing that nanoparticles penetrate deeper. We have shown that TiO particles can pass through the human cornea and reach the epidermis and even the dermis. The stretching action of normal skin has been shown to facilitate penetration of micron-sized fluorescent beads into the dermis [53]. They demonstrated the penetration of various nanoparticles into the dermis and their translocation into systemic vessels via lymphatic and regional lymphatics [54].

They recently investigated possible interactions for *S. galactiae* and *S. aureus*. These studies showed that low concentrations of 60–100 nm polyvinyl alcohol (PVA)-ZnO nanoparticles do not cause cell damage. However, they reported that nanoparticle concentrations above 0.016 M cause cell damage [54]. They examined the toxicity of ZnO nanoparticles against Vibrio fischeri, Daphnia magna and Tamnocephalus pratulus and showed that metal oxide nanoparticles can cause cell membrane damage without penetrating cells [55]. Moreover, the release of metal ions within ZnO NPs caused by dissolution under the influence of real ions has been shown to be responsible for toxicity in lung cell lines [56].

3.   Nanoproduction

The existence of nanomaterials was just an idea a long time ago. Feynman's 1959 idea that many universes underneath ushered in this era, but no tangible progress has yet been made. The existence of such a world has been established by research based on these ideas [57].

Research on nano fabrication methods was initiated to develop designs from the nano world. It was originally designed as an idea and simulated in a digital environment. The research divided nanomanufacturing into two main methods. This; It is a top-down production route and a bottom-up production route. As the name of these main methods suggests, the top-down method is likened to machining a new part from a solid material, while the bottom-up method is likened to manufacturing a large system [58].

3.1   Top-Down Production Ways

The top-down production route is a method based on size reduction by shrinking material in batches to nanoscale sizes. Nanostructures are produced using a variety of mechanical and chemical methods. The methods involved in the top-down approach are based on breaking the material into small pieces by externally energizing the bulk material through mechanical or chemical processes up to the nanoscale.

The most common examples of how to use the top-down approach are: These include mechanical grinding, chemical etching, laser cutting, sandblasting and blasting. These techniques are also called high-energy grinding or high-speed grinding, as they consume significantly more energy than conventional grinding processes. The methods involved in top-down methods are based on the principle of breaking bulk materials into small pieces that are reduced to the nanoscale by stimulation of an external mechanical or chemical process. Examples of the most common methods that can be used for top-down methods are mechanical grinding and etching. These techniques are also known as high-energy mills or high-speed mills, as they use more energy than conventional grinding processes. In this way, production begins with larger materials until it reaches the microscopic level. A good example of this method is the dry powder milling of wheat and the associated increased water holding capacity [59].

One study reported that reducing the size of green tea particles to 1000 nm improves the digestion and absorption of green tea, thereby increasing oxygen-removing enzymes and, more scientifically, increasing its antioxidant activity [60].

3.2   Bottom-Up Production Ways

This method uses atoms or molecules to create organic or inorganic structures. Carbon nanotubes can be designed using the self-assembly properties of biological forces (such as DNA) to attach them to nanostructures. Applications of this method are defined as the creation of particles using chemical reactions to form atomic or molecular structures. Gas condensation technology is the first method of producing nanometals and their alloys with a bottom-up approach. Chemical vapor deposition, chemical vapor condensation, sol-gel and spray pyrolysis are other well-known methods of this method [61]. Many classical physical and chemical methods have been used to synthesize nanoparticles in solution medium. Today, "green nanotechnology" has emerged, which includes low-cost, environmentally friendly and non-toxic biological techniques [62]. In addition, various methods have been developed within the classical method. These include techniques such as electrochemical synthesis, reverse micelle/microemulsion method, hydrothermal synthesis, sonochemical precipitation, and chemical reduction. The nanostructures obtained in most of these studies should have certain dimensions and morphologies. Although classical synthetic methods can be used to synthesize nanoparticles of desired size and morphology, these methods have disadvantages such as using green nanotechnology for research on more economical, simpler and non-toxic materials [63],[64].

3.3   Chemical Vapor Condensation Method

Chemical Vapor Concentration (CVC) was first developed in Germany in 1994 and is an ideal method to produce nanoparticles in large quantities [65]. In this method, the starting materials are metal oxides, carbonyl compounds, chlorides and hydrides, which tend to transition to the gas phase [66]. The biggest advantage of the CVC method is that it can produce almost any material with a wide range of chemical compositions from raw materials with different chemical contents available in the market [65]. This method is mainly based on the conversion of gaseous substances into particles by pyrolysis. Process flow; that is, a gas stream is introduced into the area where the starting materials are evaporated and the steam is fed into a furnace, also called a reactor, where pyrolysis takes place. Inert gases such as He, Ar and N_2 are used as the carrier gas, but in addition to the carrier gas, gases such as H_2, CO and CH_4 can also be used to reduce the compound. Atomic clusters or nanoparticles formed by pyrolysis can be collected at the furnace exit by various dust collection methods. The most common of these is the accumulation of particles in the sealed chamber of the rod through which liquid nitrogen flows.

3.4   Hydrogen Reduction Method

This is a method of producing metal nanoparticles by reducing the gas phase. Studies have shown that nanoparticles of iron group metals (Fe, Ni, Co) are specially used for laboratory-scale synthesis. Methods; particle formation, particle collection and gas cleaning. The starting solution used in the first step of the process is evaporated and particles are formed by being transported together with the carrier and/or reducing gas to a preheated zone and then to the reduction hot zone. Hydrogen can be used as both a reducing agent and carrier, but inert gases such as nitrogen and argon can also be used as carriers. Reagent concentration, reaction temperature, preheat zone temperature, and vapor/particle residence time are the main factors controlling particle size, size distribution, and crystallinity [67].

3.5   Green Synthesis

New techniques are needed because the traditional physical and chemical methods used for many years to produce nanoparticles have undesirable properties. This is the most environmentally friendly approach currently used in nanoparticle production.

It is the ultimate plant/microbe mediated biosynthesis that can be produced by biological organisms such as plants and fungi in the presence of suitable substrates with desirable properties under environmental conditions. Plants and microorganisms are exposed to high concentrations of heavy metals and damage that affects their morphology, membranes, enzymes and DNA leads to cell death. Despite these adverse effects, organisms exposed to toxic concentrations of heavy metals have developed resistance mechanisms by enzymatically oxidizing heavy metals or reducing them to less toxic forms or covalent modifications [68]. This naturally occurring phenomenon has inspired scientists to use naturally occurring materials and biological structures to create perfectly engineered nanoscale materials. Green synthesis is one of the most environmentally friendly techniques in the synthesis of nanoparticles. Generally, organic compounds obtained from plants and microorganisms are used. These natural resources are generally abundant in nature and allow for the biosynthesis of large quantities of nanoparticles [64]. Among the various natural materials used for nanoparticle synthesis, plant extracts seem to be the best candidates. Nanoparticles from plant extracts are more stable, vary in size and shape, and are produced more quickly compared to microbial production processes [69]. In this sense, many plants and extracts used for the production of various metal nanoparticles, especially copper, silver and zinc, have been reported in the literature [70].

A 2015 study showed that plants and plant products are inexpensive and renewable resources for nanomaterial production. In recent years, the use of plant extracts has replaced physical and chemical methods and is widely used in the field of health [71]. At the same time, its biocompatibility and commercial production capabilities offer many potential uses, particularly in pharmaceutical and medical applications [64].

4.   Nano-Scale Measurement And Characterization Methods

It is very important to be able to express what is produced as much as nano capacity and development in technology development. Therefore, the developed technology undergoes intensive research on imaging tools to solve scientific mysteries, expand the knowledge pool and discover new things. The methods used to characterize nanomaterials can be classified into three main groups. These are microscopy, spectroscopy and spectrometry.

Microscopy; represents the morphological information of the sample under consideration. In other words, it is used to analyze the shape and size of nanoparticles. In addition to the physical properties of objects, our eyes are healthy and need the right amount of light. Therefore, imaging of smaller dimensions was first achieved by optical microscopy. Lenses were designed to see objects and objects too small to be seen with the naked eye, and these lenses made the advancement of microscopes possible. The development of lenses and advances in optical technology contributed to and accelerated the discovery of the nanoworld. However, scientists are turning to new researches with the idea that there are structures that cannot be visualized even with a light microscope [72]. As a result of their new research, they built the first prototype of the electron microscope in 1933, with the idea that lenses could focus light and electrons in a magnetic field to enable imaging. The electron microscope produced provided a three-dimensional image in addition to the two-dimensional image of the light microscope. In order to obtain the first image in an electron microscope, the electron beam must be directed at the sample to be imaged at high speed. When accelerated electrons form an image, they either pass through a material or are reflected from its surface. A scanning electron microscope (SEM) is a microscope that reflects a beam of electrons from the surface of an object, and a TEM transmission electron microscope is a microscope that creates an image by passing an electron beam through a sample.

SEM – Scanning Electron Microscope: Obtaining images with SEM is essential. It is based on the analysis of the collection of signals resulting from the physical interaction of the electron beam with the sample surface. Electron energies providing images in SEM range from 200 to 300 Ev, 100 keV. The electron beam is focused by the focusing electromagnetic lens and objective lens, and the sample surface is scanned by the electromagnetic deflection coil. In the interaction of electron beams with matter, x-rays with keV energy also provide a signal. When the electrons collide with the sample, they leave the inner orbit of the electrons in the sample and the electrons in the upper orbit enter this orbit for energy balance and emit X-rays as they pass. A 10 mm diameter Si detector is sufficient to detect this X-ray radiation. X-ray signals are sent to an amplifier, then to a multi-channel analyzer and then to the computer of the SEM system. The characteristic X-ray SEM obtained by these processes enables SEM measurement of the elemental content of the imaged sample.

Conductive materials are visualized using SEM. The conductive material is also grounded and screened. Particularly, the sample surface, which consists of heavy atoms, can be observed in detail and sharply. To obtain an accurate image, the sample surface must be very clean and free of contamination in the vacuum system.

Therefore, SEM provides topographic information in a vacuum environment. The lateral resolution is typically about 0.5 nm and the instrumentation required for this imaging modality is intermediate to advanced.

TEM transmission electron microscope: The most distinctive feature of this microscope is its extremely sensitive technique, which can obtain images at the atomic level. The difference between the methods used for imaging in microscopy and the methods used in SEM is the way the electron method works. In a TEM, an electron beam passes through the sample under examination. After leaving the source, the electron beam that generates the signal in the TEM is focused on the material using a lens. An electron beam striking the sample under examination penetrates the sample and creates an image that provides structural information about the sample.

The microscope requires that the thickness of the sample be no more than a few hundred nanometers in order to create a beam of electrons that carries the signal that produces the image. For this reason, the samples to be imaged with TEM should be carefully prepared. However, examining very thin samples can cause problems such as: Absorption of the electron beam passing through the sample in a very short time without reflecting the basic properties of the material. To overcome these problems, the high-voltage electron microscope was developed. These microscopes use very high voltages to achieve the desired quality, with electron beam energies ranging from 100 to 500 kilovolts in high voltage microscopes. The high-energy electron beam passes through several different magnetic lens systems before focusing on the material under investigation. Just after the electrons pass through the sample, they pass through the magnetic lens system again and are reflected on the screen.

TEM microscopes are the most powerful electron microscopes in use today. TEM microscopes are the first choice in many laboratories due to their important advantages such as ease of use, image stability and phase range of 100-500 kilovolts. This microscope can provide images with a resolution of up to 0.14 nanometers. This provides information on morphology, crystal structure, and defect distribution in TEM imaging. Although the lateral resolution varies by about 0.1 nm, imaging is done in an ultra-high vacuum environment. Equipment costs are high.

Spectroscopy; Information is given about the components of the analyzed sample and the determination of its chemical structure.

Spectroscopy: A general term for methods that use light rays and particles to study the properties of substances and objects in various ways. Spectroscopy is a method of examining the properties of matter using absorbed particles, light or sound [73]. This is how we determine the "quantized" energy levels of molecules, ions, and nuclei. It is not based on chemical analysis. Spectral analysis is not chemical analysis either. Physics is a technology based on phonetics and optics. Experimentally, it only includes frequency measurements. It can be used in suitable items and places. Spectroscopy can be used in many fields such as astrophysics, medicine, electrochemistry, nuclear physics, nuclear chemistry, analytical chemistry and molecular biology. In other words, spectroscopy is the measurement and interpretation of light-matter interactions in the form of absorption and emission of light or radiation by materials [74].

Spectroscopic analysis; Provides quantitative and depth profile information. It is a tool for measuring the properties of light in specific parts of the electromagnetic spectrum and is widely used in spectroscopic analysis to identify materials. The variable being measured is usually the intensity of the light, but it can also be the polarization state, for example. The independent variable is usually the wavelength of light or units directly proportional to the photon energy, such as the inverse centimeter or electron volt, which is inversely proportional to the wavelength. Spectrometers are used in many fields. It is used, for example, in astronomy to analyze the radiation of celestial bodies and to draw conclusions about their chemical composition. A spectrometer uses a prism or diffraction grating to scatter light from a distant object into a spectrum. This would allow astronomers to identify many chemical elements with their characteristic spectral fingerprints. If the object is self-luminous, you will see spectral lines caused by the luminous gas itself.

These lines are named for the elements that cause them, such as hydrogen alpha, beta, and gamma lines. Chemical compounds can also be identified by absorption. These are usually dark bands located at certain points in the spectrum, caused by energy absorbed as light from other objects passes through gas clouds. Most of our knowledge of the chemical structure of the universe comes from spectra.

5. Conclusion

Synthesis of biogenic nanomaterials can be done by many methods, namely physical, chemical and biological or a combination of these. There are situations where one method is superior or weaker than another method in processes. Nowadays, it is more cost-effective, when left to nature, any it does not cause destruction, does not harm the living organism, is easily applicable and methods of producing the sample that will provide the desired properties are preferred. Therefore, the selection of biodegradable and biodegradable materials no waste as a result, materials that will not pose a risk of adverse reactions with the environment should be preferred.

Physical, chemical, optical, electronic and morphological properties of nanomaterials it is very important to obtain accurate data when reviewing; Therefore, appropriate techniques must be used. The device used, the structure of nanomaterials, morphological properties, wear rate of the material, three-dimensional mapping, should provide a better understanding of its chemical and physical properties. The characterization techniques that are tried to be explained in this review article are the results of nanostructure and NPs.

It is understood that it is not enough on its own to obtain its properties. More than one nanostructure characterization technique can be used simultaneously to elucidate a nanostructure. Moreover, every technique is also used in every nanostructure analysis. In scientific publications/literature reviews, more existing techniques appear to be used; but nanostructures with new properties the design and development of new nanostructure characterization techniques as they are scientifically derived, development will be inevitable.

References

[1]   D.S. Bhagat, W.B. Gurnule, G.S. Bumbrah, P. Koinkar, P.A. Chawla, Recent advances in biomedical applications of biogenic nanomaterials, Curr. Pharm. Biotechnol. 24 (2023) 86–100.

[2]   A. Hojjati-Najafabadi, S. Salmanpour, F. Sen, P.N. Asrami, M. Mahdavian, M.A. Khalilzadeh, A Tramadol Drug Electrochemical Sensor Amplified by Biosynthesized Au Nanoparticle Using Mentha aquatic Extract and Ionic Liquid, Top. Catal. 65 (2022) 587–594. https://doi.org/10.1007/s11244-021-01498-x

[3]   A. Aygun, F. Gulbagca, E.E. Altuner, M. Bekmezci, T. Gur, H. Karimi-Maleh, F. Karimi, Y. Vasseghian, F. Sen, Highly Active PdPt Bimetallic Nanoparticles Synthesized By One Step Bioreduction Method: Characterizations, Anticancer, Antibacterial Activities and Evaluation of Their Catalytic Effect for Hydrogen Generation, Int. J. Hydrogen Energy 48 (2023) 6666–6679. https://doi.org/10.1016/j.ijhydene.2021.12.144

[4]   N. Bazancir, I. Meydan, Characterization of Zn nanoparticles of Platonus orientalis plant,

investigation of DPPH radical extinquishing and antimicrobial activity, East. J. Med. 27 (2022).

[5]   A. Şavk, K. Cellat, K. Arıkan, F. Tezcan, S.K. Gülbay, S. Kızıldağ, E.Ş. Işgın, F. Şen, Highly monodisperse Pd-Ni nanoparticles supported on rGO as a rapid, sensitive, reusable and selective enzyme-free glucose sensor, Sci. Rep. 9 (2019) 19228. https://doi.org/10.1038/s41598-019-55746-y

[6]   N.H. Khand, A.R. Solangi, S. Ameen, A. Fatima, J.A. Buledi, A. Mallah, S.Q. Memon, F. Sen, F. Karimi, Y. Orooji, A new electrochemical method for the detection of quercetin in onion, honey and green tea using Co3O4 modified GCE, J. Food Meas. Charact. 2021 154 15 (2021) 3720–3730. https://doi.org/10.1007/S11694-021-00956-0

[7]   A. Cherif, R. Nebbali, J.W. Sheffield, N. Doner, F. Sen, Numerical investigation of hydrogen production via autothermal reforming of steam and methane over Ni/Al2O3 and Pt/Al2O3 patterned catalytic layers, Int. J. Hydrogen Energy (2021). https://doi.org/10.1016/j.ijhydene.2021.04.032

[8]   R. Ulus, Y. Yıldız, S. Eriş, B. Aday, F. Şen, M. Kaya, Functionalized Multi-Walled Carbon Nanotubes (f-MWCNT) as Highly Efficient and Reusable Heterogeneous Catalysts for the Synthesis of Acridinedione Derivatives, ChemistrySelect 1 (2016) 3861–3865. https://doi.org/10.1002/SLCT.201600719

[9]   J.T. Abrahamson, B. Sempere, M.P. Walsh, J.M. Forman, F. Şen, S. Şen, S.G. Mahajan, G.L.C. Paulus, Q.H. Wang, W. Choi, M.S. Strano, Excess Thermopower and the Theory of Thermopower Waves, ACS Nano 7 (2013) 6533–6544. https://doi.org/10.1021/nn402411k

[10]   F. Şen, Ö. Demirbaş, M.H. Çalımlı, A. Aygün, M.H. Alma, M.S. Nas, The dye Removal from Aqueous Solution Using Polymer Composite Films, Appl. Water Sci. 8 (2018) 206. https://doi.org/10.1007/s13201-018-0856-x.

[11]   Y. Wu, E.E. Altuner, R.N. El Houda Tiri, M. Bekmezci, F. Gulbagca, A. Aygun, C. Xia, Q. Van Le, F. Sen, H. Karimi-Maleh, Hydrogen generation from methanolysis of sodium borohydride using waste coffee oil modified zinc oxide nanoparticles and their photocatalytic activities, Int. J. Hydrogen Energy 48 (2023) 6613–6623. https://doi.org/10.1016/J.IJHYDENE.2022.04.177

[12]   H. Seçkin, İ. Meydan, Synthesis and Characterization of Veronica beccabunga Green Synthesized Silver Nanoparticles for The Antioxidant and Antimicrobial Activity, Turkish J. Agric. Res. 8 (2021) 49–55. https://doi.org/10.19159/TUTAD.805463

[13]   H. Seçkin, I. Meydan, Synthesis and characterization of *Sophora alopecuroides* L. green synthesized of Ag nanoparticles for the antioxidant, antimicrobial and DNA damage prevention activity, Brazilian J. Pharm. Sci. 58 (2022) e20992. https://doi.org/10.1590/S2175-97902022E20992

[14]   A. Van Dijken, E.A. Meulenkamp, D. Vanmaekelbergh, A. Meijerink, Identification of the transition responsible for the visible emission in ZnO using quantum size effects, J. Lumin. 90 (2000) 123–128.

[15]   O. V Kharissova, H.V.R. Dias, B.I. Kharisov, B.O. Pérez, V.M.J. Pérez, The greener synthesis of nanoparticles, Trends Biotechnol. 31 (2013) 240–248.

[16]   A. UMAZ, A. KOÇ, Yeşil Yolla Sentezlenmiş Metal Nanopartiküllerinin Biyomedikal Uygulamaları, Res. Net (n.d.).

[17]   M. Rafique, A.J. Shaikh, R. Rasheed, M.B. Tahir, H.F. Bakhat, M.S. Rafique, F. Rabbani, A review on synthesis, characterization and applications of copper nanoparticles using green method, Nano 12 (2017) 1750043.

[18]   L. Pereira, F. Mehboob, A.J.M. Stams, M.M. Mota, H.H.M. Rijnaarts, M.M. Alves, Metallic nanoparticles: microbial synthesis and unique properties for biotechnological applications, bioavailability and biotransformation, Crit. Rev. Biotechnol. 35 (2015) 114–128.

[19]   Z. Deng, M. Chen, G. Gu, L. Wu, A facile method to fabricate ZnO hollow spheres and their photocatalytic property, J. Phys. Chem. B 112 (2008) 16–22.

[20]   A. Yadav, V. Prasad, A.A. Kathe, S. Raj, D. Yadav, C. Sundaramoorthy, N. Vigneshwaran,

Functional finishing in cotton fabrics using zinc oxide nanoparticles, Bull. Mater. Sci. 29 (2006) 641–645.

[21]   K. Deepti, Precursor-controlled synthesis of hierarchical ZnO nanostructures, using oligoaniline-coated Au nanoparticle seeds/D. Krishnan, T. Pradeep, J. Cryst. Growth 311 (2009) 3889–3897.

[22]   J.P. Rao, K.E. Geckeler, Polymer nanoparticles: Preparation techniques and size-control parameters, Prog. Polym. Sci. 36 (2011) 887–913.

[23]   E. Cohen-Sela, M. Chorny, N. Koroukhov, H.D. Danenberg, G. Golomb, A new double emulsion solvent diffusion technique for encapsulating hydrophilic molecules in PLGA nanoparticles, J. Control. Release 133 (2009) 90–95.

[24]   A. Haleem, M. Javaid, R.P. Singh, S. Rab, R. Suman, Applications of nanotechnology in medical field: a brief review, Glob. Heal. J. 7 (2023) 70–77. https://doi.org/10.1016/J.GLOHJ.2023.02.008

[25]   C.-W. Lam, J.T. James, R. McCluskey, R.L. Hunter, Pulmonary toxicity of single-wall carbon nanotubes in mice 7 and 90 days after intratracheal instillation, Toxicol. Sci. 77 (2004) 126–134.

[26]   J.W. Stansbury, M.J. Idacavage, 3D printing with polymers: Challenges among expanding options and opportunities, Dent. Mater. 32 (2016) 54–64.

[27]   A. Aygun, R.N.E. Tiri, R. Bayat, F. Sen, Hydrothermal synthesis of BCQD@g-C3N4 nanocomposites supporting environmental sustainability: Organic dye removal and bacterial inactivation, J. Hazard. Mater. Adv. 16 (2024) 100464. https://doi.org/10.1016/J.HAZADV.2024.100464

[28]   C. Zhai, H. Zhang, N. Du, B. Chen, H. Huang, Y. Wu, D. Yang, One-pot synthesis of biocompatible CdSe/CdS quantum dots and their applications as fluorescent biological labels, Nanoscale Res Lett 6 (2011) 1–5.

[29]   L. Kouwenhoven, C. Marcus, Quantum dots, Phys. World 11 (1998) 35.

[30]   Y. Ghasemi, P. Peymani, S. Afifi, Quantum dot: magic nanoparticle for imaging, detection and targeting, Acta Biomed 80 (2009) 156–165.

[31]   P. Kumar, D. Kukkar, A. Deep, S.C. Sharma, L.M. Bharadwaj, Synthesis of mercaptopropionic acid stabilized CDS quantum dots for bioimaging in breast cancer, Adv. Mater. Lett 3 (2012) 471–475.

[32]   A. Aygun, I. Cobas, R.N.E. Tiri, F. Sen, Hydrothermal synthesis of B, S, and N-doped carbon quantum dots for colorimetric sensing of heavy metal ions, RSC Adv. 14 (2024) 10814–10825. https://doi.org/10.1039/D4RA00397G

[33]   S. Nazir, T. Hussain, A. Ayub, U. Rashid, A.J. MacRobert, Nanomaterials in combating cancer: therapeutic applications and developments, Nanomedicine Nanotechnology, Biol. Med. 10 (2014) 19–34.

[34]   L. Mu, S.S. Feng, A novel controlled release formulation for the anticancer drug paclitaxel (Taxol®): PLGA nanoparticles containing vitamin E TPGS, J. Control. Release 86 (2003) 33–48.

[35]   P. Singh, Y.-J. Kim, D. Zhang, D.-C. Yang, Biological Synthesis of Nanoparticles from Plants and Microorganisms, Trends Biotechnol. 34 (2016) 588–599. https://doi.org/10.1016/j.tibtech.2016.02.006

[36]   L. Sintubin, W. Verstraete, N. Boon, Biologically produced nanosilver: current state and future perspectives, Biotechnol. Bioeng. 109 (2012) 2422–2436.

[37]   P.P. Deshpande, S. Biswas, V.P. Torchilin, Current trends in the use of liposomes for tumor targeting, Nanomedicine 8 (2013) 1509–1528.

[38]   X. Cai, Y. Luo, W. Zhang, D. Du, Y. Lin, PH-Sensitive ZnO Quantum Dots-Doxorubicin Nanoparticles for Lung Cancer Targeted Drug Delivery, ACS Appl. Mater. Interfaces 8 (2016) 22442–22450. https://doi.org/10.1021/ACSAMI.6B04933

[39]   Q. Yuan, S. Hein, R.D.K. Misra, New generation of chitosan-encapsulated ZnO quantum dots loaded with drug: Synthesis, characterization and in vitro drug delivery response, Acta Biomater. 6 (2010) 2732–2739. https://doi.org/10.1016/J.ACTBIO.2010.01.025.

[40]   S. Kaviya, J. Santhanalakshmi, B. Viswanathan, [Retracted] Green Synthesis of Silver

Nanoparticles Using Polyalthia longifolia Leaf Extract along with D-Sorbitol: Study of Antibacterial Activity, J. Nanotechnol. 2011 (2011) 152970.

[41]   K.A. Khalil, H. Fouad, T. Elsarnagawy, F.N. Almajhdi, Preparation and Characterization of Electrospun PLGA/silver Composite Nanofibers for Biomedical Applications, Int. J. Electrochem. Sci. 8 (2013) 3483–3493. https://doi.org/10.1016/S1452-3981(23)14406-3

[42]   A. Ahmad, P. Mukherjee, S. Senapati, D. Mandal, M.I. Khan, R. Kumar, M. Sastry, Extracellular biosynthesis of silver nanoparticles using the fungus Fusarium oxysporum, Colloids Surfaces B Biointerfaces 28 (2003) 313–318. https://doi.org/10.1016/S0927-7765(02)00174-1

[43]   P. Kaur, R. Luthra, Silver nanoparticles in dentistry: An emerging trend, SRM J. Res. Dent. Sci. 7 (2016) 162. https://doi.org/10.4103/0976-433x.188808

[44]   S. Prabhu, E.K. Poulose, Silver nanoparticles: mechanism of antimicrobial action, synthesis, medical applications, and toxicity effects, Int. Nano Lett. 2 (2012) 32. https://doi.org/10.1186/2228-5326-2-32

[45]   H. Seckin, R.N.E. Tiri, I. Meydan, A. Aygun, M.K. Gunduz, F. Sen, An environmental approach for the photodegradation of toxic pollutants from wastewater using Pt–Pd nanoparticles: Antioxidant, antibacterial and lipid peroxidation inhibition applications, Environ. Res. 208 (2022) 112708. https://doi.org/10.1016/J.ENVRES.2022.112708

[46]   A. Fariq, T. Khan, A. Yasmin, Microbial synthesis of nanoparticles and their potential applications in biomedicine, J. Appl. Biomed. 15 (2017) 241–248.

[47]   H. Barabadi, S. Honary, P. Ebrahimi, M.A. Mohammadi, A. Alizadeh, F. Naghibi, Microbial mediated preparation, characterization and optimization of gold nanoparticles, Brazilian J. Microbiol. 45 (2014) 1493–1501. https://doi.org/10.1590/S1517-83822014000400046

[48]   V.L. Colvin, The potential environmental impact of engineered nanomaterials, Nat. Biotechnol. 21 (2003) 1166–1170. https://doi.org/10.1038/nbt875

[49]   S. American, N. America, S. American, Less is more in Medicine Author (s): A . PAUL ALIVISATOS Source : Scientific American , Vol . 285 , No . 3 (SEPTEMBER 2001), pp . 66-73 Published by : Scientific American , a division of Nature America , Inc . Stable URL : https://www.jstor.org/stable, 285 (2001) 66–73

[50]   P. Alivisatos, The use of nanocrystals in biological detection, Nat. Biotechnol. 22 (2004) 47–52.

[51]   S.G. Penn, L. He, M.J. Natan, Nanoparticles for bioanalysis, Curr. Opin. Chem. Biol. 7 (2003) 609–615.

[52]   Y. Pan, S. Neuss, A. Leifert, M. Fischler, F. Wen, U. Simon, G. Schmid, W. Brandau, W. Jahnen-Dechent, Size-dependent cytotoxicity of gold nanoparticles, Small 3 (2007) 1941–1949.

[53]   S. Arora, J.M. Rajwade, K.M. Paknikar, Nanotoxicology and in vitro studies: the need of the hour, Toxicol. Appl. Pharmacol. 258 (2012) 151–165.

[54]   M.H. Huang, S. Mao, H. Feick, H. Yan, Y. Wu, H. Kind, E. Weber, R. Russo, P. Yang, Room-temperature ultraviolet nanowire nanolasers, Science (80-.). 292 (2001) 1897–1899.

[55]   M. Heinlaan, A. Ivask, I. Blinova, H.-C. Dubourguier, A. Kahru, Toxicity of nanosized and bulk ZnO, CuO and TiO2 to bacteria Vibrio fischeri and crustaceans Daphnia magna and Thamnocephalus platyurus, Chemosphere 71 (2008) 1308–1316.

[56]   T.J. Brunner, P. Wick, P. Manser, P. Spohn, R.N. Grass, L.K. Limbach, A. Bruinink, W.J. Stark, In vitro cytotoxicity of oxide nanoparticles: comparison to asbestos, silica, and the effect of particle solubility, Environ. Sci. Technol. 40 (2006) 4374–4381.

[57]   S. Bayda, M. Adeel, T. Tuccinardi, M. Cordani, F. Rizzolio, The History of Nanoscience and Nanotechnology: From Chemical–Physical Applications to Nanomedicine, Molecules 25 (2020). https://doi.org/10.3390/MOLECULES25010112

[58]   C.C. Koch, Top-Down Synthesis Of Nanostructured Materials: Mechanical And Thermal Processing Methods., Rev. Adv. Mater. Sci. 5 (2003) 91–99.

[59]   N. Shibata, T. Saitoh, Y. Tadokoro, Y. Okawa, The cell wall galactomannan antigen from Malassezia furfur and Malassezia pachydermatis contains β-1, 6-linked linear galactofuranosyl residues and its detection has diagnostic potential, Microbiology 155 (2009) 3420–3429.

[60]   S. Machado, S.L. Pinto, J.P. Grosso, H.P.A. Nouws, J.T. Albergaria, C. Delerue-Matos, Green production of zero-valent iron nanoparticles using tree leaf extracts, Sci. Total Environ. 445 (2013) 1–8.

[61]   E. Halvacı, Ö. Özdemir, M. Kaya, Y.E. Serin, G. Tekkanat, T. Kozak, A. Aygün, Small particles, big changes; synthesis, characterization of nanomaterials and an overview of their application areas, J. Sci. Reports-B (2023) 16–38.

[62]   H. Bar, D.K. Bhui, G.P. Sahoo, P. Sarkar, S.P. De, A. Misra, Green synthesis of silver nanoparticles using latex of Jatropha curcas, Colloids Surfaces A Physicochem. Eng. Asp. 339 (2009) 134–139. https://doi.org/10.1016/j.colsurfa.2009.02.008

[63]   W.A. Goddard III, D. Brenner, S.E. Lyshevski, G.J. Iafrate, Handbook of nanoscience, engineering, and technology, CRC press, 2002.

[64]   I. Meydan, H. Seckin, H. Burhan, T. Gür, B. Tanhaei, F. Sen, Arum italicum mediated silver nanoparticles: Synthesis and investigation of some biochemical parameters, Environ. Res. 204 (2022) 112347. https://doi.org/10.1016/J.ENVRES.2021.112347

[65]   Z. Wang, H. Zhang, L. Zhang, J. Yuan, S. Yan, C. Wang, Low-temperature synthesis of ZnO nanoparticles by solid-state pyrolytic reaction, Nanotechnology 14 (2002) 11.

[66]   C.E. Knapp, C.J. Carmalt, Solution based CVD of main group materials, Chem. Soc. Rev. 45 (2016) 1036–1064. https://doi.org/10.1039/C5CS00651A

[67]   Y.J. Suh, H.D. Jang, H. Chang, W.B. Kim, H.C. Kim, Size-controlled synthesis of Fe–Ni alloy nanoparticles by hydrogen reduction of metal chlorides, Powder Technol. 161 (2006) 196–201.

[68]   C.A.M. Bonilla, V. V Kouznetsov, "Green" Quantum Dots: Basics, Green Synthesis, and Nanotechnological Applications, Green Nanotechnology-Overview Furth. Prospect. 1 (2016) 174–192.

[69]   F. Karimi, N. Rezaei-savadkouhi, M. Uçar, A. Aygun, R.N. Elhouda Tiri, I. Meydan, E. Aghapour, H. Seckin, D. Berikten, T. Gur, F. Sen, Efficient green photocatalyst of silver-based palladium nanoparticles for methyle orange photodegradation, investigation of lipid peroxidation inhibition, antimicrobial, and antioxidant activity, Food Chem. Toxicol. 169 (2022) 113406. https://doi.org/10.1016/J.FCT.2022.113406

[70]   I. Meydan, H. Burhan, T. Gür, H. Seçkin, B. Tanhaei, F. Sen, Characterization of Rheum ribes with ZnO nanoparticle and its antidiabetic, antibacterial, DNA damage prevention and lipid peroxidation prevention activity of in vitro, Environ. Res. 204 (2022) 112363. https://doi.org/10.1016/J.ENVRES.2021.112363

[71]   F. Nematollahi, Silver nanoparticles green synthesis using aqueous extract of Salvia limbata CA Mey., (2015).

[72]   B. Alberts, A. Johnson, J. Lewis, M. Raff, K. Roberts, P. Walter, Looking at the Structure of Cells in the Microscope, (2002). https://www.ncbi.nlm.nih.gov/books/NBK26880/ (accessed February 14, 2025).

[73]   V.S. ANDERSEN, [Infrared spectroscopy]., Dan. Tidsskr. Farm. 31 (1957) 93–106.

[74]   T. OBA, G. KAWATA, [TECHNIC OF INFRARED SPECTROMETRY]., Eisei Shikenjo Hokoku. 81 (1963) 41–2.

Biogenic Nanomaterials: Synthesis, Characterization, Applications, and Future Remarks   Materials Research Forum LLC
Materials Research Foundations 180 (2025)   https://doi.org/21741/9781644903759

Chapter 4

Properties of Biogenic Nanomaterials

S. Çibuk[1*], A. Ozengul[2], F. Sen[2*]

[1]Vocational School of Health Care, Van Yuzuncu Yil University, Van, Turkiye

[2]Sen Research Group, Department of Biochemistry, Kutahya Dumlupinar University, Kutahya, Turkiye

salihcibuk@yyu.edu.tr; fatih.sen@dpu.edu.tr

Abstract

Biogenic nanomaterials are synthesized by living organisms and exhibit unique properties that make them attractive for various applications in the fields of medicine, environment and energy. The properties of biogenic nanomaterials are influenced by their composition, area, size, shape and surface chemistry. In this chapter, we will discuss the properties of biogenic nanomaterials in detail, including their synthesis, characterization, and functionalization. We will also explore the various methods of synthesis, characterization, and functionalization of these materials. Additionally, we will discuss the challenges and future prospects of biogenic nanomaterials in various fields.

Keywords

Biogenic Nanomaterials, Properties, Synthesis, Characterization, Functionalization, Applications

1. Introduction

Biogenic nanomaterials are nanomaterials that are synthesized by living plants, animals, including bacteria and organisms [1–5]. These materials have unique features that make them attractive for a variety of applications, including tissue engineering, drug delivery and biosensors. The features of biogenic nanomaterials are influenced by their size, surface area, surface chemistry, shape and composition. Therefore, understanding these properties is crucial for developing efficient and effective applications of biogenic nanomaterials. In this chapter, we will discuss the properties of biogenic nanomaterials in detail [6,7]

The size and shape of biogenic nanomaterials are critical factors that influence their properties and applications. The size of biogenic nanomaterials typically ranges from a few nanometers to several hundred nanometers, depending on the organism that synthesizes them. For example, bacterial cells can synthesize nanoparticles with a size range of 10-100 nm, while plants can produce nanoparticles with a size range of 10-500 nm [3–6]. The shape of biogenic nanomaterials can also vary, depending on the organism that synthesizes them. For example, bacterial cells can synthesize spherical or rod-shaped nanoparticles, while plants can produce

50

nanoplates or nanorods. The size and shape of biogenic nanomaterials can influence their properties, such as their surface area, stability, and toxicity [7,8].

The composition of biogenic nanomaterials can also vary, depending on the organism that synthesizes them. Biogenic nanomaterials can be composed of organic or inorganic materials, or a combination of both. For example, bacterial cells can synthesize silver nanoparticles using silver ions in the presence of reducing agents [14]. These nanoparticles are composed of silver and organic molecules produced by the bacteria. Plants can produce gold nanoparticles using plant extracts, which contain various organic molecules, such as flavonoids, polyphenols, and terpenoids. The composition of biogenic nanomaterials can influence their properties, such as their stability, solubility, and toxicity [10,11].

The surface area and surface chemistry of biogenic nanomaterials are also critical factors that influence their properties and applications. The surface area of biogenic nanomaterials is usually large, which allows for increased interaction with the surrounding environment. The surface chemistry of biogenic nanomaterials can also vary, depending on the organism that synthesizes them. For example, bacterial cells can produce nanoparticles with a capping layer of organic molecules, which can influence their surface chemistry and stability. Plants can produce nanoparticles with various surface functional groups, such us carboxyl, amine groups and hydroxyl, which can influence their solubility and reactivity [17].

The synthesis of biogenic nanomaterials can be obtained through various methods, including chemical, biological and physical methods. Biological methods involve using living organisms, such as fungi, plants and bacteria.

2.   Biogenic Nanomaterials Produced by Bacteria: Synthesis, Characterization, and Applications

Bacteria are capable of synthesizing nanomaterials with unique features that make them attractive for a variety of applications in biomedicine, environmental remediation, and energy. These biogenic nanomaterials can be produced using simple and eco-friendly methods, and they offer several advantages over conventionally synthesized nanomaterials, including low toxicity, biocompatibility, and easy functionalization. In this chapter, we review the synthesis, characterization, and applications of biogenic nanomaterials produced by bacteria, with a focus on metallic nanoparticles. We discuss the mechanisms of nanoparticle synthesis by bacteria, the factors affecting nanoparticle size, shape, and composition, and the methods of nanoparticle characterization. We also highlight the applications of bacterial nanomaterials, including drug delivery, biosensing, and catalysis, and the challenges and future prospects of using bacterial nanomaterials in various fields.

Bacteria are microorganisms that have the ability to synthesize nanomaterials with unique properties. The synthesis of these biogenic nanomaterials is a complex process that involves the reduction of metal ions to form nanoparticles, which are then stabilized by various biomolecules produced by the bacteria [18]. The properties of these nanoparticles, such as shape, size and composition can be controlled by altering the reaction conditions, such us the temperature, pH and concentration of metal ions. Biogenic nanomaterials produced by bacteria offer several advantages over conventionally synthesized nanomaterials, including low toxicity, biocompatibility, and ease of functionalization [14,15].

The synthesis of biogenic nanomaterials by bacteria involves the reduction of metal ions to form nanoparticles, which are then stabilized by various biomolecules produced by the bacteria. The mechanisms of nanoparticle synthesis by bacteria can vary depending on the species of bacteria and the metal ion used [21]. For example, some bacteria, such as *Geobacter sulfurreducens* and *Shewanell oneidensis*, use extracellular electron transfer mechanisms to reduce metal ions and form nanoparticles, while others, such as *Bacillus subtilis* and *Pseudomonas aeruginosa*, use intracellular mechanisms involving enzymes and proteins to synthesize nanoparticles [19].

The factors affecting the shape, size and composition of biogenic nanoparticles produced by bacteria include the concentration of metal ions, the reaction pH, temperature, and the presence of reducing agents and stabilizing agents. The shape and size of biogenic nanoparticles can be controlled by adjusting the reaction conditions, and the composition can be altered by using different metal ions or modifying the bacterial culture conditions [17,18].

The characterization of biogenic nanoparticles produced by bacteria is essential for optimizing their applications and understanding their properties. The methods of nanoparticle characterization include scanning electron microscopy, UV-Vis spectroscopy, transmission electron microscopy, Fourier transform infrared spectroscopy and X-ray diffraction. These techniques allow for the determination of nanoparticle shape, composition, crystallinity and size, as well as the identification of biomolecules responsible for nanoparticle stabilization [18–22].

Biogenic nanoparticles produced by bacteria offer several advantages over conventionally synthesized nanoparticles in various applications, catalysis, including drug delivery and biosensing. For example, silver nanoparticles produced by *Pseudomonas stutzeri* have been shown to exhibit antibacterial activity against a range of bacteria and are being investigated for their use in wound healing and medical implants [23,24]. Gold nanoparticles produced by *Bacillus licheniformis* have been used as biosensors [25,26].

3. Biogenic Nanomaterials Produced by Fungi: Synthesis, Properties, and Applications

Biogenic nanomaterials produced by fungi have gained significant attention due to their unique properties and potential applications in various fields, including medicine, biotechnology, and environmental remediation. This chapter provides a comprehensive review of the synthesis, properties, and applications of biogenic nanomaterials produced by fungi. Fungi have been found to synthesize a large variety of nanomaterials, metal oxide, gold, palladium and platinium nanoparticles [9,23]. The properties of these biogenic nanomaterials are influenced by their composition, size, shape and surface chemistry. Various factors, such as temperature, pH and concentration of precursor materials, affect the synthesis of biogenic nanomaterials by fungi. The potential applications of these biogenic nanomaterials include antimicrobial agents, catalysts, biosensors, and environmental remediation agents [32]. This review highlights the recent progress in the field of biogenic nanomaterials produced by fungi and the challenges that need to be addressed for their efficient and effective utilization [14,28].

Nanotechnology has revolutionized various fields, including medicine, biotechnology, and environmental science, by providing efficient and effective solutions to many challenges. Biogenic nanomaterials are nanomaterials synthesized by living organisms, including bacteria, fungi, plants, and animals. Biogenic nanomaterials have gained significant attention due to their potential applications and unique properties. Fungi have been found to synthesize a large variety of nanomaterials, including platinum, gold, metal oxide, palladium and silver nanoparticles.

These biogenic nanomaterials exhibit unique properties, such as biocompatibility, high stability and low toxicity, which make them attractive for various applications. Therefore, understanding the synthesis, properties, and applications of biogenic nanomaterials produced by fungi is crucial for their efficient and effective utilization [33].

Fungi have been found to synthesize a wide range of nanomaterials using various mechanisms, including extracellular and intracellular biosynthesis. The extracellular biosynthesis of biogenic nanomaterials by fungi involves the secretion of enzymes and organic molecules that reduce the precursor materials into nanoscale particles. In contrast, the intracellular biosynthesis of biogenic nanomaterials by fungi involves the uptake of precursor materials into the fungal cells, where they are reduced to nanoscale particles. Various factors, such us temperature, concentration and pH of precursor materials, of, affect the synthesis of biogenic nanomaterials by fungi. For example, the synthesis of silver nanoparticles by fungi is influenced by the pH of the reaction mixture, with alkaline conditions promoting the synthesis of smaller particles. Similarly, the synthesis of gold nanoparticles by fungi is influenced by the concentration of precursor materials, with lower concentrations promoting the synthesis of smaller particles [6,34,35].

The properties of biogenic nanomaterials produced by fungi are influenced by their composition, shape, size, and surface chemistry. The size of biogenic nanomaterials produced by fungi typically ranges from a few nanometres to several hundred nanometres, depending on the organism and the synthesis conditions. The shape of biogenic nanomaterials produced by fungi can also vary, including spherical, rod-shaped, and flower-shaped nanoparticles. The composition of biogenic nanomaterials produced by fungi can include various metals and metal oxides, such as gold, titanium dioxide, zinc oxide, silver and palladium [6,33].

4.   Biogenic Nanomaterials Produced by Plants: Synthesis, Characterization

Plants have been recognized as an excellent source of biogenic nanomaterials due to their unique ability to synthesize a wide range of nanoparticles with desirable properties. These biogenic nanomaterials are eco-friendly, cost-effective, and have lower toxicity compared to chemically synthesized nanomaterials. In this book chapter, we provide an overview of the synthesis and characterization of biogenic nanomaterials produced by plants, highlighting the different plant species used, the mechanisms of nanoparticle formation, and the factors that influence their properties. Additionally, we discuss the various applications of plant-based biogenic nanomaterials, including biomedical, environmental, and agricultural applications [28,36].

Nanotechnology has become a rapidly growing field in last years due to its practical applications in variety of fields, including biomedical, environmental, and agricultural applications. However, the widespread use of chemically synthesized nanomaterials has raised concerns regarding their toxicity and environmental impact. Biogenic nanomaterials produced by plants have emerged as a potential alternative due to their eco-friendliness, cost-effectiveness, and low toxicity. Plant-based biogenic nanomaterials have specific features that make them appealing for variety of applications. In this book chapter, we provide an overview of the synthesis, characterization, and applications of biogenic nanomaterials produced by plants [36].

Plants can synthesize a wide range of biogenic nanomaterials, including silver, iron oxide, zinc, copper and gold nanoparticles, using various parts of the plant, such as leaves, stems, seeds, and roots. The synthesis of biogenic nanomaterials produced by plants is typically achieved through the reduction of metal ions in the presence of phytochemicals, such as phenolic compounds, flavonoids and terpenoids, which act as stabilizing and reducing agents. The reduction of metal ions can be achieved through either intracellular or extracellular mechanisms, depending on the plant species and the part of the plant used for synthesis. In the intracellular mechanism, metal ions are reduced within the plant cells, resulting in the formation of nanoparticles inside the cell. In the extracellular mechanism, nanoparticles are formed outside the cell, typically in the presence of plant extracts [37–40].

The properties of biogenic nanomaterials produced by plants are influenced by various factors, including the plant species, the part of the plant used for synthesis, the concentration of the phytochemicals present and metal ions. Therefore, understanding the characterization of these nanomaterials is crucial for developing efficient and effective applications. The characterization of biogenic nanomaterials produced by plants typically involves the use of various analytical techniques, such as UV-Vis spectroscopy, dynamic light scattering (DLS), X-ray diffraction XRD), Fourier-transform infrared spectroscopy, transmission electron microscopy (TEM) and transform infrared spectroscopy (FTIR). These techniques allow for the determination of the size, shape, composition, stability, and surface properties of biogenic nanomaterials produced by plants [26,41–44].

Biogenic nanomaterials produced by plants have numerous potential applications in various fields. In the biomedical field, plant-based biogenic nanomaterials have been used for drug delivery, imaging, and diagnosis. For example, gold nanoparticles produced by plant extracts have been used for cancer therapy, while silver nanoparticles have been used as antibacterial agents [45]. In the environmental field, plant-based biogenic nanomaterials have been used for water treatment [31,46,47].

5.   Biogenic Nanomaterials Produced by Animals

Biogenic nanomaterials produced by animals have recently gained significant attention due to their potential applications and unique properties in various fields. In this review, we present a comprehensive overview of the biogenic nanomaterials produced by different animals, including marine animals, insects, and mammals. We discuss the synthesis, characterization, applications and properties of these biogenic nanomaterials. Additionally, we highlight the challenges and future prospects of animal-based biogenic nanomaterials.

Biogenic nanomaterials are synthesized by living organisms and exhibit specific features that make them appealing for variety of applications in energy, environment, and medicine. Animals are one of the most diverse groups of organisms that can produce biogenic nanomaterials. These materials have been found in various parts of the animal body, including the bones, teeth, skin, and exoskeleton. Animal-based biogenic nanomaterials have shown great potential in biomedical, environmental, and industrial applications due to their unique properties. In this review, we present an overview of the biogenic nanomaterials produced by different animals and their potential applications [48].

Marine Animals: Marine animals have been found to produce a variety of biogenic nanomaterials, including silica, calcium carbonate, and magnetite nanoparticles. For example, diatoms, which are a type of single-celled algae, produce silica-based nanoparticles that are used in various applications, such as drug delivery, biosensors, and water filtration. Marine sponges produce silica-based nanoparticles that are used in bone tissue engineering. Some marine animals, such as fish and sharks, produce magnetite nanoparticles that are used in magnetic resonance imaging (MRI) and drug delivery [49,50].

Insects: Insects are also known to produce biogenic nanomaterials, including chitin and melanin nanoparticles. Chitin is a polysaccharide has been used in various applications and that makes up the exoskeleton of insects, such as wound healing and drug delivery. Melanin is a pigment that gives color to the hair, eyes and skin of animals and has been found to exhibit antioxidant, antibacterial, and anticancer properties. Some insects, such as butterflies and moths, produce photonic crystals that exhibit structural coloration and have potential applications in optics and sensors [51,52].

Mammals: Mammals produce biogenic nanomaterials in various parts of their bodies, including the bones, teeth, and skin. For example, bones and teeth contain apatite nanoparticles that exhibit excellent biocompatibility and are used in bone tissue engineering and drug delivery. Skin and hair contain keratin nanoparticles that exhibit excellent mechanical and thermal properties and have potential applications in cosmetics and textiles. Some mammals, such as whales and dolphins, produce magnetite nanoparticles in their brains that are used in magnetic navigation [53,54].

Synthesis and Characterization: The synthesis of animal-based biogenic nanomaterials can be achieved through various mechanisms, including biomineralization, enzymatic and elf-assembly reactions. Characterization of these materials involves various techniques, such us transmission electron microscopy, Fourier-transform infrared spectroscopy and X-ray diffraction. These techniques allow for the determination of the composition, shape, size and surface chemistry of the biogenic nanomaterials [53].

Properties and Applications: The properties of animal-based biogenic nanomaterials, such as their shape, composition, size and surface chemistry, influence their properties and potential applications. These materials have shown great potential in various fields, such as environment, industry and medicine. For example, biogenic nanom [53,54].

6.   Conclusion

In conclusion, biogenic nanomaterials have become a captivating area of research due to their specific features and potential applications in variety of fields, biotechnology, including environmental science and medicine. These materials have evolved over millions of years of natural selection, and their properties are optimized for specific functions, such as strength, flexibility, and self-repair.

One of the most intriguing features of biogenic nanomaterials is their hierarchical structure, which allows for the integration of different properties at multiple length scales. This structure is often difficult to replicate synthetically, which has led researchers to investigate the biological synthesis mechanisms in order to develop new methods for producing these materials.

Recent advancements in nanotechnology have enabled the manipulation and characterization of biogenic nanomaterials, providing new insights into their structure and properties. This has allowed researchers to develop new applications for these materials, such as drug delivery systems, tissue engineering scaffolds, and environmental remediation agents.

Despite these exciting developments, there is still much to be learned about biogenic nanomaterials. Future research will undoubtedly uncover new and unexpected properties, leading to even more exciting applications in the years to come. Overall, the study of biogenic nanomaterials is an important area of research with great potential for improving our lives and the world around us.

References

[1]   A. Hojjati-Najafabadi, A. Aygun, R.N.E. Tiri, F. Gulbagca, M.I. Lounissaa, P. Feng, F. Karimi, F. Sen, Bacillus thuringiensis Based Ruthenium/Nickel Co-Doped Zinc as a Green Nanocatalyst: Enhanced Photocatalytic Activity, Mechanism, and Efficient H_2Production from Sodium Borohydride Methanolysis, Ind. Eng. Chem. Res. 62 (2023) 4655–4664. https://doi.org/10.1021/ACS.IECR.2C03833

[2]   Y. Wu, E.E. Altuner, R.N. El Houda Tiri, M. Bekmezci, F. Gulbagca, A. Aygun, C. Xia, Q. Van Le, F. Sen, H. Karimi-Maleh, Hydrogen generation from methanolysis of sodium borohydride using waste coffee oil modified zinc oxide nanoparticles and their photocatalytic activities, Int. J. Hydrogen Energy 48 (2023) 6613–6623. https://doi.org/10.1016/J.IJHYDENE.2022.04.177

[3]   K. Nesrin, C. Yusuf, K. Ahmet, S.B. Ali, N.A. Muhammad, S. Suna, Ş. Fatih, Biogenic silver nanoparticles synthesized from Rhododendron ponticum and their antibacterial, antibiofilm and cytotoxic activities, J. Pharm. Biomed. Anal. 179 (2020) 112993. https://doi.org/10.1016/J.JPBA.2019.112993

[4]   F. Gol, A. Aygun, C. Ture, R.N.E. Tiri, Z.G. Sarıtaş, E. Kaçar, M. Arslan, F. Sen, Environmentally Friendly Synthesis of 2D Cu2O Nanoleaves: Morphological Evaluation, Their Photocatalytic Activity Against Azo Dye And Antibacterial Activity For Ceramic Structures, Bionanoscience (2024) 1–10. https://doi.org/10.1007/S12668-024-01440-X

[5]   C. Demir, A. Aygun, M.K. Gunduz, B.Y. Altınok, T. Karahan, I. Meydan, E. Halvaci, R.N.E. Tiri, F. Sen, Production of plant-based ZnO NPs by green synthesis; anticancer activities and photodegradation of methylene red dye under sunlight, Biomass Convers. Biorefinery 2024 (2024) 1–16. https://doi.org/10.1007/S13399-024-06172-2

[6]   M. Popescu, A. Velea, A. Lorinczi, Biogenic production of nanoparticles, Dig. J. Nanomater. Biostructures 5 (2010) 1035–1040.

[7]   S. Patil, R. Chandrasekaran, Biogenic nanoparticles: a comprehensive perspective in synthesis, characterization, application and its challenges, J. Genet. Eng. Biotechnol. 2020 181 18 (2020) 1–23. https://doi.org/10.1186/S43141-020-00081-3

[8]   A. Agi, R. Junin, A. Gbadamosi, Tailoring of nanoparticles for chemical enhanced oil recovery activities: a review, Int. J. Nanomanuf. 16 (2020) 107. https://doi.org/10.1504/IJNM.2020.106236

[9]   Z. Zheng, Y. Xiao, F. Zhao, J. Ulstrup, J. Zhang, Bacterially Generated Nanocatalysts and Their Applications, in: 2020: pp. 97–122. https://doi.org/10.1021/bk-2020-1342.ch005

[10]   T. Gur, I. Meydan, H. Seckin, M. Bekmezci, F. Sen, Green synthesis, characterization and bioactivity of biogenic zinc oxide nanoparticles, Environ. Res. 204 (2022) 111897. https://doi.org/10.1016/J.ENVRES.2021.111897

[11]   I. Meydan, H. Burhan, T. Gür, H. Seçkin, B. Tanhaei, F. Sen, Characterization of Rheum ribes with ZnO nanoparticle and its antidiabetic, antibacterial, DNA damage prevention and lipid peroxidation prevention activity of in vitro, Environ. Res. 204 (2022) 112363. https://doi.org/10.1016/J.ENVRES.2021.112363

[12]   A. Singh, P.K. Gautam, A. Verma, V. Singh, P.M. Shivapriya, S. Shivalkar, A.K. Sahoo, S.K. Samanta, Green synthesis of metallic nanoparticles as effective alternatives to treat antibiotics resistant bacterial infections: A review, Biotechnol. Reports 25 (2020) e00427. https://doi.org/10.1016/J.BTRE.2020.E00427

[13]   F.A.-Z. Sayed, N.G. Eissa, Y. Shen, D.A. Hunstad, K.L. Wooley, M. Elsabahy, Morphologic design of nanostructures for enhanced antimicrobial activity, J. Nanobiotechnology 20 (2022) 536. https://doi.org/10.1186/s12951-022-01733-x

[14]   A. Aygün, S. Özdemir, M. Gülcan, K. Cellat, F. Şen, Synthesis and Characterization of Reishi Mushroom-mediated Green Synthesis of Silver Nanoparticles for the Biochemical Applications, J. Pharm. Biomed. Anal. (2019) 112970. https://doi.org/10.1016/j.jpba.2019.112970

[15]   G. Gnanajobitha, K. Paulkumar, M. Vanaja, S. Rajeshkumar, C. Malarkodi, G. Annadurai, C. Kannan, Fruit-mediated synthesis of silver nanoparticles using Vitis vinifera and evaluation of their antimicrobial efficacy, J. Nanostructure Chem. 3 (2013) 67. https://doi.org/10.1186/2193-8865-3-67.

[16]   R. Mariselvam, A.J.A. Ranjitsingh, A. Usha Raja Nanthini, K. Kalirajan, C. Padmalatha, P. Mosae Selvakumar, Green synthesis of silver nanoparticles from the extract of the inflorescence of Cocos nucifera (Family: Arecaceae) for enhanced antibacterial activity, Spectrochim. Acta Part A Mol. Biomol. Spectrosc. 129 (2014) 537–541. https://doi.org/10.1016/j.saa.2014.03.066

[17]   B.D. Mattos, B.L. Tardy, M. Pezhman, T. Kämäräinen, M. Linder, W.H. Schreiner, W.L.E. Magalhães, O.J. Rojas, Controlled biocide release from hierarchically-structured biogenic silica: surface chemistry to tune release rate and responsiveness, Sci. Rep. 8 (2018) 5555. https://doi.org/10.1038/s41598-018-23921-2

[18]   K. Siva Kumar, G. Kumar, E. Prokhorov, G. Luna-Bárcenas, G. Buitron, V.G. Khanna, I.C. Sanchez, Exploitation of anaerobic enriched mixed bacteria (AEMB) for the silver and gold nanoparticles synthesis, Colloids Surfaces A Physicochem. Eng. Asp. 462 (2014) 264–270. https://doi.org/10.1016/j.colsurfa.2014.09.021

[19]   E.I. Alarcon, M. Griffith, K.I. Udekwu, eds., Silver Nanoparticle Applications, Springer International Publishing, Cham, 2015. https://doi.org/10.1007/978-3-319-11262-6

[20]   S.I. Tsekhmistrenko, V.S. Bityutskyy, O.S. Tsekhmistrenko, L.P. Horalskyi, N.O. Tymoshok, M.Y. Spivak, Bacterial synthesis of nanoparticles: A green approach, Biosyst. Divers. 28 (2020) 9–17. https://doi.org/10.15421/012002

[21]   A. Saravanan, P.S. Kumar, S. Karishma, D.-V.N. Vo, S. Jeevanantham, P.R. Yaashikaa, C.S. George, A review on biosynthesis of metal nanoparticles and its environmental applications, Chemosphere 264 (2021) 128580. https://doi.org/10.1016/j.chemosphere.2020.128580

[22]   M. Gomathy, K.G. Sabarinathan, Microbial Mechanisms of Heavy Metal Tolerance- a Review, Agric. Rev. 31 (2010) 133–138.

[23]   ABHILASH, K. REVATI, B.D. PANDEY, Microbial synthesis of iron-based nanomaterials—A review, Bull. Mater. Sci. 34 (2011) 191–198. https://doi.org/10.1007/s12034-011-0076-6

[24]   S. Rajeshkumar, C. Malarkodi, M. Vanaja, G. Gnanajobitha, K. Paulkumar, C. Kannan, G. Annadurai, Antibacterial activity of algae mediated synthesis of gold nanoparticles from turbinaria conoides, Der Pharma Chem. 5 (2013) 224–229.

[25]   H. Sowani, P. Mohite, H. Munot, Y. Shouche, T. Bapat, A.R. Kumar, M. Kulkarni, S. Zinjarde, Green synthesis of gold and silver nanoparticles by an actinomycete Gordonia amicalis HS-11: Mechanistic aspects and biological application, Process Biochem. 51 (2016) 374–383. https://doi.org/10.1016/j.procbio.2015.12.013

[26]   X. Zhang, Y. Qu, W. Shen, J. Wang, H. Li, Z. Zhang, S. Li, J. Zhou, Biogenic synthesis of gold nanoparticles by yeast Magnusiomyces ingens LH-F1 for catalytic reduction of nitrophenols, Colloids Surfaces A Physicochem. Eng. Asp. 497 (2016) 280–285. https://doi.org/10.1016/j.colsurfa.2016.02.033

[27]   F. Karimi, N. Rezaei-savadkouhi, M. Uçar, A. Aygun, R.N. Elhouda Tiri, I. Meydan, E. Aghapour, H. Seckin, D. Berikten, T. Gur, F. Sen, Efficient green photocatalyst of silver-based palladium nanoparticles for methyle orange photodegradation, investigation of lipid peroxidation inhibition, antimicrobial, and antioxidant activity, Food Chem. Toxicol. 169 (2022) 113406. https://doi.org/10.1016/J.FCT.2022.113406

[28]   R. Darabi, F.E.D. Alown, A. Aygun, Q. Gu, F. Gulbagca, E.E. Altuner, H. Seckin, I. Meydan, G. Kaymak, F. Sen, H. Karimi-Maleh, Biogenic platinum-based bimetallic nanoparticles: Synthesis,

characterization, antimicrobial activity and hydrogen evolution, Int. J. Hydrogen Energy (2023). https://doi.org/10.1016/j.ijhydene.2022.12.072

[29]   Y. Kocak, G. Oto, I. Meydan, H. Seckin, T. Gur, A. Aygun, F. Sen, Assessment of therapeutic potential of silver nanoparticles synthesized by Ferula Pseudalliacea rech. F. plant, Inorg. Chem. Commun. (2022). https://doi.org/10.1016/j.inoche.2022.109417

[30]   S. He, Z. Guo, Y. Zhang, S. Zhang, J. Wang, N. Gu, Biosynthesis of gold nanoparticles using the bacteria Rhodopseudomonas capsulata, Mater. Lett. 61 (2007) 3984–3987. https://doi.org/10.1016/j.matlet.2007.01.018

[31]   B. Syed, N.N. Prasad, S. Satisha, Endogenic mediated synthesis of gold nanoparticles bearing bactericidal activity, J. Microsc. Ultrastruct. 4 (2016) 162. https://doi.org/10.1016/j.mau.2016.01.004

[32]   A. Aygun, F. Gülbagca, L.Y. Ozer, B. Ustaoglu, Y.C. Altunoglu, M.C. Baloglu, M.N. Atalar, M.H. Alma, F. Sen, Biogenic platinum nanoparticles using black cumin seed and their potential usage as antimicrobial and anticancer agent, J. Pharm. Biomed. Anal. (2019) 112961. https://doi.org/10.1016/j.jpba.2019.112961

[33]   K.S. Uma Suganya, K. Govindaraju, V. Ganesh Kumar, T. Stalin Dhas, V. Karthick, G. Singaravelu, M. Elanchezhiyan, Blue green alga mediated synthesis of gold nanoparticles and its antibacterial efficacy against Gram positive organisms, Mater. Sci. Eng. C 47 (2015) 351–356. https://doi.org/10.1016/j.msec.2014.11.043

[34]   N. Durán, P.D. Marcato, M. Durán, A. Yadav, A. Gade, M. Rai, Mechanistic aspects in the biogenic synthesis of extracellular metal nanoparticles by peptides, bacteria, fungi, and plants, Appl. Microbiol. Biotechnol. 90 (2011) 1609–1624. https://doi.org/10.1007/s00253-011-3249-8

[35]   K. Gudikandula, P. Vadapally, M.A. Singara Charya, Biogenic synthesis of silver nanoparticles from white rot fungi: Their characterization and antibacterial studies, OpenNano 2 (2017) 64–78. https://doi.org/10.1016/j.onano.2017.07.002

[36]   S.S. Emmanuel, A.A. Adesibikan, Bio-fabricated green silver nano-architecture for degradation of methylene blue water contaminant: A mini-review, Water Environ. Res. 93 (2021) 2873–2882. https://doi.org/10.1002/wer.1649

[37]   F. Dumur, A. Guerlin, E. Dumas, D. Bertin, D. Gigmes, C.R. Mayer, Controlled spontaneous generation of gold nanoparticles assisted by dual reducing and capping agents, Gold Bull. 2011 442 44 (2011) 119–137. https://doi.org/10.1007/S13404-011-0018-5

[38]   J. Sarkar, S. Ray, D. Chattopadhyay, A. Laskar, K. Acharya, Mycogenesis of gold nanoparticles using a phytopathogen Alternaria alternata, Bioprocess Biosyst. Eng. 35 (2012) 637–643. https://doi.org/10.1007/s00449-011-0646-4

[39]   S. Rajeshkumar, C. Malarkodi, K. Paulkumar, M. Vanaja, G. Gnanajobitha, G. Annadurai, Algae Mediated Green Fabrication of Silver Nanoparticles and Examination of Its Antifungal Activity against Clinical Pathogens, Int. J. Met. 2014 (2014) 1–8. https://doi.org/10.1155/2014/692643

[40]   A. Yasmin, K. Ramesh, S. Rajeshkumar, Optimization and stabilization of gold nanoparticles by using herbal plant extract with microwave heating, Nano Converg. 1 (2014) 12. https://doi.org/10.1186/s40580-014-0012-8

[41]   M.O. Montes, A. Mayoral, F.L. Deepak, J.G. Parsons, M. Jose-Yacamán, J.R. Peralta-Videa, J.L. Gardea-Torresdey, Anisotropic gold nanoparticles and gold plates biosynthesis using alfalfa extracts, J. Nanoparticle Res. 13 (2011) 3113–3121. https://doi.org/10.1007/s11051-011-0230-5

[42]   M. Noruzi, Biosynthesis of gold nanoparticles using plant extracts, Bioprocess Biosyst. Eng. 38 (2015) 1–14. https://doi.org/10.1007/s00449-014-1251-0

[43]   T. Bennur, Z. Khan, R. Kshirsagar, V. Javdekar, S. Zinjarde, Biogenic gold nanoparticles from the Actinomycete Gordonia amarae: Application in rapid sensing of copper ions, Sensors Actuators B Chem. 233 (2016) 684–690. https://doi.org/10.1016/j.snb.2016.04.022

[44]   N. Bazancir, I. Meydan, Characterization of Zn nanoparticles of Platonus orientalis plant, investigation of DPPH radical extinquishing and antimicrobial activity, East. J. Med. 27 (2022).

[45]   F. Göl, A. Aygün, A. Seyrankaya, T. Gür, C. Yenikaya, F. Şen, Green synthesis and characterization of Camellia sinensis mediated silver nanoparticles for antibacterial ceramic applications, Mater. Chem. Phys. 250 (2020) 123037. https://doi.org/10.1016/j.matchemphys.2020.123037

[46]   A. Mewada, G. Oza, S. Pandey, M. Sharon, W. Ambernath, Extracellular Biosynthesis of Gold Nanoparticles Using Pseudomonas denitiricans and Comprehending its Stability, J. Microbiol. Biotechnol. Res. 2 (2012) 493–499.

[47]   M. Apte, P. Chaudhari, A. Vaidya, A.R. Kumar, S. Zinjarde, Application of nanoparticles derived from marine Staphylococcus lentus in sensing dichlorvos and mercury ions, Colloids Surfaces A Physicochem. Eng. Asp. 501 (2016) 1–8. https://doi.org/10.1016/j.colsurfa.2016.04.055

[48]   S.F. Ahmed, M. Mofijur, N. Rafa, A.T. Chowdhury, S. Chowdhury, M. Nahrin, A.B.M.S. Islam, H.C. Ong, Green approaches in synthesising nanomaterials for environmental nanobioremediation: Technological advancements, applications, benefits and challenges, Environ. Res. 204 (2022) 111967. https://doi.org/10.1016/J.ENVRES.2021.111967

[49]   R.L. Brutchey, D.E. Morse, Silicatein and the Translation of its Molecular Mechanism of Biosilicification into Low Temperature Nanomaterial Synthesis, Chem. Rev. 108 (2008) 4915–4934. https://doi.org/10.1021/cr078256b

[50]   D. Sharma, S. Kanchi, K. Bisetty, Biogenic synthesis of nanoparticles: A review, Arab. J. Chem. (2019). https://doi.org/10.1016/j.arabjc.2015.11.002

[51]   N. Maroufpour, M. Mousavi, B. Asgari Lajayer, M. Ghorbanpour, Biogenic Nanoparticles in the Insect World: Challenges and Constraints, in: Biog. Nano-Particles Their Use Agro-Ecosystems, Springer Singapore, Singapore, 2020: pp. 173–185. https://doi.org/10.1007/978-981-15-2985-6_10

[52]   A. Khayrova, S. Lopatin, V. Varlamov, Obtaining chitin, chitosan and their melanin complexes from insects, Int. J. Biol. Macromol. 167 (2021) 1319–1328. https://doi.org/10.1016/j.ijbiomac.2020.11.086

[53]   N. Ciftcioglu, D.S. McKay, G. Mathew, O.E. Kajander, Nanobacteria: Fact or Fiction? Characteristics, Detection, and Medical Importance of Novel Self-Replicating, Calcifying Nanoparticles, J. Investig. Med. 54 (2006) 385–394. https://doi.org/10.2310/6650.2006.06018

[54]   S. Lo, M.B. Fauzi, Current Update of Collagen Nanomaterials—Fabrication, Characterisation and Its Applications: A Review, Pharmaceutics 13 (2021) 316. https://doi.org/10.3390/pharmaceutics13030316

Biogenic Nanomaterials: Synthesis, Characterization, Applications, and Future Remarks   Materials Research Forum LLC
Materials Research Foundations 180 (2025)   https://doi.org/21741/9781644903759

Chapter 5

Effects of Biogenic Nanomaterials on Human and Animal Health

G. Sargin[1*], E. Halvaci[2], F. Sen[2*]

[1] Van YYU Vocational School of Health Service, Van Yuzuncu Yıl University, Van, Turkiye.

[2]Sen Research Group, Department of Biochemistry, Kutahya Dumlupinar University, Kutahya, Turkiye.

gulumsargin@yyu.edu.tr; fatih.sen@dpu.edu.tr

Abstract

The field of nanotechnology is on the way to becoming a field of science that increases its importance day by day as a result of the increase in nano production and the spread of usage limits. Nanotechnology can be used in medicine, environment, biotechnology, medicine, agriculture, and food to facilitate human life and animal nutrition and breeding. However, the disadvantages of these uses have created many areas of study. The aim here is to discuss the advantages of nanomaterials over their counterparts, and the various classes of nanoparticles, and to report on the extraordinary role of nanotechnology in veterinary medicine, its applications, and the adverse effects that may arise.

Keywords

Nanotechnology, Nanoparticles, Environmental Effects, Health Effects, Toxic Effects

1.   Introduction

The characteristic properties of nanoparticles (NPs) are in the direction of benefiting and convenience in many parts of our lives. However, the lack of possible effects on human health may raise concerns about its toxicological evaluation. Research at the molecular or macromolecular level consists of particles of less than 100 nm. With the use of these small particles, new materials and structures with extraordinary physical, chemical, and biological properties have emerged. This has enabled new applications in various fields [1]. Nanotechnology can be used in various sectors such as computers, electronics, medicine, environment, energy, materials, biotechnology, pharmaceuticals, textiles, agriculture, and food [2]. Nanotechnology is an ever-advancing field of technology that has a wide range of fields and is involved in the production of many products that make life easier, including cosmetics, sunscreens, paints, self-cleaning glasses, and stain-resistant clothes [3–5]. Reported as the most common material in use, according to the Nanotechnology Consumer Products Inventory, is carbon, which includes fullerenes and nanotubes. Another most referenced product is silver [6]. Then came silica, zinc oxide, titanium dioxide, and cerium oxide. Among the environmental uses of NPs, the most favoured area of use is the treatment of groundwater contaminated with iron-containing nanomaterials [7,8]. NPs such as titanium dioxide, zinc oxide, and silver are found in many products that are indispensable in daily life such as personal care and hygiene products, most cosmetic products, especially sunscreens used in summer, and many textile products. However, silver NPs, which are the most demanded of these nanoparticles, are used as hygiene additives in cleaning products, food packaging, socks, and underwear [9–30]. Recently, nanotechnology has taken place in many sectors of humanity, as well as its various applications have started to find a place for itself in the veterinary sector. The use and treatment of drugs in animals, disease diagnosis, production of animal vaccines, especially farm sanitizers, animal reproduction, and even animal farming have become increasingly involved in animal nutrition. Substitutions for commonly used antibiotics have direct implications for public health. The use of nanoparticles can create opportunities in many areas such as eliminating the problem of drug resistance development in both human and veterinary medicine, eliminating the problems caused by drug residues accumulated in animal milk and meat. In this way, it is aimed to minimise the amount of idle milk and the number of dead calves and to minimise the economic losses that may occur in dairy farming. Nanotechnology has been used to develop pet care products and hygienic products [31].

Although different classifications can be made according to their different properties, nanoparticles can be classified as follows most simply [32];

- Carbon-based nanoparticles (fullerenes, multi-walled carbon nanotubes, etc.),

- Metal-based nanoparticles (gold colloids, titanium oxide, copper oxide, superparamagnetic iron oxide nanoparticles, etc.),

- Semiconductor-based nanoparticles (quantum dots, etc.).

With the development of nanotechnology, thousands of products are being developed and presented on the market shelves. We, as buyers, buy these products and we begin to hear the negative effects as they are used. Generally, the test times of these products are not taken into account. However, even if we do not buy these nanotechnological products ourselves with money, the nanoparticles emitted into the environment during their production enter our bodies either through respiration or by consuming contaminated food and beverages. Nanotechnology laboratory wastes are generally transported to the waters, then to the soil, and from there to our food [33].

Much of our knowledge of NP interaction comes from studies of human exposure to these ultrafine particulate biomaterials (UFP), and epidemiological studies have shown that there may be a link between respiratory and cardiac health effects, particularly when inhaled exposure to UFP (<100 nm diameter) [34]. A variety of different properties of nanoparticles, not only their size but also their shape, surface charge, chemical properties, and resolution, can reveal their effects on biological systems and public health and determine the likelihood of danger when inhaled [1].

The diffusion of nanoparticles over the ecological system differs. However, one of the risks that may arise is that the proteins in the living structure change their properties. Changes that occur in the living body with the emergence of risks (such as redness, swelling, and inflammation) cause effects that can damage cells and tissues. Human uptake of nanoparticles can occur by inhalation, trans-dermal, or ingestion. In addition, exposure to living organisms for a long time will lead to the possibility of nanoparticles accumulating in tissues and organs. This may cause some risks to emerge in the coming years. This means that it is necessary to develop and implement different analyses in addition to the tests that must be performed before use. In fact, biomaterials to be used in living tissues enter the field of use after undergoing some tests. However, despite the tests, biomaterials containing nanoparticles may appear in the body after a period of allergic, inflammatory (inflammation), mutagenic carcinogenic, and immune effects [35,36].

Occupational and public exposure to NPs and their potential release into the environment is expected to grow significantly in the coming years, in proportion to the increase in the production of NPs by synthesis. It is important to identify the problems that may arise from this increase holistically and as early as possible to prevent adverse effects. A holistic perspective is necessary, as the potential harms of a new technology may be overlooked if the stages of use of a product are not taken into account. In addition, while the use of nanoparticles is increasing rapidly, tests to determine their long-term effects in vivo are very limited. Since very little research has focused on the risks associated with nanotechnology until recently and studies on nanogenotoxicity are still limited in the literature, the mechanism of action of nanoparticles on genetic material more competently inside the cell has not been elucidated in detail [37].

The first studies investigating the toxicity of nanoparticles were studies that determined that heterogeneous mixtures of humans and a large number of environmentally produced fine particulate particles (<100 nm diameter) were present in the air. With these studies, pulmonary toxicity was investigated since the particulate matter accumulation of organisms is in the respiratory tract [38,39]. Recent studies investigating the effects of exposure to particles formed as a by-product of combustion events that occur as a result of air pollution and exhaust fumes in cities, especially in winter months, show that there is a link between the level of particulate matter and the amount of morbidity and mortality [40,41]. Several researchers have reported an improved risk of asthma in children as well as adults as a link between the concentration of ultrafine particles in the air and environmental exposure [42,43].

Many laboratory studies have been conducted in different animal models and cell culture-based in vitro trials. These studies have routinely demonstrated elevated oxidative stress and visceral uptake as well as pulmonary inflammation following inhalation exposure to ultrafine particulate matter with tissue ingestion. Tissue examination and cell culture analyses in all these animal models have been able to determine the physiological changes that occur after exposure to ultrafine particles. Data show an increased incidence of oxidative stress, inflammatory cytokine production, and apoptosis. It also provided insight into gene expression activated as a result of exposure to ultrafine particles of various types, ranging from carbon-based combustion to transition metals [44,45].

The cellular toxicity of nanomaterials can differ significantly from the conventional chemicals from which they are derived depending on factors such as size, shape, stability, mode of synthesis, and surface chemistry. Since smaller particles will form a larger surface area, the more particles per unit mass, the greater the biological interaction. This leads us to conclude that the smaller the size of the particles, the higher the biological activity [46,47]. Recent research has focused on the biological effects of nanoparticles on ecosystems and mammals. However, some recent studies have been directed towards the phytotoxic properties of nano-materials in the ecosystem [48].

Magnetic nanoparticles have more applications in biomedical fields. This has resulted in greater exposure of people and their environment to magnetic nanoparticles (MNPs). In applications of magnetic nanoparticles, even if the target tissue is liver and spleen, other organs should also be evaluated in terms of toxicity. Therefore, there is a great need to investigate and analyse MNP. Potential impacts, both negative and beneficial, on human health and the ecological system are important considerations [49].

Once NPs enter the body, their elimination from the body is also very difficult due to their long half-lives. In particular, metal oxide NPs are not easily degraded and decompose very slowly. In this case, the accumulation of NPs increases oxidation on the one hand and weakens the antioxidant system on the other, initiating stress reactions that cause inflammation [50–52]. Carbon nanoparticles, which are thought to be able to cross the blood-brain barrier, have been examined due to their chemical effects. It is thought that high exposure to anionic and cationic nanoparticles may destroy the structure of the blood-brain barrier. When exposed at the same rates, anionic nanoparticles were found to have higher brain take-up rates than cationic formulations. Studies have shown that both neutral nanoparticles and anionic nanoparticles with low concentrations facilitate the direct passage of chemicals into the brain by moving toward the brain like a carrier molecule. Cationic nanoparticles have a more toxic effect on the blood-brain barrier [53]. Data from ecotoxicity tests on vertebrates, fish, artemia, and algae indicate that nanoparticles pose a low hazard potential in aquatic species [54–56]. Studies on the brains of sea bass fish exposed to carbon-based fullerene C60 also showed that there were no life-threatening effects, only oxidative destruction [57] When $TiO2$ nanoparticles were applied, lipid reoxidation was also observed in the gills, brains, and livers of carp and rainbow trout, indicating oxidative stress disease in these tissues [58,59]. Nanoparticles are taken up by fish eggs and deposited in the gills and intestines of adult fish, and nanoparticles have also been detected in the brain, blood, liver, and testes [60]. In addition, in a study examining the effects of long-term exposure to Zn and ZnO nanoparticles on artemia, oxidative stress and deposition were detected in exposed artemia [61].

Many nano-sized particles can accumulate anywhere in the lungs and respiratory tract. Nanoparticles are likely to travel to other parts of the body as they may pass into cells or be found in the intercellular fluid. As shown in Figure 1, studies are reporting that inhaled nanoparticles can travel to many internal organs and tissues, such as the brain, after entering the circulation [39,62].

Biogenic Nanomaterials: Synthesis, Characterization, Applications, and Future Remarks   Materials Research Forum LLC
Materials Research Foundations 180 (2025)   https://doi.org/21741/9781644903759

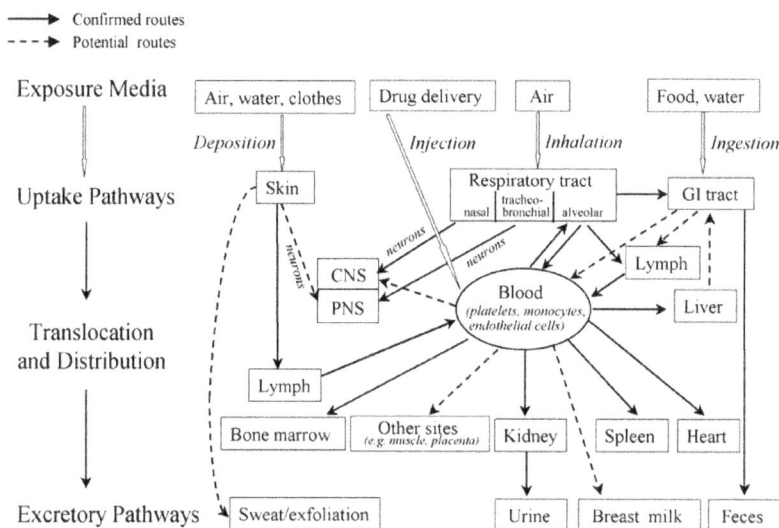

Figure 1. *The direct risk to human health presented by many nanomaterials designed with nanostructure properties is related to the possibility of exposure and the materials entering the body (Reprinted with permission from [39], Copyright EHP Publishing).*

Interactions in biological systems occur with the quick solubilization of water- or fat-soluble particles of the nanomaterial. These possible physicochemical changes should be carefully evaluated. Direct effects such as the following may occur:

- NPs can puncture and damage cell membranes;

- NPs damage the cytoskeletal components of the cell, disrupting intracellular transport and cell proliferation;

- NPs increase mutagenicity as a result of damage to DNA

- NP's damage to mitochondria and disruption of cell metabolism cause energy imbalance;

- NPs trigger apoptosis by acting on lysosome formation, inhibiting autophagy formation and degradation of macromolecules (As in Figure 2).

- Since NPs cause structural changes in cell wall proteins, they impair the transfer of substances in and out of the cell;

- They can activate inflammatory mediators, impairing tissue and organ function. However, for this effect to occur, the total number of particles in the organism must be high.

- NPs can cause oxidation;
- There are studies indicating that nanoparticles can cause serious toxicity in tissues with their small size [63].

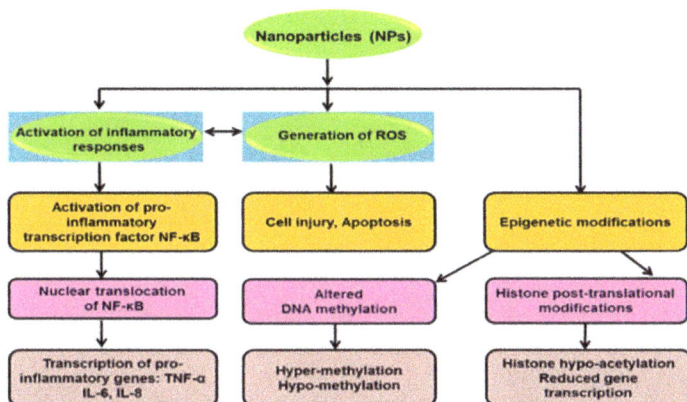

Figure 2. *Toxic effects of NPs on cell structure (Reprinted with permission from* [64]*, Copyright NIH).*

Important parameters determining the toxicity of nanoparticles are their size, number, and total localization. The smaller the size of a particle, the greater the surface area and the larger the proportion of molecules present. This leads to an increase in particle surface energy [65].

Air pollution, which directly affects human health and also affects the life of organisms, is an important factor in environmental quality. It has been statistically proven that ambient air pollution plays an active role in the formation of cardiovascular diseases [66,67]. Nanoparticles can migrate to secondary organs with particles larger than 10 μm or 2.5 μm crossing the respiratory tract and settling directly in the lungs. Such a situation may pose a health risk [68]. It has been reported that harmful NPs inhaled by macrophage cells in the lungs cannot function fully compared to larger particles [69,70]. The important factor that plays a role in the positive or negative effects of nanoparticles after they enter the body is particle size. However, besides this; factors such as chemical structure, form, magnetic, mechanical, or optical features, structure, surface forms, and surface charge or agglomeration potential may also cause negativity [71].

In a comprehensive study conducted in 2007, some diseases thought to be related to inhaled nanoparticles, lung cancer, emphysema, asthma, bronchitis, Parkinson's, and Alzheimer's. Nanoparticles that settle in the digestive tract have been associated with colon cancer and Crohn's disease. Nanoparticles that can settle in the circulatory system have been reported to cause heart diseases such as arteriosclerosis, arrhythmia, embolism, and cardiac death. As a result, diseases in organs such as liver or spleen may be observed [35]. Size-Toxicity Scale Relationship Sketch.

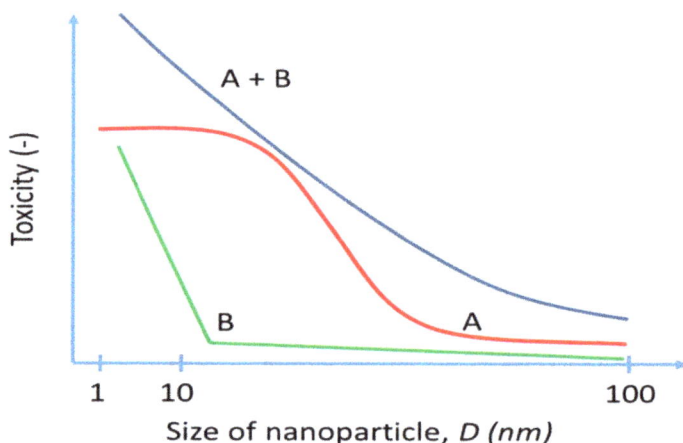

Figure 3. *Prediction plot of size and toxicity relationship. A: sigmoidal curve; B: threshold curve; A + B: synthesis of both approaches; Toxicity increases with decreasing particle size. Note: NP diameter and toxicity axis have no units in this graph. Because it's an ad hoc approach (Reprinted with permission from [72], Copyright MDPI).*

With the use of nanotechnology in energy applications, it is possible to obtain cleaner energy that will reduce environmental pollution. In this case, factors that may be harmful to health will be eliminated and the emergence of health problems will be prevented. The availability of decentralized replenishable energy in developing countries can help raise the overall general quality of life and can facilitate decentralized health care. All innovators have once again emphasized this emerging area of technology, with hazard-release nanoparticles showing superior properties, especially in the production and recovery of energy equipment. However, it is clearly stated that adverse health effects are unknown, and risks cannot be ruled out. In one embodiment, how nanomaterials are fixed is thought to play an influential role in their release into the environment. Nanoparticles placed permanently in a matrix were considered unlikely to be released and lead to human exposure [73]. Gur et al. stated that zinc oxide nanoparticles can serve many areas such as production and consumption areas; for example, it is offered to humanity with various products such as environmental and biomedical fields and can provide many advantages [8]. The destructive effect of ZnO on pathogenic microorganisms that threaten health is quite remarkable. It is thought that its use in cancer and diabetes treatment also creates positive effects [74].

There are many areas where humanity can benefit greatly from nanotechnology. However, the possibility of potential health effects of nanomaterials (NMs) should never be ruled out. This idea initiated the field of nanotoxicology, which is tasked with promoting the safe design and use of NMs as well as evaluating their toxicological potential. Although there has been no human discomfort in the use of NMs until recently, some in vivo and in vitro studies indicate that these materials may have toxic effects or induce negative biological reactions. One of the reactions that can occur is the formation of reactive oxygen species and the occurrence of oxidant damage. Oxidant damage, in particular, should be taken into account when analyzing ambient particles, as this is the mechanism by which very small particles in the environment can cause health hazards [75]. It is stated that nanomaterials like silver, magnesium, and zinc-oxide incorporated into the packaging materials of nanotechnology-based food wrappers may contaminate foodstuffs in direct contact and may cause health risks for human health when these products are consumed [76]. The ease of binding and transport of toxic chemical pollutants is due to the large surface area of nanoparticles. It is reported that the ability of nanoparticles to enter the body and cells may cause the spread of toxic substances within the body and as a result, cell and tissue damage and defence mechanism failures may occur. It is also reported that inhalation of nanosized particles may cause lung diseases in humans and mammals [77].

Since the magnetic structure in nanoparticles creates the force acting on any moving electric charge, it may cause the balance in the ecological system to be adversely affected. Dust storms can cause inflammation and inflammation of the lung surface and tissue, as well as diseases like asthma and chronic lung disease. Volcanic activities and eruptions occurring in nature can cause the emergence of nano-scale particles. This may pose a risk to the ecological system. In fact, the main danger is that there is not enough information about the volcano and the risks that its activity will bring. Because knowing these risks means taking the necessary precautions to reduce the damage they may cause. As a result of a volcanic eruption, approximately 30 million tons of ash is released. Viruses, bacteria, fungi, and chemical pollutants can be transported from one continent to another with the help of dust storms, thanks to the adhesion of nanoparticles to the ashes released as a result of volcanic activity and eruptions. They can survive under the high ultraviolet light released during this transport process. As a result of a single volcanic activity, heavy metals (Lead, Mercury, Chromium, Cadmium) that are known to be highly toxic to living things can be released into the ecological system. While nanoparticles formed as a result of volcanic activity show themselves in the respiratory organs, eyes, and skin as a short-term effect, they can cause various diseases such as lymphatic system origin diseases, elephantiasis (lymphedema) and Kaposi's sarcoma as a long-term effect. Such undesirable situations negatively affect people's quality of life and social life. The application of products obtained as a result of the application of nanoparticles in cosmetics, which is included in nanotechnological developments, is not a new method. Research has shown that black soot and some mineral powders were used for cosmetic purposes thousands of years ago. Today, nanoparticles have many applications such as sun protection creams, anti-aging creams, moisturizers, eyeshadows, mascara, and mineral makeup products [78].

The skin is the outer wall of the body and acts as an effective barrier to external agents from the environment. However, the respiratory tract, which we call inhalation, is a very easy way for particles to enter the body. Depending on their particle composition or size, they can affect the area they come into contact with. Oxidative stress and organelle damage is one of the types of irreversible cellular damage encountered as a result, which is often controversial. The most important feature of these nanoparticles, which are very mobile in nanoscales, is their ability to have a toxic effect on cells. While some of the nanotechnology products produced are fearlessly handed over, the testing process that has yet to be completed is ignored. The real frightening thing is that Nanotechnology laboratory wastes are mostly transported to the waters, then to the soil through the waters, and from there to our food. Even if you don't spend money and buy nanotechnology products directly, the nanoparticles scattered in nature during production can enter our body either by inhalation or through food and drink.

Nanoparticles pass into body tissues mainly employing respiration, nutrition, and skin. Nanoparticles taken into the body easily pass into the blood [79]. Nanomaterials have been proven to cause responses in the body against irritation, injury, or infection. However, the factors that determine the severity of the reactions are not fully understood. Also, the long-term health consequences of many cases of repeated exposures are unknown. It has been observed that there is an increase in the number of nanomaterials whose biological effects are not fully known, and their use is increasing rapidly. It should not be overlooked that many of these products may be carcinogenic or have a permanent effect on living cells. The impact of nanoparticles on living cells and genetics is a wide area that needs to be analyzed. Because it is seen that living things are exposed to more and more nanoparticles every day. As a result, it can cause cells to die, stop their functions, mutate their genes, or cause toxic effects, which are among the possible effects on living things. It is necessary to carry out studies to control different diseases, especially cancer, and many reproductive and developmental disorders. In addition to the differences in the stages of spreading nanoparticles into the ecological system, there is also the risk of changing the properties of proteins in the living structure. Changes that occur in the living body as a result of the emergence of risks, such as redness, and inflammation, cause effects that can damage cells and tissues. There is also the possibility of accumulation in living organisms. This will bring some risks in the following years. For this reason, in addition to the required tests, different tests should be developed and put into practice. Thus, the effects of nanoparticles on living things are examined in depth and it is easier to take precautions against risks.

Metal nanoparticle structures can be easily synthesized. Since it can be changed chemically, it has wide applications in consumer products, industrial products, the machinery industry, the military field, and especially medicine. Metal nanoparticles and alloys, which exhibit superior mechanical properties thanks to their crystal structures and very strong metallic bonds, are among the materials used for the musculoskeletal system [80]. Steel is the leading biomaterial used. Steel, which was used extensively in the early 1900s, was later replaced by stainless steel, cobalt, chromium, and cobalt-chromium alloys due to its oxidation feature. In fact, the biomaterials to be used in the living body enter the field of use after passing some tests. However, despite the tests performed, the allergic, immune, inflammatory, mutagenic, and carcinogenic effects of biomaterials in the body may occur over time [36]. For this reason, the allergic reaction properties, biological compatibility, and genetic toxicity of nanoparticles against tissues should be investigated and put into use.

Due to their size, they go deep into the skin layer and carry some nutritional components to these points, ensuring that the skin is nourished and regenerated, that nanoparticles help the skin look younger by showing antioxidant properties, and hide wrinkles due to their specific dimensions and optical properties are just a few of these advantages. It is stated that nanoparticle absorption may increase in damaged, injured, sensitive skin with acne and eczema, can reach the circulatory system, and cause unwanted complications. Liposomes in sunscreen can be detected as foreign matter. This may cause an allergic reaction by activating the body's immune system. There is not much information about the toxicities of nanoparticles used in the cosmetic industry. In addition, it is known that the most preferred ZnO metal in this sector has toxic properties at the nanoscale [41].

In regions where settlement is rapidly increasing, it is normal for old buildings to be demolished and replaced by new ones to make room for the growing population. During the demolition of a sufficiently large building, particles smaller than 10 μ were detected to be scattered into the ecological system. These ruins, which contain asbestos, lead, glass, and other toxic particles (metals with heavy metals in the majority), can easily reach the neighboring settlements around them. This means that as the number of nanoparticles released into the ecological system increases every day, the risks that arise will increase [42].

2. Toxic Effects of Biogenic Nanomaterials on Tissues

The most significant pathway and target organ of nanoparticles in the body is the lungs. Most nano-sized rigid particles can easily reach the alveoli of the lungs. When vaporized or gasified nanoparticles are taken into the lungs by breathing from the air, they reach the outer membrane of the chest wall (parietal pleura). It is destroyed by phagocytosis by helical macrophages. However, nanoparticles with larger sizes are similar to asbestos fibers and accumulate in these areas by forming a stoma of the internal organ against the skin. Since macrophage cells that perform phagocytosis and pinocytosis cannot phagocytize nanotubes due to their fibrous structures, pro-inflammatory, genotoxic (chemicals and radioactive elements that can cause mutation and cancer by having toxic effects on DNA and genes), mutagenic mediators are released by mesothelial cells [81]. It has been suggested that the toxic effect of vaporized or gaseous nanoparticles begins with their reaching the lungs through both the respiratory system and other systems and that the effect is pulmonary and systemic inflammation. According to this view, there may be important modifications that may result in pulmonary endothelial dysfunction, platelet activation, stimulation of thrombotic factors, atherosclerotic plaque lesion and rupture, vascular endothelial dysfunction, stimulation of lung and liver reflexes, deterioration in heart rate and rhythm, and even sudden cardiac arrest resulting from inflammation. has been reported [82,83]. As a result of the study, in which the genetic toxicity tests of some nanoparticles were demonstrated by the micronucleus (MN) test in the process of growing certain cells under their control, it was determined that nanoparticles such as iron oxide, aluminium oxide, titanium dioxide, cobalt, cobalt-chromium and silver cause DNA damage. In a different study, in addition to various cell 374 poisoning and genetic toxicity tests, it was stated that copper oxide, zinc oxide, aluminum oxide, silver, gold, chromium, cobalt, and titanium dioxide nanoparticles, whose effects were investigated in different cell cultures, had chemotherapy and genotoxic potentials [80]. 21 genetic toxicity tests were carried out on some substances containing metal nanoparticles, and 16 of them were observed to pose a risk [81]. It enters the systemic circulation by inhalation of gaseous or vaporous nanoparticles. Nanoparticles that pass into the circulatory system can reach various parts of the body. It was concluded that macrophages in many organs and tissues such as the liver, kidneys, heart, nervous system, bone marrow, and spleen are responsible for the translocation of particles into the blood through phagocytosis of epithelial and endothelial cells [84]. As long as these particles are not exposed to themselves or a different factor from the lungs, cleaning can be done with cleaning mechanisms. Maximum exposure probably increases the effects induced by these particles. Irrespective of particle size, parameters such as the loads carried by the particle in contact with cell membranes and the chemical reactivity of the particle also play an important role [85]. In another work, Sadeghi et al stated that as a result of Fe_2O_3 given to rats, nanoparticles rapidly entered the circulation and caused severe infection in the lung and liver tissues [86]. With the inhalation of ultra-fine particles, pulmonary inflammatory processes are inevitable, with particles entering the systemic circulation and cardiovascular disorders. Although we have been exposed to nano-sized particles in the air throughout life, these particles can enter the body through breathing, ingestion, and skin with the developing nanotechnology today [87]. However, since there are mechanisms that act as a series of filters, starting from the respiratory tract, nose, or mouth, from the airways of various diameters to the alveoli, regional accumulations may occur. This shows that particles of various sizes are deposited differently in the airways and also in the alveolar region. As a result of increased diffusion mobility, it means that particle deposition in all regions of the lung may increase and have different effects. This may be most important in children especially in asthma and COPD patients [88].

Nanoparticles may tend to stay longer in vital organs such as the liver, heart, spleen, kidney, and lung for long-term toxicity. In a study investigating the prothrombotic effects of nanoparticles, it was observed that intravenously administered high-dose quantum deposited on the surfaces of lungs, liver, and blood cells of 375 mice, changed the coagulation time and caused pulmonary vascular thrombosis. Based on the information obtained from the results, it is concluded that exposure to nanoparticles may increase the events of thromboembolism in the vessels [89]. There is a high percentage of nanoparticles in the air, and if people breathe it, the risk of cardiovascular disease increases. Cardiac arrhythmia is the most common finding related to the heart. The liver is the first among the organs affected in the digestive system. Intestinal spread is dependent on the rate of particle translocation, diffusion, and accessibility through mucus, initial enterocyte contact, cellular trafficking, and post-translocation events. The smaller the particle diameter, the greater the diffusion power to enterocytes through the mucus layer. Particles found in the submucosal tissue can enter both the lymphatic and capillaries. While they can reach different organs systemically through capillaries, particles via lymphatic vessels are probably important in inducing secretory immune responses. It has been suggested that disruption of epithelial barrier function is an effective factor for mucosal inflammation [86,90].

The skin layer surrounding the body acts as an important shield against microorganisms or harmful particles that may come from outside. To cross this protective layer, the nanoparticles must be very small in size, so crossing the skin barrier is entirely size-dependent. Secondly, different types of nanoparticles can be found in the innermost layers of the skin. TiO_2 particles are often used in sunscreens and penetrate the skin directly. However, there is no clear information that the particles also enter the systemic circulation. Particles in the skin may likely be phagocytized by macrophages, Langerhans cells, or other cells [91]. It is stated that there is a possibility that nanoparticles can cross the blood-brain barrier by carrier endocytosis or passive diffusion. It is thought to have adverse effects on the central nervous system [92,93]. It is known that materials made of metal nanoparticles and their alloys can be used comfortably by adapting to the skeletal-muscular system. Steel is the leading biomaterial used. Stainless steel, chromium, cobalt, and cobalt-chromium alloys were preferred instead of steel, which was widely used at the beginning of the 19th century, due to its oxidation feature [80].

A study by Oberdörster showed that C60 fullerenes cause changes in the brain of fish. Significant lipid peroxidation has been reported in the brains of sea bream and sea bass exposed to 0.5 mg/l uncoated C60 fullerenes for 48 hours [62]. In most of the studies, an increase in DNA damage was observed in parallel with the increase in nanoparticle concentration. Due to their small size, metal nanoparticles can enter directly into the cell and nucleus, which can lead to the release of free radicals in the cell. It can also cause genetic damage by binding to DNA [94]. It has been seen in some studies that nano titanium dioxide was given under the skin during pregnancy of mice, and it was seen that it passed to the fetus. It has been determined that there are some problems with the brain damage and reproductive system in male offspring in puppies followed after birth [94,95] In a different study, fish were exposed to nanoparticles for 48 hours and brain damage was observed. In addition, since nanoparticles are highly likely to be taken up by cells, they can also enter the food chain through bacteria and pose serious health threats. For example, mercury in fish, pesticides in vegetables, and hormones in meat. The popular carbon nanotube (20 times stronger and lighter than steel) has a similarity to asbestos fiber. When carbon nanotubes are released into the air, since they are of carbon origin, the defence mechanism cannot recognize them after they enter the body and it becomes difficult to detect [96].

Translocation/accumulation of NPs: Looking at the health opportunities of NT for human health, the treatment of drinking water and improving its safety, a practice seen as particularly important for developing countries, has become very important. Exposure to free nanoparticles will be greatest for those working in manufacturing facilities using nanomaterials and consumers using NT-based products. Nanoparticles can reach the ganglion and central nervous system structures via the sensory nerves ending in the systemic circulation and airway epithelium or with the help of the olfactory nerve. As a result of research, it has been seen that nanoparticles have harmful effects on the central nervous system. Titanium dioxide nanoparticle used in sunscreens has been shown to cause brain damage in mice [97]. In a more recent study investigating the prothrombotic effects of nanoparticles, it was observed that high-dose quantum dots administered intravenously accumulated on the surfaces of the lungs, liver, and blood cells of mice, causing activation in the coagulation process and causing pulmonary vascular thrombosis. Results from these studies raise the suspicion that exposure to nanoparticles may increase the incidence of arterial and/or venous thromboembolism (VTE) in humans [89]. Nanoparticle toxicity can result in oxidative stress, inflammation, and thus damage to proteins, membranes, and DNA. Vapor or gaseous nanoparticles may enter the bloodstream by inhalation or ingestion. Some nanomaterials can penetrate the skin, depending on how they are used. Conditions such as broken skin, acne, eczema, shaving scars, or severe sunburn can be effective in making nanomaterials more easily absorbed into the skin. Once biogenic nanomaterials have entered through any of the dermal, pulmonary, and gastrointestinal tracts, NPs may enter the systemic circulation. Generally, nanoparticles that pass into the blood are distributed in tissues [98,99]. Animal studies have shown that NPs given by inhalation, ingestion, or injection are first distributed to the site of exposure and then migrate to organs such as the liver, spleen, kidneys, ovary, testes, and brain [82 As NPs enter tissues and cells, they can bind to transport proteins in the circulatory system and increase the rate of phagocytosis [100]. As a result of animal and human studies, it can be said that macrophage clearance is the reason for the displacement of NPs into many tissues and organs through the circulatory and lymphatic systems [35]. NPs can accumulate in different tissues through circulation. For example, they can accumulate in some biological membranes and reproductive organs [101].

Since 2018, significant progress has been made in the implementation of nanotechnology in different areas of the water treatment, nutrition food packaging, and construction sectors [102,103]. In one study, Pd NPs obtained as a result of green synthesis may be used as antibacterial agents in biomedical fields and water hygiene in the future [104]. The increasing application of NMs in everyday products raises new questions regarding chemical and product safety.

3.   Conclusion

Nanomaterials used today should not pose a threat to health. In the research carried out to reveal the negative effects of nanotechnology in terms of the health of living things, the effects on human biological structures and organs could not be revealed in sufficient detail. The approaches and methods used to arrive at conclusions regarding the effects of produced nanomaterials and ultrafine particles have led to different results. This discrepancy suggests that general testing is needed to obtain comparable results for the potential adverse effects of nanomaterials. As the place of biogenic nanomaterials used in the field of nanotoxicology increases, it is thought that the tests used to determine the toxic effects of nanomaterials will be made more common better comparisons will be made and healthy results will be obtained. While the nanostructure-related properties of many engineered nanomaterials may classify them as potential hazards, the direct hazard they present to human health will depend on the probability of exposure and the degree of behaviour associated with the materials entering the body. Pharmacokinetic features of various nanoparticle types should be analysed in terms of health threats and databases on different nanoparticle types should be developed. Living tissues should be used to study the physicochemical properties of nanoparticles. To understand whether there is a link between these properties and toxic effects, it is necessary to know the uptake and accumulation of nanoparticles in living tissues very well. For this reason, it will be of great importance for health to determine different doses, durations of use, and exposure routes by conducting in vivo studies on a large number of various species. As a result of the studies, it has been seen that the number of nanomaterials, whose biological effects are not fully known, is increasing day by day and its use has increased in this process. It should not be ignored that many of these products may have a permanent effect on the cells in life or may be carcinogenic. Demir et al. In his study, he mentioned that chemical molecules are decision markers in some diseases. It is also possible that it may have an effect on genetics, and serious studies should be conducted on this issue. Lipids, proteins, carbohydrates, and other cell components undergo significant oxidation in the cell's structure. Oxidative stress occurs with the formation of this damage. Because it can cause cells to die, stop their functions, mutate their genes, or cause toxic effects, which are among the possible toxic effects on living things. It is necessary to carry out studies before it is too late to control different diseases, especially cancer, and many reproductive and developmental disorders. While NMs have many beneficial uses to facilitate life, they can have adverse effects on human and animal health and also have negative impacts on the environment. It is a fact that many advantages and conveniences are provided with the introduction of nanotechnology. However, toxicity studies also bring along many uncertainties such as the harmful effects of nanoparticles on human health and what kind of negativities they will create in the environment in which they live.

References

[1]   A.J. Ferreira, J. Cemlyn-Jones, C. Robalo Cordeiro, Nanoparticles, nanotechnology and pulmonary nanotoxicology, Rev. Port. Pneumol. 19 (2013) 28–37. https://doi.org/10.1016/j.rppneu.2012.09.003

[2]   A. Dağ, Nanoteknolojinin Gıdalara Uygulanması ve Sağlık Üzerine Etkisi, Beslenme ve Diyet Derg. 42 (2014) 168–174.

[3]   F. Şen, G. Gökağaç, Improving Catalytic Efficiency in the Methanol Oxidation Reaction by Inserting Ru in Face-Centered Cubic Pt Nanoparticles Prepared by a New Surfactant, tert -Octanethiol, Energy & Fuels 22 (2008) 1858–1864. https://doi.org/10.1021/ef700575t

[4]   K. Arikan, H. Burhan, R. Bayat, F. Sen, Glucose Nano Biosensor With Non-Enzymatic Excellent Sensitivity Prepared With Nickel–Cobalt Nanocomposites On F-MWCNT, Chemosphere 291 (2022) 132720. https://doi.org/10.1016/j.chemosphere.2021.132720

[5]   H. Goksu, Y. Yıldız, B. Çelik, M. Yazici, B. Kilbas, F. Sen, Eco-friendly hydrogenation of aromatic aldehyde compounds by tandem dehydrogenation of dimethylamine-borane in the presence of a reduced graphene oxide furnished platinum nanocatalyst, Catal. Sci. Technol. 6 (2016) 2318–2324. https://doi.org/10.1039/C5CY01462J

[6]   M.E. Vance, T. Kuiken, E.P. Vejerano, S.P. McGinnis, M.F. Hochella, D. Rejeski, M.S. Hull, Nanotechnology in the real world: Redeveloping the nanomaterial consumer products inventory, Beilstein J. Nanotechnol. 6 (2015) 1769–1780. https://doi.org/10.3762/bjnano.6.181

[7]   P.G. Tratnyek, R.L. Johnson, Nanotechnologies for environmental cleanup, Nano Today 1 (2006) 44–48. https://doi.org/10.1016/S1748-0132(06)70048-2

[8]   T. Gur, I. Meydan, H. Seckin, M. Bekmezci, F. Sen, Green Synthesis, Characterization and Bioactivity of Biogenic Zinc Oxide Nanoparticles, Environ. Res. 204 (2022) 111897. https://doi.org/10.1016/j.envres.2021.111897.

[9]   T. Yuranova, R. Mosteo, J. Bandara, D. Laub, J. Kiwi, Self-cleaning cotton textiles surfaces modified by photoactive SiO2/TiO2 coating, J. Mol. Catal. A Chem. 244 (2006) 160–167. https://doi.org/10.1016/j.molcata.2005.08.059

[10]   I. Meydan, A. Aygun, R.N.E. Tiri, T. Gur, Y. Kocak, H. Seckin, F. Sen, Chitosan/PVA Supported Silver Nanoparticles for Azo Dyes Removal: Fabrication, Characterization, and Assessment of Antioxidant Activity, Environ. Sci. Adv. (2024). https://doi.org/10.1039/D3VA00224A

[11]   N. Lolak, E. Kuyuldar, H. Burhan, H. Goksu, S. Akocak, F. Sen, Composites of Palladium–Nickel Alloy Nanoparticles and Graphene Oxide for the Knoevenagel Condensation of Aldehydes with Malononitrile, ACS Omega 4 (2019) 6848–6853. https://doi.org/10.1021/acsomega.9b00485

[12]   R. Ayranci, G. Başkaya, M. Güzel, S. Bozkurt, F. Şen, M. Ak, Carbon Based Nanomaterials for High Performance Optoelectrochemical Systems, ChemistrySelect 2 (2017) 1548–1555. https://doi.org/10.1002/slct.201601632

[13]   F. Şen, G. Gökağaç, S. Şen, High performance Pt nanoparticles prepared by new surfactants for C1 to C3 alcohol oxidation reactions, J. Nanoparticle Res. 15 (2013) 1979. https://doi.org/10.1007/s11051-013-1979-5

[14]   F. Sen, A.A. Boghossian, S. Sen, Z.W. Ulissi, J. Zhang, M.S. Strano, Observation of Oscillatory Surface Reactions of Riboflavin, Trolox, and Singlet Oxygen Using Single Carbon Nanotube Fluorescence Spectroscopy, ACS Nano 6 (2012) 10632–10645. https://doi.org/10.1021/nn303716n

[15]   R. Ayranci, G. Baskaya, M. Guzel, S. Bozkurt, M. Ak, A. Savk, F. Sen, Enhanced optical and electrical properties of PEDOT via nanostructured carbon materials: A comparative investigation, Nano-Structures & Nano-Objects 11 (2017) 13–19. https://doi.org/10.1016/j.nanoso.2017.05.008

[16]   F. Şen, G. Gökağaç, Pt Nanoparticles Synthesized With New Surfactants: Improvement in C1–C3 Alcohol Oxidation Catalytic Activity, J. Appl. Electrochem. 44 (2014) 199–207. https://doi.org/10.1007/s10800-013-0631-5

[17]   N. Korkmaz, Y. Ceylan, P. Taslimi, A. Karadağ, A.S. Bülbül, F. Şen, Biogenic nano silver: Synthesis, characterization, antibacterial, antibiofilms, and enzymatic activity, Adv. Powder Technol. 31 (2020) 2942–2950. https://doi.org/10.1016/j.apt.2020.05.020

[18]   Y. Yıldız, İ. Esirden, E. Erken, E. Demir, M. Kaya, F. Şen, Microwave (Mw) Assisted Synthesis of 5-Substituted 1 H -Tetrazoles via [3+2] Cycloaddition Catalyzed by Mw-Pd/Co Nanoparticles Decorated on Multi-Walled Carbon Nanotubes, ChemistrySelect 1 (2016) 1695–1701. https://doi.org/10.1002/slct.201600265

[19]   E. Demir, A. Savk, B. Sen, F. Sen, A novel monodisperse metal nanoparticles anchored graphene

oxide as Counter Electrode for Dye-Sensitized Solar Cells, Nano-Structures & Nano-Objects 12 (2017) 41–45. https://doi.org/10.1016/j.nanoso.2017.08.018

[20]   A. Aygun, F. Gulbagca, E.E. Altuner, M. Bekmezci, T. Gur, H. Karimi-Maleh, F. Karimi, Y. Vasseghian, F. Sen, Highly Active PdPt Bimetallic Nanoparticles Synthesized By One Step Bioreduction Method: Characterizations, Anticancer, Antibacterial Activities and Evaluation of Their Catalytic Effect for Hydrogen Generation, Int. J. Hydrogen Energy 48 (2023) 6666–6679. https://doi.org/10.1016/j.ijhydene.2021.12.144

[21]   H. Göksu, H. Burhan, S.D. Mustafov, F. Şen, Oxidation of Benzyl Alcohol Compounds in the Presence of Carbon Hybrid Supported Platinum Nanoparticles (Pt@CHs) in Oxygen Atmosphere, Sci. Rep. 10 (2020) 5439. https://doi.org/10.1038/s41598-020-62400-5

[22]   N. Zare, F. Ameen, M. Bekmezci, M. Akin, R. Bayat, T. Kozak, E. Halvaci, I. Isik, F. Sen, Electrochemical sensing of vitamin C using PtNi nanomaterials supported on carbon nanotubes produced by arc discharge method, J. Food Meas. Charact. 18 (2024) 6759–6769. https://doi.org/10.1007/s11694-024-02689-2

[23]   R. Bayat, E. Halvaci, T. Kozak, M. Bekmezci, F. Sen, Electrocatalytic performance of Shape-Controlled synthesized PdNi-rGO nano Cube on sodium borohydride and methanol electrooxidation, Fuel 366 (2024) 131248. https://doi.org/10.1016/j.fuel.2024.131248

[24]   A. Şavk, K. Cellat, K. Arıkan, F. Tezcan, S.K. Gülbay, S. Kızıldağ, E.Ş. Işgın, F. Şen, Highly Monodisperse Pd-Ni Nanoparticles Supported On Rgo As A Rapid, Sensitive, Reusable And Selective Enzyme-Free Glucose Sensor, Sci. Rep. 9 (2019) 19228. https://doi.org/10.1038/s41598-019-55746-y

[25]   R. Ulus, Y. Yıldız, S. Eriş, B. Aday, F. Şen, M. Kaya, Functionalized Multi-Walled Carbon Nanotubes (f -MWCNT) as Highly Efficient and Reusable Heterogeneous Catalysts for the Synthesis of Acridinedione Derivatives, ChemistrySelect 1 (2016) 3861–3865. https://doi.org/10.1002/slct.201600719

[26]   F. Gulbagca, A. Aygün, M. Gülcan, S. Ozdemir, S. Gonca, F. Şen, Green Synthesis of Palladium Nanoparticles: Preparation, Characterization, and Investigation of Antioxidant, Antimicrobial, Anticancer, and DNA Cleavage Activities, Appl. Organomet. Chem. 35 (2021). https://doi.org/10.1002/aoc.6272

[27]   J.T. Abrahamson, B. Sempere, M.P. Walsh, J.M. Forman, F. Şen, S. Şen, S.G. Mahajan, G.L.C. Paulus, Q.H. Wang, W. Choi, M.S. Strano, Excess Thermopower and the Theory of Thermopower Waves, ACS Nano 7 (2013) 6533–6544. https://doi.org/10.1021/nn402411k

[28]   H. Göksu, Y. Yıldız, B. Çelik, M. Yazıcı, B. Kılbaş, F. Şen, Highly Efficient and Monodisperse Graphene Oxide Furnished Ru/Pd Nanoparticles for the Dehalogenation of Aryl Halides via Ammonia Borane, ChemistrySelect 1 (2016) 953–958. https://doi.org/10.1002/slct.201600207

[29]   S. Günbatar, A. Aygun, Y. Karataş, M. Gülcan, F. Şen, Carbon-nanotube-based rhodium nanoparticles as highly-active catalyst for hydrolytic dehydrogenation of dimethylamineborane at room temperature, J. Colloid Interface Sci. 530 (2018) 321–327. https://doi.org/10.1016/j.jcis.2018.06.100

[30]   H. Goksu, Y. Yıldız, B. Çelik, M. Yazıcı, B. Kılbas, F. Sen, Eco Friendly Hydrogenation of Aromatic Aldehyde Compounds By Tandem Dehydrogenation of Dimethylamine Borane in the Presence of a Reduced Graphene Oxide Furnished Platinum Nanocatalyst, Catal. Sci. Technol. 6 (2016) 2318–2324. https://doi.org/10.1039/C5CY01462J

[31]   A. El-Sayed, M. Kamel, Advanced applications of nanotechnology in veterinary medicine, Environ. Sci. Pollut. Res. 27 (2020) 19073–19086. https://doi.org/10.1007/s11356-018-3913-y

[32]   N. Lewinski, V. Colvin, R. Drezek, Cytotoxicity of Nanoparticles, Small 4 (2008) 26–49. https://doi.org/10.1002/smll.200700595

[33]   W. Hannah, P.B. Thompson, Nanotechnology, risk and the environment: a review, J. Environ. Monit. 10 (2008) 291–300. https://doi.org/10.1039/B718127M

[34]   C. Mühlfeld, B. Rothen-Rutishauser, F. Blank, D. Vanhecke, M. Ochs, P. Gehr, Interactions of

nanoparticles with pulmonary structures and cellular responses, Am. J. Physiol. Cell. Mol. Physiol. 294 (2008) L817–L829. https://doi.org/10.1152/ajplung.00442.2007

[35]   C. Buzea, I.I. Pacheco, K. Robbie, Nanomaterials and Nanoparticles: Sources and Toxicity, Biointerphases 2 (2007) MR17–MR71. https://doi.org/10.1116/1.2815690

[36]   S. Pfuhler, R. Elespuru, M.J. Aardema, S.H. Doak, E. Maria Donner, M. Honma, M. Kirsch-Volders, R. Landsiedel, M. Manjanatha, T. Singer, J.H. Kim, Genotoxicity of nanomaterials: Refining strategies and tests for hazard identification, Environ. Mol. Mutagen. 54 (2013) 229–239. https://doi.org/10.1002/em.21770

[37]   Z. Atli Sekeroglu, From nanotechnology to nanogenotoxicology: genotoxic effect of cobalt-chromium nanoparticles, Turkish Bull. Hyg. Exp. Biol. 70 (2013) 33–42. https://doi.org/10.5505/TurkHijyen.2013.70298

[38]   Y.S. Cheng, G.K. Hansen, Y.F. Su, H.C. Yeh, K.T. Morgan, Deposition of ultrafine aerosols in rat nasal molds, Toxicol. Appl. Pharmacol. 106 (1990) 222–233. https://doi.org/10.1016/0041-008X(90)90242-M

[39]   G. Oberdörster, E. Oberdörster, J. Oberdörster, Nanotoxicology: An Emerging Discipline Evolving from Studies of Ultrafine Particles, Environ. Health Perspect. 113 (2005) 823–839. https://doi.org/10.1289/ehp.7339

[40]   D. W, MacNeeK, How can ultrafine particles be responsible for increased mortality?, Arch. Monaldi Mal. Torace 55 (2000) 135–139.

[41]   Y. Zhu, W.C. Hinds, M. Krudysz, T. Kuhn, J. Froines, C. Sioutas, Penetration of freeway ultrafine particles into indoor environments, J. Aerosol Sci. 36 (2005) 303–322. https://doi.org/10.1016/j.jaerosci.2004.09.007

[42]   K. Donaldson, M. Ian Gilmour, W. MacNee, Asthma and PM10, Respir. Res. 1 (2000) 12–15. https://doi.org/10.1186/rr5

[43]   M. Guarnieri, J.R. Balmes, Outdoor air pollution and asthma, Lancet 383 (2014) 1581–1592. https://doi.org/10.1016/S0140-6736(14)60617-6

[44]   P.G. Barlow, K. Donaldson, J. MacCallum, A. Clouter, V. Stone, Serum exposed to nanoparticle carbon black displays increased potential to induce macrophage migration, Toxicol. Lett. 155 (2005) 397–401. https://doi.org/10.1016/j.toxlet.2004.11.006

[45]   V. Stone, M. Tuinman, J.E. Vamvakopoulos, J. Shaw, D. Brown, S. Petterson, S.P. Faux, P. Borm, W. MacNee, F. Michaelangeli, K. Donaldson, Increased calcium influx in a monocytic cell line on exposure to ultrafine carbon black, Eur. Respir. J. 15 (2000) 297. https://doi.org/10.1034/j.1399-3003.2000.15b13.x

[46]   F. Cassee, H. Muijser, E. Duistermaat, J. Freijer, K. Geerse, J. Marijnissen, J. Arts, Particle size-dependent total mass deposition in lungs determines inhalation toxicity of cadmium chloride aerosols in rats. Application of a multiple path dosimetry model, Arch. Toxicol. 76 (2002) 277–286. https://doi.org/10.1007/s00204-002-0344-8

[47]   L. Yang, D.J. Watts, Particle surface characteristics may play an important role in phytotoxicity of alumina nanoparticles, Toxicol. Lett. 158 (2005) 122–132. https://doi.org/10.1016/j.toxlet.2005.03.003

[48]   A. Awaad, Histopathological and immunological changes induced by magnetite nanoparticles in the spleen, liver and genital tract of mice following intravaginal instillation, J. Basic Appl. Zool. 71 (2015) 32–47. https://doi.org/10.1016/j.jobaz.2015.03.003

[49]   M.-L. Kung, S.-L. Hsieh, C.-C. Wu, T.-H. Chu, Y.-C. Lin, B.-W. Yeh, S. Hsieh, Enhanced reactive oxygen species overexpression by CuO nanoparticles in poorly differentiated hepatocellular carcinoma cells, Nanoscale 7 (2015) 1820–1829. https://doi.org/10.1039/C4NR05843G

[50]   B. Fahmy, S.A. Cormier, Copper oxide nanoparticles induce oxidative stress and cytotoxicity in airway epithelial cells, Toxicol. Vitr. 23 (2009) 1365–1371. https://doi.org/10.1016/j.tiv.2009.08.005

[51]   Z. Wang, N. Li, J. Zhao, J.C. White, P. Qu, B. Xing, CuO Nanoparticle Interaction with Human Epithelial Cells: Cellular Uptake, Location, Export, and Genotoxicity, Chem. Res. Toxicol. 25 (2012) 1512–1521. https://doi.org/10.1021/tx3002093

[52]   S. Naseem, M.A. Gatoo, A.M. Dar, M.Y. Arfat, K. Qasim, in Vivo Toxicity of Nanoparticles: Modalities and Treatment, Eur. Chem. Bull. 3 (2014) 992–1000. https://doi.org/10.17628/ECB.2014.3.992

[53]   P.R. Lockman, J.M. Koziara, R.J. Mumper, D.D. Allen, Nanoparticle Surface Charges Alter Blood–Brain Barrier Integrity and Permeability, J. Drug Target. 12 (2004) 635–641. https://doi.org/10.1080/10611860400015936

[54]   S.B. Lovern, R. Klaper, Daphnia magna mortality when exposed to titanium dioxide and fullerene (C 60) nanoparticles, Environ. Toxicol. Chem. 25 (2006) 1132–1137. https://doi.org/10.1897/05-278R.1

[55]   S.B. Lovern, J.R. Strickler, R. Klaper, Behavioral and Physiological Changes in Daphnia magna when Exposed to Nanoparticle Suspensions (Titanium Dioxide, Nano-C 60 , and C 60 HxC 70 Hx), Environ. Sci. Technol. 41 (2007) 4465–4470. https://doi.org/10.1021/es062146p

[56]   M. Ates, J. Daniels, Z. Arslan, I.O. Farah, Effects of aqueous suspensions of titanium dioxide nanoparticles on Artemia salina: assessment of nanoparticle aggregation, accumulation, and toxicity, Environ. Monit. Assess. 185 (2013) 3339–3348. https://doi.org/10.1007/s10661-012-2794-7

[57]   E. Oberdörster, S. Zhu, T.M. Blickley, P. McClellan-Green, M.L. Haasch, Ecotoxicology of carbon-based engineered nanoparticles: Effects of fullerene (C60) on aquatic organisms, Carbon N. Y. 44 (2006) 1112–1120. https://doi.org/10.1016/j.carbon.2005.11.008

[58]   G. FEDERICI, B. SHAW, R. HANDY, Toxicity of titanium dioxide nanoparticles to rainbow trout (Oncorhynchus mykiss): Gill injury, oxidative stress, and other physiological effects, Aquat. Toxicol. 84 (2007) 415–430. https://doi.org/10.1016/j.aquatox.2007.07.009

[59]   L. HAO, Z. WANG, B. XING, Effect of sub-acute exposure to TiO2 nanoparticles on oxidative stress and histopathological changes in Juvenile Carp (Cyprinus carpio), J. Environ. Sci. 21 (2009) 1459–1466. https://doi.org/10.1016/S1001-0742(08)62440-7

[60]   B. Nowack, T.D. Bucheli, Occurrence, behavior and effects of nanoparticles in the environment, Environ. Pollut. 150 (2007) 5–22. https://doi.org/10.1016/j.envpol.2007.06.006

[61]   M. Ates, J. Daniels, Z. Arslan, I.O. Farah, H.F. Rivera, Comparative evaluation of impact of Zn and ZnO nanoparticles on brine shrimp (Artemia salina) larvae: effects of particle size and solubility on toxicity, Environ. Sci. Process. Impacts 15 (2013) 225–233. https://doi.org/10.1039/C2EM30540B

[62]   G. Oberdörster, Z. Sharp, V. Atudorei, A. Elder, R. Gelein, W. Kreyling, C. Cox, Translocation of Inhaled Ultrafine Particles to the Brain, Inhal. Toxicol. 16 (2004) 437–445. https://doi.org/10.1080/08958370490439597

[63]   K. Zhang, Integration Of ER Stress, Oxidative Stress And The Inflammatory Response In Health And Disease., Int. J. Clin. Exp. Med. 3 (2010) 33.

[64]   W.N. Missaoui, R.D. Arnold, B.S. Cummings, Toxicological status of nanoparticles: What we know and what we don't know, Chem. Biol. Interact. 295 (2018) 1–12. https://doi.org/10.1016/j.cbi.2018.07.015

[65]   G. Oberdörster, A. Maynard, K. Donaldson, V. Castranova, J. Fitzpatrick, K. Ausman, J. Carter, B. Karn, W. Kreyling, D. Lai, S. Olin, N. Monteiro-Riviere, D. Warheit, H. Yang, Principles for characterizing the potential human health effects from exposure to nanomaterials: elements of a screening strategy, Part. Fibre Toxicol. 2 (2005) 8. https://doi.org/10.1186/1743-8977-2-8

[66]   C.A. Pope, Epidemiology of fine particulate air pollution and human health: biologic mechanisms and who's at risk?, Environ. Health Perspect. 108 (2000) 713–723. https://doi.org/10.1289/ehp.108-1637679

[67]   B. Groneberg-Kloft, T. Kraus, A. van Mark, U. Wagner, A. Fischer, Analysing the causes of

chronic cough: relation to diesel exhaust, ozone, nitrogen oxides, sulphur oxides and other environmental factors, J. Occup. Med. Toxicol. 1 (2006) 6. https://doi.org/10.1186/1745-6673-1-6

[68]   D. Li, Y. Li, G. Li, Y. Zhang, J. Li, H. Chen, Fluorescent reconstitution on deposition of PM 2.5 in lung and extrapulmonary organs, Proc. Natl. Acad. Sci. 116 (2019) 2488–2493. https://doi.org/10.1073/pnas.1818134116

[69]   D.E. Schraufnagel, The health effects of ultrafine particles, Exp. Mol. Med. 52 (2020) 311–317. https://doi.org/10.1038/s12276-020-0403-3

[70]   E. Fröhlich, S. Salar-Behzadi, Toxicological Assessment of Inhaled Nanoparticles: Role of in Vivo, ex Vivo, in Vitro, and in Silico Studies, Int. J. Mol. Sci. 15 (2014) 4795–4822. https://doi.org/10.3390/ijms15034795

[71]   P. Khanna, C. Ong, B. Bay, G. Baeg, Nanotoxicity: An Interplay of Oxidative Stress, Inflammation and Cell Death, Nanomaterials 5 (2015) 1163–1180. https://doi.org/10.3390/nano5031163

[72]   M. Macko, J. Antoš, F. Božek, J. Konečný, J. Huzlík, J. Hegrová, I. Kuřitka, Development of New Health Risk Assessment of Nanoparticles: EPA Health Risk Assessment Revised, Nanomaterials 13 (2022) 20. https://doi.org/10.3390/nano13010020

[73]   A.R. Köhler, C. Som, A. Helland, F. Gottschalk, Studying the potential release of carbon nanotubes throughout the application life cycle, J. Clean. Prod. 16 (2008) 927–937. https://doi.org/10.1016/j.jclepro.2007.04.007

[74]   I. Meydan, H. Burhan, T. Gür, H. Seçkin, B. Tanhaei, F. Sen, Characterization of Rheum ribes with ZnO nanoparticle and its antidiabetic, antibacterial, DNA damage prevention and lipid peroxidation prevention activity of in vitro, Environ. Res. 204 (2022) 112363. https://doi.org/10.1016/j.envres.2021.112363

[75]   T. Xia, N. Li, A.E. Nel, Potential Health Impact of Nanoparticles, Annu. Rev. Public Health 30 (2009) 137–150. https://doi.org/10.1146/annurev.publhealth.031308.100155

[76]   H. Bouwmeester, S. Dekkers, M.Y. Noordam, W.I. Hagens, A.S. Bulder, C. de Heer, S.E.C.G. ten Voorde, S.W.P. Wijnhoven, H.J.P. Marvin, A.J.A.M. Sips, Review of health safety aspects of nanotechnologies in food production, Regul. Toxicol. Pharmacol. 53 (2009) 52–62. https://doi.org/10.1016/j.yrtph.2008.10.008

[77]   M.N. Moore, Do nanoparticles present ecotoxicological risks for the health of the aquatic environment?, Environ. Int. 32 (2006) 967–976. https://doi.org/10.1016/j.envint.2006.06.014

[78]   L. Taylor, H. Schmitt, W. Carrier, M. Nakagawa, Lunar Dust Problem: From Liability to Asset, in: 1st Sp. Explor. Conf. Contin. Voyag. Discov., American Institute of Aeronautics and Astronautics, Reston, Virigina, 2005. https://doi.org/10.2514/6.2005-2510

[79]   C. Medina, M.J. Santos-Martinez, A. Radomski, O.I. Corrigan, M.W. Radomski, Nanoparticles: pharmacological and toxicological significance, Br. J. Pharmacol. 150 (2007) 552–558. https://doi.org/10.1038/sj.bjp.0707130

[80]   H. Xie, M.M. Mason, J.P. Wise, Genotoxicity of metal nanoparticles, Rev. Environ. Health 26 (2011). https://doi.org/10.1515/REVEH.2011.033

[81]   K. Donaldson, C.A. Poland, R.P.F. Schins, Possible genotoxic mechanisms of nanoparticles: Criteria for improved test strategies, Nanotoxicology 4 (2010) 414–420. https://doi.org/10.3109/17435390.2010.482751

[82]   K. BéruBé, D. Balharry, K. Sexton, L. Koshy, T. Jones, COMBUSTION-DERIVED NANOPARTICLES: MECHANISMS OF PULMONARY TOXICITY, Clin. Exp. Pharmacol. Physiol. 34 (2007) 1044–1050. https://doi.org/10.1111/j.1440-1681.2007.04733.x

[83]   R. Duffin, N.L. Mills, K. Donaldson, Nanoparticles-A Thoracic Toxicology Perspective, Yonsei Med. J. 48 (2007) 561. https://doi.org/10.3349/ymj.2007.48.4.561

[84]   H. Erdem, E. Canakci, A. Karatas, M. Akcay Celik, A. Kilinc, Evaluation of distant organ effect

of renal ischemia and reperfusion with claudin-5, J. Crit. Intensive Care 11 (2020) 3–7. https://doi.org/10.37678/dcybd.2020.2155

[85]   T. Kato, T. Yashiro, Y. Murata, D. Herbert, K. Oshikawa, M. Bando, S. Ohno, Y. Sugiyama, Evidence that exogenous substances can be phagocytized by alveolar epithelial cells and transported into blood capillaries, Cell Tissue Res. 311 (2003) 47–51. https://doi.org/10.1007/s00441-002-0647-3.

[86]   L. Sadeghi, V.Y. Babadi, H.R. Espanani, Toxic effects of the Fe2O3 nanoparticles on the liver and lung tissue, Bratislava Med. J. 116 (2015) 373–378. https://doi.org/10.4149/BLL_2015_071

[87]   K. Donaldson, F.A. Murphy, R. Duffin, C.A. Poland, Asbestos, carbon nanotubes and the pleural mesothelium: a review and the hypothesis regarding the role of long fibre retention in the parietal pleura, inflammation and mesothelioma, Part. Fibre Toxicol. 7 (2010) 5. https://doi.org/10.1186/1743-8977-7-5

[88]   C. Demir, D. Tasdemir, S. Ilhan, A.F. Isik, H. Bayram, Toxicological and inflammatory effects of engineered nanoparticles and diesel exhaust particles on human mesothelial cells, in: Airw. Cell Biol. Immunopathol., European Respiratory Society, 2017: p. PA3918. https://doi.org/10.1183/1393003.congress-2017.PA3918.

[89]   J. Geys, A. Nemmar, E. Verbeken, E. Smolders, M. Ratoi, M.F. Hoylaerts, B. Nemery, P.H.M. Hoet, Acute Toxicity and Prothrombotic Effects of Quantum Dots: Impact of Surface Charge, Environ. Health Perspect. 116 (2008) 1607–1613. https://doi.org/10.1289/ehp.11566

[90]   N. Hussain, Recent advances in the understanding of uptake of microparticulates across the gastrointestinal lymphatics, Adv. Drug Deliv. Rev. 50 (2001) 107–142. https://doi.org/10.1016/S0169-409X(01)00152-1

[91]   M.-F. Song, Y.-S. Li, H. Kasai, K. Kawai, Metal nanoparticle-induced micronuclei and oxidative DNA damage in mice, J. Clin. Biochem. Nutr. 50 (2012) 211–216. https://doi.org/10.3164/jcbn.11-70

[92]   R.N. Alyaudtin, A. Reichel, R. Löbenberg, P. Ramge, J. Kreuter, D.J. Begley, Interaction of Poly(butylcyanoacrylate) Nanoparticles with the Blood-Brain Barrier in vivo and in vitro, J. Drug Target. 9 (2001) 209–221. https://doi.org/10.3109/10611860108997929

[93]   M. Khalid, M. Asad, P. Henrich-Noack, M. Sokolov, W. Hintz, L. Grigartzik, E. Zhang, A. Dityatev, B. van Wachem, B. Sabel, Evaluation of Toxicity and Neural Uptake In Vitro and In Vivo of Superparamagnetic Iron Oxide Nanoparticles, Int. J. Mol. Sci. 19 (2018) 2613. https://doi.org/10.3390/ijms19092613

[94]   A. Dhawan, V. Sharma, Toxicity assessment of nanomaterials: methods and challenges, Anal. Bioanal. Chem. 398 (2010) 589–605. https://doi.org/10.1007/s00216-010-3996-x

[95]   X. Wang, Y. Wang, Z.G. Chen, D.M. Shin, Advances of Cancer Therapy by Nanotechnology, Cancer Res. Treat. 41 (2009) 1. https://doi.org/10.4143/crt.2009.41.1.1

[96]   S. Syed, A. Zubair, M. Frieri, Immune Response to Nanomaterials: Implications for Medicine and Literature Review, Curr. Allergy Asthma Rep. 13 (2013) 50–57. https://doi.org/10.1007/s11882-012-0302-3.

[97]   T.C. Long, J. Tajuba, P. Sama, N. Saleh, C. Swartz, J. Parker, S. Hester, G. V. Lowry, B. Veronesi, Nanosize Titanium Dioxide Stimulates Reactive Oxygen Species in Brain Microglia and Damages Neurons in Vitro, Environ. Health Perspect. 115 (2007) 1631–1637. https://doi.org/10.1289/ehp.10216

[98]   D.P.K. Lankveld, A.G. Oomen, P. Krystek, A. Neigh, A. Troost – de Jong, C.W. Noorlander, J.C.H. Van Eijkeren, R.E. Geertsma, W.H. De Jong, The kinetics of the tissue distribution of silver nanoparticles of different sizes, Biomaterials 31 (2010) 8350–8361. https://doi.org/10.1016/j.biomaterials.2010.07.045

[99]   E. Sadauskas, N.R. Jacobsen, G. Danscher, M. Stoltenberg, U. Vogel, A. Larsen, W. Kreyling, H. Wallin, Biodistribution of gold nanoparticles in mouse lung following intratracheal instillation, Chem. Cent. J. 3 (2009) 16. https://doi.org/10.1186/1752-153X-3-16

[100]   N. Barkalina, C. Charalambous, C. Jones, K. Coward, Nanotechnology in reproductive medicine: Emerging applications of nanomaterials, Nanomedicine Nanotechnology, Biol. Med. 10 (2014) e921–e938. https://doi.org/10.1016/J.NANO.2014.01.001

[101]   R. Wang, B. Song, J. Wu, Y. Zhang, A. Chen, L. Shao, Potential adverse effects of nanoparticles on the reproductive system, Int. J. Nanomedicine Volume 13 (2018) 8487–8506. https://doi.org/10.2147/IJN.S170723

[102]   T.M. Benn, P. Westerhoff, Nanoparticle Silver Released into Water from Commercially Available Sock Fabrics, Environ. Sci. Technol. 42 (2008) 4133–4139. https://doi.org/10.1021/es7032718

[103]   J. Bott, R. Franz, Investigations into the Potential Abrasive Release of Nanomaterials due to Material Stress Conditions—Part B: Silver, Titanium Nitride, and Laponite Nanoparticles in Plastic Composites, Appl. Sci. 9 (2019) 221. https://doi.org/10.3390/app9020221

[104]   Y. Liang, H. Demir, Y. Wu, A. Aygun, R.N. Elhouda Tiri, T. Gur, Y. Yuan, C. Xia, C. Demir, F. Sen, Y. Vasseghian, Facile synthesis of biogenic palladium nanoparticles using biomass strategy and application as photocatalyst degradation for textile dye pollutants and their in-vitro antimicrobial activity, Chemosphere 306 (2022) 135518. https://doi.org/10.1016/j.chemosphere.2022.135518

Biogenic Nanomaterials: Synthesis, Characterization, Applications, and Future Remarks   Materials Research Forum LLC
Materials Research Foundations 180 (2025)   https://doi.org/21741/9781644903759

Chapter 6

Anticancer Effects of Biogenic Nanomaterials

M. Ermaya[1], H. Demir[2*], C. Demir[3], E. Halvaci[4], F. Sen[4*]

[1]Van Regional Training and Research Hospital, Van, Turkiye

[2]Department of Biochemistry, Faculty of Science, Van Yuzuncu Yıl University, Van, Turkiye

[3]Vocational School of Health Care, Van Yuzuncu Yıl University, Van, Turkiye

[4]Sen Research Group, Department of Biochemistry, Kutahya Dumlupinar University, Kutahya, Turkiye

halitdemir@yyu.edu.tr; fatih.sen@dpu.edu.tr

Abstract

Nanotechnology is used in many sectors, from cosmetics, pharmaceuticals, paints and biotechnology to cancer treatment. Some studies determined that cancer cells can be killed by nanoparticles. Thanks to specific transporter systems improved using nanotechnology, applications that do not influence healthful cells but only kill cancerous cells are now probable. The technology in question; Tumor-specific ligands, antibodies, anticancer drugs, moreover imaging catheters have led to the improvement of nanoparticles that can be simultaneously conjugated with multiple all-duty molecules and applied in oncology. These nanoparticles can be with ease transported through blood vessels moreover interact with the tumor-specific proteins they target on the surface in addition within of cancer cells.

Keywords

Anticancer Effect, Cancer, Nanoparticles, Tumor Imaging

1.   Introduction

In recent years, drug delivery systems; It has come to the fore in many studies in cancer studies, thanks to their properties such as targeting tumor sites, increasing the bioavailability of low-soluble anticancer drugs, preventing systemic toxicity, and eliminating multidrug resistance [1–13]. Among the many drug delivery systems developed to date, it is obvious that inorganic-based nanoparticles with magnetic properties are promising in that they can be used in both diagnostic and therapeutic applications. It has become possible for these particles to be used as smart nanodrug deliverers due to their ease of synthesis and the ability to easily integrate different cell-specific targeting, imaging and therapeutic functions into the system [14–29]. In latest years, there have been many researches focusing on the intratumoral accumulation and anticancer activity effects of scales of transmitter systems [30]. Because, in order to overcome the abnormal bio barriers that occur in tumor areas, the characterization of synthesized nanoparticle scales is of great importance in terms of understanding the anticancer effectiveness of delivery systems and improving their capacities. While the scale of the nanoparticles determines their capability to attach to the cell membrane, the surfaces of the transmitter systems can be actively increased by modifying the surfaces of the transmitter systems with ligands specific to the receptors overexpressed by cancer cells [9,31,32]. Thanks to this, while drug concentrations in cancer cells are increased through nanoparticles, systemic toxicity that anticancer drugs may cause against normal cells is also minimized. In research conducted on the reticuloendothelial system (RES), which is among the body's protection mechanisms and responsible for clearing nanoparticles, it has been observed that nanoparticles with diameters varying between 10 and 100 nm can be prevented from being retained by the RES, while on the other hand, they have longer blood circulation times [33]. However, when the studies were examined, no significant overlap could be detected on the anticancer activities of nanoparticles at the same scales, because in addition to the unique pathological features of each tumor type, various important behaviours were observed against various regions within the same tumor [34,35]. Therefore, nanoparticle designs reduced to a single scale may not be the most effective approach against all types of cancer. Therefore, it was deemed necessary to design the nanoparticle according to the exact location and conditions (such as primary or metastatic tumor, tumor scene moreover aggressiveness, host organ, local vascular properties in addition hemodynamic). In addition, nanoparticle scale may also affect drug loading and release into delivery systems. While particles with smaller diameters have higher drug release rates because their surface area/volume ratio is larger, larger particles allow more drug loading thanks to their large core contents and show a slower drug release character [36,37]. Nanotechnology, which emerges as a brand new branch of science in today's world, is developing rapidly. When we look at the fields of nanotechnology, it is observed that it has spread to energy, production, health, medicine, environment, defence, and many other areas. Since nanoparticles have various properties, their usage areas seem to cover a wide range of areas [38]. Nanoparticles used in health-related fields are used in biomedical targeted technologies to provide more specific analyses, for molecular imaging in many radiological fields such as count up tomography, magnetic resonance, fluorescence moreover ultrasound, for many purposes such as targeted treatment, drug development systems and vaccine development. has become a situation. In addition to these common and beneficial areas of use, nanoparticles have become the subject of scientifically study to examine their possible toxic effects on the respiratory system, nervous system, blood, gastrointestinal system and skin due to their molecular properties. Solid colloidal polymeric particulate systems prepared with polymers of natural or synthetic nature, which release the active substance in the dissolved, trapped or adsorbed state, whose particle sizes vary between 10-1000 nm, are called "nanoparticles" [39]. Nanotechnology aims to synthesize new nanostructures after designing them or to give new extraordinary properties to nanostructures and to use these properties in new functions [40]. In latest years, there has been an acceleration in research on the usage of nano biotechnology, especially in biomedicine, for the benefit of humans. It has enabled the production of more effective and targeted drugs in the medical field and the development of methods to destroy cancerous tissues without causing any harm to healthy cells [41]. Nanotechnology is the genesis moreover use of tools or systems by controlling matter at the nanometer size. When the natural-sized functional components of living cells are taken into consideration, the application of nanotechnology in the biotechnology environment has become mandatory and the expression "Nano biotechnology", that is, " implementation of nanotechnology in living sciences" has emerged. Nano medicine is describing as the appeal of nano biotechnology in medicine moreover is dependent on the usage of nano sized equipment moreover apparatus for diagnostic moreover treatment, as well as the improvement of new pharmaceuticals called nano-pharmaceuticals. Applications of nanotechnology in the field of cancer are also called nano-oncology [42]. Malignant cells can develop resistance to the anticancer efficacy of chemotherapeutic drugs as a result of differences in their content at the molecular level. They also have the ability to hinder hypoxic circumstances moreover assist nutritional intake by developing tumor survival tactics that activate the antigenic mechanism. Timings associated with apoptosis or cell proliferation cause differences in the signalling pathways liable for chemotherapy strength moreover drug toleration in the cancer cell. Therefore, major therapeutic approaches specifically aim to inhibit tumor angiogenesis [43,44]. Cancer treatment strategies: It is in the form of different combinations of radiotherapy, adjuvant / neo-adjuvant surgery and chemotherapy. The most important drawback of chemotherapy is the undesirable side effects of the drugs. Hence, study is being conducted to improve new therapeutic formulations using targeted tissue-specific nanoparticles to refrain cytotoxic efficacy on healthful cells. Nano-oncology research has pioneered purpose-based drug improvement approaches that increase the survival ratio of cancer patients. In research conducted for this purpose, therapeutic agents become highly special in addition have a high closeness because they target different molecules depending on the genotype and phenotype of the tumor tissue [45,46]. Moreover, nanoparticle-based drug handing over systems attract attention because of their capability to create a controlled-extricate reservoir that can safely transmit therapeutic substances to intended areas or certain cells. In order to provide safe usage in the medical field, nanoparticles; It must be biocompatible, meaning that the structure must be able to combine into a biological system outside reason of an immune reaction or negative side efficacy when released directly into the tumor or the bloodstream [47]. Nano-particulate drug transmit systems have many advantages over free drugs. The basic ones are protecting the drug from early deterioration moreover biological elements, boosting the sorption of the drug by goal tissues, to check the pharmacokinetic features of the drug to further its distribution in tissues, boosting intracellular penetration further reducing systemic toxicity. Thanks to all the advantages in question, the number of nanoformulations confirmed for usage in cancer therapy is boosting over time [48,49] (Table 1).

Table 1. *Nano formulations that have been confirmed for usage in cancer treatment in addition are in phase (since 2016) [25, 26].*

Trade Name	Formulation	Indication	Approval Status
Abraxane®	Paclitaxel loaded albumin	Breast, lung and pancreatic cancer	FDA 2005 EMA 2008
Adcertis®	Brentuximab vedotin	Lymphoma	FDA 2011
Aurimune	Colloidal gold for tumor necrosis factor release	Solid tumors	Faz II
AuroLase	Gold-coated silica nanoparticles	Photothermal ablation in head and neck cancers	Faz I
BIND-014	Docetaxel–poly(styrene-alt-maleic acid) conjugate	Non-small cell lung cancer	Phase II completed
CPX-1	Liposome loaded with irinotecan and floxuridine	colorectal cancer	Phase II completed
CriPec®	Docetaxel	Solid tumors	Faz I
CRLX101	Camptothecin conjugated cyclodextrin-PEG copolymer	Non-small cell lung cancer	Faz II

DepoCyt®	Cytarabine loaded liposome	Glioblastoma, leukemia	FDA 2007
Doxil®	PEGylated liposomal Doxorubicin	Myeloma, ovarian cancer	FDA 1995 EMA 1996
DaunoXome	Liposomal daunorubicin	HIV-associated Kaposi's sarcoma	FDA 1996
Eligard	leuprolide acetate	Prostate cancer	FDA 2002
Gendicine®	Recombinant adenovirus	Head and neck cancer	Çin 2003
Genexol-PM®	Paclitaxel-loaded monomethoxy poly(ethylene glycol)-block-poly(D,L-lactide) (mPEG-PDLLA) micelle	Breast and pancreatic cancer	South Korea 2006,
Kadcyla®	Emtanin-bound trastuzumab	Metastatic breast cancer	FDA 2013
Marqibo®	Liposomal Vincristine sulfate	Acute lymphoblastic leukemia	FDA 2012
Megace®	Megestrol Acetate	Breast cancer	FDA 2005
Mepact®	Mifamurtide loaded liposome	Osteosarcoma	EMA 2009
MM-302	Antibody (HER-2) and doxorubicin liposomal conjugate	HER2 positive breast cancer	Faz III
Myocet®	Doxorubicin loaded liposome	Metastatic breast cancer	EMA 2000
NanoTherm™	Aminosilane coated iron oxide nanoparticle	Local ablation in glioblastoma, prostate and pancreatic cancer	EMA 2010
NanoXray	Hafnium oxide nanoparticle	Solid tumor radiotherapy	Faz I
NC-6004	Polyethylene glycol poly(glutamic acid) block copolymer containing cisplatin	Pancreatic cancer	Faz II
Oncaspar®	PEGylated L-asparaginase	Acute lymphoblastic leukemia	FDA 1994
Ontak®	Recombinant DNA created from IL-2 and diphtheria toxin sequence	T-cell lymphoma	FDA 1999
Onyvide®	Irinotecan loaded liposome	Pancreatic cancer	FDA 2015
Opaxio™	Paclitaxel conjugated α-poly(L)-glutamic acid	Lung and ovarian cancer	Faz III
ThermoDox®	Doxorubicin loaded liposome	Hepatocellular carcinoma	Faz III
Vyxeos™	Liposome loaded with cytarabine and daunorubicin	Acute myeloid leukemia	Faz III
Zinostatin Stimalamer®	Neocarzinostatin conjugated Styrene-maleic acid copolymer	Hepatocellular carcinoma	Japan 1994

Nano-Sized drug propellant systems; Its condition in the body may vary depending on surface charges, particle diameter, shape and polymer/material used. Particles with hydrodynamic diameters under 10 nm are quickly removed from the body by the kidney. Most of the particles given to the body through injection are retained by the liver further spleen. To avoid liver further spleen elimination, the scale of nanoparticles must be below 200 nm [50,51]. It has been observed that the zeta potency value, which is an indicator of the surface charge of nanoparticles, affects the direction of nanoparticles within the tumor. Increased cellular uptake because of the electrostatic coaction between positively charged nanoparticles further cell membranes has been observed in different studies. It has been observed that nanoparticles with neutral superficial charge at the same scale have lower cellular uptake compared to cationic nanoparticles [49]. To create an effective and reliable nano-carrier system, several important features need to be taken into account. Nanoscale systems must be made of a material that is well-defined, soluble, has the ability to circulate in the blood for a long time without causing aggregation, has high uptake by target cells, is biologically adaptable and can be easily functionalized. Functionalization of the surfaces of nanoparticulate systems is very important for both active and passive targeting strategies. In this study, the anticancer effects of nanoparticles were examined in detail.

2.   Cancer Nanotechnology

Cancer nanotechnology: is an interdisciplinary field of science, engineering moreover medical sciences that provides new tools to combat cancer. Cancer nanotechnology, a specific area of nano medicine, is based on nanoparticles that display useful-unlikely properties in medical (oncological) applications. The 5–200 nm scales of nanoparticles used in medicine allow for an unprecedented interaction with biological systems at the molecular grade. The usefulness of nanotechnology in cancer disease is molecular tumor imaging, mutational diagnosis, molecular diagnostics, targeted therapy, in addition cancer bioinformatics [52].

3.   Targeted Delivery of Therapeutic Agents with Nanoparticles

To submit of cancer drugs with nanoparticles tumor tissues can be reached with passive or active aiming. The tumor gaoling tactic with nano medicine-based therapeutics has occurred as a hopeful approach to come through the deficit specificity of conventional chemotherapeutic agents moreover to eliminate the troubles run across in customary cancer treatment in today's conditions [53,54]. Drug handing over systems can be usage especially to goal tumors moreover advance the therapeutic impact further pharmacokinetic profile of anticancer drugs. Nanoparticles, which are researched for this purpose, are nanometer-scale transmitter structures that aim to develop the biodistribution of systemically administered chemotherapeutic drugs. The main purpose of nanoscale drug handing over systems is to develop the equilibrium between the effectiveness moreover toxicity of the chemotherapeutic molecule after systemic management [55]. With this aim, it is intended to goal nano particulate systems to the tumor tissue further therefore obstruct systemic toxicity. Tumor gaoling tactics are splinted into passive targeting further active targeting.

3.1   Passive Aiming

Angiogenesis is very important for growth of the tumor. Antigenic blood vessels in tissues with tumors have 600–800 nm wide spaces between adjoining endothelial cells, distinct from those in normal tissues [56,57]. This defective vascular structure allows nanoparticles to pass through these spaces, nanoparticles can collect inside tumor tissues [58]. Passive targeting is the ability of nanoparticles to arrive at cancerous tissue from the circulatory system, thanks to their particle size, by taking advantage of the "EPR efficacy", beside known as the "increased permeability in addition retention efficacy" [59]. The EPR efficacy was first explained by Maeda in 1986. The basic feature of EPR physiology is; The vascular construction lets particles such as proteins, macromolecules, liposomes further micelles to pass within the cell [60]. When the tumor tissue arrives an exact size, the initially existing vessels are not sufficient to supply the needed oxygen in addition nutrients. In conclusion, cancer cells begin to lose their necrotic core, which causes the secretion of growth factors that detent angiogenesis in the tumor tissue. In the tumor tissue, new capillaries are formed from the circumambient capillaries. Angiogenesis in tumors; It is a processing in which new erratic blood vessels rapidly develop, presenting a discontinuous epithelium moreover devoid of the basement membrane of normal vascular constructions [41]. During the processing, there are spaces called fenestration between endothelial cells in newly created blood vessels, and the lengths of these fenestrations vary between 200-2000 nm relying on the tumor type [49]. Nanoparticles in the cyclic system can beside reach the tumor tissue by leaking through these gaps between endothelial cells, relying on their size. In addition, in healthy tissues, extracellular fluids are regularly discharged within lymphatic vessels. This lets the renovation of interstitial liquid further the turn back of colloids to the cycle. Nevertheless, in tumor tissue, lymphatic system function is moreover impaired in addition therefore retention in interstitial liquid is very low [61]. The success of the EPR effect bases on the lymphatic drainage rate and the degree of blood flow further capillary disorder, which are different in different tumor types. The number of tumor targeting studies aimed at utilizing the EPR effect using nano particulate drug delivery systems is increasing. However, there is still no product developed with this strategy and confirmed for clinical usage. However, it has been observed that FDA-approved Abraxane® and Doxil® cannot benefit from the EPR effect as intended in clinical practice. While the clinical therapeutic impact of Doxil® is upper to conventional treatments in ovarian cancer further AIDS- relating to Kaposi's sarcoma, it has been shown to be equivalent in multiple myeloma further metastatic breast cancer [62,63]. But, from the beginning, the intended treatment strategy for Doxil® in the cure of solid tumors was to stay behind in the cycle for a long time thanks to the PEG chains present on its surface and to ensure tumoral accumulation by taking advantage of the EPR impact. However, in clinical applications, the tumoral accumulation of Doxil was not as expected and it was observed that it caused toxicity on the skin because of its long-term retention in the blood. The lack of discrimination of tumoral accumulation of Doxil from free doxorubicin means that the EPR effect considered in theory varies in the clinic (Figure 1). The EPR hypothesis, known in theory, is not seen in all vessels in the tumor, as it cannot be the same for all tumors in the biological system. On the contrary what is known, EPR is a highly heterogeneous formation in the natural tumor context. It is affected by many environmental agents further varies greatly rely on the tumor type. Therefore, in studies aimed at benefiting from the EPR effect, tumor content and morphology need to be investigated in detail.

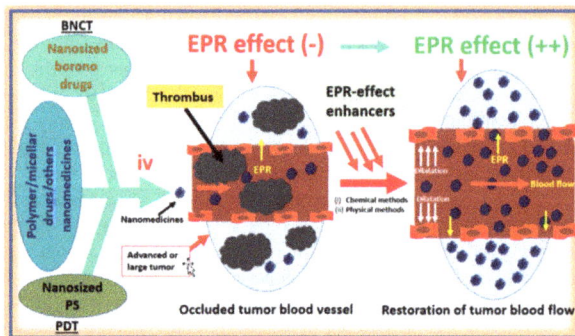

Figure 1. *Features that enable and prevent the usage of the EPR efficacy inherent in the tumor tissue (Reprinted with permission from [64], Copyright MDPI).*

3.2   Active Targeting

In passive aiming, high osmotic pressure can cause the drug out of the cells. Therefore, it may be necessary to reintroduce the drug to some parts of tumor tissue. Another strategy to achieve this limitation is the conjugation of antibody to nanoparticles [56]. In active targeting, the superficial of nanoparticles can be modified with peptides [65,66] proteins [67,68], and antibodies [69,70] specific to the targeted organ or/and tumor tissue. These molecules supply specific gaoling to molecules that are over expressed in cancer cells. Physicochemical features of nanoparticles, such as particle dimension, target molecule selection moreover intensity, play a significant role in the effectiveness of active targeting in vitro further in vivo environments [49]. Active coupling can ensure the release of the drug into or outside the cell in a targeted manner. Along with, the targeting tactic usage in active targeting also makes it easier the cellular uptake of nanoparticles [71]. In the active aiming strategy, there are two types of cell targets. The first of these is direct gaoling of the cancer cell. For this aim, specific molecules such as transferrin, folate, epidermal growth agent receptor or glycoprotein receptors, which are overexpressed in exact types of cancer, can be aimed. Other approach is to goal the tumor endothelium. There is excessive expression of VEGF, integrin, vascular cell adhering molecules further matrix metalloproteinase to ensure angiogenesis in tumor tissue [60]. It is possible to target tumor-specific nanoparticles with ligands suitable for these molecules [72,73].

4.   Nanoparticles Usage in Drug to Submit to Tumor Cells

Drug submission is important in cancer treatment so as to optimize the efficacy of drugs usage in cancer cure in addition reduce their side effects. Different nanotechnologies depended on nanoparticles can facilitate drug transmit, especially towards tumors themselves and increase drug efficacy [74,75]. Nanoparticles used in drug transmit in cancer treatment, hydrogels, micelles, liposomes, dendrimers, nano cells and nanotubes. Nanoparticles are composed of polymer, metal, ceramic, etc. materials. They can be in different shapes and scales because of the various methods further materials used in their production. These enable nanoparticles to have various properties [57,76].

4.1   Hydrogels

Hydrogels is a technology that uses hydrophobic polysaccharides in drug and therapeutic protein delivery. New systems using cholesterol pullulan (pullulan, a polysaccharide polymer containing malt triose units and derived from starch by Aureobasidium pullulans) show great promise. With pullulan on the outside, 4 cholesterol molecules come together spontaneously to form the hydrophobic core [77].

Table 2. Formulations of Polymeric Micelles Evaluated in Clinical Studies [78].

Formulation	Polymers That Make Up Nanoparticles	The Medicine She Carries
Genexol-Pm	Poly (ethylene glycol)- poly (D, L lactide)	Paclitaxel
NK105	Poly (ethylene glycol)-poly(aspartic acid)	Paklitaksel
NC-6004	Poly (ethylene glycol)-poly(glutamic acid) (Cisplatin)	Cisplatin
NC-4016	Poly (ethylene glycol)- poly(glutamic acid)(dichloro(1,2 diaminocydohexane)platinum II	Dichloro(1,2diaminocyd ohexane) Platinum ii
NK012	Poly (ethylene glycol)- poly(glutamic acid)(7-methyl-10 hydroxy camptothecin)	7-methyl-10 hydroxy camptothecin
NK911	Poly (ethylene glycol)-poly(aspartic acid) (doxorubicin)	Doxorubicin
SP1049C	Pluronic L61, F127/ 2-methyloxirane, poly (ethylene oxide)-block-poly (propylene oxide) block poly (ethylene oxide)	Doxorubicin

It has been revealed that pH sensitive biodegradable passive targeted hydrogels that transmit curcumin, which is in the structure of polyethylene glycol cross-linked to acrylic polymers, are more uptake by cervical cancer cells (HeLa). Thus, it has been observed that they have increased anti-proliferative effect and stimulate more apoptosis. It was also observed that the solubility of hydrogel-delivered curcumin in water is higher than that of independent curcumin in dimethyl sulfoxide (DMSO) [79]. It has been observed that passive targeted hydrogels in the poly (ethylene glycol)-b-poly (ε- caprolactone) make cyclodextrin moreover delivering doxorubicin inhibit the growth of human bladder cancer (EJ) cells [75]. Active targeted hydrogels with covalently bonded polyacrylamide structure to Coomassie Blue molecules conjugated with polyethylene glycol and F3 peptides; have been found to have a long in vivo circulation time. It has also been observed that they can be used in the imaging of brain tumors [80]. It has been observed that passive targeted hydrogels, which are gluconate, carboxylic acid poly (organophosphazene), transmitting camptothecin, have cytotoxic effects in cell lines of human diagnosed with lung cancer (A549) and 3 types of cell lines of human colon cancer (DLD-1, HCT 116 and HT-29). In vivo studies have shown that they inhibit colon tumor growth more effectively than free camptothecin [57]. Paclitaxel-conducting passively targeted hydrogels, which are in the structure of 5-methyl-2-(2,4,6-trimethoxyphenyl)- [1,3]-5-dioxanemethyl methacrylate, has a cytotoxic effect on tumor cells in humans diagnosed with lung cancer (A549) and empty hydrogels have no cytotoxic effect, has not been shown. Thus, it has been demonstrated that cytotoxicity caused by drug-loaded hydrogels is entirely related with drug release [81].

4.2   Micelles

Micelles are spherical molecular bundles of amphiphilic polymers. While the nucleus of the micelles find room to hydrophobic drugs, the shells make them water-soluble. It is thus in a hydrophilic form which allows distribution of the poorly soluble content. Biocompatible (no reaction occurs between the micelle and the cell-tissue), stabilized, camptothecin-delivering passively targeted micelles surrounded by polyethylene glycol provide high drug concentration in tumors. At the same time, it causes less drug toxicity than free camptothecin in healthy tissues [82]. It has been shown that passive targeted micelles that transmit cisplatin in the structure of methoxypoly (ethylene glycol)-block-poly (L-glutamic acid) inhibit cell proliferation in cervical cancer cells (HeLa) moreover lung cancer cells (A549). Increased blood circulation times have been observed in vivo compared to free cisplatin [83]. It has been revealed that micelles in the Tat peptide-linked methoxypoly (ethylene glycol) (MPEG) / polycaprolactone (PCL) dib lock copolymer structure that deliver small interfering RNA (siRNA) actively targeted to Vascular Endothelial Growth Factor have high antitumor effects in vivo. Thus, it has been shown that they can be used as therapeutic siRNA deliverers in gene silencing [83]. Poly (ethylene glycol)-b-poly(N-(2-hydroxypropyl) meth acrylamide diacetate co-(N-(2-hydroxypropyl)-meth acrylamide co-histidine) moreover poly(ethylene glycol)-b-poly(D, L) It has been observed that passively targeted micelles that deliver rapamycin in the -lactase)) dib lock copolymer structure have an antitumor effect in mice with colon tumors, a cytotoxic effect in colorectal carcinoma cells (HCT 116) and also induce high rates of apoptosis. However, it has been observed that free rapamycin does not have such an effect [84]. Micelles such as SP1049C, NC6004, NC4016, NK105, NK1012, NK911 and Genexol-PM have been evaluated in clinical studies (Table 2) [78]. Micelles have advantages such as lower toxicity, increased drug solubility and circulation half-life compared to other nanoparticles used in drug delivery.

4.3   Liposomes

Liposomes are referred to as self-assembled spherical, closed colloidal, lipid bilayers surrounding an aqueous central cavity. Liposomes are a more studied formulation due to the use of nanoparticles for drug to send. Early liposomes were being detected by macrophages because they had unmodified phospholipid surfaces. This feature caused them to be quickly cleared from circulation. Therefore, delivery of liposomal drugs to solid tumors was difficult [85].

Table 3. *Commercially Available Liposomal Formulations [56,85,86].*

Component	İsim
Liposomal daunorobicin	DaunoXome
Latent liposomal doxorubicin	Doxil/Caelyx
Liposomal doxorubicin	Myocet

In today's world, attention is paid to the fact that the liposomal drugs that are being improved have the ability to escape macrophage recognition. Surface-changed liposomes usually have hydrophilic carbohydrates or polymers added to the liposome superficies. This superficies modification resolved the problem of rapid clearance from the cycle and provided the liposomes with an importantly incremented half-life in the blood. It has been observed that liposomes in cholesterol and hyaluronic structure liposomes active targeted to CD44 antigen that transmit doxorubicin have a more cytotoxic effect than free doxorubicin in melanoma cells [87]. Liposomes actively targeted (with the Epidermal Growth Element Receptor Antibody) to the Epidermal Growth Factor Receptor transmitting siRNA, which is in the structure of 1,2-Dioleoyl-3-trimethylammonium propane cholesterol, were more effective in MDA-MB-231 breast cancer cells, which overexpress EGFR, than untargeted liposomes. effective gene silencing has been observed [85]. It has been observed that passive targeted liposomes that transmit daunorubicin, which is in the structure of polyethylene glycol-distearoyl phosphatidylethanolamine, are more uptake by breast cancer stem cells (MCF-7) than free daunorubicin. In addition, it has been observed that they activate proapoptotic Bax protein abolishes mitochondrial membrane potential, opens mitochondrial transition pores, initiate the caspase 9 and 3 cascades by causing the affranchise of cytochrome c and stimulate apoptosis [88].

4.4   Dendrimers

Drug delivery systems originating from Dendrimer are focused on drug encapsulation. But it is difficult to control drugs delivered via dendrimers in a released form. Because they contain linear polymers, their behaviour is different. Dendrimers have high circulation, which provides an advantage in drug transport. Doxorubicin-transmitting passively targeted dendrimers in oxidized polyethylene structure were found to have 10 times more cytotoxic effect than free doxorubicin in colon cancer cells. In vivo studies have shown that they are taken up to 9 times more by cancer cells, causing tumor size reduction and complete survival of mice after 60 days [87].

It has been observed that active paclitaxel-delivering dendrimers containing folate-dextran and galactose-linked poly (propylene imine) have a cytotoxic effect on HeLa and SiHa (human cervical cancer cells) cells rather than free paclitaxel. It has been observed that gold-conducting dendrimers containing 5 polys (amido amine) have a long cycle time moreover can be used as molecular imaging probes in computed tomography [89].

4.5   Nano Cells

Bacteria normally divide from the canter of their cytoplasm. The working group of a scientist named MacDiarmid announced that they had found a way to force bacteria to divide from the ends of their cytoplasm. Thus, they managed to produce 400 nm nano cells from small buds formed from the cytoplasm of bacteria at each division (Figure 2) [90]. Nano cells are non-chromosomal non-living structures resembling bacteria. Active targeting of nano cells with antibodies against specific receptors membranes cancerous cells results endocytosis of the nano cells with the release of the drug they contain. Since nano cells have a hard membrane, they do not deteriorate when injected and successfully transport the drugs they carry to the target region [57]. To promote progressive tumor growth, the dosages of drugs to submitted through nano cells are thereabout almost 1000 times less than the dosage of free drug. It has been observed that doxorubicin significantly regresses and inhibits tumor growth in canine lymphoma, despite loading a very small amount of doxorubicin into nano cells active on the CD3 T cell surface receptor derived from Salmonella typhimurium (S. typhimurium). Reduced dose is an important factor to limit toxicity [90]. siRNA-delivering nano cells have been observed to silence drug resistance genes in Caco2 colon cancer cells that overexpress the MDR1 gene resistant to both 5-fluorouracil and irinotecan (Composer). Thus, it has been demonstrated that drug resistance is reversed with these nano cells [91]. The reliability of paclitaxel-delivering EGFR-active nano cells made from bacteria has been tested in phase I studies, and it has been stated that this research will continue with phase II studies. With this research, nanocells were tested on humans for the first time [92].

4.6 Nanotubes

It is possible to add drug molecules straight to antibodies. However, the chemical bonds used in the addition of more than one drug molecule to an antibody importantly limit the ability of the antibodies. The many analogs of tumor-special antibodies can be covalently attached to carbon nanotubes targeting the tumor [93].

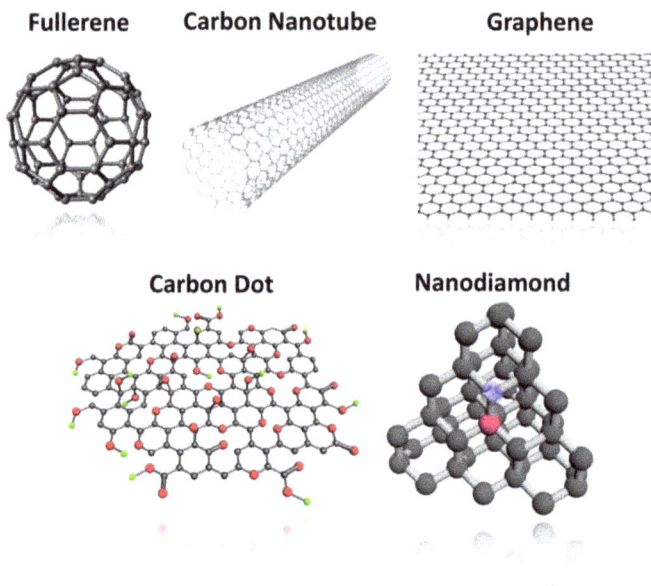

Figure 2. *Fulleran, carbon nanotube, graphene, carbon dot, nanodiamond structure and conformation (Reprinted with permission from [94], Copyright ACS).*

Tumor-targeting antibodies can bind to fullerenes (C60), which won the Nobel Prize in Chemistry for research scientists led by Richard Smalley in 1996. However, they can also be loaded with many cancer drugs such as Taxol (Figure 2) [95]. It has been revealed that fullerenes, which actively deliver docetaxel to the folate receptor contained in polyethyleneimine, do not have a toxic effect on healthy organs. In addition, it has been observed that they are 7.5 times more taken up by prostate tumors and have a high antitumor effect. These results have shown that these fullerenes may provide hope for cancer treatment in the future, with their high treatment effectiveness and very few side effects [96]. It has been observed that passively targeted carbon nanotubes surrounded by polyethylene glycol and delivering paclitaxel are taken up by breast tumor cells moreover reduce the tumor size [97].

Biogenic Nanomaterials: Synthesis, Characterization, Applications, and Future Remarks   Materials Research Forum LLC
Materials Research Foundations 180 (2025)   https://doi.org/21741/9781644903759

5.   Nanoparticles Used for Tumor Imaging

Molecular imaging tactic applied to identify cancerous cells; are imaging probes guided by ligands that can identify further interact with targeted molecules, particularly biomarker molecules manufactured by cancer cells. Nanoparticles of 10–100 nm size have an extra circulation time that does not belong to small molecular imaging agents [57].

5.1   Quantum Dot Nanoparticles

Half-conducting quantum dots (QDs) are nanometer-scale particles with enhanced signal brightness moreover stability and unmatched optical in addition electronic features. It is likely to simultaneously display and monitor large amounts of tumor markers with quantum dots emitting separate wavelengths. This adds to the uniqueness of detecting cancerous cells and increases its sensitivity (Figure 4) [57]. Cadmium is known to be the key component of quantum dot nanoparticles. A number of concerns over the toxicity of cadmium make future clinical application of quantum dot nanoparticles uncertain. Quantum dot nanoparticles actively targeted to MUC1 mucin protein in imaging of ovarian cancer cells [98], quantum dot nanoparticles actively targeted to HER2 receptors in imaging of HER2/neu positive breast tumor cells [99], Actively targeted to tumor-united glycoprotein 72 (TAG-72) It has been observed that specifically designed quantum dot nano particles can be used to image gastric cancer cells [100].

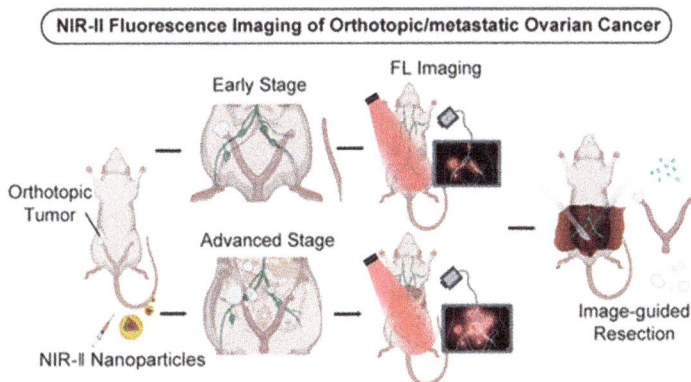

Figure 3. Reported NIR-II imaging studies of ovarian cancer, fluorescence imaging with late-stage metastasis detection through an intraperitoneal xenograft model (Reprinted with permission from [101], Copyright ACS).

5.2   Superparamagnetic Iron Oxide Nanoparticles

When iron oxide nanoparticles which are superparamagnetic, image cells that are cancerous, magnetic resonance imaging is applied in the form of contrast medium. Iron oxide nanoparticles with low toxicity can be retained blood retention time may be long moreover being biodegradable [102,103]. It has been observed that iron oxide nanoparticles actively aimed to the folate receptor can be used in tumor imaging [104]. Different forms of iron oxide nanoparticles have been tested in clinical studies further proven to be secure for human usage (Table 4) [104,105].

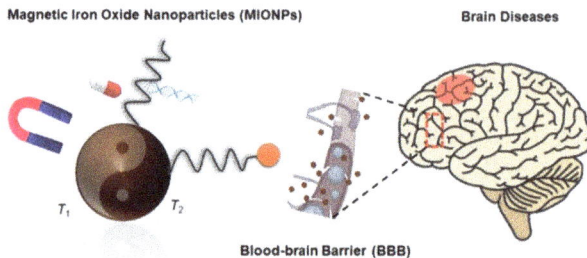

Magnetic Iron Oxide Nanoparticles (MIONPs)

Brain Diseases

T_1 T_2

Blood-brain Barrier (BBB)

Figure 4. Brain Tumor Imaging Using Superparamagnetic Iron Oxide Nanoparticle Actively Targeted to the Epidermal Growth Factor Receptor (Reprinted with permission from [106], Copyright ELSEVIER).

Table 4. Superparamagnetic Iron Oxide Nanoparticle Formulations Available in the Market [107].

Compound	Trade name	Coating agent
Ferristene (OMP)	Abdoscan	Stiren divinilbenzen
Ferumoxsil (AMI-121)	GastroMARK Lumirem	Siloksan
Ferumoxide (AMI-25)	Feridex Endorem	Dekstran

6. Conclusion

Rapid and promising developments in cancer nanotechnology are attracting attention. Physicists, chemists, engineers, biologists, and clinicians keep moving forward new and effective nano systems that can be used in cancer diagnosis and treatment. Thanks to this technology, we will soon be one step closer to developing approaches for the definitive cure of such fatal illnesses. In research, the anticancer activities of two different sized nanoparticle-based drug send systems against human colon cancer Caco-2 in addition HCT-116 cells were investigated. With cellular uptake experiments, it was determined that nanoparticles of both sizes could target cancer cells at increasing rates depending on time. In a study evaluating the cytotoxic impact of AgNPs on MCF-7 breast cancer cells, it has been resulted that AgNPs synthesis is a good option moreover that AgNPs synthesized by applying rosemary extract can be a potential agent of choice in the treatment of breast cancer. Gur et al. in research of, it has been explained that the protective effect formed zinc oxide nanoparticles (ZnO NPs) in various ways the protective effect against DNA damage caused by Thymbra Spicata L. plant by applying green synthesis increases depending on the concentration. In another study investigating a docetaxel (DTX) loaded nanoparticle that can dually target osteoclasts moreover bone metastatic tumor cells to cure bone metastases of lung cancer, DTX@AHP not only significantly inhibited the proliferation of bone metastases of lung cancer, however moreover inhibited osteolysis in tumor-bearing mice in the process. It has been reported to decrease. Research shows that both the number and use of nanomaterials are increasing. Thanks to this technology, it will be one step closer to progress approaches for the definitive cure of such deadly illness in the near future.

Reference

[1]   R.. Müller, C. Jacobs, O. Kayser, Nanosuspensions as Particulate Drug Formulations in Therapy, Adv. Drug Deliv. Rev. 47 (2001) 3–19. https://doi.org/10.1016/S0169-409X(00)00118-6

[2]   A. Shapira, Y.D. Livney, H.J. Broxterman, Y.G. Assaraf, Nanomedicine for Targeted Cancer Therapy: Towards the Overcoming of Drug Resistance, Drug Resist. Updat. 14 (2011) 150–163. https://doi.org/10.1016/j.drup.2011.01.003

[3]   F. Gulbagca, A. Aygün, M. Gülcan, S. Ozdemir, S. Gonca, F. Şen, Green Synthesis of Palladium Nanoparticles: Preparation, Characterization, and Investigation of Antioxidant, Antimicrobial, Anticancer, and DNA Cleavage Activities, Appl. Organomet. Chem. 35 (2021). https://doi.org/10.1002/aoc.6272

[4]   J.T. Abrahamson, B. Sempere, M.P. Walsh, J.M. Forman, F. Şen, S. Şen, S.G. Mahajan, G.L.C. Paulus, Q.H. Wang, W. Choi, M.S. Strano, Excess Thermopower and the Theory of Thermopower Waves, ACS Nano 7 (2013) 6533–6544. https://doi.org/10.1021/nn402411k

[5]   H. Göksu, Y. Yıldız, B. Çelik, M. Yazıcı, B. Kılbaş, F. Şen, Highly Efficient and Monodisperse Graphene Oxide Furnished Ru/Pd Nanoparticles for the Dehalogenation of Aryl Halides via Ammonia Borane, ChemistrySelect 1 (2016) 953–958. https://doi.org/10.1002/slct.201600207.

[6]   F. Şen, G. Gökağaç, Improving Catalytic Efficiency in the Methanol Oxidation Reaction by Inserting Ru in Face-Centered Cubic Pt Nanoparticles Prepared by a New Surfactant, tert -Octanethiol, Energy & Fuels 22 (2008) 1858–1864. https://doi.org/10.1021/ef700575t

[7]   K. Arikan, H. Burhan, R. Bayat, F. Sen, Glucose Nano Biosensor With Non-Enzymatic Excellent Sensitivity Prepared With Nickel–Cobalt Nanocomposites On F-MWCNT, Chemosphere 291 (2022) 132720. https://doi.org/10.1016/j.chemosphere.2021.132720

[8]   H. Goksu, Y. Yıldız, B. Çelik, M. Yazici, B. Kilbas, F. Sen, Eco-friendly hydrogenation of aromatic aldehyde compounds by tandem dehydrogenation of dimethylamine-borane in the presence of a reduced graphene oxide furnished platinum nanocatalyst, Catal. Sci. Technol. 6 (2016) 2318–2324. https://doi.org/10.1039/C5CY01462J

[9]   T. Gur, I. Meydan, H. Seckin, M. Bekmezci, F. Sen, Green Synthesis, Characterization and Bioactivity of Biogenic Zinc Oxide Nanoparticles, Environ. Res. 204 (2022) 111897. https://doi.org/10.1016/j.envres.2021.111897

[10]   I. Meydan, A. Aygun, R.N.E. Tiri, T. Gur, Y. Kocak, H. Seckin, F. Sen, Chitosan/PVA Supported Silver Nanoparticles for Azo Dyes Removal: Fabrication, Characterization, and Assessment of Antioxidant Activity, Environ. Sci. Adv. (2024). https://doi.org/10.1039/D3VA00224A

[11]   H. Göksu, H. Burhan, S.D. Mustafov, F. Şen, Oxidation of Benzyl Alcohol Compounds in the Presence of Carbon Hybrid Supported Platinum Nanoparticles (Pt@CHs) in Oxygen Atmosphere, Sci. Rep. 10 (2020) 5439. https://doi.org/10.1038/s41598-020-62400-5

[12]   A. Şavk, K. Cellat, K. Arıkan, F. Tezcan, S.K. Gülbay, S. Kızıldağ, E.Ş. Işgın, F. Şen, Highly Monodisperse Pd-Ni Nanoparticles Supported On Rgo As A Rapid, Sensitive, Reusable And Selective Enzyme-Free Glucose Sensor, Sci. Rep. 9 (2019) 19228. https://doi.org/10.1038/s41598-019-55746-y

[13]   R. Ulus, Y. Yıldız, S. Eriş, B. Aday, F. Şen, M. Kaya, Functionalized Multi-Walled Carbon Nanotubes (f -MWCNT) as Highly Efficient and Reusable Heterogeneous Catalysts for the Synthesis of Acridinedione Derivatives, ChemistrySelect 1 (2016) 3861–3865. https://doi.org/10.1002/slct.201600719

[14]   C. Sun, J.S.H. Lee, M. Zhang, Magnetic Nanoparticles in MR Imaging and Drug Delivery, Adv. Drug Deliv. Rev. 60 (2008) 1252–1265. https://doi.org/10.1016/j.addr.2008.03.018.

[15]   J. Park, J. von Maltzahn, E. Ruoslahti, S.N. Bhatia, M.J. Sailor, Micellar Hybrid Nanoparticles for Simultaneous Magnetofluorescent Imaging and Drug Delivery, Angew. Chemie Int. Ed. 47 (2008) 7284–7288. https://doi.org/10.1002/anie.200801810

[16]   N. Korkmaz, Y. Ceylan, P. Taslimi, A. Karadağ, A.S. Bülbül, F. Şen, Biogenic nano silver:

Synthesis, characterization, antibacterial, antibiofilms, and enzymatic activity, Adv. Powder Technol. 31 (2020) 2942–2950. https://doi.org/10.1016/j.apt.2020.05.020

[17] Y. Yıldız, İ. Esirden, E. Erken, E. Demir, M. Kaya, F. Şen, Microwave (Mw) Assisted Synthesis of 5-Substituted 1 H -Tetrazoles via [3+2] Cycloaddition Catalyzed by Mw-Pd/Co Nanoparticles Decorated on Multi-Walled Carbon Nanotubes, ChemistrySelect 1 (2016) 1695–1701. https://doi.org/10.1002/slct.201600265

[18] E. Demir, A. Savk, B. Sen, F. Sen, A novel monodisperse metal nanoparticles anchored graphene oxide as Counter Electrode for Dye-Sensitized Solar Cells, Nano-Structures & Nano-Objects 12 (2017) 41–45. https://doi.org/10.1016/j.nanoso.2017.08.018

[19] A. Aygun, F. Gulbagca, E.E. Altuner, M. Bekmezci, T. Gur, H. Karimi-Maleh, F. Karimi, Y. Vasseghian, F. Sen, Highly Active PdPt Bimetallic Nanoparticles Synthesized By One Step Bioreduction Method: Characterizations, Anticancer, Antibacterial Activities and Evaluation of Their Catalytic Effect for Hydrogen Generation, Int. J. Hydrogen Energy 48 (2023) 6666–6679. https://doi.org/10.1016/j.ijhydene.2021.12.144

[20] N. Zare, F. Ameen, M. Bekmezci, M. Akin, R. Bayat, T. Kozak, E. Halvaci, I. Isik, F. Sen, Electrochemical sensing of vitamin C using PtNi nanomaterials supported on carbon nanotubes produced by arc discharge method, J. Food Meas. Charact. 18 (2024) 6759–6769. https://doi.org/10.1007/s11694-024-02689-2

[21] R. Bayat, E. Halvaci, T. Kozak, M. Bekmezci, F. Sen, Electrocatalytic performance of Shape-Controlled synthesized PdNi-rGO nano Cube on sodium borohydride and methanol electrooxidation, Fuel 366 (2024) 131248. https://doi.org/10.1016/j.fuel.2024.131248

[22] S. Günbatar, A. Aygun, Y. Karataş, M. Gülcan, F. Şen, Carbon Nanotube Based Rhodium Nanoparticles as Highly Active Catalyst for Hydrolytic Dehydrogenation of Dimethylamineborane at Room Temperature, J. Colloid Interface Sci. 530 (2018) 321–327. https://doi.org/10.1016/j.jcis.2018.06.100

[23] H. Goksu, Y. Yıldız, B. Çelik, M. Yazici, B. Kilbas, F. Sen, Eco Friendly Hydrogenation of Aromatic Aldehyde Compounds By Tandem Dehydrogenation of Dimethylamine Borane in the Presence of a Reduced Graphene Oxide Furnished Platinum Nanocatalyst, Catal. Sci. Technol. 6 (2016) 2318–2324. https://doi.org/10.1039/C5CY01462J

[24] N. Lolak, E. Kuyuldar, H. Burhan, H. Goksu, S. Akocak, F. Sen, Composites of Palladium–Nickel Alloy Nanoparticles and Graphene Oxide for the Knoevenagel Condensation of Aldehydes with Malononitrile, ACS Omega 4 (2019) 6848–6853. https://doi.org/10.1021/acsomega.9b00485

[25] R. Ayranci, G. Başkaya, M. Güzel, S. Bozkurt, F. Şen, M. Ak, Carbon Based Nanomaterials for High Performance Optoelectrochemical Systems, ChemistrySelect 2 (2017) 1548–1555. https://doi.org/10.1002/slct.201601632

[26] F. Şen, G. Gökağaç, S. Şen, High performance Pt nanoparticles prepared by new surfactants for C1 to C3 alcohol oxidation reactions, J. Nanoparticle Res. 15 (2013) 1979. https://doi.org/10.1007/s11051-013-1979-5

[27] F. Sen, A.A. Boghossian, S. Sen, Z.W. Ulissi, J. Zhang, M.S. Strano, Observation of Oscillatory Surface Reactions of Riboflavin, Trolox, and Singlet Oxygen Using Single Carbon Nanotube Fluorescence Spectroscopy, ACS Nano 6 (2012) 10632–10645. https://doi.org/10.1021/nn303716n

[28] R. Ayranci, G. Baskaya, M. Guzel, S. Bozkurt, M. Ak, A. Savk, F. Sen, Enhanced optical and electrical properties of PEDOT via nanostructured carbon materials: A comparative investigation, Nano-Structures & Nano-Objects 11 (2017) 13–19. https://doi.org/10.1016/j.nanoso.2017.05.008

[29] F. Şen, G. Gökağaç, Pt Nanoparticles Synthesized With New Surfactants: Improvement in C1–C3 Alcohol Oxidation Catalytic Activity, J. Appl. Electrochem. 44 (2014) 199–207. https://doi.org/10.1007/s10800-013-0631-5

[30]   E.A. Sykes, J. Chen, G. Zheng, W.C.W. Chan, Investigating the Impact of Nanoparticle Size on Active and Passive Tumor Targeting Efficiency, ACS Nano 8 (2014) 5696–5706. https://doi.org/10.1021/nn500299p

[31]   F. Danhier, O. Feron, V. Préat, To Exploit the Tumor Microenvironment: Passive and Active Tumor Targeting of Nanocarriers for Anti Cancer Drug Delivery, J. Control. Release 148 (2010) 135–146. https://doi.org/10.1016/j.jconrel.2010.08.027

[32]   I. Meydan, H. Seckin, H. Burhan, T. Gür, B. Tanhaei, F. Sen, Arum Italicum Mediated Silver Nanoparticles: Synthesis and Investigation of Some Biochemical Parameters, Environ. Res. 204 (2022) 112347. https://doi.org/10.1016/j.envres.2021.112347

[33]   S. Laurent, D. Forge, M. Port, A. Roch, C. Robic, L. Vander Elst, R.N. Muller, Magnetic Iron Oxide Nanoparticles: Synthesis, Stabilization, Vectorization, Physicochemical Characterizations, and Biological Applications, Chem. Rev. 108 (2008) 2064–2110. https://doi.org/10.1021/cr068445e

[34]   K. Uchiyama, A. Nagayasu, Y. Yamagiwa, T. Nishida, H. Harashima, H. Kiwada, Effects of the Size and Fluidity of Liposomes on Their Accumulation in Tumors: A Presumption of Their Interaction With Tumors, Int. J. Pharm. 121 (1995) 195–203. https://doi.org/10.1016/0378-5173(95)00015-B

[35]   A. Nagayasu, K. Uchiyama, H. Kiwada, The size of Liposomes: A Factor Which Affects Their Targeting Efficiency to Tumors and Therapeutic Activity of Liposomal Antitumor Drugs, Adv. Drug Deliv. Rev. 40 (1999) 75–87. https://doi.org/10.1016/S0169-409X(99)00041-1

[36]   J. Gao, H. Gu, B. Xu, Multifunctional Magnetic Nanoparticles: Design, Synthesis, and Biomedical Applications, Acc. Chem. Res. 42 (2009) 1097–1107. https://doi.org/10.1021/ar9000026.

[37]   J.E. Lee, N. Lee, H. Kim, J. Kim, S.H. Choi, J.H. Kim, T. Kim, I.C. Song, S.P. Park, W.K. Moon, T. Hyeon, Uniform Mesoporous Dye-Doped Silica Nanoparticles Decorated with Multiple Magnetite Nanocrystals for Simultaneous Enhanced Magnetic Resonance Imaging, Fluorescence Imaging, and Drug Delivery, J. Am. Chem. Soc. 132 (2010) 552–557. https://doi.org/10.1021/ja905793q

[38]   C. Medina, M.J. Santos-Martinez, A. Radomski, O.I. Corrigan, M.W. Radomski, Nanoparticles: Pharmacological and Toxicological Significance, Br. J. Pharmacol. 150 (2007) 552–558. https://doi.org/10.1038/sj.bjp.0707130

[39]   S.-T. C. Tuba; HASCİCEK, Polimerik Nanopartikuler Ilaç Taşıyıcı Sistemlerde Yüzey Modifikasyonu, Ankara Univ. Eczac. Fak. Derg. 38 (2009) 137–154. https://doi.org/10.1501/Eczfak_0000000522

[40]   P. Bajpai, Nanotechnology in Forest Industry, 2 (2016) 258.

[41]   K. Jain, Advances in the Field of Nanooncology, BMC Med. 8 (2010) 83. https://doi.org/10.1186/1741-7015-8-83

[42]   C. Jin, K. Wang, A. Oppong-Gyebi, J. Hu, Application of Nanotechnology in Cancer Diagnosis and Therapy, Int. J. Med. Sci. 17 (2020) 2964–2973. https://doi.org/10.7150/ijms.49801

[43]   C. Braicu, R. Chiorean, A. Irimie, S. Chira, C. Tomuleasa, E. Neagoe, A. Paradiso, P. Achimas-Cadariu, V. Lazar, I. Berindan-Neagoe, Novel Insight Into Triple Negative Breast Cancers the Emerging Role of Angiogenesis, and Antiangiogenic Therapy, Expert Rev. Mol. Med. 18 (2016) e18. https://doi.org/10.1017/erm.2016.17

[44]   S.A. Rosenzweig, Acquired Resistance to Drugs Targeting Tyrosine Kinases, in: 2018: pp. 71–98. https://doi.org/10.1016/bs.acr.2018.02.003.

[45]   M. Herbrink, B. Nuijen, J.H.M. Schellens, J.H. Beijnen, Variability in Bioavailability of Small Molecular Tyrosine Kinase Inhibitors, Cancer Treat. Rev. 41 (2015) 412–422. https://doi.org/10.1016/j.ctrv.2015.03.005

[46]   C. Vlad, P. Kubelac, D. Vlad, A. Irimie, P.A. Cadariu, Evaluation of Clinical, Morphopathological and Therapeutic Prognostic Factors in Rectal Cancer. Experience of a Tertiary Oncology Center, J. B.U.ON. 20 (2015) 92–99.

[47]   A.Z. Wilczewska, K. Niemirowicz, K.H. Markiewicz, H. Car, Nanoparticles as Drug Delivery Systems, Pharmacol. Reports 64 (2012) 1020–1037. https://doi.org/10.1016/S1734-1140(12)70901-5

[48]   A.C. Anselmo, S. Mitragotri, Nanoparticles in the Clinic: An Update, Bioeng. Transl. Med. 4 (2019). https://doi.org/10.1002/btm2.10143

[49]   M. Rocha, N. Chaves, S. Bao, Nanobiotechnology for Breast Cancer Treatment, in: Breast Cancer - From Biol. to Med., InTech, 2017. https://doi.org/10.5772/66989

[50]   J.D. Heidel, M.E. Davis, Clinical Developments in Nanotechnology for Cancer Therapy, Pharm. Res. 28 (2011) 187–199. https://doi.org/10.1007/s11095-010-0178-7

[51]   S.D. Perrault, C. Walkey, T. Jennings, H.C. Fischer, W.C.W. Chan, Mediating Tumor Targeting Efficiency of Nanoparticles Through Design, Nano Lett. 9 (2009) 1909–1915. https://doi.org/10.1021/nl900031y

[52]   Y.-E.L. Koo, W. Fan, H. Hah, H. Xu, D. Orringer, B. Ross, A. Rehemtulla, M.A. Philbert, R. Kopelman, Photonic Explorers Based on Multifunctional Nanoplatforms for Biosensing and Photodynamic Therapy, Appl. Opt. 46 (2007) 1924. https://doi.org/10.1364/AO.46.001924

[53]   R.K. Jain, T. Stylianopoulos, Delivering Nanomedicine to Solid Tumors, Nat. Rev. Clin. Oncol. 7 (2010) 653–664. https://doi.org/10.1038/nrclinonc.2010.139

[54]   A.I. Minchinton, I.F. Tannock, Drug Penetration in Solid Tumours, Nat. Rev. Cancer 6 (2006) 583–592. https://doi.org/10.1038/nrc1893

[55]   T. Lammers, F. Kiessling, W.E. Hennink, G. Storm, Drug Targeting to Tumors: Principles, Pitfalls and (pre-) Clinical Progress, J. Control. Release 161 (2012) 175–187. https://doi.org/10.1016/j.jconrel.2011.09.063

[56]   T.M. Allen, P.R. Cullis, Drug Delivery Systems: Entering the Mainstream, Science (80-.). 303 (2004) 1818–1822. https://doi.org/10.1126/science.1095833

[57]   V. Torchilin, Tumor Delivery of Macromolecular Drugs Based on the EPR Effect, Adv. Drug Deliv. Rev. 63 (2011) 131–135. https://doi.org/10.1016/j.addr.2010.03.011

[58]   V.P. Torchilin, Passive and Active Drug Targeting: Drug Delivery to Tumors as an Example, in: 2010: pp. 3–53. https://doi.org/10.1007/978-3-642-00477-3_1

[59]   J.W. Nichols, Y.H. Bae, Epr: Evidence and Fallacy, J. Control. Release 190 (2014) 451–464. https://doi.org/10.1016/j.jconrel.2014.03.057

[60]   F. Danhier, To Exploit the Tumor Microenvironment: Since the EPR Effect Fails in the Clinic, What is the Future of Nanomedicine?, J. Control. Release 244 (2016) 108–121. https://doi.org/10.1016/j.jconrel.2016.11.015

[61]   M.A. Swartz, M.E. Fleury, Interstitial Flow and Its Effects in Soft Tissues, Annu. Rev. Biomed. Eng. 9 (2007) 229–256. https://doi.org/10.1146/annurev.bioeng.9.060906.151850

[62]   M.J. Bissell, D. Radisky, Putting Tumours in Context, Nat. Rev. Cancer 1 (2001) 46–54. https://doi.org/10.1038/35094059

[63]   E.L. da Rocha, L.M. Porto, C.R. Rambo, Nanotechnology Meets 3D in Vitro Models: Tissue Engineered Tumors and Cancer Therapies, Mater. Sci. Eng. C 34 (2014) 270–279. https://doi.org/10.1016/j.msec.2013.09.019

[64]   W. Islam, T. Niidome, T. Sawa, Enhanced Permeability and Retention Effect as a Ubiquitous and Epoch-Making Phenomenon for the Selective Drug Targeting of Solid Tumors, J. Pers. Med. 12 (2022) 1964. https://doi.org/10.3390/jpm12121964

[65]   J. Zong, S.L. Cobb, N.R. Cameron, Peptide Functionalized Gold Nanoparticles: Versatile Biomaterials for Diagnostic and Therapeutic Applications, Biomater. Sci. 5 (2017) 872–886. https://doi.org/10.1039/C7BM00006E

[66]   C.-J. Lin, C.-H. Kuan, L.-W. Wang, H.-C. Wu, Y. Chen, C.-W. Chang, R.-Y. Huang, T.-W.

Wang, Integrated Self Assembling Drug Delivery System Possessing Dual Responsive and Active Targeting for Orthotopic Ovarian Cancer Theranostics, Biomaterials 90 (2016) 12–26. https://doi.org/10.1016/j.biomaterials.2016.03.005

[67]   M. Beck, T. Mandal, C. Buske, M. Lindén, Serum Protein Adsorption Enhances Active Leukemia Stem Cell Targeting of Mesoporous Silica Nanoparticles, ACS Appl. Mater. Interfaces 9 (2017) 18566–18574. https://doi.org/10.1021/acsami.7b04742

[68]   J. Lee, J.A. Kang, Y. Ryu, S.-S. Han, Y.R. Nam, J.K. Rho, D.S. Choi, S.-W. Kang, D.-E. Lee, H.-S. Kim, Genetically Engineered and Self Assembled Oncolytic Protein Nanoparticles for Targeted Cancer Therapy, Biomaterials 120 (2017) 22–31. https://doi.org/10.1016/j.biomaterials.2016.12.014.

[69]   V.H. Shargh, H. Hondermarck, M. Liang, Antibody Targeted Biodegradable Nanoparticles for Cancer Therapy, Nanomedicine 11 (2016) 63–79. https://doi.org/10.2217/nnm.15.186

[70]   Z. Wang, R. Qiao, N. Tang, Z. Lu, H. Wang, Z. Zhang, X. Xue, Z. Huang, S. Zhang, G. Zhang, Y. Li, Active Targeting Theranostic Iron Oxide Nanoparticles for MRI and Magnetic Resonance Guided Focused Ultrasound Ablation of Lung Cancer, Biomaterials 127 (2017) 25–35. https://doi.org/10.1016/j.biomaterials.2017.02.037

[71]   O. Veiseh, J.W. Gunn, M. Zhang, Design and Fabrication of Magnetic Nanoparticles for Targeted Drug Delivery and Imaging, Adv. Drug Deliv. Rev. 62 (2010) 284–304. https://doi.org/10.1016/j.addr.2009.11.002

[72]   I. Meydan, H. Burhan, T. Gür, H. Seçkin, B. Tanhaei, F. Sen, Characterization of Rheum ribes with ZnO nanoparticle and its antidiabetic, antibacterial, DNA damage prevention and lipid peroxidation prevention activity of in vitro, Environ. Res. 204 (2022) 112363. https://doi.org/10.1016/j.envres.2021.112363

[73]   H. Seckin, R.N.E. Tiri, I. Meydan, A. Aygun, M.K. Gunduz, F. Sen, An Environmental Approach for the Photodegradation of Toxic Pollutants From Wastewater Using Pt–Pd Nanoparticles: Antioxidant, Antibacterial and Lipid Peroxidation Inhibition Applications, Environ. Res. 208 (2022) 112708. https://doi.org/10.1016/j.envres.2022.112708

[74]   T. Lammers, W.E. Hennink, G. Storm, Tumour Targeted Nanomedicines: Principles and Practice, Br. J. Cancer 99 (2008) 392–397. https://doi.org/10.1038/sj.bjc.6604483

[75]   W. Zhu, Y. Li, L. Liu, Y. Chen, F. Xi, Supramolecular Hydrogels as a Universal Scaffold for Stepwise Delivering Dox and Dox Cisplatin Loaded Block Copolymer Micelles, Int. J. Pharm. 437 (2012) 11–19. https://doi.org/10.1016/j.ijpharm.2012.08.007

[76]   S. Bai, D. Liu, Y. Cheng, H. Cui, M. Liu, M. Cui, B. Zhang, Q. Mei, S. Zhou, Osteoclasts and Tumor Cells Dual Targeting Nanoparticle to Treat Bone Metastases of Lung Cancer, Nanomedicine Nanotechnology, Biol. Med. 21 (2019) 102054. https://doi.org/10.1016/j.nano.2019.102054

[77]   R. Singh, J.W. Lillard, Nanoparticle Based Targeted Drug Delivery, Exp. Mol. Pathol. 86 (2009) 215–223. https://doi.org/10.1016/j.yexmp.2008.12.004

[78]   J. Gong, M. Chen, Y. Zheng, S. Wang, Y. Wang, Polymeric Micelles Drug Delivery System in Oncology, J. Control. Release 159 (2012) 312–323. https://doi.org/10.1016/j.jconrel.2011.12.012

[79]   G.V. Kumar, Deepa, A.K.T. Thulasidasan, R.J. Anto, Jisha Pillai, Cross Linked Acrylic Hydrogel for the Controlled Delivery of Hydrophobic Drugs in Cancer Therapy, Int. J. Nanomedicine (2012) 4077. https://doi.org/10.2147/IJN.S30149

[80]   G. Nie, H.J. Hah, G. Kim, Y.K. Lee, M. Qin, T.S. Ratani, P. Fotiadis, A. Miller, A. Kochi, D. Gao, T. Chen, D.A. Orringer, O. Sagher, M.A. Philbert, R. Kopelman, Hydrogel Nanoparticles with Covalently Linked Coomassie Blue for Brain Tumor Delineation Visible to the Surgeon, Small 8 (2012) 884–891. https://doi.org/10.1002/smll.201101607

[81]   K.A. V. Zubris, Y.L. Colson, M.W. Grinstaff, Hydrogels as Intracellular Depots for Drug Delivery, Mol. Pharm. 9 (2012) 196–200. https://doi.org/10.1021/mp200367s

[82]   F. Karimi, N. Rezaei-savadkouhi, M. Uçar, A. Aygun, R.N. Elhouda Tiri, I. Meydan, E. Aghapour, H. Seckin, D. Berikten, T. Gur, F. Sen, Efficient green photocatalyst of silver-based palladium nanoparticles for methyle orange photodegradation, investigation of lipid peroxidation inhibition, antimicrobial, and antioxidant activity, Food Chem. Toxicol. 169 (2022) 113406. https://doi.org/10.1016/j.fct.2022.113406.

[83]   T. Kanazawa, K. Sugawara, K. Tanaka, S. Horiuchi, Y. Takashima, H. Okada, Suppression of Tumor Growth by Systemic Delivery of Anti Vegf siRNA With Cell Penetrating Peptide Modified MPEG–PCL Nanomicelles, Eur. J. Pharm. Biopharm. 81 (2012) 470–477. https://doi.org/10.1016/j.ejpb.2012.04.021

[84]   Y.-C. Chen, C.-L. Lo, Y.-F. Lin, G.-H. Hsiue, Rapamycin Encapsulated in Dual Responsive Micelles for Cancer Therapy, Biomaterials 34 (2013) 1115–1127. https://doi.org/10.1016/j.biomaterials.2012.10.034

[85]   P. Sapra, P. Tyagi, T. Allen, Ligand Targeted Liposomes for Cancer Treatment, Curr. Drug Deliv. 2 (2005) 369–381. https://doi.org/10.2174/156720105774370159

[86]   G.J.. Charrois, T.M. Allen, Drug Release Rate Influences the Pharmacokinetics, Biodistribution, Therapeutic Activity, and Toxicity of Pegylated Liposomal Doxorubicin Formulations in Murine Breast Cancer, Biochim. Biophys. Acta - Biomembr. 1663 (2004) 167–177. https://doi.org/10.1016/j.bbamem.2004.03.006

[87]   C.C. Lee, E.R. Gillies, M.E. Fox, S.J. Guillaudeu, J.M.J. Fréchet, E.E. Dy, F.C. Szoka, A Single Dose of Doxorubicin Functionalized Bow Tie Dendrimer Cures Mice Bearing C-26 Colon Carcinomas, Proc. Natl. Acad. Sci. 103 (2006) 16649–16654. https://doi.org/10.1073/pnas.0607705103

[88]   L. Zhang, H.-J. Yao, Y. Yu, Y. Zhang, R.-J. Li, R.-J. Ju, X.-X. Wang, M.-G. Sun, J.-F. Shi, W.-L. Lu, Mitochondrial Targeting Liposomes Incorporating Daunorubicin and Quinacrine for Treatment of Relapsed Breast Cancer Arising From Cancer Stem Cells, Biomaterials 33 (2012) 565–582. https://doi.org/10.1016/j.biomaterials.2011.09.055

[89]   T. Huang, G. Li, Y. Guo, G. Zhang, D. Shchabin, X. Shi, M. Shen, Recent Advances in PAMAM Dendrimer Based CT Contrast Agents for Molecular Imaging and Theranostics of Cancer, Sensors & Diagnostics 2 (2023) 1145–1157. https://doi.org/10.1039/D3SD00101F

[90]   J.A. MacDiarmid, N.B. Mugridge, J.C. Weiss, L. Phillips, A.L. Burn, R.P. Paulin, J.E. Haasdyk, K.-A. Dickson, V.N. Brahmbhatt, S.T. Pattison, A.C. James, G. Al Bakri, R.C. Straw, B. Stillman, R.M. Graham, H. Brahmbhatt, Bacterially Derived 400 nm Particles for Encapsulation and Cancer Cell Targeting of Chemotherapeutics, Cancer Cell 11 (2007) 431–445. https://doi.org/10.1016/j.ccr.2007.03.012

[91]   C. DAĞLIOĞLU, İnsan Kolon Kanseri Hücrelerine Karşı İnorganik Nanopartikül Temelli İlaç Taşıyıcı Sistemlerin Kullanılması: Partikül Büyüklüğünün Antikanser Aktivitesine Etkisi, Politek. Derg. 23 (2020) 171–179. https://doi.org/10.2339/politeknik.496351

[92]   Ç. AYDIN ACAR, S. Pehlivanoğlu, Gümüş Nanopartiküllerin Biberiye Özütü ile Biyosentezi ve MCF-7 Meme Kanseri Hücrelerinde Sitotoksik Etkisi, Süleyman Demirel Üniversitesi Sağlık Bilim. Derg. 10 (2019) 172–176. https://doi.org/10.22312/sdusbed.543053

[93]   M.R. McDevitt, D. Chattopadhyay, B.J. Kappel, J.S. Jaggi, S.R. Schiffman, C. Antczak, J.T. Njardarson, R. Brentjens, D.A. Scheinberg, Tumor Targeting with Antibody-Functionalized, Radiolabeled Carbon Nanotubes, J. Nucl. Med. 48 (2007) 1180–1189. https://doi.org/10.2967/jnumed.106.039131

[94]   G. Hong, S. Diao, A.L. Antaris, H. Dai, Carbon Nanomaterials for Biological Imaging and Nanomedicinal Therapy, Chem. Rev. 115 (2015) 10816–10906. https://doi.org/10.1021/acs.chemrev.5b00008

[95]   J.M. Ashcroft, D.A. Tsyboulski, K.B. Hartman, T.Y. Zakharian, J.W. Marks, R.B. Weisman, M.G. Rosenblum, L.J. Wilson, Fullerene (C60) Immunoconjugates: Interaction of Water Soluble C60

Derivatives With the Murine Anti-gp240 MelanomaAantibody, Chem. Commun. (2006) 3004. https://doi.org/10.1039/b601717g

[96]   J. Shi, H. Zhang, L. Wang, L. Li, H. Wang, Z. Wang, Z. Li, C. Chen, L. Hou, C. Zhang, Z. Zhang, PEI Derivatized Fullerene Drug Delivery Using Folate as a Homing Device Targeting to Tumor, Biomaterials 34 (2013) 251–261. https://doi.org/10.1016/j.biomaterials.2012.09.039

[97]   Z. Liu, J.T. Robinson, S.M. Tabakman, K. Yang, H. Dai, Carbon Materials for Drug Delivery Amp; Cancer Therapy, Mater. Today 14 (2011) 316–323. https://doi.org/10.1016/S1369-7021(11)70161-4

[98]   R. Savla, O. Taratula, O. Garbuzenko, T. Minko, Tumor Targeted Quantum Dot Mucin 1 Aptamer Doxorubicin Conjugate for Imaging and Treatment of Cancer, J. Control. Release 153 (2011) 16–22. https://doi.org/10.1016/j.jconrel.2011.02.015

[99]   I. V. Balalaeva, T.A. Zdobnova, I. V. Krutova, A.A. Brilkina, E.N. Lebedenko, S.M. Deyev, Passive and Active Targeting of Quantum Dots for Whole Body Fluorescence Imaging of Breast Cancer Xenografts, J. Biophotonics 5 (2012) 860–867. https://doi.org/10.1002/jbio.201200080

[100]   Y.-P. ZHANG, P. SUN, X.-R. ZHANG, W.-L. YANG, In Vitro Gastric Cancer Cell Imaging Using Near Infrared Quantum Dot Conjugated CC49, Oncol. Lett. 4 (2012) 996–1002. https://doi.org/10.3892/ol.2012.870

[101]   T. Pu, Y. Liu, Y. Pei, J. Peng, Z. Wang, M. Du, Q. Liu, F. Zhong, M. Zhang, F. Li, C. Xu, X. Zhang, NIR-II Fluorescence Imaging for the Detection and Resection of Cancerous Foci and Lymph Nodes in Early-Stage Orthotopic and Advanced-Stage Metastatic Ovarian Cancer Models, ACS Appl. Mater. Interfaces 15 (2023) 32226–32239. https://doi.org/10.1021/acsami.3c04949

[102]   Y.F. Tan, P. Chandrasekharan, D. Maity, C.X. Yong, K.-H. Chuang, Y. Zhao, S. Wang, J. Ding, S.-S. Feng, Multimodal Tumor Imaging by Iron Oxides and Quantum Dots Formulated in Poly (Lactic Acid) Dalpha Tocopheryl Polyethylene Glycol 1000 Succinate Nanoparticles, Biomaterials 32 (2011) 2969–2978. https://doi.org/10.1016/j.biomaterials.2010.12.055

[103]   C. Fan, W. Gao, Z. Chen, H. Fan, M. Li, F. Deng, Z. Chen, Tumor Selectivity of Stealth Multi Functionalized Superparamagnetic Iron Oxide Nanoparticles, Int. J. Pharm. 404 (2011) 180–190. https://doi.org/10.1016/j.ijpharm.2010.10.038

[104]   B. Hamm, T. Staks, M. Taupitz, R. Maibauer, A. Speidel, A. Huppertz, T. Frenzel, R. Lawaczeck, K.J. Wolf, L. Lange, Contrast Enhanced MR Imaging of Liver and Spleen: First Experience in Humans With a New Superparamagnetic Iron Oxide, J. Magn. Reson. Imaging 4 (1994) 659–668. https://doi.org/10.1002/jmri.1880040508

[105]   K. Maier-Hauff, R. Rothe, R. Scholz, U. Gneveckow, P. Wust, B. Thiesen, A. Feussner, A. von Deimling, N. Waldoefner, R. Felix, A. Jordan, Intracranial Thermotherapy using Magnetic Nanoparticles Combined with External Beam Radiotherapy: Results of a Feasibility Study on Patients with Glioblastoma Multiforme, J. Neurooncol. 81 (2007) 53–60. https://doi.org/10.1007/s11060-006-9195-0

[106]   R. Qiao, C. Fu, H. Forgham, I. Javed, X. Huang, J. Zhu, A.K. Whittaker, T.P. Davis, Magnetic Iron Oxide Nanoparticles for Brain Imaging and Drug Delivery, Adv. Drug Deliv. Rev. 197 (2023) 114822. https://doi.org/10.1016/j.addr.2023.114822

[107]   J.E. Rosen, L. Chan, D.-B. Shieh, F.X. Gu, Iron Oxide Nanoparticles for Targeted Cancer Imaging and Diagnostics, Nanomedicine Nanotechnology, Biol. Med. 8 (2012) 275–290. https://doi.org/10.1016/j.nano.2011.08.017

Biogenic Nanomaterials: Synthesis, Characterization, Applications, and Future Remarks   Materials Research Forum LLC
Materials Research Foundations 180 (2025)   https://doi.org/21741/9781644903759

Chapter 7

Antimicrobial Effects of Biogenic Nanomaterials

D. Doğan[1], I. Meydan[1*], M. S. Erdoğan[2], F. Sen[3]

[1]Vocational School of Health Care, Van Yuzuncu Yil University, Van, Turkey

[2]Altintas Vocational School, Kutahya Dumlupinar University, Kutahya, Turkey

[3]Sen Research Group, Department of Biochemistry, Kutahya Dumlupınar University, Kutahya, Turkiye

ismetmeydan@yyu.edu.tr; fatih.sen@dpu.edu.tr

Abstract

Nanotechnology has revolutionized antimicrobial research, particularly through biogenic nanoparticles, which offer eco-friendly and effective antimicrobial solutions. This chapter explores the synthesis of nanoparticles using chemical, physical, and biogenic methods, underlining the advantages of biogenic approaches. Various metallic and metal oxide nanoparticles, including gold, silver, copper, platinum, zinc oxide, iron, palladium, and titanium dioxide, are discussed in terms of their antimicrobial efficacy. The mechanisms of action, such as cell membrane disruption and oxidative stress induction, are also highlighted. The chapter provides insights into the potential of biogenic nanomaterials for combating microbial resistance in diverse applications.

Keywords

Nanotechnology, Nanoparticle Synthesis, Biogenic Nanomaterials, Antimicrobial Effect, Microbial Resistance

1.   Introduction

In recent years, nanotechnology, which has brought a new dimension to research, has become the focus of attention. Nanotechnology, a field of study with the potential to create the next industrial revolution on the economy and society, offers promising solutions [1,2]. Each of the nano-sized materials obtained by intensive studies of nanotechnology has different properties. The resulting nanoparticles have increased their production due to their impressive properties and aroused interest as a comprehensive research area.

These microscopic particles, which are invisible to the human eye but can be measured with high resolution and advanced devices, are the main target point in chemistry with both particle size and geometric shape [3]. Nanoparticles also create a transition between bulk materials and atomic-molecular structures. To achieve this, metal nanoparticles such as silver, gold, copper, platinum, zinc, titanium, palladium and zinc oxide, titanium oxide are commonly used. It shows itself with its applicability in different fields such as medicine, catalysis, electrochemistry, biotechnology, and pharmacology [4–11].

Different methods are used to synthesize nanoparticles of different contents and sizes. Nanoparticles, which can be produced physically and chemically, have been open to development and less preferred due to the use of harmful chemicals during their synthesis, being toxic, costly, and needing high energy. The most important issue in these continuously developed methods is to synthesize products that do not harm the environment. As a result, biological systems have been the best approach to produce the metal and metal oxide particles at nanometric scale [4,12,13].

It is more advantageous than chemical methods with its affordable cost and less toxicity. Another reason why it is preferred is that nanoparticles become more compatible with the coating of biological molecules on the surface. This biocompatibility has received much attention, especially in the medical field [12].

It has a pH value close to neutral at room temperature, and due to these features, it becomes reliable in terms of low energy consumption and environment. Bacterias, fungies, plants are environmentally friendly steps for the synthesis of nanoparticles [14].

These assets used in the biological synthesis process not only lead to the formation of nanoparticles, but also increase their biocompatibility by coating the surface of the nanoparticle. In order to synthesize metallic nanoparticles, it can bring an additional feature on nanoparticles as an antimicrobial compared to nanoparticles synthesized by chemical and physical methods through these biological entities [15]. Searching for harmless alternatives to antimicrobial agents has an important place in overcoming the resistance of pathogens. When the surface area of nanoparticles increases, their interaction with pathogenic bacteria increases. Thus, these nanoparticles become an antimicrobial agent. Bacteria easily adhere to the surfaces of these small-sized nanoparticles [14,16]. Since there are many different biological entities for synthesizing metal nanoparticles, easily scalable and the most compatible with green synthesis are more preferable. For this reason, it provides advantages such as the use of plants in synthesis, low cost, and the presence of active metabolites [17,18].

Plant extracts obtained from plant parts such as stems, seeds, roots, and leaves can determine the shape, size, and physicochemical properties of the nanoparticles obtained [19].

For example; The antimicrobial effect of silver nanoparticles obtained biogenically from the leaf of the *Cassia fistula* plant was seen at the end of the study [20].

In another study, silver nanoparticles were developed using the flower extract of the *Couroupita guianensis* tree. It has been proven that these developed nanoparticles have strong antibacterial effects against many pathogens [21].

In addition, microorganisms can form nanoparticles through green synthesis, thanks to the reductase enzyme found in metals [22].

Molecules that must be formed outside the cell in order for microorganisms to survive can increase due to environmental conditions. They produce the desired metal nanoparticles by both intracellular and extracellular mechanisms. In the intracellular mechanism, ions are transported to the cell and subsequently form nanoparticles within the cell by electrostatic and enzymatic interaction. The extracellular mechanism, on the other hand, occurs when this interaction takes place on the surface of the cell [23].

Gold nanoparticles obtained from the prokaryotic *Phormidium* and the eukaryotic *Coelastrella photosynthetic* microorganisms were synthesized. It was observed that these synthesized nanoparticles showed particularly strong antioxidant properties [24].

Apart from that, fungi, bacteria and algae have the capacity to produce nanoparticles with antimicrobial effect. Gold nanoparticles were produced biogenically from *Cladosporium cladosporioides*, an endophytic fungus of seaweed, by Joshi C.G et al. It showed an impressive antimicrobial effect against pathogenic microorganisms [25].

In particular, algae are actively used in the biogenic synthesis of metal nanoparticles, as they have good metal-holding capacity with both living and dead biomass. In addition, the aqueous extract of algae contains secondary metabolites such as polysaccharides, proteins, and tannins. Silver and copper nanoparticles obtained from the extract of *Botryococcus braunii*, a green algae species, showed that they are highly toxic to gram-negative and gram-positive bacteria [26][27].

2.   Nanotechnology and Nanoparticles

The concept of nano refers to one billionth of any measure. Nobel laureate Richard P Feynman first mentioned the concept of nanotechnology at the meeting of the American Physical Society. Subsequently, he contributed enormously to the study of the atom at this nanoscale in many fields. Tokyo University of Science professor Norio Taniguchi coined the term nanotechnology to describe finer dimensions [28].

Nanotechnology is used in different fields such as physics, chemistry, biology, drug development, water treatment, information and communication technologies, and the production of stronger and lighter materials [29,30]. Nanoparticles synthesized by refining from atomic groups or materials form the backbone of Nanotechnology. The production of nanoparticles by both humans and nature dates back to ancient times. For example, the Lycurgus cup, which the Romans used to make glass, and which has nanometric metals in different sizes, contains silver and gold nanoparticles. Faraday's synthesis of colloidal gold is one of the most important examples. Since 1975, it has been supported by studies that magnetic nanoparticles are pioneers in medicine and biology. Biogenic magnetite is used to learn the earth's magnetic field [31,32]. The nanostructured crystal papillae found on the upper side of the epidermal cells of the lotus leaf are one of the best examples of natural nanoparticles. They reduce the contact area of water with the leaf. With the discovery of this structure, superhydrophobic surfaces could be produced for various products such as self-cleaning paint, cooling liquid, grafted porous surfaces [15].

Nanoparticles are also used to improve the pharmacological and therapeutic properties of anticancer drugs through targeted drug delivery [33].

The properties of nanoparticles such as electrical, magnetic, physicochemical, size control, and increased surface area / volume ratio distinguish them from many other materials [34].

The electrons trapped in these particles change the band gap energy, causing the particle size and surface-to-volume ratio to change. This gives nanoparticles their superior properties [35]. When the materials used are converted to the nanoscale, they show unpredictable properties such as great strengths, chemical reactivity, electrical conductivity, superparamagnetic behavior, and other properties that the same material does not have at the micro or macro scale [36].

These uniquely-proper nanomaterials are used in many different areas such as electronic batteries, fuel additives, solar cells, catalysts, cosmetics, petroleum, textiles, pharmaceuticals, food additives, agriculture, biosensor production, diagnostic imaging, vaccine production, and the development of antimicrobial and chemotherapy drugs [37],[38]. As a result, many different nanoparticle examples have been presented in the scientific literature.

3.   Nanoparticle Synthesis

There are two basic approaches to nanoparticle synthesis: top-down and bottom-up. Top-down approaches involve breaking and cutting bulk materials into nanoscale fragments. Methods such as mechanical grinding, etching, electro-explosion, sputtering, and laser cutting are included. The bottom-up approach includes the formation of nanoparticles by combining atom-atom, molecule-molecule or cluster-cluster. Supercritical fluid synthesis, spinning, use of templates, and laser pyrolysis are important examples of the bottom-up approach [39,40].

The top-down method can produce large amounts of nanoparticles. However, in this method, nanoparticles with surface defects can be formed and nanoparticles can be damaged [41]. The optical and physicochemical properties of nanoparticles are closely related to their surface morphology. For these reasons, the production of nanoparticles with the top-down method is limited [42]. Techniques such as self-assembly of monomer-polymer molecules, chemical or electrochemical nanostructural precipitation, chemical vapor deposition, plasma or flame spray synthesis, and bioassisted synthesis are used for the production of nanoparticles. The bottom-up approach is a suitable technique for a variety of applications.

Biogenic Nanomaterials: Synthesis, Characterization, Applications, and Future Remarks   Materials Research Forum LLC
Materials Research Foundations 180 (2025)   https://doi.org/21741/9781644903759

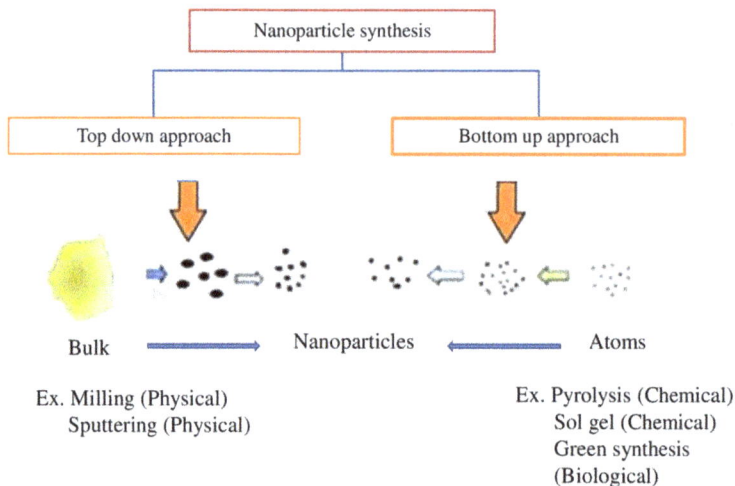

Figure 1. *Top-down and bottom-up approach to the synthesis of nanoparticles. Reprinted with permission from [15], (Copyright Springer Nature).*

4.   Synthesis Methods of Nanoparticles

4.1   *Physical Methods of Synthesis of Nanoparticles*

The production of nanoparticles using physical methods includes deposition and sputtering plasma based techniques [43]. In these methods, the synthesis rate of metal nanoparticles is very slow. For example, in the sputtering technique, the particle size distribution is very large and it has been noted that only 6-8% of the sputtered material is smaller than 100nm. High energy is required for the plasma technique. Slow production rate, waste by-products, high energy consumption, and very wide size distribution made the physical method undesirable [44].

4.2   *Chemical Methods of Synthesis of Nanoparticles*

Many commonly used chemical methods have been proposed. Chemical reduction, pyrolysis, sol-gel method, microemulsion, hydrothermal synthesis and chemical vapor deposition are some of the methods [45].

However, the use of hazardous chemicals and reagents during the synthesis process can lead to fatal situations for humans and the environment. Therefore, the use of this method is limited [46]. For example, in the chemical synthesis of silver nanoparticles, borohydride is required as a reducing agent, a sealant (starch, polyethylene glycol), and different substances to stabilize it [3].

4.3   Biogenic Synthesis of Nanoparticles

Although the mechanism of nanoparticles obtained by biogenic synthesis is not fully understood, it is preferred because it is environmentally friendly and non-toxic. Seen as a new technology, biogenic synthesis has proven to be cheaper than chemical and physical methods. Mechanisms in biogenic synthesis emerging studies for these biological processes try to explain. Thus, new abilities of unique microorganisms were discovered in this way [47].

Studies have shown that reducing agents of microorganisms and plants are used through biological mass or extract to synthesize biogenic nanoparticles. Substances such as carbohydrates, sugar, proteins are secreted and synthesis occurs through redox reactions of metal ions [48]. Metabolites, which are involved in the reduction process, make it difficult to elucidate the mechanism of action due to their chemical diversity. Parameters such as pH, reaction time, reactant concentration, temperature play an important role in the physical and chemical properties of the obtained nanoparticles [35].

Biogenic synthesis is based on the process established by plants and microorganisms to resist or adapt to the high concentrations of toxic metals found in the environment [12]. The use of plants to synthesize nanoparticles has advantages such as low cost, scaling up synthesis, and formation of more active metabolites compared to other sources [17]. By controlling biological synthesis, nano-structured products with desired geometry and structure can be obtained. Microorganisms such as plants, bacteria, fungi, yeast, viruses, algae are widely used in nanoparticle formation. Microorganisms have the capacity to produce many nanoparticles. Considering these reasons, it has become inevitable to work on nanoparticle synthesis [49]. There are different types of microbes that react with metals. For example, Bacteria and fungi can produce intracellular and extracellular production. There are different ways to produce both. In intracellular synthesis, positively charged ions interact with the negatively charged wall, transporting metal ions across the cell wall. Enzymes in the cell reduce these ions to metal nanoparticles [50]. In the biogenic synthesis of nanoparticles, the biological material acts not only as a reducing agent on the nanoparticles, but also as the capping agent of the synthesized nanoparticles. These biological substances avoid the aggregation of nanoparticles and the growth of nanoparticles in size. In addition, it can provide additional properties such as antioxidant and anti-inflammatory effects for the nanoparticle [12]. For example; Gold nanoparticles were obtained by using *Lactobacillus kimchicus* bacteria. (Figure 2.) While nanoclusters are forming, HAuCl ions first begin to nucleate in an electrostatic way. Subsequently, these formed nanoclusters are gradually transferred across the cell wall.

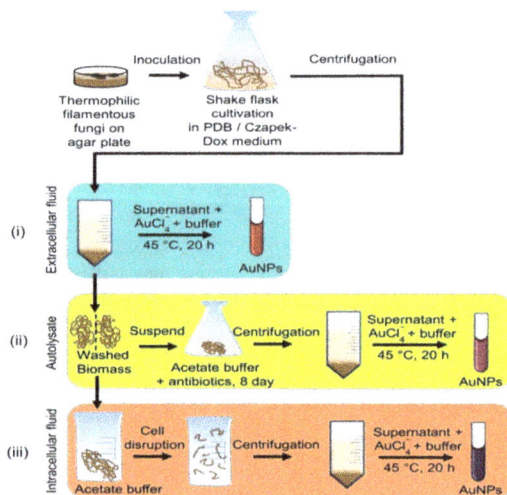

Figure 2. *Extracellular, autolysate, intracellular synthesis mechanism of gold nanoparticles using Thermophilic filamentous fungi (Reprinted with permission from [42], Copyright Springer Nature)..*

In extracellular synthesis, however, metal ions accumulate on the cell surface and occur with the incorporation of reducing ions by enzymes. In a study, *Cupriavidas sp.* in the extracellular synthesis of silver nanoparticles, it was concluded that silver ions were attached to the surface of the cell and nitrate reductase was used as an enzyme and reduced them to atomic silver nanoparticles [51]. Biogenically produced nanoparticles are a superior synthesis compared to chemical methods because they have less chemical use and higher catalytic activity. *Penicillium rugulosum* was used to synthesize more useful uniform-size Au nanoparticles than bacteria and yeast [52]. Leaf extracts of *Jasminum sambac* were used to prepare stable Au, Ag and Au-Ag alloy nanoparticles [53].

In general, bacteria, fungi and yeasts are tolerant to metals and thrive in extreme environmental conditions. Silver nanoparticles were synthesized by the action of *Fusarium oxysporum* nitrate reductase and anthraquinones in fungi [54]. There are plants and microbes that are abundant biological resources in the world. There are some disadvantages in using microbes; It is expensive, creating limited space for the creation of nanoparticles due to molecular mutation, barriers to mass cultivation, and other reasons. Especially for microorganisms involved in biogenic nanoparticle synthesis, extensive tools are needed in the process for the production, maintenance and purification of nanoparticles. There should be freezers to preserve microorganisms, centrifuges for the purification of nanoparticles, and temperature and shaking controlled incubators for the production of nanoparticles. Therefore, these materials required for production and purification make the process expensive. Using plants as a source for synthesis, they are more effective on nanoparticles than other sources, thanks to the active ingredients (phenols, flavonoids, tannins, saponins, polysaccharides, polyphenols, etc.) Considering the advantages of biogenic synthesis, we can count the production of stable nanoparticles, the biocompatible coating on the surface of the nanoparticles that provide the most active surface area, the absence of dangerous by-products, the absence of reducing and stabilizing needs. In addition, its low cost and easy scalability makes this synthesis preferable [55–57].

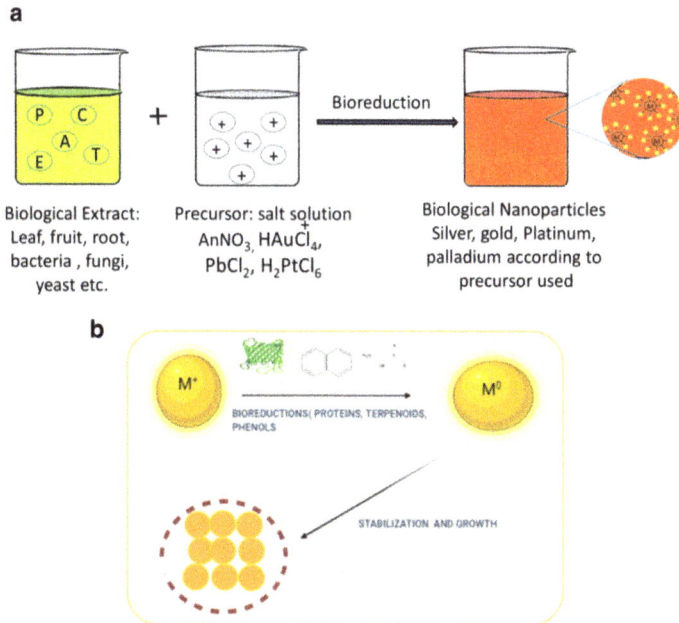

Figure 3. *a) Biological synthesis of nanoparticles, (b) Mechanism of plant-mediated metallic nanoparticle synthesis (Reprinted with permission from [8], Copyright Springer Nature)..*

Biological extracts are reduced to metal ions and their by-products by reacting with heat, temperature and salt at a certain pH. The synthesis of metallic nanoparticles depends on the action of phytocomponents in plant extracts. Plants with too many active ingredients synthesize metal nanoparticles by activating the reaction mechanism [58].

Biogenic Nanomaterials: Synthesis, Characterization, Applications, and Future Remarks   Materials Research Forum LLC
Materials Research Foundations 180 (2025)   https://doi.org/21741/9781644903759

5.   Biogenic Metallic Nanoparticles Used for Antimicrobial Action

These metal nanoparticles, which are preferred due to the properties mentioned above, have become important in many biomedical applications. Metal-based nanoparticles such as silver, copper, gold, zinc, platinum, iron palladium, titanium oxide is used against microorganisms due to their antimicrobial effect. Many studies have investigated the antibacterial effect of green nanoparticles against pathogenic microorganisms. In a study, copper nanoparticles obtained from *Sida acuta* showed antibacterial activity against *Escherichia coli, Proteus vulgaris* and *Staphylococcus aureus*. [59–61].

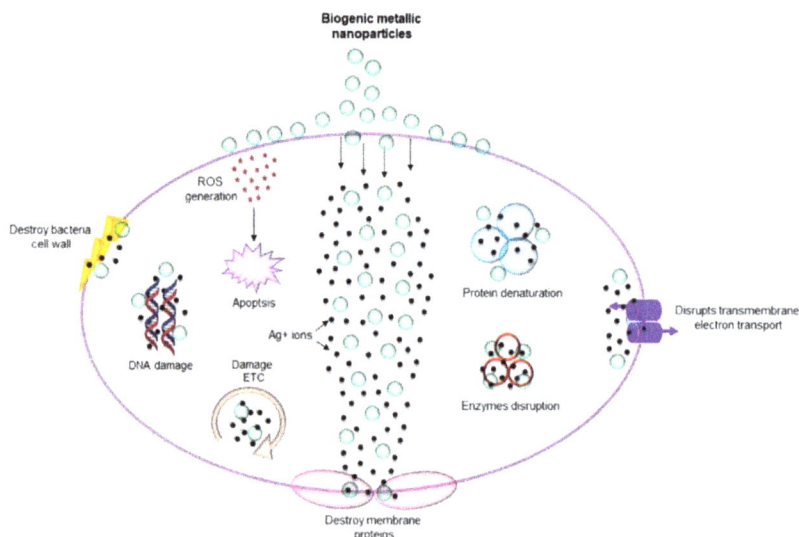

Figure 4. *Different antimicrobial action mechanisms of biogenic metal nanoparticles. ROS: Reactive oxygen species(Reprinted with permission from [3] Copyright MDPI).*

In addition, when these biogenic nanoparticles are applied together with antibiotics such as lincomycin, oleandomycin, vancomycin, novobiocin, penicillin G and rifampicin, an increase in the antimicrobial effect of antibiotics has been observed [62]. In line with these findings, it was concluded that biogenic metal nanoparticles would increase their antimicrobial effect when combined with antibiotics. Nanoparticles focus on the bacterial cell, target the cell membrane and impair the permeability of the membrane. They inhibit cell division and gene transfer by reacting with components such as proteins, nucleic acids [63,64].

Most of the metallic nanoparticles show their effect through multiple mechanisms in response to the development of resistance by microorganisms. To resist these nanoparticles, the microbial cell would have to show multiple gene mutations at the same time, which seems unlikely. Let's talk about some metal nanoparticles that show these effects:

6.   Silver Nanoparticles

Silver nanoparticles are increasing their applicability in the medical field thanks to their low toxicity and high bactericidal properties. In particular, the most important application of silver and silver nanoparticles in medicine is used as an anti-infective ointment for burns and open wounds. It is thought that the antibacterial effect of silver nanoparticles occurs through the release of silver ions within the cell. When silver nanoparticles touch bacteria, free radicals are formed, and these free radicals damage the cell membrane and make it porous [65].

Silver ions of nanoparticles can interact with the thiol groups of vital enzymes and inactivate them. As another approach, there are sulfur and phosphorus in the structure of DNA. Nanoparticles can act on these bases in DNA, causing cell death and destroying DNA. In this way, this interaction of silver nanoparticles and DNA can destroy microbes. Silver nanoparticles provide better contact due to their large surface area[66]. Therefore, they show better antimicrobial properties compared to other metal nanoparticles. It has been seen that silver nanoparticles obtained by green synthesis using *Syzygium aromaticum* can be used as effective growth inhibitors against microorganisms commonly found in the oral cavity [67]. In another study, *Escherichia coli* L-cysteine was used for the synthesis of silver nanoparticles. These synthesized silver nanoparticles showed antimicrobial activity against *E. coli* and *S. aureus* [68]. In their research, Pasupuleti V.R et al. synthesized silver nanoparticles using *Rhinacanthus nasutus* leaf extract. They have seen the antimicrobial effects of these nanoparticles on various bacteria [69].

Silver nanoparticles obtained from the perspective of green chemistry were synthesized from the leaf of *Psidium guajava*. The antimicrobial effect of these particles on *Pseudomonas aeruginosa* was examined and it was found that they have a potential to inhibit bacterial growth [70]. It has been observed that silver nanoparticles obtained from *Arum italicum*, which is rich in content and used in different health areas, have a stronger effect than antibiotics [18]. In a study reviewed, silver nanoparticles with an effective antibacterial mechanism were pretreated at a non-lethal level.

Bacteria exposed to ampicillin in this way showed higher levels of intracellular ATP, lower levels of membrane damage and oxidative stress. Accordingly, it provides information that pretreated bacterial cells with silver nanoparticles increase their resistance to antibiotics [71].

7.   Gold Nanoparticles

Gold nanoparticles: Apart from being used in optoelectronics, magnetic devices, biosensors, drug release, DNA labeling, it is used as antibacterial, anti-HIV, antitumor, antibiofilm agents. They can also be easily synthesized with high chemical and thermal stability. Gold nanoparticles are synthesized biogenically from a whole plant or from a combination of various components [72,73].

Using *Jasminum auriculatum* leaf extract, environmentally harmless biogenic gold nanoparticles were produced. In the study, it was seen that gold nanoparticles are a versatile candidate for heterogeneous catalysis. Outstandingly produced, these gold nanoparticles have demonstrated antimicrobial activity against pathogenic bacteria [74]. In their study, Sharma T.S. K. et al. synthesized gold nanoparticles from *Camellia japonica* leaf at room temperature without using any other chemicals. Gold nanoparticles showed tremendous antibacterial activity against seven different pathogens. Gold nanoparticles can also be obtained in different ways. It has been reported that there are different forms of gold nanoparticles such as spherical, octahedral, ten-sided, tetrahedral, nano-triangle, and nano-rods. In the research, nanocubic gold particles of gelidium amansii were obtained. The antibacterial potential of these particles has been tested on human pathogens [75].

As a result, it was observed that the gold nanoparticles were biocompatible and showed inhibition on cancer cells at high concentrations. It was stated that it needs to be discovered in order to become a bioresource in the biomedical and biosensor fields [76]. It has been observed that gold nanoparticles synthesized from the leaf extract of *Anacardium occidentale* have an increasing effect against pathogens, and it has been proven that they can be applied safely [77].

8.   Copper Nanoparticles

Copper nanoparticles have become a potential focus in the fields of electronics, optics, and biomedicine. The obtained materials are used in conductive films, catalysis, and nanofluids [78].

These nanoparticles are very reactive due to their high surface-to-volume ratio. For this reason, they can interact very easily with other particles and act as antimicrobials. Copper is very sensitive to air and differs from gold and silver in that its oxides are extremely thermodynamically stable. It has been difficult to produce stable copper oxide nanoparticles due to its high oxidation tendency [79]. Although the mechanism underlying the antimicrobial effect of copper nanoparticles has not been fully determined, it has been reported that Cu^{+2} ions help cell death by damaging enzymes and DNA because of contact with the cell membrane [80]. Using the bark extract of *Punica granatum*, copper nanoparticles were biologically synthesized from copper sulfate solution. It was observed that these synthesized particles showed a high antibacterial effect against pathogens such as *Micrococcus luteus* [81]. In a study, *Bauhinia tomentosa* leaf extract was a good bioreducer to produce copper oxide nanoparticles. It was observed that these formed nanoparticles showed strong antibacterial activity against *E. coli* and *P. Aeruiginosa* [82]. Green synthesized copper nanoparticles using Magnolia kobus, which is used as a reducing agent, showed a high antibacterial effect [83]. It was observed that copper oxide nanoparticles obtained from the leaves of *Tabernaemontana divaricate* by green chemistry method contained a maximum inhibition zone against urinary tract pathogen [84]. In line with the studies examined, these nanoparticles have become an environmentally friendly alternative for biomedical applications.

9.　Zinc Oxide Nanoparticles

Zinc oxide nanoparticles, which have a wide scope such as photocatalytic, water treatment, antimicrobial, are much different from the known zinc oxide particle. Zinc oxide nanoparticles can be produced by the chemical reaction of zinc metal with alcohol, precipitation in aqueous solution, chemical reaction of zinc acetate dihydrate and NaOH.

These nanoparticles, by using photocatalyst to remove heavy metal ions and break down electronic devices, drop UV rays on the surfaces on which they are used and have a wide band area that produces electron holes. The main reason for its use for human health is due to its UV-reducing effect [1,38]. It is known that zinc oxide nanoparticles at room temperature have high binding energy that can be used in photoelectronics, semiconductors and solar cells [85].

Zinc oxide nanoparticles obtained through green synthesis are preferred as they are more economical and more environmentally friendly than chemical and physical synthesis. Zinc oxide nanoparticles formed by plant extract (stem, root, leaf); It was observed that it showed higher photocatalytic, antioxidant and antimicrobial properties compared to nanoparticles synthesized by fungi, bacteria, algae, and yeast [86,87].

It has been observed that zinc oxide nanoparticles obtained from *Helichrysum arenarium* have inhibitory properties against six different pathogens. These biogenic nanoparticles have been proven to have potent antimicrobial effects [88].

Biogenically obtained zinc-oxide nanoparticles were synthesized using the aqueous root extract of *Ruta graveolens*. It has been subjected to measurements for properties as antioxidant and antibacterial. It was observed that it exhibited significant activity against gram-positive and gram-negative pathogens. In addition, 1-diphenyl-2-picrylhydrazil showed an antioxidant effect by destroying free radicals [89].

Rajashekara S et al., zinc oxide nanoparticles obtained from leaf extracts of *Calotropis gigantea* have been characterized as antimicrobial and anticancer. As a result of the study, it has been shown that it is a good anticancer agent and can be used as an antibacterial [90].

In another study, it was observed that zinc oxide nanoparticles synthesized from *Aquilegia pubiflora* extract showed a high antimicrobial effect against the pathogens studied, and it was discovered that the formed particle was extraordinarily biocompatible [91].

Zinc oxide nanoparticles were formed by microbial synthesis from the soil bacterium *Pseudomonas putida* and showed great antimicrobial activity against gram(+) and gram(-) microbes [92]. Zinc nanoparticles obtained from the *Rheum ribes* plant strongly inhibited gram(+) and gram(-) microorganisms [88].

10.   Platinum Nanoparticles

It is known that biogenically prepared platinum nanoparticles are used in medical diagnosis, catalysts, chemical sensors, pharmacology, environment, and biomedical fields [93].

It can be used as a heterogeneous and homogeneous catalyst due to its large surface/volume ratios and high surface energy. It can easily take place in aqueous media with a catalytic reaction with plant extracts without ligand. The efficiency of platinum nanoparticles formed even in very small proportions is quite high [94]. The most important feature that distinguishes platinum nanoparticles from other nanoparticles is that they can be used in proton membrane exchange fuel cells [95]. It has been shown that platinum nanoparticles obtained biogenically using black cumin seeds are a potential antibacterial in the pharmaceutical industry by showing high zone diameters against gram-positive and gram-negative bacteria, especially at certain concentrations [96]. In another study, platinum nanoparticles were obtained using an extract of a green algae, *Botryococcus braunii*, and showed promising antimicrobial results [27]. In their study, Eltaweil A.S et al., synthesized platinum nanoparticles from *Atriplex halimus* leaves as a reducing agent. It showed high antimicrobial activity against gram-negative bacteria [97]. It has been observed that platinum nanoparticles synthesized green from the leaf pulp of *Combretum erythrophyllum* have strong antibacterial effects against pathogenic gram positive *Staphylococcus epidermidis* and gram *Klebsiella oxytoca*, *Klebsiella aerogenes* [98].

11.   Iron Nanoparticles

Iron nanoparticles have unique properties such as catalytic activity, optical, electronic, magnetic and antibacterial [99]. Iron and iron oxide nanoparticles; Catalysis and magnetic materials are widely used in biology and medicine, cosmetics, and drugs. Because these nanoparticles increase the surface area, it seems to reduce the negative effect of the drug on normal cells [100]. In particular, it has been seen that iron oxides have a very important place in applications such as catalysis, magnetic resonance imaging (MRI), cancer treatment, and DNA analysis [101].

Alam T et al. obtained biocompatible iron oxide nanoparticles were from the extract of *Skimmia laureola* leaf [102]. It was observed that these obtained nanoparticles inhibited the growth of *Ralstonia solanacearum* pathogen and showed an inhibitory effect. In another study, the antimicrobial effect of two iron oxide nanoparticles synthesized using two different seaweed species, *Colpomenia sinuosa* and *Pterocladia capillacea*, collected from the Mediterranean and Egypt were investigated. As a result of the study, it was seen that it showed a broad antibacterial effect against the growth of gram bacteria [103].

It was observed that iron oxide nanoparticles synthesized from *Borassus flabellifer* seed coat showed antimicrobial activity in direct proportion to the increase in concentrations [104].

Iron oxide nanoparticles obtained environmentally friendly by using *Ruellia tuberosa* as a reducing agent showed lethal effects against gram-negative and gram-positive pathogens [105].

12.   Palladium Nanoparticles

Palladium nanoparticles can act as catalysts in many organic processes including carbon-carbon cross-coupling, oxidation, and reduction reactions [106].

Palladium nanoparticles are widely used in many applications such as hydrogen sensor, chemical modifier, automotive catalytic converter, optical limitation devices, as well as catalyst use [107]. In addition to its various applications, palladium nanoparticles have been shown to have biomedical diagnostic and therapeutic effects with minimal side effects [108]. Palladium nanoparticles synthesized in an environmentally friendly manner using *Filicium decipiens* leaf extract have been shown to have antimicrobial activity against both gram-positive and gram-negative bacteria.[109].

In another study, palladium nanoparticles were obtained by using *Santalum album* extract. It has been observed that these created nanoparticles have an improved bactericidal effect against the pathogens used [110].

Sivamaruthi B.S et al. investigated both the anticancer and antimicrobial effects of silver-palladium bimetallic nanoparticles formed from the fruit pulp of *Terminalia chebula*. The resulting silver-palladium nanoparticles induced the formation of ROS in lung cancer cells and showed antimicrobial effect against resistant *Pseudomonas aeruginosa* [111].

Palladium nanoparticles biogenically synthesized using *Couroupita guianensis* fruit extract have shown antimicrobial activity against different pathogens and have been a reliable choice for biomedical studies [112].

13.   Titanium Oxide-Dioxide Nanoparticles

Titanium dioxide nanoparticles among metal oxides are inert, non-toxic, environmentally friendly catalysts with a high refractive index, capable of absorbing UV light. In addition, it is seen that titanium dioxide nanoparticles react with oxygen and hydroxide adsorbed on the surface to obtain oxygen free radicals and hydroxyl free radicals that can directly attack the cell wall and cell membrane [113].

It is used in sunscreen lotions, paints, papers, plastics, medical applications, toothpastes, food colorants to whiten and provide opacity [114].

It has been observed that titanium nanoparticles obtained using a microalga, *Phaeodactylum tricornutum*, have an antimicrobial effect [115].

In another study, titanium dioxide nanoparticles were synthesized from *Trichoderma citrinoviridae* extract, which was used as a reducing agent. Antibacterial activity of these particles has been observed against *Pseudomonas aeuginosa*, which is extremely drug-resistant [116]. Ahmad W et al., in their study, showed that titianium dioxide nanoparticles obtained from *Mentha arvensis* leaf extract were synthesized using various methods, these particles were synthesized as spherical and exhibited antibacterial and antifungal activity against microorganisms [117]. Titanium oxide nanoparticles were biosynthesized using *Trigonella foenum-graecum*, and these particles, examined by Kirby-Bauer method, showed impressive antimicrobial activity against all tested microorganisms [118].

14. Antimicrobial Effect of Nanoparticles

Antibiotics are chemical compounds that can inhibit the growth or kill microorganisms. Antibiotics are used to treat diseases and support medical procedures.

Antibiotics show their antibacterial effects by inhibiting enzymes, preventing DNA, RNA and protein synthesis, and disrupting membrane structure [119]. Antimicrobial resistance is defined as the ability to resist a drug that can successfully treat the microbe. Bacterial resistance is seen because of a natural or acquired mechanism. If bacteria can inhibit the effect of an antibiotic, natural resistance emerges.

An antibiotic targeted to bind to a receptor on a bacterial cell will not work if the bacterium does not have the receptor. In an acquired resistance, the genetic structure changes and the effect of antibiotics decreases [16]. However, the resistance of microorganisms to drugs causes problems such as the use of high-dose drugs and high-dose toxicity treatments [120]. As a result of the mechanisms created by antibiotic resistance, it causes the drugs to remain in the cell for a shorter time due to less accumulation of drugs in the cells or the inability to reach the therapeutic levels of drugs easily. As a result, it is exposed to high doses and repeated applications, which causes adverse effects in humans and animals [121,122].

To find solutions to these problems, scientists continued their studies to develop the drugs used. Metals such as silver, copper, titanium, and zinc, which are antimicrobial, are at the forefront of research. Unlike antibiotics, these metals are known to act against microorganisms in different ways, such as breaking the membrane, damaging cellular components such as DNA and protein, forming reactive oxygen species and so on [123].

The geometric diversity of nanoparticles is one of the most important reasons for their effectiveness in the antimicrobial field. Because the antibacterial effect of nanoparticles is directly proportional to the surface area of these components [121]. Munmi Hazarika et al. studied on biogenically obtained palladium nanoparticles and showed excellent antimicrobial activity against *Cronobacter sakazaki* [106].

Gold nanoparticles biogenically synthesized from the *Crescentia cuteje* tree showed impressive antimicrobial effects against both gram-positive and gram-negative bacterias [124].

Silver nanoparticles synthesized from *Fusarium oxysporum* solution were investigated as an inhibitor of microbial propagation. It has shown potent activity against some human and orange tree pathogens (*Candida parapsilosis* and *Xanthomonas axonopodis*) [125].

Garibo D et al. studied *Lysiloma acapulcensis*, a perennial tree in Mexico, which is rich in antimicrobial content [126]. Silver nanoparticles obtained from the extract were found to increase the antimicrobial effect, and the effect of biogenic silver nanoparticles was greater than silver nanoparticles obtained by chemical method.

Selenium nanoparticles obtained from *Spirulina platensis* extract showed antimicrobial activity against gram-positive and gram-negative pathogens depending on the dose used. It is seen that the formed selenium nanoparticles are a reliable and successful candidate for use in the medical field [127]. Kamran et al. produced manganese nanoparticles by *Cinnamomum verum* bark. It has been determined that these biosynthesized nanoparticles are a potential antimicrobial against *Staphlyococcus aureus* and *Escherichia coli*. This also serves to obtain antimicrobial substances naturally [128].

Silver nanoparticles obtained by using the extracellular filtrate from *Bionectria ochroleuca*, an epiphytic fungus species, offer promising approaches by showing strong antimicrobial effect on clinically potent *Candida albicans*, *Candida glabrata* and *Candida parabsilosis* [129]. Zinc nanoparticles were obtained from *Pichia kudriavzeii*, a yeast species, by the green method. It showed maximum inhibition on various pathogens according to the increase in the doses used [130].

Green synthesized cobalt oxide nanoparticles were formed from *Geranium wallichianum* leaves as both a reducing and stabilizing agent. It exhibited remarkable toxic potential especially on *Bacillus subtilis* as antibacterial against different bacterial pathogens [131].

Iron oxide nanoparticles obtained from *Mikania mikrantha* have been promising to intervene in the field of medicine by showing very high antibacterial, antifungal and antimicrobial effects against different bacterial and fungal pathogens by combining the phenolic acids, flavonoids, terpenoids and proteins in the leaf extract [132].

15. Conclusions

Microorganisms have started to develop resistance to antibiotics due to the use of drugs in excessive doses and the inadequacy of drugs. This situation causes a serious health problem. Metal nanoparticles synthesized with a green and sustainable approach have proven to have strong effects against pathogens, either alone or in combination. These more environmentally friendly, more economical, less toxic substances can be used not only in the biomedical field, but also in many applications such as agriculture, cosmetics, and textiles. These nanoparticles obtained from biological entities can have different properties with different mechanisms. As a result of the research, these metal nanoparticles were found to have a significant amount of antimicrobial activity. Especially against gram-positive and gram-negative strains, it has had multifaceted important effects. We believe that the fascinating and rapidly developing microbial nanotechnology will be groundbreaking for the future. Although we have presented the latest developments in this study, we wish that more research on this subject will be conducted and that it will be an inspiration for this research.

References

[1]   R. Sharma, R. Garg, A. Kumari, A review on biogenic synthesis, applications and toxicity aspects of zinc oxide nanoparticles, EXCLI J. 19 (2020) 1325.

[2]   I. Meydan, A. Aygun, R.N.E. Tiri, T. Gur, Y. Kocak, H. Seckin, F. Sen, Chitosan/PVA Supported Silver Nanoparticles for Azo Dyes Removal: Fabrication, Characterization, and Assessment of Antioxidant Activity, Environ. Sci. Adv. (2024). https://doi.org/10.1039/D3VA00224A

[3]   P. Singh, A. Garg, S. Pandit, V.R.S.S. Mokkapati, I. Mijakovic, Antimicrobial Effects of Biogenic Nanoparticles, Nanomater. (Basel, Switzerland) 8 (2018). https://doi.org/10.3390/NANO8121009

[4]   S. Menon, R. S., V.K. S., A review on biogenic synthesis of gold nanoparticles, characterization, and its applications, Resour. Technol. 3 (2017) 516–527. https://doi.org/10.1016/j.reffit.2017.08.002

[5]   A. Hojjati-Najafabadi, S. Salmanpour, F. Sen, P.N. Asrami, M. Mahdavian, M.A. Khalilzadeh, A Tramadol Drug Electrochemical Sensor Amplified by Biosynthesized Au Nanoparticle Using Mentha aquatic Extract and Ionic Liquid, Top. Catal. 65 (2022) 587–594. https://doi.org/10.1007/s11244-021-01498-x

[6]   A. Şavk, K. Cellat, K. Arıkan, F. Tezcan, S.K. Gülbay, S. Kızıldağ, E.Ş. Işgın, F. Şen, Highly monodisperse Pd-Ni nanoparticles supported on rGO as a rapid, sensitive, reusable and selective enzyme-free glucose sensor, Sci. Rep. 9 (2019) 19228. https://doi.org/10.1038/s41598-019-55746-y

[7]   A. Cherif, R. Nebbali, J.W. Sheffield, N. Doner, F. Sen, Numerical investigation of hydrogen production via autothermal reforming of steam and methane over Ni/Al2O3 and Pt/Al2O3 patterned catalytic layers, Int. J. Hydrogen Energy (2021). https://doi.org/10.1016/j.ijhydene.2021.04.032

[8]   F. Şen, G. Gökağaç, Improving catalytic efficiency in the methanol oxidation reaction by inserting Ru in face-centered cubic Pt nanoparticles prepared by a new surfactant, tert-octanethiol, Energy and Fuels 22 (2008) 1858–1864. https://doi.org/10.1021/ef700575t

[9]   F. Gulbagca, A. Aygün, M. Gülcan, S. Ozdemir, S. Gonca, F. Şen, Green Synthesis of Palladium Nanoparticles: Preparation, Characterization, and Investigation of Antioxidant, Antimicrobial, Anticancer, and DNA Cleavage Activities, Appl. Organomet. Chem. 35 (2021). https://doi.org/10.1002/aoc.6272

[10]   E. Demir, B. Sen, F. Sen, Highly efficient Pt nanoparticles and f-MWCNT nanocomposites based counter electrodes for dye-sensitized solar cells, Nano-Structures & Nano-Objects 11 (2017) 39–45. https://doi.org/10.1016/J.NANOSO.2017.06.003

[11]   B. Sen, E. Kuyuldar, A. Şavk, H. Calimli, S. Duman, F. Sen, Monodisperse ruthenium–copper alloy nanoparticles decorated on reduced graphene oxide for dehydrogenation of DMAB, Int. J. Hydrogen Energy 44 (2019) 10744–10751. https://doi.org/10.1016/j.ijhydene.2019.02.176

[12]   G. Tortella, O. Rubilar, P. Fincheira, J.C. Pieretti, P. Duran, I.M. Lourenço, A.B. Seabra, Bactericidal and virucidal activities of biogenic metal-based nanoparticles: Advances and perspectives, Antibiotics 10 (2021) 783.

[13]   Y. Kocak, R.N.E. Tiri, A. Aygun, I. Meydan, N. Bennini, T. Karahan, F. Sen, Microwave-Assisted Fabrication of AgRuNi Trimetallic NPs with Their Antibacterial vs Photocatalytic Efficiency for Remediation of Persistent Organic Pollutants, Bionanoscience (2024). https://doi.org/10.1007/s12668-023-01237-4

[14]   Z. He, Z. Zhang, S. Bi, Nanoparticles for organic electronics applications, Mater. Res. Express 7 (2020) 012004. https://doi.org/10.1088/2053-1591/ab636f

[15]   S. Patil, R. Chandrasekaran, Biogenic nanoparticles: a comprehensive perspective in synthesis, characterization, application and its challenges, J. Genet. Eng. Biotechnol. 2020 181 18 (2020) 1–23. https://doi.org/10.1186/S43141-020-00081-3

[16]   E.T. Bekele, B.A. Gonfa, O.A. Zelekew, H.H. Belay, F.K. Sabir, Synthesis of titanium oxide

nanoparticles using root extract of Kniphofia foliosa as a template, characterization, and its application on drug resistance bacteria, J. Nanomater. 2020 (2020) 2817037.

[17]   P. Khandel, R.K. Yadaw, D.K. Soni, L. Kanwar, S.K. Shahi, Biogenesis of metal nanoparticles and their pharmacological applications: present status and application prospects, J. Nanostructure Chem. 8 (2018) 217–254. https://doi.org/10.1007/s40097-018-0267-4

[18]   I. Meydan, H. Seckin, H. Burhan, T. Gür, B. Tanhaei, F. Sen, Arum italicum mediated silver nanoparticles: Synthesis and investigation of some biochemical parameters, Environ. Res. 204 (2022) 112347. https://doi.org/10.1016/J.ENVRES.2021.112347.

[19]   A.M. El Shafey, Green synthesis of metal and metal oxide nanoparticles from plant leaf extracts and their applications: A review, Green Process. Synth. 9 (2020) 304–339.

[20]   Y.K. Mohanta, S.K. Panda, K. Biswas, A. Tamang, J. Bandyopadhyay, D. De, D. Mohanta, A.K. Bastia, Biogenic synthesis of silver nanoparticles from Cassia fistula (Linn.): In vitro assessment of their antioxidant, antimicrobial and cytotoxic activities, IET Nanobiotechnology 10 (2016) 438–444. https://doi.org/10.1049/iet-nbt.2015.0104

[21]   R. Singh, C. Hano, F. Tavanti, B. Sharma, Biogenic Synthesis and Characterization of Antioxidant and Antimicrobial Silver Nanoparticles Using Flower Extract of Couroupita guianensis Aubl., Materials (Basel). 14 (2021) 6854. https://doi.org/10.3390/ma14226854

[22]   P. Singh, Y.J. Kim, C. Wang, R. Mathiyalagan, D.C. Yang, Weissella oryzae DC6-facilitated green synthesis of silver nanoparticles and their antimicrobial potential, Artif. Cells, Nanomedicine, Biotechnol. 44 (2016) 1569–1575. https://doi.org/10.3109/21691401.2015.1064937

[23]   M.E. Cueva, L.E. Horsfall, The contribution of microbially produced nanoparticles to sustainable development goals, Microb. Biotechnol. 10 (2017) 1212–1215. https://doi.org/10.1111/1751-7915.12788

[24]   D. MubarakAli, J. Arunkumar, K.H. Nag, K.A. SheikSyedIshack, E. Baldev, D. Pandiaraj, N. Thajuddin, Gold nanoparticles from Pro and eukaryotic photosynthetic microorganisms—Comparative studies on synthesis and its application on biolabelling, Colloids Surfaces B Biointerfaces 103 (2013) 166–173.

[25]   C.G. Joshi, A. Danagoudar, J. Poyya, A.K. Kudva, B.L. Dhananjaya, Biogenic synthesis of gold nanoparticles by marine endophytic fungus-Cladosporium cladosporioides isolated from seaweed and evaluation of their antioxidant and antimicrobial properties, Process Biochem. 63 (2017) 137–144.

[26]   N. Salvador, A. Gómez Garreta, L. Lavelli, M.A. Ribera, Antimicrobial activity of Iberian macroalgae, Sci. Mar. 71 (2007) 101–114. https://doi.org/10.3989/scimar.2007.71n1101

[27]   A. Arya, K. Gupta, T.S. Chundawat, D. Vaya, Biogenic synthesis of copper and silver nanoparticles using green alga Botryococcus braunii and its antimicrobial activity, Bioinorg. Chem. Appl. 2018 (2018) 7879403.

[28]   E. Drexler, Engines of creation: The coming era of nanotechnology, Anchor, 1987.

[29]   K. Nesrin, C. Yusuf, K. Ahmet, S.B. Ali, N.A. Muhammad, S. Suna, Ş. Fatih, Biogenic silver nanoparticles synthesized from Rhododendron ponticum and their antibacterial, antibiofilm and cytotoxic activities, J. Pharm. Biomed. Anal. 179 (2020) 112993. https://doi.org/10.1016/J.JPBA.2019.112993

[30]   Y. Wu, E.E. Altuner, R.N. El Houda Tiri, M. Bekmezci, F. Gulbagca, A. Aygun, C. Xia, Q. Van Le, F. Sen, H. Karimi-Maleh, Hydrogen generation from methanolysis of sodium borohydride using waste coffee oil modified zinc oxide nanoparticles and their photocatalytic activities, Int. J. Hydrogen Energy 48 (2023) 6613–6623. https://doi.org/10.1016/J.IJHYDENE.2022.04.177

[31]   K.M. Krishnan, Biomedical Nanomagnetics: A Spin Through Possibilities in Imaging, Diagnostics, and Therapy, IEEE Trans. Magn. 46 (2010) 2523–2558. https://doi.org/10.1109/TMAG.2010.2046907

[32]   R. Blakemore, Magnetotactic Bacteria, Science (80-.). 190 (1975) 377–379. https://doi.org/10.1126/science.170679

[33]   M. Prabaharan, Chitosan-based nanoparticles for tumor-targeted drug delivery, Int. J. Biol. Macromol. 72 (2015) 1313–1322. https://doi.org/10.1016/j.ijbiomac.2014.10.052

[34]   S. Iravani, Green synthesis of metal nanoparticles using plants, Green Chem. 13 (2011) 2638–2650. https://doi.org/10.1039/c1gc15386b

[35]   Q. Wu, W. Miao, Y. Zhang, H. Gao, D. Hui, Mechanical properties of nanomaterials: A review, Nanotechnol. Rev. 9 (2020) 259–273. https://doi.org/10.1515/ntrev-2020-0021

[36]   C. Ostiguy, B. Roberge, C. Woods, B. Soucy, Engineered nanoparticles: current knowledge about OHS risks and prevention measures, (2010).

[37]   I. Khan, K. Saeed, I. Khan, Nanoparticles: Properties, applications and toxicities, Arab. J. Chem. 12 (2019) 908–931. https://doi.org/10.1016/J.ARABJC.2017.05.01

[38]   T. Gur, I. Meydan, H. Seckin, M. Bekmezci, F. Sen, Green synthesis, characterization and bioactivity of biogenic zinc oxide nanoparticles, Environ. Res. 204 (2022) 111897. https://doi.org/10.1016/J.ENVRES.2021.111897

[39]   R.M. Tripathi, S.J. Chung, Biogenic nanomaterials: Synthesis, characterization, growth mechanism, and biomedical applications, J. Microbiol. Methods 157 (2019) 65–80. https://doi.org/10.1016/J.MIMET.2018.12.008

[40]   M. Bayat, M. Zargar, E. Chudinova, T. Astarkhanova, E. Pakina, In Vitro Evaluation of Antibacterial and Antifungal Activity of Biogenic Silver and Copper Nanoparticles: The First Report of Applying Biogenic Nanoparticles against Pilidium concavum and Pestalotia sp. Fungi, Molecules 26 (2021). https://doi.org/10.3390/molecules26175402

[41]   K.N. Thakkar, S.S. Mhatre, R.Y. Parikh, Biological synthesis of metallic nanoparticles, Nanomedicine Nanotechnology, Biol. Med. 6 (2010) 257–262. https://doi.org/10.1016/j.nano.2009.07.002

[42]   D. Decarolis, Y. Odarchenko, J.J. Herbert, C. Qiu, A. Longo, A.M. Beale, Identification of the key steps in the self-assembly of homogeneous gold metal nanoparticles produced using inverse micelles, Phys. Chem. Chem. Phys. 22 (2020) 18824–18834.

[43]   C. Dhand, N. Dwivedi, X.J. Loh, A.N. Jie Ying, N.K. Verma, R.W. Beuerman, R. Lakshminarayanan, S. Ramakrishna, Methods and Strategies for the Synthesis of Diverse Nanoparticles and Their Applications: A Comprehensive Overview, RSC Adv. 5 (2015) 105003–105037. https://doi.org/10.1039/C5RA19388E

[44]   P.K. Seetharaman, R. Chandrasekaran, S. Gnanasekar, G. Chandrakasan, M. Gupta, D.B. Manikandan, S. Sivaperumal, Antimicrobial and larvicidal activity of eco-friendly silver nanoparticles synthesized from endophytic fungi Phomopsis liquidambaris, Biocatal. Agric. Biotechnol. 16 (2018) 22–30. https://doi.org/10.1016/j.bcab.2018.07.006

[45]   M. Darroudi, Z. Sabouri, R. Kazemi Oskuee, A. Khorsand Zak, H. Kargar, M.H.N.A. Hamid, Sol–gel synthesis, characterization, and neurotoxicity effect of zinc oxide nanoparticles using gum tragacanth, Ceram. Int. 39 (2013) 9195–9199. https://doi.org/10.1016/j.ceramint.2013.05.021

[46]   M. Zhang, J. Yang, Z. Cai, Y. Feng, Y. Wang, D. Zhang, X. Pan, Detection of engineered nanoparticles in aquatic environments: current status and challenges in enrichment, separation, and analysis, Environ. Sci. Nano 6 (2019) 709–735. https://doi.org/10.1039/C8EN01086B

[47]   T. Mabey, D. Andrea Cristaldi, P. Oyston, K.P. Lymer, E. Stulz, S. Wilks, C. William Keevil, X. Zhang, Bacteria and nanosilver: the quest for optimal production, Crit. Rev. Biotechnol. 39 (2019) 272–287. https://doi.org/10.1080/07388551.2018.1555130

[48]   S.S. Salem, A. Fouda, Green Synthesis of Metallic Nanoparticles and Their Prospective Biotechnological Applications: an Overview, Biol. Trace Elem. Res. 199 (2021) 344–370. https://doi.org/10.1007/S12011-020-02138-3

[49]   P. Moitra, M. Alafeef, K. Dighe, M.B. Frieman, D. Pan, Selective Naked-Eye Detection of

SARS-CoV-2 Mediated by N Gene Targeted Antisense Oligonucleotide Capped Plasmonic Nanoparticles, ACS Nano 14 (2020) 7617–7627. https://doi.org/10.1021/acsnano.0c03822

[50]   K. Mathivanan, R. Selva, J.U. Chandirika, R.K. Govindarajan, R. Srinivasan, G. Annadurai, P.A. Duc, Biologically synthesized silver nanoparticles against pathogenic bacteria: Synthesis, calcination and characterization, Biocatal. Agric. Biotechnol. 22 (2019) 101373.

[51]   B. Mughal, S.Z.J. Zaidi, X. Zhang, S.U. Hassan, Biogenic nanoparticles: Synthesis, characterisation and applications, Appl. Sci. 11 (2021) 2598.

[52]   A. Mishra, S.K. Tripathy, S.-I. Yun, Fungus mediated synthesis of gold nanoparticles and their conjugation with genomic DNA isolated from Escherichia coli and Staphylococcus aureus, Process Biochem. 47 (2012) 701–711.

[53]   S. Yallappa, J. Manjanna, B.L. Dhananjaya, Phytosynthesis of stable Au, Ag and Au–Ag alloy nanoparticles using J. Sambac leaves extract, and their enhanced antimicrobial activity in presence of organic antimicrobials, Spectrochim. Acta Part A Mol. Biomol. Spectrosc. 137 (2015) 236–243. https://doi.org/10.1016/j.saa.2014.08.030

[54]   N. Durán, P.D. Marcato, O.L. Alves, G.I. De Souza, E. Esposito, Mechanistic aspects of biosynthesis of silver nanoparticles by several Fusarium oxysporum strains, J. Nanobiotechnology 3 (2005) 8. https://doi.org/10.1186/1477-3155-3-8

[55]   P. Singh, S. Pandit, J. Garnæs, S. Tunjic, V. Mokkapati, A. Sultan, A. Thygesen, A. Mackevica, R.V. Mateiu, A.E. Daugaard, A. Baun, I. Mijakovic, Green synthesis of gold and silver nanoparticles from Cannabis sativa (industrial hemp) and their capacity for biofilm inhibition, Int. J. Nanomedicine Volume 13 (2018) 3571–3591. https://doi.org/10.2147/IJN.S157958

[56]   P. Singh, Y.-J. Kim, D. Zhang, D.-C. Yang, Biological Synthesis of Nanoparticles from Plants and Microorganisms, Trends Biotechnol. 34 (2016) 588–599. https://doi.org/10.1016/j.tibtech.2016.02.006

[57]   J.H. Jo, P. Singh, Y.J. Kim, C. Wang, R. Mathiyalagan, C.-G. Jin, D.C. Yang, Pseudomonas deceptionensis DC5-mediated synthesis of extracellular silver nanoparticles, Artif. Cells, Nanomedicine, Biotechnol. 44 (2016) 1576–1581. https://doi.org/10.3109/21691401.2015.1068792

[58]   J. Singh, T. Dutta, K.H. Kim, M. Rawat, P. Samddar, P. Kumar, "Green" synthesis of metals and their oxide nanoparticles: Applications for environmental remediation, J. Nanobiotechnology 16 (2018). https://doi.org/10.1186/s12951-018-0408-4

[59]   J.S. Fernandez-Moure, M. Evangelopoulos, K. Colvill, J.L. Van Eps, E. Tasciotti, Nanoantibiotics: A New Paradigm for the Treatment of Surgical Infection, Nanomedicine 12 (2017) 1319–1334. https://doi.org/10.2217/nnm-2017-0401

[60]   H. Seçkin, İ. Meydan, Synthesis and Characterization of Veronica beccabunga Green Synthesized Silver Nanoparticles for The Antioxidant and Antimicrobial Activity, Turkish J. Agric. Res. 8 (2021) 49–55. https://doi.org/10.19159/TUTAD.805463

[61]   S. Sathiyavimal, S. Vasantharaj, D. Bharathi, M. Saravanan, E. Manikandan, S.S. Kumar, A. Pugazhendhi, Biogenesis of copper oxide nanoparticles (CuONPs) using Sida acuta and their incorporation over cotton fabrics to prevent the pathogenicity of Gram negative and Gram positive bacteria, J. Photochem. Photobiol. B Biol. 188 (2018) 126–134. https://doi.org/10.1016/j.jphotobiol.2018.09.014

[62]   J. Pasquet, Y. Chevalier, J. Pelletier, E. Couval, D. Bouvier, M.-A. Bolzinger, The contribution of zinc ions to the antimicrobial activity of zinc oxide, Colloids Surfaces A Physicochem. Eng. Asp. 457 (2014) 263–274. https://doi.org/10.1016/j.colsurfa.2014.05.057

[63]   T.C. Dakal, A. Kumar, R.S. Majumdar, V. Yadav, Mechanistic Basis of Antimicrobial Actions of Silver Nanoparticles, Front. Microbiol. 7 (2016). https://doi.org/10.3389/fmicb.2016.01831

[64]   Y.N. Slavin, J. Asnis, U.O. Häfeli, H. Bach, Metal nanoparticles: understanding the mechanisms

behind antibacterial activity, J. Nanobiotechnology 15 (2017) 65. https://doi.org/10.1186/s12951-017-0308-z

[65]   T. Kim, G.B. Braun, Z. She, S. Hussain, E. Ruoslahti, M.J. Sailor, Composite Porous Silicon–Silver Nanoparticles as Theranostic Antibacterial Agents, ACS Appl. Mater. Interfaces 8 (2016) 30449–30457. https://doi.org/10.1021/acsami.6b09518

[66]   S. Prabhu, E.K. Poulose, Silver nanoparticles: mechanism of antimicrobial action, synthesis, medical applications, and toxicity effects, Int. Nano Lett. 2 (2012) 32. https://doi.org/10.1186/2228-5326-2-32

[67]   E.A. Jardón-Romero, E. Lara-Carrillo, M.G. González-Pedroza, V. Sánchez-Mendieta, E.N. Salmerón-Valdés, V.H. Toral-Rizo, O.F. Olea-Mejía, S. López-González, R.A. Morales-Luckie, Antimicrobial activity of biogenic silver nanoparticles from Syzygium aromaticum against the five most common microorganisms in the oral cavity, Antibiotics 11 (2022) 834.

[68]   S. Perni, V. Hakala, P. Prokopovich, Biogenic synthesis of antimicrobial silver nanoparticles capped with l-cysteine, Colloids Surfaces A Physicochem. Eng. Asp. 460 (2014) 219–224. https://doi.org/10.1016/j.colsurfa.2013.09.034

[69]   V.R. Pasupuleti, TNVKV Prasad, Rayees Ahmad Sheikh, Satheesh Krishna Balam, Ganapathi Narasimhulu, Siew Hua Gan, C. Reddy, I. Rahman, Biogenic silver nanoparticles using Rhinacanthus nasutus leaf extract: synthesis, spectral analysis, and antimicrobial studies, Int. J. Nanomedicine (2013) 3355. https://doi.org/10.2147/IJN.S49000

[70]   D. Bose, S. Chatterjee, Biogenic synthesis of silver nanoparticles using guava (Psidium guajava) leaf extract and its antibacterial activity against Pseudomonas aeruginosa, Appl. Nanosci. 6 (2016) 895–901.

[71]   C. Kaweeteerawat, P. Na Ubol, S. Sangmuang, S. Aueviriyavit, R. Maniratanachote, Mechanisms of antibiotic resistance in bacteria mediated by silver nanoparticles, J. Toxicol. Environ. Heal. Part A 80 (2017) 1276–1289

[72]   S.G. Ali, M.A. Ansari, M.A. Alzohairy, M.N. Alomary, S. AlYahya, M. Jalal, H.M. Khan, S.M.M. Asiri, W. Ahmad, A.A. Mahdi, A.M. El-Sherbeeny, M.A. El-Meligy, Biogenic Gold Nanoparticles as Potent Antibacterial and Antibiofilm Nano-Antibiotics against Pseudomonas aeruginosa, Antibiotics 9 (2020) 100. https://doi.org/10.3390/antibiotics9030100

[73]   R.M. Tripathi, B.R. Shrivastav, A. Shrivastav, Antibacterial and catalytic activity of biogenic gold nanoparticles synthesised by Trichoderma harzianum, IET Nanobiotechnology 12 (2018) 509–513. https://doi.org/10.1049/iet-nbt.2017.0105

[74]   S. Balasubramanian, S.M.J. Kala, T.L. Pushparaj, Biogenic synthesis of gold nanoparticles using Jasminum auriculatum leaf extract and their catalytic, antimicrobial and anticancer activities, J. Drug Deliv. Sci. Technol. 57 (2020) 101620. https://doi.org/10.1016/j.jddst.2020.101620

[75]   T.S.K. Sharma, K. Selvakumar, K.Y. Hwa, P. Sami, M. Kumaresan, Biogenic fabrication of gold nanoparticles using Camellia japonica L. leaf extract and its biological evaluation, J. Mater. Res. Technol. 8 (2019) 1412–1418. https://doi.org/10.1016/j.jmrt.2018.10.006

[76]   P.S. Murphin Kumar, D. MubarakAli, R.G. Saratale, G.D. Saratale, A. Pugazhendhi, K. Gopalakrishnan, N. Thajuddin, Synthesis of nano-cuboidal gold particles for effective antimicrobial property against clinical human pathogens, Microb. Pathog. 113 (2017) 68–73. https://doi.org/10.1016/j.micpath.2017.10.032

[77]   V. Sunderam, D. Thiyagarajan, A.V. Lawrence, S.S.S. Mohammed, A. Selvaraj, In-vitro antimicrobial and anticancer properties of green synthesized gold nanoparticles using Anacardium occidentale leaves extract, Saudi J. Biol. Sci. 26 (2019) 455–459. https://doi.org/10.1016/j.sjbs.2018.12.001

[78]   O. Rubilar, M. Rai, G. Tortella, M.C. Diez, A.B. Seabra, N. Durán, Biogenic nanoparticles:

copper, copper oxides, copper sulphides, complex copper nanostructures and their applications, Biotechnol. Lett. 35 (2013) 1365–1375. https://doi.org/10.1007/s10529-013-1239-x

[79]   M.F. Al-Hakkani, Biogenic copper nanoparticles and their applications: A review, SN Appl. Sci. 2 (2020) 505. https://doi.org/10.1007/s42452-020-2279-1

[80]   U. Bogdanović, V. Lazić, V. Vodnik, M. Budimir, Z. Marković, S. Dimitrijević, Copper nanoparticles with high antimicrobial activity, Mater. Lett. 128 (2014) 75–78. https://doi.org/10.1016/j.matlet.2014.04.106

[81]   P. Kaur, R. Thakur, A. Chaudhury, Biogenesis of copper nanoparticles using peel extract of *Punica granatum* and their antimicrobial activity against opportunistic pathogens, Green Chem. Lett. Rev. 9 (2016) 33–38. https://doi.org/10.1080/17518253.2016.1141238

[82]   G. Sharmila, R.S. Pradeep, K. Sandiya, S. Santhiya, C. Muthukumaran, J. Jeyanthi, N.M. Kumar, M. Thirumarimurugan, Biogenic synthesis of CuO nanoparticles using Bauhinia tomentosa leaves extract: characterization and its antibacterial application, J. Mol. Struct. 1165 (2018) 288–292.

[83]   H. Lee, J.Y. Song, B.S. Kim, Biological synthesis of copper nanoparticles using Magnolia kobus leaf extract and their antibacterial activity, J. Chem. Technol. Biotechnol. 88 (2013) 1971–1977.

[84]   R. Sivaraj, P.K.S.M. Rahman, P. Rajiv, H.A. Salam, R. Venckatesh, Biogenic copper oxide nanoparticles synthesis using Tabernaemontana divaricate leaf extract and its antibacterial activity against urinary tract pathogen, Spectrochim. Acta Part A Mol. Biomol. Spectrosc. 133 (2014) 178–181. https://doi.org/10.1016/j.saa.2014.05.048

[85]   N. Matinise, X.G. Fuku, K. Kaviyarasu, N. Mayedwa, M. Maaza, ZnO nanoparticles via Moringa oleifera green synthesis: Physical properties & mechanism of formation, Appl. Surf. Sci. 406 (2017) 339–347. https://doi.org/10.1016/j.apsusc.2017.01.219

[86]   N. Supraja, T. Prasad, T.G. Krishna, E. David, Synthesis, characterization, and evaluation of the antimicrobial efficacy of Boswellia ovalifoliolata stem bark-extract-mediated zinc oxide nanoparticles, Appl. Nanosci. 6 (2016) 581–590.

[87]   N. Bazancir, I. Meydan, Characterization of Zn nanoparticles of Platonus orientalis plant, investigation of DPPH radical extinquishing and antimicrobial activity, East. J. Med. 27 (2022).

[88]   I. Meydan, H. Burhan, T. Gür, H. Seçkin, B. Tanhaei, F. Sen, Characterization of Rheum ribes with ZnO nanoparticle and its antidiabetic, antibacterial, DNA damage prevention and lipid peroxidation prevention activity of in vitro, Environ. Res. 204 (2022) 112363. https://doi.org/10.1016/J.ENVRES.2021.112363

[89]   K. Lingaraju, H. Raja Naika, K. Manjunath, R.B. Basavaraj, H. Nagabhushana, G. Nagaraju, D. Suresh, Biogenic synthesis of zinc oxide nanoparticles using Ruta graveolens (L.) and their antibacterial and antioxidant activities, Appl. Nanosci. 6 (2016) 703–710.

[90]   S. Rajashekara, A. Shrivastava, S. Sumhitha, S. Kumari, Biomedical Applications of Biogenic Zinc Oxide Nanoparticles Manufactured from Leaf Extracts of Calotropis gigantea (L.) Dryand., Bionanoscience 10 (2020) 654–671. https://doi.org/10.1007/s12668-020-00746-w

[91]   H. Jan, M. Shah, H. Usman, M.A. Khan, M. Zia, C. Hano, B.H. Abbasi, Biogenic Synthesis and Characterization of Antimicrobial and Antiparasitic Zinc Oxide (ZnO) Nanoparticles Using Aqueous Extracts of the Himalayan Columbine (Aquilegia pubiflora), Front. Mater. 7 (2020) 516743. https://doi.org/10.3389/FMATS.2020.00249

[92]   J. Jayabalan, G. Mani, N. Krishnan, J. Pernabas, J.M. Devadoss, H.T. Jang, Green biogenic synthesis of zinc oxide nanoparticles using Pseudomonas putida culture and its In vitro antibacterial and anti-biofilm activity, Biocatal. Agric. Biotechnol. 21 (2019) 101327. https://doi.org/10.1016/j.bcab.2019.101327

[93]   A. Naseer, A. Ali, S. Ali, A. Mahmood, H.S. Kusuma, A. Nazir, M. Yaseen, M.I. Khan, A. Ghaffar, M. Abbas, M. Iqbal, Biogenic and eco-benign synthesis of platinum nanoparticles (Pt NPs) using

plants aqueous extracts and biological derivatives: environmental, biological and catalytic applications, J. Mater. Res. Technol. 9 (2020) 9093–9107. https://doi.org/10.1016/J.JMRT.2020.06.013

[94]   K.S. Siddiqi, A. Husen, Green Synthesis, Characterization and Uses of Palladium/Platinum Nanoparticles, Nanoscale Res. Lett. 11 (2016) 1–13. https://doi.org/10.1186/S11671-016-1695-Z

[95]   M. Hosny, M. Fawzy, E.M. El-Fakharany, A.M. Omer, E.M.A. El-Monaem, R.E. Khalifa, A.S. Eltaweil, Biogenic synthesis, characterization, antimicrobial, antioxidant, antidiabetic, and catalytic applications of platinum nanoparticles synthesized from Polygonum salicifolium leaves, J. Environ. Chem. Eng. (2022). https://doi.org/10.1016/j.jece.2021.106806

[96]   A. Aygun, F. Gülbagca, L.Y. Ozer, B. Ustaoglu, Y.C. Altunoglu, M.C. Baloglu, M.N. Atalar, M.H. Alma, F. Sen, Biogenic platinum nanoparticles using black cumin seed and their potential usage as antimicrobial and anticancer agent, J. Pharm. Biomed. Anal. (2019) 112961. https://doi.org/10.1016/j.jpba.2019.112961

[97]   A.S. Eltaweil, M. Fawzy, M. Hosny, E.M. Abd El-Monaem, T.M. Tamer, A.M. Omer, Green synthesis of platinum nanoparticles using Atriplex halimus leaves for potential antimicrobial, antioxidant, and catalytic applications, Arab. J. Chem. (2022). https://doi.org/10.1016/j.arabjc.2021.103517

[98]   O.T. Fanoro, S. Parani, R. Maluleke, T.C. Lebepe, R.J. Varghese, N. Mgedle, V. Mavumengwana, O.S. Oluwafemi, Biosynthesis of Smaller-Sized Platinum Nanoparticles Using the Leaf Extract of Combretum erythrophyllum and Its Antibacterial Activities, Antibiotics (2021). https://doi.org/10.3390/antibiotics10111275

[99]   M. Harshiny, C.N. Iswarya, M. Matheswaran, Biogenic synthesis of iron nanoparticles using Amaranthus dubius leaf extract as a reducing agent, Powder Technol. 286 (2015) 744–749. https://doi.org/10.1016/j.powtec.2015.09.021

[100]   K.S. Siddiqi, A. Husen, Fabrication of metal nanoparticles from fungi and metal salts: scope and application, Nanoscale Res. Lett. 11 (2016) 1–15.

[101]   A.B. Seabra, P. Haddad, N. Duran, Biogenic synthesis of nanostructured iron compounds: applications and perspectives, IET Nanobiotechnology 7 (2013) 90–99.

[102]   T. Alam, R.A.A. Khan, A. Ali, H. Sher, Z. Ullah, M. Ali, Biogenic synthesis of iron oxide nanoparticles via Skimmia laureola and their antibacterial efficacy against bacterial wilt pathogen Ralstonia solanacearum, Mater. Sci. Eng. C 98 (2019) 101–108. https://doi.org/10.1016/j.msec.2018.12.117

[103]   D.M.S.A. Salem, M.M. Ismail, M.A. Aly-Eldeen, Biogenic synthesis and antimicrobial potency of iron oxide (Fe3O4) nanoparticles using algae harvested from the Mediterranean Sea, Egypt, Egypt. J. Aquat. Res. 45 (2019) 197–204.

[104]   J. Sandhya, S. Kalaiselvam, Biogenic synthesis of magnetic iron oxide nanoparticles using inedible borassus flabellifer seed coat: characterization, antimicrobial, antioxidant activity and in vitro cytotoxicity analysis, Mater. Res. Express 7 (2020) 015045. https://doi.org/10.1088/2053-1591/ab6642

[105]   S. Vasantharaj, S. Sathiyavimal, P. Senthilkumar, F. LewisOscar, A. Pugazhendhi, Biosynthesis of iron oxide nanoparticles using leaf extract of Ruellia tuberosa: Antimicrobial properties and their applications in photocatalytic degradation, J. Photochem. Photobiol. B Biol. 192 (2019) 74–82. https://doi.org/10.1016/j.jphotobiol.2018.12.025

[106]   M. Hazarika, D. Borah, P. Bora, A.R. Silva, P. Das, Biogenic synthesis of palladium nanoparticles and their applications as catalyst and antimicrobial agent, PLoS One 12 (2017) e0184936. https://doi.org/10.1371/journal.pone.0184936

[107]   A.J. Kora, L. Rastogi, Green synthesis of palladium nanoparticles using gum ghatti (Anogeissus latifolia) and its application as an antioxidant and catalyst, Arab. J. Chem. 11 (2018) 1097–1106. https://doi.org/10.1016/j.arabjc.2015.06.024

[108]   Y. Liang, H. Demir, Y. Wu, A. Aygun, R.N. Elhouda Tiri, T. Gur, Y. Yuan, C. Xia, C. Demir, F.

Sen, Y. Vasseghian, Facile synthesis of biogenic palladium nanoparticles using biomass strategy and application as photocatalyst degradation for textile dye pollutants and their in-vitro antimicrobial activity, Chemosphere 306 (2022) 135518. https://doi.org/10.1016/J.CHEMOSPHERE.2022.135518

[109]   G. Sharmila, M. Farzana Fathima, S. Haries, S. Geetha, N. Manoj Kumar, C. Muthukumaran, Green synthesis, characterization and antibacterial efficacy of palladium nanoparticles synthesized using Filicium decipiens leaf extract, J. Mol. Struct. 1138 (2017) 35–40. https://doi.org/10.1016/J.MOLSTRUC.2017.02.097

[110]   G. Sharmila, S. Haries, M. Farzana Fathima, S. Geetha, N. Manoj Kumar, C. Muthukumaran, Enhanced catalytic and antibacterial activities of phytosynthesized palladium nanoparticles using Santalum album leaf extract, Powder Technol. 320 (2017) 22–26. https://doi.org/10.1016/J.POWTEC.2017.07.026

[111]   B.S. Sivamaruthi, V.S. Ramkumar, G. Archunan, C. Chaiyasut, N. Suganthy, Biogenic synthesis of silver palladium bimetallic nanoparticles from fruit extract of Terminalia chebula – In vitro evaluation of anticancer and antimicrobial activity, J. Drug Deliv. Sci. Technol. 51 (2019) 139–151. https://doi.org/10.1016/j.jddst.2019.02.024

[112]   S. Gnanasekar, J. Murugaraj, B. Dhivyabharathi, V. Krishnamoorthy, P.K. Jha, P. Seetharaman, R. Vilwanathan, S. Sivaperumal, Antibacterial and cytotoxicity effects of biogenic palladium nanoparticles synthesized using fruit extract of Couroupita guianensis Aubl., J. Appl. Biomed. 16 (2018) 59–65. https://doi.org/10.1016/j.jab.2017.10.001

[113]   G. Rajakumar, A.A. Rahuman, B. Priyamvada, V.G. Khanna, D.K. Kumar, P.J. Sujin, Eclipta prostrata leaf aqueous extract mediated synthesis of titanium dioxide nanoparticles, Mater. Lett. 68 (2012) 115–117.

[114]   E. Tilahun Bekele, B.A. Gonfa, F.K. Sabir, Use of different natural products to control growth of titanium oxide nanoparticles in green solvent emulsion, characterization, and their photocatalytic application, Bioinorg. Chem. Appl. 2021 (2021) 6626313.

[115]   G. Caliskan, T. Mutaf, H.C. Agba, M. Elibol, Green synthesis and characterization of titanium nanoparticles using microalga, Phaeodactylum tricornutum, Geomicrobiol. J. 39 (2022) 83–96.

[116]   S. Arya, H. Sonawane, S. Math, P. Tambade, M. Chaskar, D. Shinde, Biogenic titanium nanoparticles (TiO2NPs) from Tricoderma citrinoviride extract: synthesis, characterization and antibacterial activity against extremely drug-resistant Pseudomonas aeruginosa, Int. Nano Lett. 11 (2021) 35–42. https://doi.org/10.1007/s40089-020-00320-y

[117]   W. Ahmad, K.K. Jaiswal, S. Soni, Green synthesis of titanium dioxide (TiO 2) nanoparticles by using Mentha arvensis leaves extract and its antimicrobial properties, Inorg. Nano-Metal Chem. 50 (2020) 1032–1038. https://doi.org/10.1080/24701556.2020.1732419

[118]   S. Subhapriya, P. Gomathipriya, Green synthesis of titanium dioxide (TiO2) nanoparticles by Trigonella foenum-graecum extract and its antimicrobial properties, Microb. Pathog. 116 (2018) 215–220. https://doi.org/10.1016/j.micpath.2018.01.027

[119]   M.A. Kohanski, D.J. Dwyer, J.J. Collins, How antibiotics kill bacteria: from targets to networks, Nat. Rev. Microbiol. 8 (2010) 423–435. https://doi.org/10.1038/nrmicro2333

[120]   A. Ray, Challenges in Antimicrobial Resistance: An Update, EC Pharmacol. Toxicol. 6 (2018) 865–877.

[121]   A.J. Huh, Y.J. Kwon, "Nanoantibiotics": a new paradigm for treating infectious diseases using nanomaterials in the antibiotics resistant era, J. Control. Release 156 (2011) 128–145.

[122]   M.C. Teixeira, E. Sanchez-Lopez, M. Espina, A.C. Calpena, A.M. Silva, F.J. Veiga, M.L. Garcia, E.B. Souto, Advances in antibiotic nanotherapy, in: Emerg. Nanotechnologies Immunol., Elsevier, 2018: pp. 233–259. https://doi.org/10.1016/B978-0-323-40016-9.00009-9.

[123]   P. Singh, H. Singh, S. Ahn, V. Castro-Aceituno, Z. Jiménez, S.Y. Simu, Y.J. Kim, D.C. Yang,

Pharmacological importance, characterization and applications of gold and silver nanoparticles synthesized by Panax ginseng fresh leaves, Artif. Cells, Nanomedicine, Biotechnol. 45 (2017) 1415–1424. https://doi.org/10.1080/21691401.2016.124354

[124]   P. Seetharaman, R. Chandrasekaran, S. Gnanasekar, I. Mani, S. Sivaperumal, Biogenic gold nanoparticles synthesized using Crescentia cujete L. and evaluation of their different biological activities, Biocatal. Agric. Biotechnol. 11 (2017) 75–82. https://doi.org/10.1016/j.bcab.2017.06.004

[125]   D. Ballottin, S. Fulaz, F. Cabrini, J. Tsukamoto, N. Duran, O.L. Alves, L. Tasic, Antimicrobial textiles: Biogenic silver nanoparticles against Candida and Xanthomonas, Mater. Sci. Eng. C 75 (2017) 582–589.

[126]   D. Garibo, H.A. Borbón-Nuñez, J.N.D. de León, E. García Mendoza, I. Estrada, Y. Toledano-Magaña, H. Tiznado, M. Ovalle-Marroquin, A.G. Soto-Ramos, A. Blanco, Green synthesis of silver nanoparticles using Lysiloma acapulcensis exhibit high-antimicrobial activity, Sci. Rep. 10 (2020) 12805.

[127]   A.-M.E. Abdel-Moneim, M.T. El-Saadony, A.M. Shehata, A.M. Saad, S.A. Aldhumri, S.M. Ouda, N.M. Mesalam, Antioxidant and antimicrobial activities of Spirulina platensis extracts and biogenic selenium nanoparticles against selected pathogenic bacteria and fungi, Saudi J. Biol. Sci. 29 (2022) 1197–1209.

[128]   U. Kamran, H.N. Bhatti, M. Iqbal, S. Jamil, M. Zahid, Biogenic synthesis, characterization and investigation of photocatalytic and antimicrobial activity of manganese nanoparticles synthesized from Cinnamomum verum bark extract, J. Mol. Struct. (2019). https://doi.org/10.1016/j.molstruc.2018.11.006

[129]   A.G. Rodrigues, P.J. Romano de Oliveira Gonçalves, C.A. Ottoni, R. de Cássia Ruiz, M.A. Morgano, W.L. de Araújo, I.S. de Melo, A.O. De Souza, Functional textiles impregnated with biogenic silver nanoparticles from Bionectria ochroleuca and its antimicrobial activity, Biomed. Microdevices 21 (2019) 1–10.

[130]   A. Boroumand Moghaddam, M. Moniri, S. Azizi, R. Abdul Rahim, A. Bin Ariff, W. Zuhainis Saad, F. Namvar, M. Navaderi, R. Mohamad, Biosynthesis of ZnO nanoparticles by a new Pichia kudriavzevii yeast strain and evaluation of their antimicrobial and antioxidant activities, Molecules 22 (2017) 872.

[131]   J. Iqbal, B.A. Abbasi, R. Batool, A.T. Khalil, S. Hameed, S. Kanwal, I. Ullah, T. Mahmood, Biogenic synthesis of green and cost effective cobalt oxide nanoparticles using Geranium wallichianum leaves extract and evaluation of in vitro antioxidant, antimicrobial, cytotoxic and enzyme inhibition properties, Mater. Res. Express 6 (2019) 115407. https://doi.org/10.1088/2053-1591/ab4f04

[132]   A. Biswas, C. Vanlalveni, R. Lalfakzuala, S. Nath, S.L. Rokhum, Mikania mikrantha leaf extract mediated biogenic synthesis of magnetic iron oxide nanoparticles: Characterization and its antimicrobial activity study, Mater. Today Proc. 42 (2021) 1366–1373.

Biogenic Nanomaterials: Synthesis, Characterization, Applications, and Future Remarks   Materials Research Forum LLC
Materials Research Foundations 180 (2025)   https://doi.org/21741/9781644903759

Chapter 8

The Antioxidant Properties of
Green Synthesized Nanoparticles

T. Karahan[1*], E. Halvaci[2], F. Sen[2*]

[1]Vocational School of Health Services, Van Yuzuncu Yıl University, Turkiye.

[2]Sen Research Group, Department of Biochemistry, Kutahya Dumlupinar University, Kutahya, Turkiye.

tugbakarahan@yyu.edu.tr; fatih.sen@dpu.edu.tr

Abstract

Metallic nanoparticles (MNPs) are substances used in various biomedical fields, showing important physicochemical and antioxidant properties. The synthesis of MNPs can be done quickly, easily, cheaply, and sustainably, which has led to a rise in market demand. Researchers highly appreciate noble metal nanoparticles such as Platinum (Pt), gold (Au), and silver (Ag) for their superior conductivity, chemical stability, catalytic, and antioxidant abilities. In this sense, it is extremely important to develop nanoscale materials and nanostructures that offer new antioxidant properties. In this chapter, the antioxidant effects and biocompatibility of green synthesis based MNPs (AgNP, AuNP and PtNP) produced using different plant extracts will be reviewed and examined.

Keywords

Metallic Nanoparticles, Green Synthesis, Antioxidant Properties, Silver Nanoparticles

1. Introduction

Scientist with important work in the field of physics Dr. Feynman's speech at the conference in 1959 on the question of how the tools and materials used can be structured at the molecular level accelerated the emergence of nanotechnology understanding [1]. Along with Feynman, an important name in the emergence of nanotechnology, other scientists have pioneered this field. Studies utilizing nanoparticles have advanced significantly in recent years, indicating their significance as a subject of study. With the realization of the changes that may occur thanks to nanotechnology, which has become an interesting and valuable field of study today, it is foreseen that it will start a new period for future generations. The reason for the increase in interest on this subject is that they exhibit unusual properties and functionality in a certain size range of substances different from their volumetric structures. At the core of nanotechnology is to be able to work at the molecular level, to be structured at the atomic atom, and to organize comprehensive structures primarily through new molecular arrangements. Green chemistry, which is a new and developing arm of nanotechnology, is of great interest in global research studies. Green nanotechnology is seen as an excellent method to reduce the unfavorable efficacies of the generation and applying of nanomaterials and to reduce the risky aspects of nanotechnology [2]. Day by day, the development of nanotechnology affects the scientific world to a great extent and comes to an important place. Nanoparticles (NPs) are the building blocks of nanotechnology and are distinguished by their unique compositions, sizes, forms, and higher surface area/volume ratios [3]. The various techniques employed throughout the synthesis step are what give NPs their distinct characteristics. The ability to develop pharmacological, antibacterial, anticancer, anti-inflammatory, surfactant, and drug carrier compounds has been made possible by these qualities. The synthesis of NPs by physical and chemical processes is costly, and often results in the formation of hazardous byproducts. The green technique decreases the chemical load on the environment, is easier to synthesize, more affordable, and does away with pointless steps in the synthesis process [4,5]. Green synthesis is the term used to describe the synthesis process connected with creating biogenic NPs, which are NPs made from biological components. Prokaryotic/eukaryotic cells or extracted biomolecules that function as reducing agents are used in the green manufacturing of nanoparticles. Compared to other microscopic organisms, plant biomass/extract offers some significant benefits in the synthesis of NP from diverse biological components. Biomolecules with organic functional groups that are present in plant biomass are the means by which metallic nanoparticles are biosynthesized by plants. One key factor enabling NPs to be in the nanoscale range is the synthesis method and methods employed. Environmental elements that affect synthesis include temperature, pH, the concentration of reducing agents, the concentration of metal ions, interaction duration, and pressure [6]. The unique properties of NPs can be advantageous in a variety of practical applications, such as industrial and biological administrations [7]. NPs are of major interest in medicinal and pharmaceutical biotechnology [8]. Many works have showed the utility of metal nanoparticles in recent therapy procedures [9]. Noble metal nanoparticles (NMNPs) [Pt, Ag, Au] and their combinatorial counterparts have garnered significant attention among recognized MNPs because of their unique chemical, photo-chemical, electrical, and optical characteristics. For instance, because Ag ions are poisonous to many pathogenic germs, nanosilver (Ag) is used in therapeutic treatment and user production. Furthermore, because gold nanoparticles (AuNPs) exhibit antifungal, anti-diabetic, and antibacterial properties, several noteworthy applications of AuNPs in biological and environmental domains have been documented [10]. At the moment, a lot of interest is being drawn to platinum nanoparticles (PtNPs) because to their many biological applications, which include photo-ablation therapy, controlled drug release, antimicrobial properties, hyperthermia, cancer treatment, and bioimaging [11]. Many chemical and physical methods such as photochemical, radiolysis, electrochemical and radiation techniques have been applied for metal nanoparticle synthesis. High energy requirements, toxic chemicals and low efficiency have emerged as limitations in this area. But more recently, the emphasis has been on environmentally friendly ways to prepare metal nanoparticles for the benefit of humans and the environment [12]. Therefore, the improvement of a different green synthesis technique for metal nanoparticles is especially very important for practical implementations. The Environmental Protection Agency (EPA) has supported green chemistry during the past 20 years in an effort to prevent or reduce the use of harmful and hazardous chemicals due to the chemical industry's increasing need for sustainable practices [13]. The production of plant-mediated NPs does not be necessary the laborious processes as it seems. With the green synthesis approach, plant-based extracts are combined with a metallic salt, and the solution is usually finished in a matter of minutes to two hours. Therefore, unlike green synthesis methods mediated-microorganisms, plant-based production does not need complicated efforts to develop microbial cultures or carry the hazard of leaching, thus prevention hazardous efficacies on people welfare and the environment. For the production of MNPs, extracts from a variety of plant components, including flowers, latexes, fruits, bark, roots, leaves, and seeds, were utilized (Figure 1). Since AuNPs and AgNPs are safer than other MNPs, this strategy has gained traction in the last 10 years [14]. The *Cola nitida* plant's seed, seed coat, and leaf were used to create AuNPs, which had excellent results [15].

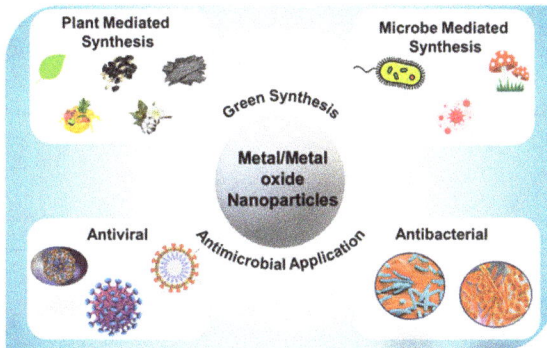

Figure 1. MNPs synthesized using various materials and their biomedical applications. (Reprinted with permission from [16], Copyright ACS).

2.   Metal Nanoparticles (MNPs)

Metallic nanoparticles are nanoscale particles with dimensions ranging from 1 to 100 nm in length, width, and thickness. Different attributes like shape, size, composition, architecture, and crystallinity define the physicochemical features of MNPs [17]. The physical, chemical, optical, electrical, thermal, catalytic, bioactive, and toxicological properties of NPs vary along with their sizes and forms. Separator agents can be used to stop particle development and create NPs of the appropriate size. Additionally, compared to NPs made up of a single type of particle, NPs made up of several types of particles may have varied catalytic, magnetic, and optical characteristics. About 40-50% of NP atoms are highly reactivated due to the fact that they are on the surface. Thanks to these properties, NP has become relatively important to other materials. Noble metals, such as gold, silver, or platinum, are typically used in a variety of chemical and physical processes to create NPs [18]. Noble metals' unparalleled qualities such as their high ionization energy, very high reduction potential, non-reactivity, high melting point, resistance to corrosion, and oxidation have sparked intense interest in designing novel MNPs for use in medicinal applications. In addition, its applications in biomedicine and pharmaceuticals, NPs of noble metals are widely utilized, especially in trading fabrications such as detergents, soaps, shampoos, toothpaste cosmetics, but these processes are not environmentally friendly [19]. Therefore, there is a necessity to improve a nonpoisonous, environmentally harmonious NP manufacture technology. The idea of security has led to improvements in safe, straightforward, affordable, repeatable, and scalable green synthesis techniques for NPs in recent years. Consequently, various biological systems like yeast, fungi, bacteria, and herb extracts are in this days widely used in green synthesis methods for the production of NPs [20,21] Because of its versatility and ease of use, plant-mediated NP green synthesis is presently regarded as the gold standard among these green biological techniques. The molecules, cells and organs of plants have been developed through bioengineering to supply novel nanomaterials with formidable sustainable benefits. Thanks to green nanotechnology, which has the potential to improve existing environmental problems, nanomaterials produced by eliminating or reducing pollution have begun to occupy an important place today. Organic dissolvent and chemical reagents are not utilized in the formation of metal nanoparticles (MNPs). MNPs have original features with their nanostructure. MNPs and MONPs; catalysts, drug delivery systems; It has many applications such as enhancing contrast agents, effective nutrition packing equipment, components indicating nano biosensor structure, disinfectants to control pathogens and pests, antibiotics, antiseptics and nano electronic constituents (Figure 2) [22,23].

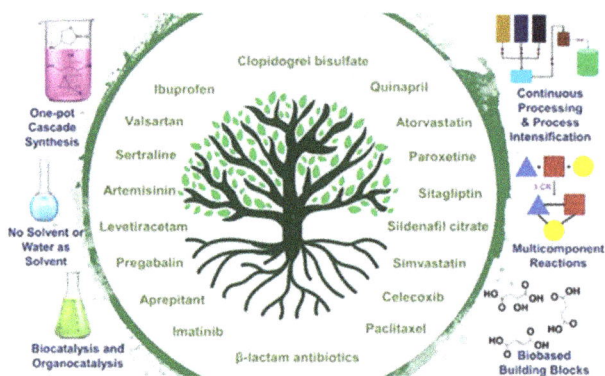

Figure 2. Effects of metallic nanomaterials mediated by green synthesis. (Reprinted with permission from [24], Copyright ACS).

3.   Synthesis of Nanoparticles

The top-down strategy, which shows the formation of materials smaller than large materials for the production of nanoparticles, and the bottom-up, which indicates the formation of materials larger than atomic or molecular-structured materials, i.e., smaller materials there are two basic approaches (Figure 3) [25]. In the top-down approach, ball grinding gives high energy to a volumetric material in its entire state by mechanical or chemical processes, mechanical-chemical processes, etching, sonication, spraying, laser ablation, lithography (printing), chemical, chemical, it is a method of obtaining nanoparticles by using thermal and natural methods, that is, by using mechanical and chemical methods such as machinery, acid and so on. With this method, generalized physical etching and mechanical grinding processes are performed in the production of nanoparticles. In this approach, high energy is used, and chemical or mechanical processes are applied. In its bottom-up approach, it is based on building organic or inorganic structures by combining atoms or molecules with chemical reactions or biological processes. In order to combine nano-structures, the self-assembly of biological systems such as DNA and living things in nature is used to control carbon nanotubes [26]. Atomic or molecular structures are combined with chemical and biological methods and increased to nano size. In this approach, the desired properties are determined and then NP is obtained, and the appropriate products are selected to create their chemical composition. The most common examples used in this method are prime gas condensation, flame synthesis, molecular buzzing, atom sheet precipitation, combustion, age chemical synthesis electro-explosion, laser ablation, chemical vapor coating, chemical vapor condensation, chemical vapor, left gel and spray are pyrolysis methods [27,28].

In addition, physical, chemical and biological synthesis stages are used in nanoparticle synthesis. Physical, chemical or microorganism (fungus, bacteria, algae, etc.) and plants (root, rhizome, lump, bark, leaf, flower, etc.), by using fruit and seed, nanoparticles can be produced by biological methods. With physical techniques, high energy is needed in nanoparticle hand. Chemical methods, on the other hand, often use high amounts of toxic chemicals. In addition, since the particle reactivity and toxicity of the synthesis product increase, it can cause different chemical molecules to be formed that can damage people healthiness and the habitat [29].

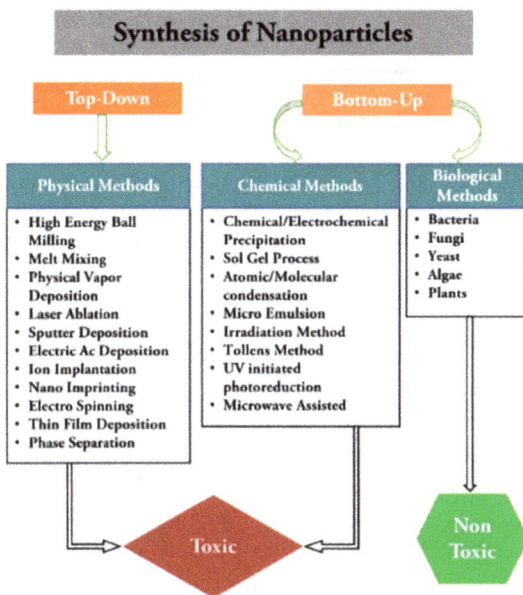

Figure 3. Overview of various methods for the synthesis of NPs. Reptinted with permission from [30], Copyright MDPI).

4.   Green Synthesis of Nanoparticles Using Plant Extracts

Different biomaterials, including bacteria, fungus, yeast, viruses, microalgae, plant biomass, and extract, are used in the green production of nanoparticles (NPs). Easements in the biological synthesis stage, where various biological microorganisms are used, further increase the success in this field. The green technique makes it possible to create novel nanoparticles (NPs) with varying sizes, contents, and physicochemical characteristics. NP is created in a single step by reducing molecules that are present in plants and microorganisms, including proteins, enzymes, phenolic compounds, amines, alkaloids, and. Microorganisms can produce NPs in natural environments free of hazardous substances and challenging circumstances. The synthesis of microorganisms is accomplished by the reduction of bioextraction and enzyme-mediated minerals, and the characteristics of NPs produced in this manner are very similar to those of NPs manufactured chemically. Biomembranes, polysaccharides, and biologically reduced proteins are involved. Microorganisms' reductases play crucial roles in several biological reduction processes, such as the production of sulfate reductase NPs. Because they may be handled relatively simply for green synthesis, bacteria have been the subject of much investigation when compared to other natural sources [31]. At ambient temperature and in moderate growing conditions, a wide variety of bacteria have the capacity to create NPs with different shapes. The places where NPs develop differ based on the types of bacteria that reducing. Intracellular formation is facilitated by the reducing enzymes or functional groups found within the cell or in the cell Wall. As a result, the NPs produced are restricted in area and occasionally challenging to extract from the cell. NPs are extracellular particles that are often produced when bacteria release their enzymes from the cell or when reducing enzymes dissolve in the solvent media.

One of the green uses is the generation of NP and algae, much as with other green techniques. Algae possess phenolic compounds and organic chemicals, which act as reducing agents in the creation of NP synthesis, making them useful in many applications [32]. Algal-mediated synthesis of NPs; it includes the incubation of algal extract, metal precursor solution and algal extract with metal precursor solution. The reaction is initiated by mixing the liquid algal extract with the metal precursor solution. The color change of the mixture is a typical indicator of the start of the reaction, after which the particles come together to form NP growth, thus forming thermodynamically stable NPs of different size and shape. Fungal NP synthesis might occur extracellularly or intracellularly. The extracellular pathway can synthesis bigger NPs than the intracellular pathway, despite the extracellular pathway being much faster. Reports suggest that this size discrepancy may be the result of particle nucleation within the fungus. Among the significant benefits are the ease of growing fungus, the creation of biomass, the use of extracellular enzymes in the manufacturing process, and the simplicity of producing huge quantities. When NPs are produced intracellularly, they develop inside the cell, but when they are produced extracellularly, they form utilizing filtrate or supernatane on the cell's outside. Because of the nature of the endeavor, obtaining NPs that are created intracellularly is more challenging. Extracellular applications benefit greatly from this using *Pleurotus citrinopileatus* (yellow oyster mushroom), they synthesized spherical AgNPs about 7 nm in size with antibacterial activity [33]. In another study, it was successfully synthesized NP with global, 17.5 nm anticancer activity with extracellular production using *Pleurotus Ostreatus* [34]. Studies on the biogenic production of metal nanoparticles (NPs) from actinomics and their therapeutic and biological uses have become accessible. In actinomics culture, proteins stabilize NP formation while enzymes released from the cell wall and membrane aid in the reduction of silver and gold ions. It has been determined that using plant biomass or extracts from biomaterials for the production of NPs is a more dependable and ecologically friendly approach. When comparing plant-mediated biosynthesis to other biomaterials, its benefits include safe use, affordable costs, an easy-to-follow process, the removal of the need for meticulous cell culture maintenance, rapid synthesis, environmental sensitivity, and more stable NPs. Additionally, NPs offer greater control over size and shape, making them more appropriate for large-scale synthesis. Because the manufactured NPs have a non-toxic structure thanks to biocompatible phytonanotechnology and the use of water as a solvent, their use in biomedical and environmental domains has increased. Various plant components, including the stem, leaf, flower, fruit, latex, seed, and seed shell, can be employed in the synthesis of NP. The fact that synthesis procedures based on plant extracts are simpler and less expensive than those based on microbiological processes, or the utilization of the entire plant is a major benefit.

Plant biomolecules that are involved in the reduction of metals include terpenoids, flavones, ketones, aldehydes, proteins, amino acids, vitamins, alkaloids, tannins, phenolics, saponins, and polysaccharides. Researchers used the fruit extract of the tomato plant, *Solanum lycopersicum*, to create AgNPs. According to *S. lycopersicum*, the fruit extract's citric acid functioned as a reducing agent and caused the physiologically reduced AgNPs of malic acid to close [35]. It was conducted a study on the synthesis of AuNPs using rose petals and found that the flower extract was the cause of the reduction of tetrachloroaurate salt to AuNPs due to its high sugar and protein content [36]. In conventional chemical and physical techniques, stabilizing agents are added to reducing agents to prevent undesired agglomeration and the danger of toxicity to the cell and environment during the reduction of metal ions and the creation of synthetic nanoparticles. The agents utilized in the green synthesis approach of producing biocompatibility nanoparticles are found naturally in the biological organisms that are employed. Studies conducted in recent years have proven that green methods are more effective in producing NPs, with lower cost, ease of characterization, and harmlessness compared to conventional methods used in recent years [37]. Plant-based synthesis is fast, economical and easier than other green techniques (bacteria and fungi). Plant-based synthesis of NPs is frequently preferred by researchers because of the negative effects on the environment caused by the toxic side effects that occur during NP production by physical and chemical methods. Green synthesis is never a laborious method. A appropriate non-toxic reducing agent, a safe chemical for the stabilization step, and an ecologically friendly solvent (such as water or ethanol) are essential for the green synthesis of nanoparticles. Under room temperature conditions, a metal salt is produced with the herb extract quite easily within minutes to a few hours, and the formation is effortlessly completed with the color transformation (Figure 4). This method is particularly prominent for silver (Ag) and gold (Au) NPs compared to other metallic NPs. Phytochemicals found in plant extracts can act as natural stabilizing and/or reducing agents in the formation of NPs. Thanks to the bioactive compounds (reducing agents) found in plants, MNPs are synthesized. Thus, the synthesis of NPs based on plant extracts has less adverse effects on humans. In addition, it offers great potential features in fields such as agriculture, food, bioengineering, cosmetics, nanomedicine, which have very important contributions to the environment. Therefore, green chemistry is considered to be a good solution to conventional chemical fabrication, which needs more complicated approaches [38].

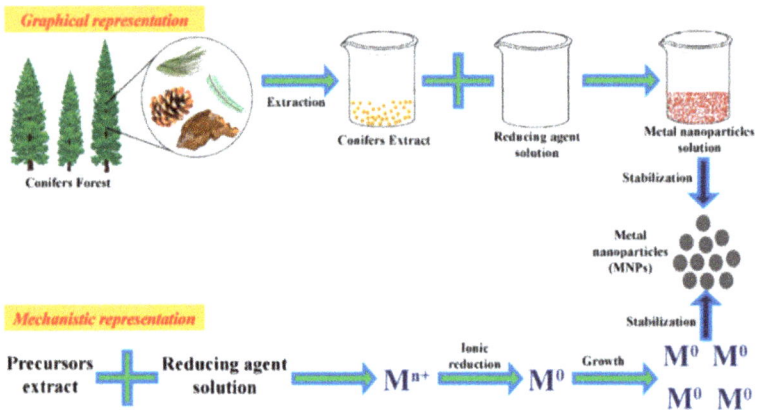

Figure 4. Biological production of MNPs mediated by green synthesis. (Reprinted with permission from [39], Copyright MDPI).

Green nanobiotechnology generally refers to the synthesis, with the aid of diverse biotechnological instruments, of nanoparticles or nanomaterials through biological pathways involving microorganisms, plants, and viruses, or their byproducts, such as proteins and lipids. In many ways, green technology-produced nanoparticles are superior to those made via physical and chemical processes. Green methods, for instance, reduce energy consumption, do away with the need for pricey chemicals, and generate goods and byproducts that are environmentally benign. The synthesis of nanoparticles using a biological system involves three main steps: choosing the solvent medium to be used, choosing a reducing agent that is both environmentally benign and friendly, and choosing a non-toxic substance to act as a capping agent to stabilize the synthesized nanoparticles. In a research examining the various antioxidant components of extracts from blackberries, blueberries, turmeric, and pomegranates, it was shown that the fruit may form Au and Ag nanoparticles (NPs) in the 20–500 nm range, which are useful for antioxidant and anticancer treatments [40]. In another work, $AgNO_3$ was converted to Ag NPs by MW irradiation using beet juice, and the generated nanomaterial revealed remarkable photocatalytic action for degradation of methyl orange (MO) dye [41].

5.   Green Synthesis of AgNPs Using Plant Extracts

In several disorders, silver (Ag) is employed as a medicinal agent. Both top-down and bottom-up methods can be used to create silver nanoparticles. A reducing agent and silver metal ion solution are needed to produce silver nanoparticles (Ag NP). Plants, vitamins, proteins, and amino acids are employed in the synthesis of Ag NP. The reducing agent and solvent contents can alter the form of the particles that are generated. The size of the nanoparticles can be significantly altered by the medium's pH. It is well known that Ag NPs are mostly utilized in the treatment of cancer. The antibacterial and antimicrobial activity of Ag NPs makes it effective for use in the medical field. In addition, Ag exhibits high catalytic activity and antioxidant properties. Much work has been published on the production of Ag NPs by the method of biological synthesis. Ag NPs are produced with different species of bacteria at sizes less than 140 nm. A different study reported that Ag nanoparticles produced using olive leaves showed antibacterial properties. A 2022 study using Japanese mint produced Ag nanoparticles and indicated that it is possible to use them in cancer treatment [42]. Again in a different study, Ag nanoparticles were produced using the leaves of the "neem" tree and reported to exhibit high antidyabetic property against gram-positive/negative bacteria [43].

AgNPs, which are produced simply, quickly and economically by using almost all plant species and their various parts, have the potential to be used in almost every field. With the help of a wide range of biomolecules, Ag ions may be reduced and stabilized to create AgNPs with ease (such as polysaccharides, vitamins, amino acids, proteins, phenolics, saponins, alkaloids and/or terpenes) [44]. AgNPs; It has a wide of variety implementations, involved their utilization as catalysts, electronic constituents, and chemical sensors in medicinal diagnosis imaging, medicinal treaphy engagements, and pharmaceutical products [45].

AgNPs interact with the cell to cause cell wall breakdowns, impacting the electron transport chain, forming radicals, interacting with ribosomes to influence protein synthesis, interacting with enzymes to affect the functioning of enzymes, and interacting with DNA to cause DNA damage [46]. A wide variety of types and numbers of phytochemicals found in plant extracts serve as reducing and covering agents, and nano-sized AgNPs [47], are produced by green synthesis method. The synthesis of acyl AgNPs takes place in a two-step process. First, Ag^+ ions are reduced to Ag^0. After stabilization and agglomeration processes, oligomeric aggregates of colloidal AgNPs are formed and the synthesis is thus completed [48]. Biological catalysts take part in the reduction process. The covering and lowering roles of phyto-compounds in *Cassia angustifolia* flower extract were validated in a research investigating the phyto-synthesis of silver nanoparticles (AgNPs) using the flower extract's aqueous form [49]. Elsewhere, metabolites and biomolecules found in the plant-mediated extract have been reported to act as reducing agents in synthesized AgNPs [50]. According to reports, a variety of plants, including *Momordica charantia, Terminalia bellirica, Eucalyptus chapmaniana, Rhynchosia suaveolens, Amaranthus retroflexus, Passiflora foetida*, and *Cochlospermum religiosum*, are employed in the synthesis of AgNPs [51,52].

6.   Antioxidant Potential Applications of Silver Nanoparticles (AgNPs)

Gold (Au) has a melting point of $1064°C$ and a boiling temperature of $2808°C$. Apart from its unique conductivity feature, it possesses other significant attributes including high scattering and absorption, low toxicity, and compatibility. The area where it is mostly used is medicine. By adding reductors to gold metal ions, stable Au nanoparticles can be obtained. Au nanoparticles are synthesized thanks to biomolecules such as flavonoids and phenols contained in plant extracts [53]. The Au nanoparticles synthesized in this way can be rubbed into different forms such as triangular, spherical, etc. Light-emitting diodes and other electrical gadgets employ gold and silver nanoparticles. Production of Au nanoparticles was provided with different studies. In one of these studies, plant extracts such as geranium and aloe vera were used [21]. Bacteria and algae are also other natural agents used in the production of Au nanoparticles. In the biological synthesis method, different materials such as Amla, tamarind and fungus were used in the production of Au nanoparticles. Nanoparticle powders produced by this method have shown antibacterial, antioxidant, anticancer properties. In addition, Au nanoparticles exhibit pain relief and are involved in determining protein-protein interaction. Furthermore, several plant extracts, including those from lemon, craton plant, and arabica coffee, are utilized to produce Au nanoparticles [54].

Various phytochemicals contained in plant extracts enable plants to show antioxidant properties. Compared to bacteria and enzymes, plants are able to absorb around 75% of the solar radiation and transform it into chemical energy. The production of this chemical energy by other methods is quite expensive. This situation, which is provided by plants, offers very convenient, sustainable and renewable resources [55]. In addition, plants have an important place in the production of metal nanoparticles, thanks to the chemicals they contain such as antioxidants and sugars [56]. The flavonoids found in leaf extracts are more effective than other phytochemicals because they have a reducing and stabilizing function of nanocomposite toxicity. In addition, flavonoids can neutralize reactive oxygen radicals (ROS) produced by metal oxides due to their antioxidant properties [57]. Flavonoids also show hepatoprotective, anticancer and antiviral properties. The creation of metal-oxide nanoparticles (MONPs) utilizing plant extracts having great amounts of flavonoids enables them to acquire more features, leading to numerous significant implementations. In one study, a wheat germ's AgNPs significantly helped to reduce the negative effects of salty stress in wheat by changing the concentration of abscisic acid, ion homeostasis, and defensive mechanisms, which include both enzymatic and non-enzymatic antioxidants [58]. In some studies, antioxidant properties of various plant-based nanoparticles such as AgNPs prepared from *C. carandas* leaf extract, AuNPs prepared from *C. inerme* leaf extract, AgNPs and NiONPs formed from stevia leaf extract have been demonstrated [59]. The phytocompounds coated on the surface of the NPs certainly have a pronounced effect on the observed antioxidant activity. A single in vitro test, like that the DPPH (2,2-diphenyl-1-picrylhydrazil) assay, is often run. On the other hand, because of the complicated character of phytocompounds, and especially since the definition of antioxidant efficiency is highly dependent on the reaction mechanism related, antioxidant efficiency should not be evaluated utilizing a only way [60]. For this reason, the effectiveness of outcomes from in vitro cell-free antioxidant experiments should be limited to explication as regards chemical reactivity, but in vivo (cellular) confirmation is necessary. In a study, the antioxidant activity, FRAP, ABTS cation radical and DPPH scavenging activity of AgNPs synthesized from *Caesalpinia pulcherrima* extract with superoxide were examined and it was observed that AgNPs were highly efficient in scavenging various reactive oxygen species [61]. Degenerative diseases and aging can be brought on by excessive oxidative stress resulting from mitochondria and other internal or external causes damaging different cell macromolecules (DNA, proteins, and membrane lipids) [62]. Antioxidants can prevent this harmful system and be utilizated to cure ageing and age-involved illnesses [63].

7.   Green Synthesis and Antioxidant Potentials of Gold Nanoparticles (Au NPs)

Data on its low toxicity, inertness and biocompatibility make gold interesting for researchers in biological and biomedical applications. Small molecules or nucleic acids, as well as covalent or non-covalent interactions, may readily functionalize and alter the surface of gold nanoparticles. Drugs or gene products can be directly attached to Au NP by ionic or covalent bonds. Studies on the use of Au NPs as drug-loaded nanocarriers, especially in cancer diagnosis and treatment, have increased [64]. Conventional cancer treatments (chemotherapy, radiation treatment) function by killing fast-growing cells, while other fast-growing cells (blood, hair cells) are harmed besides cancer cells. With the use of Au NPs in drug delivery, the desired drug can be delivered to the desired area in the desired amount. Thus, when used in cancer treatment, the risk of damage to healthy tissues is minimized. Various drugs, DNA, RNA, proteins can be transported by clinging to the particle surface and can be arranged as chemotherapeutic drug carriers for cancer treatment. The advantages of AuNPs in chemotherapeutic drug delivery can be listed as size and shape control and ease of synthesis, ease of surface modification, superiority of shape and size-dependent optical properties, biocompatibility and stability. Their limitations can be listed as non-biodegradability, lack of porous surface, biodistribution by surface modification, toxicity and alteration of pharmacokinetics, and lack of information about their toxicity [65]. In the synthesis of gold nanoparticles, flavonoids, phenols, protein, etc. contained in biogenic complexes. Different chemical components, such serve as reducing agents and aid in the production of gold nanoparticles. After 2.5 hours of reaction, gold nanoparticles were created using *Azadirachta indica* leaf extract. The surface of the nanoparticles most likely absorbed this extract, which was high in terpenoids and flavanones, and its stability was monitored for four weeks. The investigation's findings showed that nanoparticles are spherical, planar and that the majority of them are hexagonal and triangular in shape [66]. DPPH radical scavenging activity was measured with gold nanoparticles produced from *Suaeda monoica* leaf and a significant antioxidant capacity result was found by determining 43% antioxidant capacity at 1 mg/ml [67]. It was found that various concentrations of gold nanoparticles produced using the leaf extract of *Nerium oleander* showed good antioxidant activity. It was observed that antioxidant activity increased with increasing nanoparticle concentration [68].

8. Green Synthesis and Antioxidant Potentials of Palladium and Platinum Nanoparticles

Several techniques, including ion exchange, thermal degradation, and chemical degradation based on the reduction of palladium ions with the ingesting agent, are used to manufacture palladium (Pd) nanoparticles. Pd nanoparticles were produced using different extracts such as camphor tree, banana bark, tea, soy tree leaf, chlorella [69]. Additionally, coffee and tea have been used to create Pd nanoparticles. In addition, with the phenolics contained in the propeller flower extract, the palladium can also be converted to zero-valence Pd. Pd nanoparticles have a strong affinity for hydrogen and strong catalytic activity. Both palladium (Pd) and platinum (Pt) are high density and expensive metals. The delivery and manufacture of pharmaceuticals depend heavily on platinum. It is employed in the creation of chemotherapy medications, including carboplatin, cisplatin, and platinum. It is also favored in sensor applications and utilized as an antibacterial agent. Platinum and palladium are pricey, silver-white metals with a high density [70]. The fact that the biosynthesis of both nanoparticles from plants is environmentally friendly and sustainable has shed light on many researchers. They are synthesized from *Cinnamomum camphora, Gardenia jasminoides, Pinus resinosa, Anogeissus latifolia, Glycine max, Ocimun sanctum, Curcuma longa, Musa paradisica, Cinnamom zeylanicum, Pulicaria glutinosa, Doipyros kaki* and *many other plants* [71]. According to the studies, Pd nanoparticles demonstrated excellent antioxidant qualities at lower nanoparticle doses and served as a nanocatalyst for the reduction of dyes such as methyl orange, methylene blue, coomassie brilliant blue G-250, and 4-nitrophenol, demonstrating catalytic efficiency. [72]. It is known that nanoparticles produced from metals like titanium, gold, silver and platinum show very good antioxidant properties and especially PtNPs decrease the impact of reactive oxygen species. For a long time, PtNPs have just been utilized as a catalyst in medical science due to their high conductivity and reactivity features [73]. Bimetallic nanoparticles of gold and platinum can remove the effect of superoxide anion and hydrogen peroxide. Preliminary results of a mouse lung study suggested that intranasal application of PtNPs to secondhand smokers could prevent decreasing of antioxidant capacity, NFkB activation, and neutrophilic inflammation [74]. Studies on PtNPs have shown that they act as antioxidant enzymes like catalase and superoxide dismutase, which have an inhibitory effect on free radical damage.

9.   Conclusion

As a result, biological systems work like nanotechnological factories in every process. Studies on the green synthesis of nanomaterials with plants, microorganisms and other different materials are increasing day by day. In recent years, the production method made with green chemistry approach, which is environmentally friendly, has been adopted by most researchers and has become an important research area. Numerous applications, including water purification, dye degradation, textile engineering, bioengineering sciences, sensors, imaging, antioxidant and antibacterial characteristics, biotechnology, and electronics, are possible with plant-mediated nanoparticle manufacturing. In addition, it provides important contributions to technological development with its economical, non-toxic and easily accessible features. Recognizing the attitude of nanoparticles in biological fields opens up novel avenues for developing therapies, especially in the medical field, which is necessary for the improving of reliable nanotechnology. In the light of this information, more advantages than promised will continue to be obtained thanks to the advances made with nanotechnology.

References

[1]   S. Bayda, M. Adeel, T. Tuccinardi, M. Cordani, F. Rizzolio, The History of Nanoscience and Nanotechnology: From Chemical Physical Applications to Nanomedicine, Molecules. 25 (2019) 112. https://doi.org/10.3390/molecules25010112

[2]   B. Aday, Y. Yıldız, R. Ulus, S. Eris, F. Sen, M. Kaya, One Pot, Efficient and Green Synthesis of Acridinedione Derivatives Using Highly Monodisperse Platinum Nanoparticles Supported with Reduced Graphene Oxide, New J. Chem. 40 (2016) 748–754. https://doi.org/10.1039/C5NJ02098K

[3]   A.U. Khan, M. Khan, N. Malik, M.H. Cho, M.M. Khan, Recent Progress Of Algae And Blue Green Algae Assisted Synthesis Of Gold Nanoparticles For Various Applications, Bioprocess Biosyst. Eng. 42 (2019) 1–15. https://doi.org/10.1007/s00449-018-2012-2

[4]   A. Aygün, F. Gülbağça, M.S. Nas, M.H. Alma, M.H. Çalımlı, B. Ustaoglu, Y.C. Altunoglu, M.C. Baloğlu, K. Cellat, F. Şen, Biological synthesis of silver nanoparticles using Rheum ribes and evaluation of their anticarcinogenic and antimicrobial potential: A novel approach in phytonanotechnology, J. Pharm. Biomed. Anal. 179 (2020). https://doi.org/10.1016/j.jpba.2019.113012

[5]   G. Yavuz, E. Yilmaz, E. Halvaci, C. Catal, I. Turk, Nanotechnology in Medical Applications ; Recent Developments in Devices and Materials, J. Sci. Reports-C. (2023) 1–32.

[6]   A. Rana, K. Yadav, S. Jagadevan, A Comprehensive Review On Green Synthesis Of Nature Inspired Metal Nanoparticles: Mechanism, Application And Toxicity, J. Clean. Prod. 272 (2020) 122880. https://doi.org/10.1016/j.jclepro.2020.122880

[7]   K. Behdinan, R. Moradi-Dastjerdi, B. Safaei, Z. Qin, F. Chu, D. Hui, Graphene and CNT Impact on Heat Transfer Response of Nanocomposite Cylinders, Nanotechnol. Rev. 9 (2020) 41–52. https://doi.org/10.1515/ntrev-2020-0004

[8]   E. Kalantari, S.M. Naghib, A Comparative Study On Biological Properties Of Novel Nanostructured Monticellite Based Composites With Hydroxyapatite Bioceramic, Mater. Sci. Eng. C. 98 (2019) 1087–1096. https://doi.org/10.1016/j.msec.2018.12.140

[9]   S. Bharathiraja, N.Q. Bui, P. Manivasagan, M.S. Moorthy, S. Mondal, H. Seo, N.T. Phuoc, T.T.

Vy Phan, H. Kim, K.D. Lee, J. Oh, Multimodal Tumor-Homing Chitosan Oligosaccharide Coated Biocompatible Palladium Nanoparticles For Photo Based Imaging And Therapy, Sci. Rep. 8 (2018) 500. https://doi.org/10.1038/s41598-017-18966-8

[10]　A.A. Dudhane, S.R. Waghmode, L.B. Dama, V.P. Mhaindarkar, A. Sonawane, S. Katariya, Synthesis and Characterization of Gold Nanoparticles Using Plant Extract of Terminalia Arjuna With Antibacterial Activity, Int. J. Nanosci. Nanotechnol. 15 (2019) 75–82.

[11]　M. Jeyaraj, S. Gurunathan, M. Qasim, M.-H. Kang, J.-H. Kim, A Comprehensive Review on the Synthesis, Characterization, and Biomedical Application of Platinum Nanoparticles, Nanomater. 2019, Vol. 9, Page 1719. 9 (2019) 1719. https://doi.org/10.3390/NANO9121719

[12]　F. Karimi, R.N. Elhouda Tiri, A. Aygun, F. Gulbagca, S. Özdemir, S. Gonca, T. Gur, F. Sen, One Step Synthesized Biogenic Nanoparticles Using Linum Usitatissimum: Application Of Sun Light Photocatalytic, Biological Activity And Electrochemical H2O2 Sensor, Environ. Res. 218 (2023) 114757. https://doi.org/10.1016/j.envres.2022.114757

[13]　F. Behzad, S.M. Naghib, M.A.J. Kouhbanani, S.N. Tabatabaei, Y. Zare, K.Y. Rhee, An Overview of the Plant Mediated Green Synthesis of Noble Metal Nanoparticles for Antibacterial Applications, J. Ind. Eng. Chem. 94 (2021) 92–104. https://doi.org/10.1016/j.jiec.2020.12.005

[14]　Y. Kocak, G. Oto, I. Meydan, H. Seckin, T. Gur, A. Aygun, F. Sen, Assessment of therapeutic potential of silver nanoparticles synthesized by Ferula Pseudalliacea rech. F. plant, Inorg. Chem. Commun. (2022). https://doi.org/10.1016/j.inoche.2022.109417

[15]　A. Lateef, S.A. Ojo, B.I. Folarin, E.B. Gueguim-Kana, L.S. Beukes, Kolanut (Cola nitida) Mediated Synthesis of Silver Gold Alloy Nanoparticles: Antifungal, Catalytic, Larvicidal and Thrombolytic Applications, J. Clust. Sci. 27 (2016) 1561–1577. https://doi.org/10.1007/s10876-016-1019-6

[16]　A. Rana, S. Pathak, D.-K. Lim, S.-K. Kim, R. Srivastava, S.N. Sharma, R. Verma, Recent Advancements in Plant and Microbe Mediated Synthesis of Metal and Metal Oxide Nanomaterials and Their Emerging Antimicrobial Applications, ACS Appl. Nano Mater. 6 (2023) 8106–8134. https://doi.org/10.1021/acsanm.3c01351

[17]　A. Aygun, F. Gulbagca, E.E. Altuner, M. Bekmezci, T. Gur, H. Karimi-Maleh, F. Karimi, Y. Vasseghian, F. Sen, Highly Active PdPt Bimetallic Nanoparticles Synthesized By One Step Bioreduction Method: Characterizations, Anticancer, Antibacterial Activities and Evaluation of Their Catalytic Effect for Hydrogen Generation, Int. J. Hydrogen Energy. 48 (2023) 6666–6679. https://doi.org/10.1016/j.ijhydene.2021.12.144

[18]　A. Aygün, S. Özdemir, M. Gülcan, K. Cellat, F. Şen, Synthesis and Characterization of Reishi Mushroom Mediated Green Synthesis of Silver Nanoparticles for the Biochemical Applications, J. Pharm. Biomed. Anal. 178 (2020) 112970. https://doi.org/10.1016/j.jpba.2019.112970

[19]　R. Gul, H. Jan, G. Lalay, A. Andleeb, H. Usman, R. Zainab, Z. Qamar, C. Hano, B.H. Abbasi, Medicinal Plants and Biogenic Metal Oxide Nanoparticles: A Paradigm Shift to Treat Alzheimer's Disease, Coatings. 11 (2021) 717. https://doi.org/10.3390/coatings11060717.

[20]　M. Nadeem, D. Tungmunnithum, C. Hano, B.H. Abbasi, S.S. Hashmi, W. Ahmad, A. Zahir, The Current Trends in The Green Syntheses of Titanium Oxide Nanoparticles and Their Applications, Green Chem. Lett. Rev. 11 (2018) 492–502. https://doi.org/10.1080/17518253.2018.1538430

[21]　S. Jadoun, R. Arif, N.K. Jangid, R.K. Meena, Green synthesis of nanoparticles using plant extracts: a review, Environ. Chem. Lett. 19 (2020) 355–374. https://doi.org/10.1007/s10311-020-01074-x

[22]　T. Gur, I. Meydan, H. Seckin, M. Bekmezci, F. Sen, Green synthesis, characterization and bioactivity of biogenic zinc oxide nanoparticles, Environ. Res. 204 (2022) 111897. https://doi.org/10.1016/J.ENVRES.2021.111897

[23]　F. Gulbagça, A. Aygun, E.E. Altuner, M. Bekmezci, T. Gur, F. Sen, H. Karimi-Maleh, N. Zare, F.

Karimi, Y. Vasseghian, Facile bio-fabrication of Pd-Ag bimetallic nanoparticles and its performance in catalytic and pharmaceutical applications: Hydrogen production and in-vitro antibacterial, anticancer activities, and model development, Chem. Eng. Res. Des. 180 (2022) 254–264. https://doi.org/10.1016/J.CHERD.2022.02.024

[24]   S. Kar, H. Sanderson, K. Roy, E. Benfenati, J. Leszczynski, Green Chemistry in the Synthesis of Pharmaceuticals, Chem. Rev. 122 (2022) 3637–3710. https://doi.org/10.1021/acs.chemrev.1c00631

[25]   C. Dhand, N. Dwivedi, X.J. Loh, A.N. Jie Ying, N.K. Verma, R.W. Beuerman, R. Lakshminarayanan, S. Ramakrishna, Methods and Strategies for the Synthesis of Diverse Nanoparticles and Their Applications: A Comprehensive Overview, RSC Adv. 5 (2015) 105003–105037. https://doi.org/10.1039/C5RA19388E

[26]   M. BEYKAYA, A. ÇAĞLAR, An Investigation on Synthesis of Silver Nanoparticles (AgNP) and their Antimicrobial effectiveness by using Herbal Extracts, Afyon Kocatepe Univ. J. Sci. Eng. 16 (2016) 631–641. https://doi.org/10.5578/fmbd.34220

[27]   H. Göksu, B. Çelik, Y. Yıldız, F. Şen, B. Kılbaş, Superior Monodisperse CNT-Supported CoPd (CoPd@CNT) Nanoparticles for Selective Reduction of Nitro Compounds to Primary Amines with NaBH 4 in Aqueous Medium, ChemistrySelect. 1 (2016) 2366–2372. https://doi.org/10.1002/slct.201600509

[28]   N. Lolak, E. Kuyuldar, H. Burhan, H. Goksu, S. Akocak, F. Sen, Composites of Palladium–Nickel Alloy Nanoparticles and Graphene Oxide for the Knoevenagel Condensation of Aldehydes with Malononitrile, ACS Omega. 4 (2019) 6848–6853. https://doi.org/10.1021/acsomega.9b00485

[29]   M. Guilger-Casagrande, R. de Lima, Synthesis of Silver Nanoparticles Mediated by Fungi: A Review, Front. Bioeng. Biotechnol. 0 (2019) 287. https://doi.org/10.3389/FBIOE.2019.00287

[30]   S. Raj, R. Trivedi, V. Soni, Biogenic Synthesis of Silver Nanoparticles, Characterization and Their Applications, Surfaces. 5 (2021) 67–90. https://doi.org/10.3390/surfaces5010003

[31]   F. Şen, Ö. Demirbaş, M.H. Çalımlı, A. Aygün, M.H. Alma, M.S. Nas, The dye Removal from Aqueous Solution Using Polymer Composite Films, Appl. Water Sci. 8 (2018) 206. https://doi.org/10.1007/s13201-018-0856-x

[32]   I. Khan, K. Saeed, I. Khan, Nanoparticles: Properties, applications and toxicities, Arab. J. Chem. 12 (2019) 908–931. https://doi.org/10.1016/J.ARABJC.2017.05.011

[33]   A.K. Bhardwaj, A. Shukla, S. Maurya, S.C. Singh, K.N. Uttam, S. Sundaram, M.P. Singh, R. Gopal, Direct Sunlight Enabled Photo-Biochemical Synthesis of Silver Nanoparticles and Their Bactericidal Efficacy: Photon Energy As Key for Size And Distribution Control, J. Photochem. Photobiol. B Biol. 188 (2018) 42–49. https://doi.org/10.1016/j.jphotobiol.2018.08.019

[34]   A.F.M. Ismail, M.M. Ahmed, A.A.M. Salem, Biosynthesis of Silver Nanoparticles Using Mushroom Extracts: Induction of Apoptosis in HepG2 and MCF-7 Cells via Caspases Stimulation and Regulation of BAX and Bcl-2 Gene Expressions, J. Pharm. Biomed. Sci. 5 (2015) 1–9.

[35]   M. Umadevi, M.R. Bindhu, V. Sathe, A Novel Synthesis of Malic Acid Capped Silver Nanoparticles using Solanum lycopersicums Fruit Extract, J. Mater. Sci. Technol. 29 (2013) 317–322. https://doi.org/10.1016/j.jmst.2013.02.002

[36]   M. Noruzi, D. Zare, K. Khoshnevisan, D. Davoodi, Rapid Green Synthesis of Gold Nanoparticles Using Rosa Hybrida Petal Extract At Room Temperature, Spectrochim. Acta Part A Mol. Biomol. Spectrosc. 79 (2011) 1461–1465. https://doi.org/10.1016/j.saa.2011.05.001

[37]   T.M. Abdelghany, A.M.H. Al-Rajhi, M.A. Al Abboud, M.M. Alawlaqi, A. Ganash Magdah, E.A.M. Helmy, A.S. Mabrouk, Recent Advances in Green Synthesis of Silver Nanoparticles and Their Applications: About Future Directions. A Review, Bionanoscience. 8 (2018) 5–16. https://doi.org/10.1007/S12668-017-0413-3

[38]   J. Hühn, C. Carrillo-Carrion, M.G. Soliman, C. Pfeiffer, D. Valdeperez, A. Masood, I. Chakraborty, L. Zhu, M. Gallego, Z. Yue, M. Carril, N. Feliu, A. Escudero, A.M. Alkilany, B. Pelaz, P.

del Pino, W.J. Parak, Selected Standard Protocols for the Synthesis, Phase Transfer, and Characterization of Inorganic Colloidal Nanoparticles, Chem. Mater. 29 (2017) 399–461. https://doi.org/10.1021/acs.chemmater.6b04738

[39]   K. Bhardwaj, D.S. Dhanjal, A. Sharma, E. Nepovimova, A. Kalia, S. Thakur, S. Bhardwaj, C. Chopra, R. Singh, R. Verma, D. Kumar, P. Bhardwaj, K. Kuča, Conifer Derived Metallic Nanoparticles: Green Synthesis and Biological Applications, Int. J. Mol. Sci. 21 (2020) 9028. https://doi.org/10.3390/ijms21239028

[40]   M.N. Nadagouda, N. Iyanna, J. Lalley, C. Han, D.D. Dionysiou, R.S. Varma, Synthesis of silver and gold nanoparticles using antioxidants from blackberry, blueberry, pomegranate, and turmeric extracts, ACS Sustain. Chem. Eng. 2 (2014) 1717–1723. https://doi.org/10.1021/SC500237K

[41]   J. Kou, R.S. Varma, Beet Juice Induced Green Fabrication of Plasmonic AgCl/Ag Nanoparticles, ChemSusChem. 5 (2012) 2435–2441. https://doi.org/10.1002/cssc.201200477

[42]   J. Han, F. Yang, M. Wang, M. Wang, N. Xing, With Drawn: Green synthesis of Ag nanoparticles using Mentha arvensis extract: Preparation, characterization and investigation of its anti-human bladder cancer application, Inorg. Chem. Commun. (2022) 110060. https://doi.org/10.1016/j.inoche.2022.110060

[43]   Y. Iqbal, A. Raouf Malik, T. Iqbal, M. Hammad Aziz, F. Ahmed, F.A. Abolaban, S. Mansoor Ali, H. Ullah, Green Synthesis of ZnO and Ag-doped ZnO Nanoparticles Using Azadirachta Indica Leaves: Characterization and Their Potential Antibacterial, Antidiabetic, and Wound Healing Activities, Mater. Lett. 305 (2021) 130671. https://doi.org/10.1016/j.matlet.2021.130671

[44]   F. Göl, A. Aygün, A. Seyrankaya, T. Gür, C. Yenikaya, F. Şen, Green synthesis and characterization of Camellia sinensis mediated silver nanoparticles for antibacterial ceramic applications, Mater. Chem. Phys. 250 (2020) 123037. https://doi.org/10.1016/j.matchemphys.2020.123037

[45]   I. Meydan, H. Seckin, H. Burhan, T. Gür, B. Tanhaei, F. Sen, Arum italicum mediated silver nanoparticles: Synthesis and investigation of some biochemical parameters, Environ. Res. 204 (2022) 112347. https://doi.org/10.1016/J.ENVRES.2021.112347

[46]   M.P. Patil, G.-D. Kim, Eco-Friendly Approach for Nanoparticles Synthesis and Mechanism Behind Antibacterial Activity of Silver and Anticancer Activity of Gold Nanoparticles, Appl. Microbiol. Biotechnol. 101 (2017) 79–92. https://doi.org/10.1007/s00253-016-8012-8

[47]   S.H. Ghoran, M.F. Dashti, A. Maroofi, M. Shafiee, A. Zare-Hoseinabadi, F. Behzad, M. Mehrabi, A. Jangjou, K. Jamali, Biosynthesis of Zinc Ferrite Nanoparticles Using Polyphenol-Rich Extract of Citrus Aurantium Flowers, Nanomedicine Res. J. 5 (2020) 20–28. https://doi.org/10.22034/NMRJ.2020.01.003

[48]   Z.-R. Mashwani, M.A. Khan, T. Khan, A. Nadhman, Applications of Plant Terpenoids in the Synthesis of Colloidal Silver Nanoparticles, Adv. Colloid Interface Sci. 234 (2016) 132–141. https://doi.org/10.1016/j.cis.2016.04.008

[49]   D. Bharathi, V. Bhuvaneshwari, Evaluation of the Cytotoxic and Antioxidant Activity of Phyto-synthesized Silver Nanoparticles Using Cassia angustifolia Flowers, Bionanoscience. 9 (2019) 155–163. https://doi.org/10.1007/S12668-018-0577-5

[50]   Y.J. Lee, K. Song, S.-H. Cha, S. Cho, Y.S. Kim, Y. Park, Sesquiterpenoids from Tussilago farfara Flower Bud Extract for the Eco-Friendly Synthesis of Silver and Gold Nanoparticles Possessing Antibacterial and Anticancer Activities, Nanomaterials. 9 (2019) 819. https://doi.org/10.3390/nano9060819

[51]   M.S. Bethu, V.R. Netala, L. Domdi, V. Tartte, V.R. Janapala, Potential Anticancer Activity of Biogenic Silver Nanoparticles Using Leaf Extract of Rhynchosia Suaveolens : An Insight Into The Mechanism, Artif. Cells, Nanomedicine, Biotechnol. 46 (2018) 104–114. https://doi.org/10.1080/21691401.2017.1414824

[52]   B. Bahrami-Teimoori, Y. Nikparast, M. Hojatianfar, M. Akhlaghi, R. Ghorbani, H.R. Pourianfar,

Characterisation and Antifungal Activity of Silver Nanoparticles Biologically Synthesised By Amaranthus Retroflexus Leaf Extract, J. Exp. Nanosci. 12 (2017) 129–139. https://doi.org/10.1080/17458080.2017.1279355

[53]   B. Şahin, E. Demir, A. Aygün, H. Gündüz, F. Şen, Investigation of the effect of pomegranate extract and monodisperse silver nanoparticle combination on MCF-7 cell line, J. Biotechnol. 260 (2017) 79–83. https://doi.org/10.1016/j.jbiotec.2017.09.012

[54]   D.S. Bhagat, W.B. Gurnule, S.G. Pande, M.M. Kolhapure, A.D. Belsare, Biosynthesis of Gold Nanoparticles for Detection of Dichlorvos Residue from Different Samples, Mater. Today Proc. 29 (2020) 763–767. https://doi.org/10.1016/j.matpr.2020.04.589

[55]   T. Gur, Green Synthesis, Characterizations of Silver Nanoparticles Using Sumac (Rhus coriaria L.) Plant Extract and Their Antimicrobial and DNA Damage Protective Effects, Front. Chem. 10 (2022). https://doi.org/10.3389/fchem.2022.968280

[56]   M. Nasrollahzadeh, R. Akbari, Z. Issaabadi, S.M. Sajadi, Biosynthesis and Characterization of Ag/MgO Nanocomposite and Its Catalytic Performance in the Rapid Treatment of Environmental contaminants, Ceram. Int. 46 (2020) 2093–2101. https://doi.org/10.1016/j.ceramint.2019.09.191

[57]   A. Mishra, S. Kumar, A.K. Pandey, Scientific Validation of the Medicinal Efficacy of Tinospora Cordifolia, Sci. World J. 2013 (2013) 1–8. https://doi.org/10.1155/2013/292934

[58]   I. Wahid, S. Kumari, R. Ahmad, S.J. Hussain, S. Alamri, M.H. Siddiqui, M.I.R. Khan, Silver Nanoparticle Regulates Salt Tolerance in Wheat Through Changes in ABA Concentration, Ion Homeostasis, and Defense Systems, Biomolecules. 10 (2020) 1506. https://doi.org/10.3390/biom10111506

[59]   S. Srihasam, K. Thyagarajan, M. Korivi, V.R. Lebaka, S.P.R. Mallem, Phytogenic Generation of NiO Nanoparticles Using Stevia Leaf Extract and Evaluation of Their In-Vitro Antioxidant and Antimicrobial Properties, Biomolecules. 10 (2020) 89. https://doi.org/10.3390/biom10010089

[60]   D. Tungmunnithum, S. Drouet, A. Kabra, C. Hano, Enrichment in Antioxidant Flavonoids of Stamen Extracts from Nymphaea lotus L. Using Ultrasonic-Assisted Extraction and Macroporous Resin Adsorption, Antioxidants. 9 (2020) 576. https://doi.org/10.3390/antiox9070576

[61]   P. Moteriya, S. Chanda, Synthesis and Characterization of Silver Nanoparticles Using Caesalpinia Pulcherrima Flower Extract and Assessment of Their in Vitro Antimicrobial, Antioxidant, Cytotoxic, and Genotoxic Activities, Artif. Cells, Nanomedicine, Biotechnol. 45 (2017) 1556–1567. https://doi.org/10.1080/21691401.2016.1261871

[62]   C. Hano, D. Tungmunnithum, Plant Polyphenols, More than Just Simple Natural Antioxidants: Oxidative Stress, Aging and Age-Related Diseases, Medicines. 7 (2020) 26. https://doi.org/10.3390/medicines7050026

[63]   I. Meydan, H. Burhan, T. Gür, H. Seçkin, B. Tanhaei, F. Sen, Characterization of Rheum ribes with ZnO nanoparticle and its antidiabetic, antibacterial, DNA damage prevention and lipid peroxidation prevention activity of in vitro, Environ. Res. 204 (2022) 112363. https://doi.org/10.1016/J.ENVRES.2021.112363

[64]   B. Şahin, A. Aygün, H. Gündüz, K. Şahin, E. Demir, S. Akocak, F. Şen, Cytotoxic effects of platinum nanoparticles obtained from pomegranate extract by the green synthesis method on the MCF-7 cell line, Colloids Surfaces B Biointerfaces. 163 (2018) 119–124. https://doi.org/10.1016/j.colsurfb.2017.12.042

[65]   F. Gulbagca, S. Ozdemir, M. Gulcan, F. Sen, Synthesis and Characterization of Rosa Canina Mediated Biogenic Silver Nanoparticles for Anti Oxidant, Antibacterial, Antifungal, and DNA Cleavage Activities, Heliyon. 5 (2019) e02980. https://doi.org/10.1016/j.heliyon.2019.e02980

[66]   S.S. Shankar, A. Rai, A. Ahmad, M. Sastry, Rapid Synthesis of Au, Ag, and Bimetallic Au Core Ag Shell Nanoparticles Using Neem (Azadirachta Indica) Leaf Broth, J. Colloid Interface Sci. 275 (2004)

496–502. https://doi.org/10.1016/j.jcis.2004.03.003

[67]   F. Arockiya Aarthi Rajathi, R. Arumugam, S. Saravanan, P. Anantharaman, Phytofabrication of Gold Nanoparticles Assisted By Leaves of Suaeda Monoica and Its Free Radical Scavenging Property, J. Photochem. Photobiol. B Biol. 135 (2014) 75–80. https://doi.org/10.1016/j.jphotobiol.2014.03.016

[68]   K. Tahir, S. Nazir, B. Li, A.U. Khan, Z.U.H. Khan, P.Y. Gong, S.U. Khan, A. Ahmad, Nerium Oleander Leaves Extract Mediated Synthesis of Gold Nanoparticles and Its Antioxidant Activity, Mater. Lett. 156 (2015) 198–201. https://doi.org/10.1016/j.matlet.2015.05.062

[69]   I. Hussain, N.B. Singh, A. Singh, H. Singh, S.C. Singh, Green synthesis of nanoparticles and its potential application, Biotechnol. Lett. 38 (2016) 545–560. https://doi.org/10.1007/S10529-015-2026-7

[70]   J. Lin, F. Gulbagca, A. Aygun, R.N. Elhouda Tiri, C. Xia, Q. Van Le, T. Gur, F. Sen, Y. Vasseghian, Phyto-mediated synthesis of nanoparticles and their applications on hydrogen generation on NaBH4, biological activities and photodegradation on azo dyes: Development of machine learning model, Food Chem. Toxicol. 163 (2022) 112972. https://doi.org/10.1016/J.FCT.2022.112972

[71]   K.S. Siddiqi, A. Husen, Green Synthesis, Characterization and Uses of Palladium/Platinum Nanoparticles, Nanoscale Res. Lett. 11 (2016) 1–13. https://doi.org/10.1186/S11671-016-1695-Z

[72]   A.J. Kora, L. Rastogi, Green synthesis of palladium nanoparticles using gum ghatti (Anogeissus latifolia) and its application as an antioxidant and catalyst, Arab. J. Chem. 11 (2018) 1097–1106. https://doi.org/10.1016/j.arabjc.2015.06.024

[73]   H. Cheng, C. Xi, X. Meng, Y. Hao, Y. Yu, F. Zhao, Polyethylene Glycol Stabilized Platinum Nanoparticles: The Efficient and Recyclable Catalysts for Selective Hydrogenation of O-Chloronitrobenzene to O-Chloroaniline, J. Colloid Interface Sci. 336 (2009) 675–678. https://doi.org/10.1016/j.jcis.2009.04.076

[74]   S. Onizawa, K. Aoshiba, M. Kajita, Y. Miyamoto, A. Nagai, Platinum Nanoparticle Antioxidants Inhibit Pulmonary Inflammation in Mice Exposed to Cigarette Smoke, Pulm. Pharmacol. Ther. 22 (2009) 340–349. https://doi.org/10.1016/j.pupt.2008.12.015

Biogenic Nanomaterials: Synthesis, Characterization, Applications, and Future Remarks   Materials Research Forum LLC
Materials Research Foundations 180 (2025)   https://doi.org/21741/9781644903759

Chapter 9

Antiviral Effect of Biogenic Nanomaterials

N. Akman[1*], D. Ikballı[2], F. Sen[2*]

[1]Faculty of Health Sciences Van Yuzuncu Yıl University, Van, Turkiye

[2]Sen Research Group, Department of Biochemistry, Kutahya Dumlupinar University, Kutahya, Turkiye.

nurakman@yyu.edu.tr; fatih.sen@dpu.edu.tr

Abstract

About 15 million people die from infectious diseases caused by viruses, bacteria, fungi and parasites. Viral diseases such as influenza, HIV, malaria, and tuberculosis are major global health issues. The constant mutation of viruses and the ineffectiveness of some antiviral drugs have led to the development of nanotechnology-based approaches. Nanoparticles can improve drug bioavailability, helping to overcome challenges like solubility, toxicity, and resistance. Additionally, nanoparticles aim to target viruses while protecting healthy cells from damage. Research shows that nanoparticles have high potential in enhancing the effectiveness of traditional antiviral drugs and preventing viral infections.

Keywords

Nanotechnology, Nanoparticles, Antiviral Activity, Drug Delivery Systems, Bioavability

1. Introduction

1.1 Definition and Sources of Nanoparticle

The Greek word "nano", meaning dwarf, is used as a measurement unit nanometer (nm), which is one billionth. Nanoparticles are particles with a minimum size in the range of 1-100 nm, which in some special cases can also be considered 200-300 nm [1,2]. First, Nobel Prize-winning physicist Richard Feynman made his 1959's "There is plenty of room at the bottom)" In his speech entitled (There is plenty of room at the bottom)", he explains the properties of materials and devices in the nm range, the importance of nanostructured materials (nanomaterials) has been acknowledged, arguing that they will provide great and important opportunities in the future [3]. Concerns about the human health and environmental effects of nanoparticles, especially designed nanomaterials, began in the 2000's and began to affect the flow of particles in the environment and in the biological system, research on exposure sources and their toxicity remains fairly recent [4]. Due to the fact that the dimensions of materials such as fullerenes and nanotubes are 1 nm and below, the concept of nanoparticles was expanded in 2011 by the European Commission on Environmental and Human Health and Safety (European Commission, EC) and released/aggregate/it is the concept that if at least one size is between 50-100 nm in more than 50% of agglomerated nanomaterials, it can be considered nanomaterial, regardless of its source [5]. Nanoparticles can be classified in different ways according to their properties. For example; the particle shape is classified as a series of nanoparticles, tube, dendrimer, rod and prism type [6]. But the simple and commonly used classification is as follows; - Organic nanoparticles: Fullers (C60, C70 and its derivatives) and carbon nanotubes (single Walled Carbon Nanotube, single (Single Walled, SWCNT) and many (Multiple Walled Carbon Nanotube, MWCNT) [7,8].

Inorganic nanoparticles: Metals (gold and silver colloids), metal oxides (iron, zinc, titanium, cerium, cobalt, copper oxides) and quantum dots (cadmium selenide). Although they must be of ultra-small size, their modification to special shapes results in different magnetic and optical properties, gaining a high degree of robustness, flexibility, conductivity and absorption activity, nanoparticles have been widely used in many industries in recent years [9]. Nanoparticle sources can be categorized into two primary categories: anthropogenic and natural sources. Natural sources: clay minerals, colloids, hydroxides and metal oxides nanoparticles in the form of humic substances such as allophane and imogolite in the inorganic and organic structure are the main soil and/or vegetation source. Anthropogenic nanoparticle sources, as in their conscious use due to their specific properties, are unknowingly contaminated during man-made processes such as carbon nanotubes and fullerenes [10].

Nanotechnology sector, which is a rapidly growing sector in which nanoparticles are used; food, textile, chemical, materials, informatics, etc, it is used in many fields such as automobile and metal industry, and in the field of medicine from the development of vaccines and targeted drug molecules to early detection for magnetic resonance, ultrasound, fluorescence, etc, molecular imaging agent in radiological fields such as nuclear and computed tomography, and allows for many dramatic developments, from proteomic and genomic technologies to biosensor creation for more sensitive analyses [11–18]. A large number of nanotechnological products ranging from dirt-resistant baby clothes, stain-resistant tablecloths, water-resistant non-soaking paints, shaving lotions, sunscreens have also taken their place among the products used daily [19] (Figure 1). The number of nanomaterial products used as color and taste enhancer, antimicrobial, preservative as food additive and food packaging material is around 100 [20].

The production methods of nanoparticles are well known in the synthesis of nanoparticles. Gold nanoparticles were involved in the production of the drug known as "Suvarna Bhasma", which was established before the 17th century and used by Hindus, which is good for rheumatoid arthritis. Gold nanoparticles produced by biological methods, also known as top-down methods, were produced by Hindus, while Michael Faraday searched for the chemical method for the first time in 1857 [21]. There are two main perspectives in the synthesis of nanoparticles. Ball down and bottom up. [22], these perspectives are as follows. Giving energy to a material with a certain volume chemically or mechanically and reducing this material to nano size with this energy is called top-down approach. For example, etching and mechanical grinding are the best examples of the top-down approach. Recently, high-energy milling or high-speed mills are also called high-energy milling or high-speed mills because they consume more energy than other milling processes. In this approach, synthesis starts at micro levels. For example, converting wheat into flour to increase water retention by dry milling is a very good example of this approach. [23]. In a study on green tea, it was observed that absorption and digestion were facilitated by the reduction of green tea debris up to 1000 nanometers and thus enzyme activity increased, in other words, the antioxidant effect increased. [24]. The bottom-up approach is based on building organic or inorganic structures with atoms or molecules Carbon nanotubes can be incorporated into nanostructures by looking at the self-assembly of both forces in nature and biological systems, such as DNA. The aim of the application of this method is to enlarge atomic and molecular structures by chemical reactions to form particles. The gas condensation method is a method used in the production of nano metals and alloys and works with the participatory approach. Chemical evaporation coating, chemical evaporation densification, spray pyrolysis and sol gel methods are other elements of this approach. [25]. The basic method for the chemical and physical synthesis of nanoparticles in solution has been used for many years. Today, the method called "green nanotechnology" is used. This method is a friendly, non-toxic, environmental and cheaper biological method. [26].

2.   Physical Properties of Nanomaterials

This section discusses the physical properties of nanomaterials and the important consequences of the small size of bulk materials. The first and most important result of the small particle size is its rather large surface area, and it is necessary to talk about the surface-to-volume ratio to see the effect of the importance of this geometric variable. In the spherical particle system, the area of a single particle with a diameter of D is a=D2 and its volume is v=D3/6. [27]. Surface/volume ratio is written in R=a/v=6/D. This ratio is inversely proportional to particle size and increases with decreasing particle size. Similarly, it is valid within the surface per mole. The field a is a very important quantity for thermodynamic evaluations [28].

2.1   Surface Effects in Nanomaterials

In nanomaterials, the surface creates a sharp interface between the particle and the atmosphere surrounding it. For granular materials, these are free surfaces. Nanomaterials have large surfaces and can be represented as spherical particles. As mentioned earlier, the surface/volume ratio is inversely proportional to the particle diameter. If it is desired to be more realistic, it is considered that the thickness of the surfaces is certain, and the surface is partially effective on the volume. Based on many physical properties, the thickness of the region where the particles affected by the surface is between 0.5nm and 1.5nm [29].

2.2   Surface Energy

The origin of surface energy is based on a model that accepts that particles are formed by breaking down large solid materials into smaller pieces. To achieve the requirement of the model, it is necessary to break the bonds between neighbouring atoms. To break a single bond, we need a mesh bonding energy, a size of energy. So much energy is needed to break a large material into small pieces. Here N is the number of bonds that will break on the surface. Once the bonds are broken, two new surfaces are formed, and each bond on these surfaces carries as much energy as u/2. Thus, the energy required to break a single particle from a large material becomes nsu/2. Ns is the number of atoms on the surface. A few broken off bonds unit price zone is used to obtain the contribution of o=Nu/2 from broken bonds to surface energy.

3.   Thermodynamics of Dimensional Nanomaterials

Taking into account the concept of size effect, the basic correlations of nanoparticles can be developed within others. Due to the large surface of the material composed of nanoparticles, the energy stored as surface energy must also be taken into account when the thermodynamics of the system is considered. In many cases, the amount of energy held on the surface is on the same order as the phase conversion energy of the bulk material. Surface energy also controls the stability of multi-phase systems and Gibb's release energy $G = U - TS + \gamma A$ written as. In this sense, G, U, S and T are the energy of freedom, energy, entropy and temperature respectively, while surface energy and A are the surface area of the molar system. For small particles, the energy associated with the surface is the same value as the formation energy. Therefore, it is normal to expect a strong attachment of particle size over phase transformations.

4.   Theoretical Melting Models of Nanomaterials

Today, the dependence of the melting temperature of nano materials on the size of the nanomaterial is tried to be understood by experimental and MD simulations as well as theoretical models. First, Pawlow developed a theoretical model in 1909 that predicts a drop in the melting temperature of nanoparticles and says that this temperature is linear with the opposite particle size [30]. The experimental study confirming the results of this theoretical model was again conducted by Pawlow [31]. In the following process, many researchers examined the change of melting temperature with the size of the nanomaterial [32] and many theoretical models have been successfully applied to understand the relationship between melting temperature and the size of nanomaterial [33]. Decreasing the melting temperature is a common condition in all free nanomaterials except for some abnormal conditions. In some cases, the melting temperature of the nanomaterial was higher than the bulk material. This phenomenon observed for nanomaterials is called superheating, and there are also theoretical models literature developed to determine the melting temperature for superheating nanomaterials. Most researchers use theoretical models to grasp experimental findings and compare MD simulation results:

- Different researchers have observed differences in the change of melting temperature even for the same material. This observation can easily be foreseen with the theoretical model.

- The model can be used to understand the size-bound melting of non-spherical prism, decahedral, pyramid-shaped nanoparticles and nanowires and thin films.

- Compatible with Model MD simulations. Therefore, it is necessary to know more about theoretical models

Nanomaterials are used in many industries such as food, informatics, textiles and metals industries, allowing for many dramatic developments in the medical field from targeted drug molecules to the creation of imaging agents and biosensors for early detection. Although today they are among the indispensable parts of life as a miracle material, nanomaterials used with increasing density bring exposure risks that threaten both occupational and human health environmental. In addition to their use in the field of nanotechnology, they also pose a risk as impurity during production [34]. Metal-based nanomaterials are also used in the paint industry and glass, in buffing, in electronic goods production, in pharmaceutical and feed adding, in personal care products and cosmetics, they are materials often used in imaging in the field of biotechnology and medicine, and each person is exposed to metal-based nanomaterials in a variety of ways, from the water they drink to the air they take to the food they eat. As it is known, the skin, lungs and stomach-gut system are in constant contact with the environment and are therefore the most usual entry points for natural or anthropogenic nanoparticles. Because of their small size, nanoparticles can be transported through these entryways into circulatory and lymphatic systems and eventually into secondary organs such as body tissues and the brain, liver, and kidney [35].

5.   Nanoparticle Exposure and Toxicity

Nanomaterials, which are used with increasing intensity day by day, bring with them the exposure risks that threaten both occupational and public health. Out of exposure to nanoparticles by inhaling, ingesting or through the skin without being noticed as a drug agent and/or aide in treatment, orally as an imaging agent or biosensor in diagnosis and diagnosis, with implant application, it can be taken from the skin or by injection directly into the body through the blood [4]. Nanoparticles can be absorbed and interacted with by many organs and systems depending on particle size and surface characteristics [36]. But the main exposure pathways are the respiratory system, the gastrointestinal (GI) system, and the skin, while the secondary target organs include the kidney, heart, and brain [37].

As it is known, nanoparticles enable adsorption of various molecules with large surface areas [38]. Proteins with particularly charged polar and hydrophobic groups tend to form ionic and hydrophobic interactions and hydrogen bonds. The interaction of the nanoparticle protein, also called protein corona, with the dynamic surface blood plasma proteins, can cause changes in the kinetics of the particle, the formation of new antigenic properties, changes in enzyme activity. Nanomaterials used in increasing density also bring exposure risks that directly threaten both occupational and public health. In addition to its uses in the field of nanotechnology, impurity during production can affect the environment and human health. As a result of many studies investigating their release to the environment, the recent effects on human and environmental health in the receiving environment, the suspicions of some of them being toxic, they can accumulate in organisms and their undesirable effects on various organisms have increased significantly. Nanomaterials that can readily pass through biological layers can cause poisoned effects at the level of organs, tissues, cells and proteins, and in general, their toxic properties can include particle size, agglomeration/aggregation size distribution, shape, degree, quantity, and so on, its kinetics varies based on physical-chemical criterion for example [39]. Although the mechanisms of toxic effects of nanoparticles on cells are not yet conclusive, the findings suggest that reactive oxygen species are generally involved in the formation of reactive oxygen species, damage to mitochondrial functions, increased cell and plasma membrane permeability, it can cause cell death through apoptosis or necrosis and interactions at the level of genes and proteins.

Although in vivo studies have reported that the transition of nanomaterials to GI epithelial cells may be limited due to particularly tight intercellular connections (tight junction), it is necessary that the particles entering the digestive tract access the bloodstream through the gut and although it is possible to be absorbed from the gut depending on size, the presence of DNA damage in bone marrow cells following high-dose gavage has been detected in experimental animals [40]. The third major exposure pathway for nanoparticles is the skin pathway and is crucial for occupational exposure. Since 1986, many nanoparticles such as titanium dioxide (TiO_2) and zinc oxide (ZnO) have been used for different purposes in the production of personal care products [41]. Although nanoparticles are difficult to penetrate through the skin, the skin allows for the passage of particularly small lipophilic molecules [42]. Particles can pass more easily, especially from hair roots, wounds, and lesion areas [43].

5.1   Methods used in Nanoparticle Toxicity

Low cost, simple, with alternative procedures in accordance with the 3R principle (Placement, Reduction and Purification; 'Replacement, Reduction and Refinement') in terms of ethical, financial, political and legal regulations, in vitro tests used in 16 to achieve rapid results are the first and important step in assessing the toxic effect potentials of nanoparticles and it is possible to evaluate parameters such as cytotoxicity, genotoxicity and oxidative stress [44].

In vitro tests are also used to determine the correct dose for in vivo studies. As it is known, in vivo studies are studies that require compliance with relevant laws due to time-consuming, expensive, ethical problems. However, although in vitro results can be extrapolated to humans, evaluation of complex systems such as the immune system is not possible under in vitro conditions [45]. For this reason, it is more appropriate to support in vitro results obtained in the evaluation of nanoparticle toxicity with in vivo conditions [46].

6.   Methods Used in the Determination of Characterization of Nanoparticles

When evaluating interactions in living systems, nanoparticle characterization is very important. Important parameters in characterization detection; size, surface area, surface chemistry, size distribution, shape, reactivity, hydrophobicity, composition, crystal structure, porosity, surface charge, agglomeration/aggregate status and purity [47]. The most common methods used to detect characterization parameters are mentioned below;

Brunauer- Emmett -Teller (BET) method: In this method, where nitrogen gas is usually used, it is essential to determine the surface areas of the particles depending on the physical adsorbing capacity of the solid particles by the gas molecules sent to the environment under pressure. As a result of fluid nitrogen mono-layer absorption on the surfaces of the debris and then evaporation of this layer, the amount of free nitrogen is measured. BET represents the freely accessible surface area of surface gases

The primary particle calibre is measured the unique surface zone and density of the particles [48]. Dynamic lighter scattering (Dynamic Light Scattering, DLS): It is depend on the measurement of the intensity and change of light emitted from small particles that make Brown movements in the dilute solution [49]. Transmission (transmissive) electron microscope (Transmission Electron Microscopy, TEM): TEM, which offers the possibilityto examine the material to be examined by photonray , which is created in the hoover context to create a high-resolution picture and is thinned with electromagnetical lenses in the same environment, it is a method of camerization that can use simultaneous imaging techniques in determining the microstructure and crystal structures of the material. To create a two-dimensional image on the phosphofluorescence screen, it passes electrons through, for example, the very thin section. The brightness of a special area of the image is proportional to the number of electrons passing through it, for example. The pictures created in the microscope are created by counting the radiations or reflected back photon that arise from the interaction of the photon bar with the material [50]. Scanning photon microscope (Scanning Electron Microscopy, SEM): The photons obtained using the electron gun are directed to the sample via electromagnetic lenses with an acceleration voltage of 100-200 kV. Unlike TEM, the image is obtained on the basis of the collection of photons scattered from the sample. Samples in SEM, the simplistic and common used technique for measuring grain range and range distribution and detecting morphology, are measured in their dry state, the particles prepared as suspension are required to be vacuum-dried so that the particles change their properties [51]. X-ray diffraction method (X-Ray Diffraction Spectroscopy, XRD): Based on the specific atomic sequences of each crystal phase, it breaks the X-rays in a characteristic order. Diffraction occurs when the sent beam encounters a solid sample of atoms arranged regularly. This is a characterization technique used in the identification of solid materials and the determination of crystal structures. In the evaluation, the refractive and dispersion data of X-rays sent to the solid sample are collected [52]. Atomic force microscopy (Atomic Force Microscopy, AFM): Dimension, morphology, with three-dimensional imaging of the surface, for the help of a needle tip tapering down to atomic dimensions, it offers information about its many physical properties, such as surface texture and roughness. Imaging is performed as a result of the examination of the interaction of the needle tip with the surface. Along with imaging the sample magnetic differences, phase, electrical conductivity and surfaces can also be detected. Unlike other microscopic techniques, multiple scans with AFM can provide stronger statistical evaluation [53].

7.   Antiviral Effect of Biogenic Nanomaterials

World population, colds, influenza, chickenpox, human immunodeficiency virus, viral encephalitis, herpes keratitis, etc, it faces infections and life-threatening infections caused by viral infections caused by different viruses, such as infectious mononucleosis. In order to prevent viral diseases, great efforts were made to develop both antiviral drugs and to develop new vaccines. However, although great advances have been made in treatment from time to time, viral diseases have not been completely prevented [54]. One of the best ways to prevent diseases caused by viral infections is to reduce public health problems caused by viral diseases. In the past years, significant achievements have been achieved in preventing and rooting for viral diseases such as chickenpox, hepatitis A, polio [55]. However, the development of resistance against different pathogens involved in the viruses is shown as the main reason for the transformation of diseases that seriously affect the medical, pharmacological and biotechnological sectors and increase the mortality rates. Although efforts to develop vaccines and new drugs against viruses are in full swing, scientists are looking for an agent that will not only act as an inexpensive and broad-spectrum antiviral agent but also as a protective agent against attacks by pathogens [56]. New stages of drug development are underway along with traditional ways to prevent viral-derived diseases from turning into pandemics and reduce mortality and disease rates. Today, the development of vaccines with nanoparticles against viruses that pose a global problem such as dengue, hepatitis B, HIV, influenza virus is ongoing [57]. Now nanotechnological advances provide a new platform for the development of potential and effective agents by replacing the physicochemical properties of nanomaterials with surface areas, high surface area/volume ratio and increased reactivity [58].

Physics via wide choice technical aspects, including the characterization, structures, production, and manipulation of particles at the nanoscale, the emergence of nanotechnology, an interdisciplinary field between biology and chemistry has begun to offer a wide range of applications to various fields of technology and science [3]. Nanoparticles designed in recent years have been used in many different fields such as cosmetics, textiles, pharmaceuticals, agriculture and electronics. Significant advances have been made in the application and preparation of nanoparticles specifically designed from antiviral and antibacterial petition in medical field [59]. According to their content in the structure of nanoparticles, they are divided into two classes: organic and inorganic nanoparticles. Both organic and inorganic nanoparticles

Biogenic Nanomaterials: Synthesis, Characterization, Applications, and Future Remarks   Materials Research Forum LLC
Materials Research Foundations 180 (2025)   https://doi.org/21741/9781644903759

suppress viral reproduction by inhibiting viral replication stages in the host cell by blocking target cell receptors in drug delivery systems, including antiviral drugs (Figure 1). Nanoparticles are used not only in the development of new antiviral drugs, but also in increasing the effectiveness of existing antiviral drugs. For example, while oseltamivir is actively used in influenza A virus, its use is limited because the virus develops resistance to the drug. At this stage, oseltamivir with selenium nanoparticles was combined to prevent the development of resistance to the drug. Combined oseltamivir selenium compound has been reported to inhibit influenza A virus and have less toxic effects [60]. The type and antiviral action of some nanoparticles with antiviral properties are presented in table 1. In addition, organic and inorganic nanoparticles have begun to be extensively researched in developing new vaccines owing to their ability to cause immune reaction and form continuous antigens [61–65].

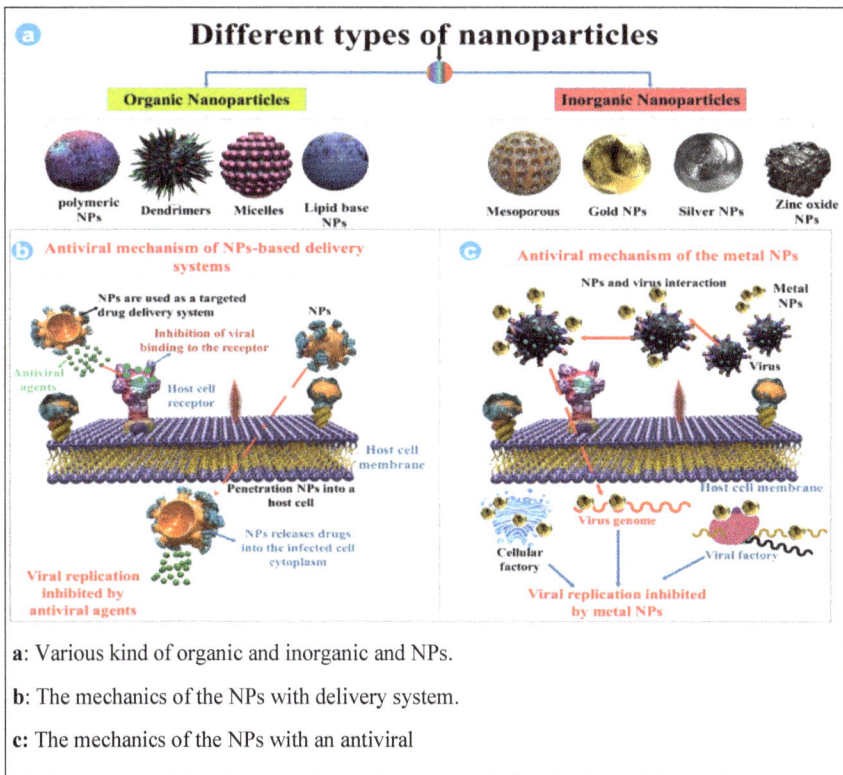

a: Various kind of organic and inorganic and NPs.

b: The mechanics of the NPs with delivery system.

c: The mechanics of the NPs with an antiviral

Figure 1: Several NPs role in treating the viral infection with antiviral aspect and delivery factors (Reprinted with permission from [66], Copyright Springer Nature).

In addition to the use of traditional antiviral and antibacterial agents to fight bacteria and viruses, nanoparticles such as zinc oxide, copper oxides, copper, iron oxide, silver and titanium oxide have been used as powerful antimicrobic proxy [67]. Several studies have shown that nanoparticles based on metal ions can cause the Coxsackievirus B4 virus, Herpes Simplex virus (HSV-1), Newcastle virus (NVD), hepatitis B and A virus, influenza virus, and, it has been proven effective against hand-to-foot oral disease virus and chikungunya virus [68]. The mechanisms of antiviral activity of metal nanoparticles have been demonstrated in figure 2 [69]. In 2019, 414 million people infected with the virus were recovered from coronavirus in the COVID-19 pandemic, which affected the entire world, while unfortunately 5.8 million people died [70]. The use of nanoparticles and nanocarriers in various fields has raised approaches that may be involved in the treatment of COVID-19 [71]. The very small size of nanoparticles means that it can effectively enter living organisms and use it to diagnose viruses can also be effective in viral-caused diseases [72].

Table 1. Type, size and antiviral activity of nanoparticles with antiviral properties

Nanoparticle Type	Inhibited virus	Result	References
Oseltamivir + Selenyum NP	İnfluenza A virüsü	İnfluenza A virüsü inhibisyonu	[73]
Polysaccharides coated with silver nanoparticles	Tacaribe ve hepatitis B virus	Invented viruses in the invitro environment	[74]
Polysaccharides coated with silver nanoparticles	monkeypox virus	Inhibition in invitro environment	[75]
Polyvinylpyrrolidone coated silver nanoparticles	HIV-1	Inhibition of the entry of the virus into the host cell	[76]
Curcumin coated AgNP	Respiratory Sinsitial Virus, COVID-19	Inhibition of Respiratory Sinsitial Virus	[75]

Human immunodeficiency virus (HIV) is a non-treatable contagious disease that currently damage 38 million citizens worldwide (WHO, 2019Studies on HIV in recent years have helped to transform HIV from a very dangerous fatal disease into a chronic disease, but these studies do not change the fact that HIV is a global health problemSince 2019, 690,000 people have died from HIV and 1.7 million people are thought to be infected with HIV [77]. South Africa has the largest and most advanced treatment facilities in the world, yet new HIV infections continue to occur [78]. To reduce the HIV virus, producing new generation treatments for patients with HIV is as effective and safe as producing vaccines [79].

Metal nanoparticles, a new antiviral drug for the treatment of HIV and other diseases, have now emerged Resistance is a concept that aims to eliminate HIV type and drug-induced side effects [80]. Nanoparticles (silver, gold, gallium, iron) synthesised by green and chemical means are shown to be an accurate and real treatment method for HIV by binding to CD4 HIV virion and gp120 protein and dampening infectivity and fusion [81–84]. (Aderibigbe, 2017). In the synthesis using plants, pharmacologically effective plant metabolites help the synthesis of nanoparticles as stabilising, reducing or capping agents [85]. Since this method is more monodispersity and stable than chemical methods, nanoparticles can be synthesised by this method. This method is of great importance as it reduces the production costs and toxicity of nanoparticles [86].

It is also known that many compounds synthesised by this method and the plants obtained from them show an antiviral effect against HIV [87]. These include *Centella asiatica (L.)* Urb (Apiaceae), a medicinal plant native to Africa, is recommended for the treatment of diseases such as HIV, syphilis, circulatory diseases, lupus, diseases, gonorrhoea, wounds and respiratory tract [88]. *Apiaceae species* (including Asiatica) are recommended to be taken in sufficient doses in the trade of herbal supplements, skin and care products and medicines [89]. In this study, the production of silver nanoparticles (AgNPs) using root extracts, stolon and aqueous leaves of *C. asiatica* is reported.

At the same time, the anti-HIV activity of artificial AgNPs can be evaluated by adding MT-4 infected cells.

Nanoparticles of zirconium, gold, silver sulphide, grapheme, silver, titanium oxide and polymeric compounds are more effective in dispersion systems relative to common built vaccines [90]. In addition, Nanoparticles have an important place in antiviral therapy by rising the transfer of hydrophobic drugs and the effectivity of drug use [91]. Nanoparticle-based drugs suppress viral replication by preventing the virus from binding to and entering the cell, neutralizing viruses directly. Different metal nanoparticles such as polylactic acid and so on are widely used in the treatment of COVID-19 [66].

Different metal nanoparticles use different ways of action mechanism against microbial activity. Although the mechanism of action of nanoparticles is not fully understood in numerous studies, it has generally been explained that metal nanoparticles damage the cell membrane by emitting metal ions [92].

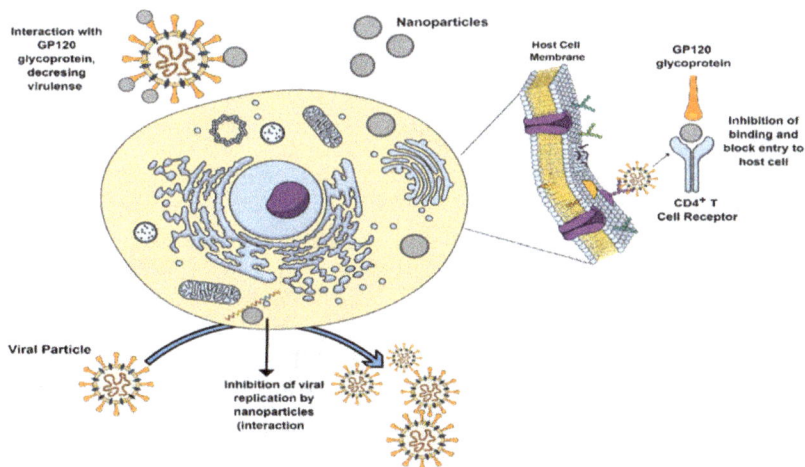

Figure 2. *Primary outlined activity mechanisms by which metal-based NPs can action versus viruses (Reprinted with permission from [69], Copyright Springer Nature).*

Although silver-containing nanoparticles (AgNPs) have been reported to be effective against viruses and bacteria, the mechanism of antiviral action has not yet been fully clarifying [93]. Although the AgNP mechanism of action has not been fully elucidated, it has been stated that it has demonstrated strong antiviral activity against the viruses hepadnaviridae, retroviridae, paramyxoviridae, herpesviridae, orthomyxoviridae, poxviridae, and arenaviridae). Several studies have reported that AgNP antiviral activitity are caused by zero-valued silver nanoparticles (nAg0). nAg0 has been reported to act by releasing Ag ions in proteins that interact with thiol groups and inhibit DNA replicas [93]. Antiviral activity of AgNPs versus monkey pox virus (MPV) and human immunodeficiency virus (HIV-1) has been confirmed in vitro studies [94]. Although the exact antiviral activity of AgNPs is unknown, scientists have focused on several mechanisms. Rogers and colleagues examined the antiviral effects of polysaccharide-coated and uncoated silver nanoparticles (10-80 nm) in various concentrations (12.5-100 µg/ml) against MPV using a plaque reduction test. Ag-PS-25 (polysaccharide coated AgNPs, 25 nm), AgNP-55 (non-coated AgNPs, 55 nm), Ag-PS-10 and whole silver nitrate concentrations except 100g/ml, it showed a considerable decline in the average number of record-forming entity. Between these silver-containing agents was the Ag-PS-10 nanoparticle, whose antiviral activity against MPV was inflamed. The mechanism of action in their antiviral activity was reported to be due to their weakening in viral replicas [95]. Along with the antiviral activity of AgNPs, it has been reported that solar ions reduce inflammation by reducing the level of cytokines, in which silver nanocrystals are involved in the wound healing process [96].

Iron oxide-based nanoparticles have been proven to have antiviral activity versus the H1N1 influenza-A virus [97]. In the treatment of infection caused by the herpes simplex virus (HSV), while testing tin, zinc oxide, silver and gold nanoparticles, [98], gold and silver nanoparticles used seaweed *Sargassum wightii* for biogenic synthesis. *Sargassum wightii* combined both gold and silver nanoparticles, and both nanoparticles combined *Sargassum wightii* showed antiviral activity against the herpes simplex virus. Silver-based nanoparticles have been reported to reduce the cytopathic effect of HSV-2 and HSV-2 at a concentration of 70 microliters, while gold-based nanoparticles have been reported to reduce the cytopathic effect at the same concentration at a higher level [99].

The antiviral activity of selenium nanoparticles was assessed against type-1 dengue virus. Increasing doses of selenium nanoparticles have been shown to reduce viral growth, even inhibiting viral growth at a maximum dose of 700 ppm, and selenium nanoparticles have strong antiviral activity [100].

Although copper has been reported to be used as an antibacterial material, copper and copper compounds have been reported to inhibit viral activation [101]. Recent studies have reported that copper materials are effective in bird flu virus [102], while copper ions with a value of +2 are effective in HIV virus [103].

It has been widely used in pharmaceutical formulation and pharmaceuticals, as the second most abundant polysaccharide in nature, nontoxicity, biodegradable, biocompatibility and ability to pass tight epithelial ports. Citosan's very good penetration into tissue has been considered a promising nanomaterial in vaccine development because it transcends frequently constructed connective tissues and can form antigen-antibody relationships [104].

8. Conclusion

the availability of new antiviral agents or the enhancement of the effectiveness of existing antiviral agents due to the wide spread of viral diseases worldwide is vital in preventing epidemic outbreaks. Rapid development of resistance to conventional antiviral drugs against infections in the fight against viral diseases leads to prolonged treatment time, increased death rates and infections spread rapidly all over the world, leading to a global health problem. When the studies are examined, the fact that nanoparticles show antiviral activity in many viruses or increase the antiviral activity of traditional drugs offers a ray of hope in facilitating treatment. With the further development of nanoparticles-based antiviral agents, antiviral activity can be increased by overcoming the problems encountered in traditional antiviral drugs, and resistance and toxicity problems can be reduced. In this case, viral diseases will cease to be a global health problem and a burden to the world economy.

Reference

[1]   C. Buzea, I.I. Pacheco, K. Robbie, Nanomaterials and Nanoparticles: Sources and Toxicity, Biointerphases 2 (2007) MR17–MR71. https://doi.org/10.1116/1.2815690

[2]   B. Sekhon, Nanotechnology in agri-food production: an overview, Nanotechnol. Sci. Appl. (2014) 31. https://doi.org/10.2147/NSA.S39406

[3]   S. Bayda, M. Adeel, T. Tuccinardi, M. Cordani, F. Rizzolio, The History of Nanoscience and Nanotechnology: From Chemical–Physical Applications to Nanomedicine, Molecules 25 (2020) 112. https://doi.org/10.3390/MOLECULES25010112

[4]   L. Xuan, Z. Ju, M. Skonieczna, P. Zhou, R. Huang, Nanoparticles-induced potential toxicity on human health: Applications, toxicity mechanisms, and evaluation models, MedComm 4 (2023). https://doi.org/10.1002/mco2.327

[5]   C.B. Golin, T.L. Bougher, A. Mallow, B.A. Cola, Toward a comprehensive framework for nanomaterials: An interdisciplinary assessment of the current Environmental Health and Safety Regulation regarding the handling of carbon nanotubes, J. Chem. Heal. Saf. 20 (2013) 9–24. https://doi.org/10.1016/j.jchas.2013.02.014

[6]   S. Khan, M.K. Hossain, Classification and properties of nanoparticles, in: Nanoparticle-Based Polym. Compos., Elsevier, 2022: pp. 15–54. https://doi.org/10.1016/B978-0-12-824272-8.00009-9

[7]   E. Demir, B. Sen, F. Sen, Highly efficient Pt nanoparticles and f-MWCNT nanocomposites based counter electrodes for dye-sensitized solar cells, Nano-Structures & Nano-Objects 11 (2017) 39–45. https://doi.org/10.1016/J.NANOSO.2017.06.003

[8]   Y. Yıldız, İ. Esirden, E. Erken, E. Demir, M. Kaya, F. Şen, Microwave (Mw)-assisted Synthesis of 5-Substituted 1 H -Tetrazoles via [3+2] Cycloaddition Catalyzed by Mw-Pd/Co Nanoparticles Decorated on Multi-Walled Carbon Nanotubes, ChemistrySelect 1 (2016) 1695–1701. https://doi.org/10.1002/slct.201600265

[9]   X. Wang, X. Zhong, J. Li, Z. Liu, L. Cheng, Inorganic nanomaterials with rapid clearance for biomedical applications, Chem. Soc. Rev. 50 (2021) 8669–8742. https://doi.org/10.1039/D0CS00461H

[10]   S. Bakshi, Z.L. He, W.G. Harris, Natural Nanoparticles: Implications for Environment and Human Health, Crit. Rev. Environ. Sci. Technol. 45 (2015) 861–904. https://doi.org/10.1080/10643389.2014.921975

[11]   R. Nagraik, A. Sharma, D. Kumar, S. Mukherjee, F. Sen, A.P. Kumar, Amalgamation of biosensors and nanotechnology in disease diagnosis: Mini-review, Sensors Int. 2 (2021) 100089. https://doi.org/10.1016/J.SINTL.2021.100089

[12]   A. Şavk, K. Cellat, K. Arıkan, F. Tezcan, S.K. Gülbay, S. Kızıldağ, E.Ş. Işgın, F. Şen, Highly

monodisperse Pd-Ni nanoparticles supported on rGO as a rapid, sensitive, reusable and selective enzyme-free glucose sensor, Sci. Rep. 9 (2019) 19228. https://doi.org/10.1038/s41598-019-55746-y

[13]   A. Cherif, R. Nebbali, J.W. Sheffield, N. Doner, F. Sen, Numerical investigation of hydrogen production via autothermal reforming of steam and methane over Ni/Al2O3 and Pt/Al2O3 patterned catalytic layers, Int. J. Hydrogen Energy (2021). https://doi.org/10.1016/j.ijhydene.2021.04.032

[14]   S. Günbatar, A. Aygun, Y. Karataş, M. Gülcan, F. Şen, Carbon-nanotube-based rhodium nanoparticles as highly-active catalyst for hydrolytic dehydrogenation of dimethylamineborane at room temperature, J. Colloid Interface Sci. 530 (2018) 321–327. https://doi.org/10.1016/J.JCIS.2018.06.100

[15]   H. Goksu, Y. Yildiz, B. Çelik, M. Yazici, B. Kilbas, F. Sen, Eco-friendly hydrogenation of aromatic aldehyde compounds by tandem dehydrogenation of dimethylamine-borane in the presence of a reduced graphene oxide furnished platinum nanocatalyst, Catal. Sci. Technol. 6 (2016) 2318–2324. https://doi.org/10.1039/C5CY01462J

[16]   F. Şen, G. Gökağaç, S. Şen, High performance Pt nanoparticles prepared by new surfactants for C1 to C3 alcohol oxidation reactions, J. Nanoparticle Res. 15 (2013) 1979. https://doi.org/10.1007/s11051-013-1979-5

[17]   B. Sen, B. Demirkan, A. Şavk, S. Karahan Gülbay, F. Sen, Trimetallic PdRuNi nanocomposites decorated on graphene oxide: A superior catalyst for the hydrogen evolution reaction, Int. J. Hydrogen Energy 43 (2018) 17984–17992. https://doi.org/10.1016/J.IJHYDENE.2018.07.122

[18]   H.K. Elaibi, F.F. Mutlag, E. Halvaci, A. Aygun, F. Sen, Review: Comparison of traditional and modern diagnostic methods in breast cancer, Measurement 242 (2025) 116258. https://doi.org/10.1016/J.MEASUREMENT.2024.116258

[19]   F. Çığ, Ç.C. Toprak, Z. Erden, Nanotechnology and The Use of Nanoparticles and Its Effect on Wheat Growing, Muş Alparslan Univ. J. Agric. Nat. 4 (2024) 23–29. https://doi.org/10.59359/maujan.1344423

[20]   S.K. Ameta, A.K. Rai, D. Hiran, R. Ameta, S.C. Ameta, Use of Nanomaterials in Food Science, in: Biog. Nano-Particles Their Use Agro-Ecosystems, Springer Singapore, Singapore, 2020: pp. 457–488. https://doi.org/10.1007/978-981-15-2985-6_24

[21]   R.D. Prasad, N.R. Prasad, R.S. Prasad, N. Prasad, S.R. Prasad, M. Shrivastav, M. Gour, V.B. Kale, Z. Guo, C.B. Desai, G.M. Nazeruddin, Y.I. Shaikh, S. Banga, P. Sane, M. Saxena, J. Kamble, A Review on Nanotechnology from Prehistoric to Modern Age, ES Gen. (2024). https://doi.org/10.30919/esg1117

[22]   N. Abid, A.M. Khan, S. Shujait, K. Chaudhary, M. Ikram, M. Imran, J. Haider, M. Khan, Q. Khan, M. Maqbool, Synthesis of nanomaterials using various top-down and bottom-up approaches, influencing factors, advantages, and disadvantages: A review, Adv. Colloid Interface Sci. 300 (2022) 102597. https://doi.org/10.1016/j.cis.2021.102597

[23]   M.A. Pagani, A. Marti, G. Bottega, Wheat Milling and Flour Quality Evaluation, in: Bak. Prod. Sci. Technol., Wiley, 2014: pp. 17–53. https://doi.org/10.1002/9781118792001.ch2

[24]   S. KOO, S. NOH, Green tea as inhibitor of the intestinal absorption of lipids: potential mechanism for its lipid-lowering effect, J. Nutr. Biochem. 18 (2007) 179–183. https://doi.org/10.1016/j.jnutbio.2006.12.005

[25]   C.D.E. Lakeman, D.A. Payne, Sol-gel processing of electrical and magnetic ceramics, Mater. Chem. Phys. 38 (1994) 305–324. https://doi.org/10.1016/0254-0584(94)90207-0

[26]   S.M. Petrovic, M.-E. Barbinta-Patrascu, Organic and Biogenic Nanocarriers as Bio-Friendly Systems for Bioactive Compounds' Delivery: State of the Art and Challenges, Materials (Basel). 16 (2023) 7550. https://doi.org/10.3390/ma16247550

[27]   K.A. Altammar, A review on nanoparticles: characteristics, synthesis, applications, and challenges, Front. Microbiol. 14 (2023) 1155622. https://doi.org/10.3389/FMICB.2023.1155622

[28]   S. Mourdikoudis, R.M. Pallares, N.T.K. Thanh, Characterization techniques for nanoparticles: comparison and complementarity upon studying nanoparticle properties, Nanoscale 10 (2018) 12871–12934. https://doi.org/10.1039/C8NR02278J

[29]   D.R. Baer, M.H. Engelhard, G.E. Johnson, J. Laskin, J. Lai, K. Mueller, P. Munusamy, S. Thevuthasan, H. Wang, N. Washton, A. Elder, B.L. Baisch, A. Karakoti, S.V.N.T. Kuchibhatla, D. Moon, Surface characterization of nanomaterials and nanoparticles: Important needs and challenging opportunities, J. Vac. Sci. Technol. A Vacuum, Surfaces, Film. 31 (2013). https://doi.org/10.1116/1.4818423

[30]   A. Safaei, M.A. Shandiz, S. Sanjabi, Z.H. Barber, Modelling the size effect on the melting temperature of nanoparticles, nanowires and nanofilms, J. Phys. Condens. Matter 19 (2007) 216216. https://doi.org/10.1088/0953-8984/19/21/216216

[31]   K.K. Nanda, Size-dependent melting of nanoparticles: Hundred years of thermodynamic model, Pramana 72 (2009) 617–628. https://doi.org/10.1007/s12043-009-0055-2

[32]   W.H. Qi, M.P. Wang, Size and shape dependent melting temperature of metallic nanoparticles, Mater. Chem. Phys. 88 (2004) 280–284. https://doi.org/10.1016/j.matchemphys.2004.04.026

[33]   H.D. Berry, Q. Zhang, Theoretical investigation of size and shape dependent melting temperature of transition metal clusters, Solid State Commun. 376 (2023) 115357. https://doi.org/10.1016/j.ssc.2023.115357

[34]   S. Malik, K. Muhammad, Y. Waheed, Nanotechnology: A Revolution in Modern Industry, Molecules 28 (2023) 661. https://doi.org/10.3390/MOLECULES28020661

[35]   S. Siddique, J.C.L. Chow, Application of Nanomaterials in Biomedical Imaging and Cancer Therapy, Nanomaterials 10 (2020) 1700. https://doi.org/10.3390/nano10091700

[36]   A.B. Jindal, The effect of particle shape on cellular interaction and drug delivery applications of micro- and nanoparticles, Int. J. Pharm. 532 (2017) 450–465. https://doi.org/10.1016/j.ijpharm.2017.09.028

[37]   N. Ali, J. Katsouli, E.L. Marczylo, T.W. Gant, S. Wright, J. Bernardino de la Serna, The potential impacts of micro-and-nano plastics on various organ systems in humans, EBioMedicine 99 (2024) 104901. https://doi.org/10.1016/j.ebiom.2023.104901

[38]   M. Manyangadze, N.H.M. Chikuruwo, T.B. Narsaiah, C.S. Chakra, M. Radhakumari, G. Danha, Enhancing adsorption capacity of nano-adsorbents via surface modification: A review, South African J. Chem. Eng. 31 (2020) 25–32. https://doi.org/10.1016/j.sajce.2019.11.003

[39]   C. Moraru, M. Mincea, G. Menghiu, V. Ostafe, Understanding the Factors Influencing Chitosan-Based Nanoparticles-Protein Corona Interaction and Drug Delivery Applications, Molecules 25 (2020) 4758. https://doi.org/10.3390/molecules25204758

[40]   Y. He, M. Cheng, R. Yang, H. Li, Z. Lu, Y. Jin, J. Feng, L. Tu, Research Progress on the Mechanism of Nanoparticles Crossing the Intestinal Epithelial Cell Membrane, Pharmaceutics 15 (2023) 1816. https://doi.org/10.3390/pharmaceutics15071816

[41]   K.-B. Kim, Y.W. Kim, S.K. Lim, T.H. Roh, D.Y. Bang, S.M. Choi, D.S. Lim, Y.J. Kim, S.-H. Baek, M.-K. Kim, H.-S. Seo, M.-H. Kim, H.S. Kim, J.Y. Lee, S. Kacew, B.-M. Lee, Risk assessment of zinc oxide, a cosmetic ingredient used as a UV filter of sunscreens, J. Toxicol. Environ. Heal. Part B 20 (2017) 155–182. https://doi.org/10.1080/10937404.2017.1290516

[42]   M. Raszewska-Famielec, J. Flieger, Nanoparticles for Topical Application in the Treatment of Skin Dysfunctions—An Overview of Dermo-Cosmetic and Dermatological Products, Int. J. Mol. Sci. 23 (2022) 15980. https://doi.org/10.3390/ijms232415980

[43]   M. Schneider, F. Stracke, S. Hansen, U.F. Schaefer, Nanoparticles and their interactions with the dermal barrier, Dermatoendocrinol. 1 (2009) 197–206. https://doi.org/10.4161/derm.1.4.9501

[44]   R.C. Hubrecht, E. Carter, The 3Rs and Humane Experimental Technique: Implementing Change,

Animals 9 (2019) 754. https://doi.org/10.3390/ani9100754

[45]   J.E. Polli, In Vitro Studies are Sometimes Better than Conventional Human Pharmacokinetic In Vivo Studies in Assessing Bioequivalence of Immediate-Release Solid Oral Dosage Forms, AAPS J. 10 (2008) 289–299. https://doi.org/10.1208/s12248-008-9027-6

[46]   B. Kong, J.H. Seog, L.M. Graham, S.B. Lee, Experimental Considerations on The Cytotoxicity of Nanoparticles, Nanomedicine 6 (2011) 929–941. https://doi.org/10.2217/nnm.11.77

[47]   N. Joudeh, D. Linke, Nanoparticle classification, physicochemical properties, characterization, and applications: a comprehensive review for biologists, J. Nanobiotechnology 2022 201 20 (2022) 1–29. https://doi.org/10.1186/S12951-022-01477-8

[48]   M. Bushell, S. Beauchemin, F. Kunc, D. Gardner, J. Ovens, F. Toll, D. Kennedy, K. Nguyen, D. Vladisavljevic, P.E. Rasmussen, L.J. Johnston, Characterization of Commercial Metal Oxide Nanomaterials: Crystalline Phase, Particle Size and Specific Surface Area, Nanomaterials 10 (2020) 1812. https://doi.org/10.3390/nano10091812

[49]   P.S. Russo, K.A. Streletzky, A. Gorman, W. Huberty, X. Zhang, Characterization of polymers by dynamic light scattering, in: Mol. Charact. Polym., Elsevier, 2021: pp. 441–498. https://doi.org/10.1016/B978-0-12-819768-4.00014-2

[50]   H. Karimi-Maleh, F. Karimi, Y. Orooji, G. Mansouri, A. Razmjou, A. Aygun, F. Sen, A new nickel-based co-crystal complex electrocatalyst amplified by NiO dope Pt nanostructure hybrid; a highly sensitive approach for determination of cysteamine in the presence of serotonin, Sci. Rep. (2020). https://doi.org/10.1038/s41598-020-68663-2

[51]   M.. Ellis, H. Hebert, Structure analysis of soluble proteins using electron crystallography, Micron 32 (2001) 541–550. https://doi.org/10.1016/S0968-4328(00)00049-4

[52]   A.A. Bunaciu, E. gabriela Udriştioiu, H.Y. Aboul-Enein, X-Ray Diffraction: Instrumentation and Applications, Crit. Rev. Anal. Chem. 45 (2015) 289–299. https://doi.org/10.1080/10408347.2014.949616

[53]   C.J.R. Sheppard, MICROSCOPY | Overview, in: Encycl. Mod. Opt., Elsevier, 2005: pp. 61–69. https://doi.org/10.1016/B0-12-369395-0/00823-X

[54]   S. Galdiero, M. Rai, A. Gade, A. Falanga, N. Incoronato, L. Russo, M. Galdiero, S. Gaikwad, A. Ingle, Antiviral activity of mycosynthesized silver nanoparticles against herpes simplex virus and human parainfluenza virus type 3, Int. J. Nanomedicine (2013) 4303. https://doi.org/10.2147/IJN.S50070

[55]   A.M. Arvin, H.B. Greenberg, New viral vaccines, Virology 344 (2006) 240–249. https://doi.org/10.1016/j.virol.2005.09.057

[56]   J. Davies, D. Davies, Origins and Evolution of Antibiotic Resistance, Microbiol. Mol. Biol. Rev. 74 (2010) 417–433. https://doi.org/10.1128/MMBR.00016-10

[57]   Y. Huang, X. Guo, Y. Wu, X. Chen, L. Feng, N. Xie, G. Shen, Nanotechnology's frontier in combatting infectious and inflammatory diseases: prevention and treatment, Signal Transduct. Target. Ther. 2024 91 9 (2024) 1–50. https://doi.org/10.1038/s41392-024-01745-z

[58]   I. Khan, K. Saeed, I. Khan, Nanoparticles: Properties, applications and toxicities, Arab. J. Chem. 12 (2019) 908–931. https://doi.org/10.1016/J.ARABJC.2017.05.011

[59]   L. V. Srinivasan, S.S. Rana, A critical review of various synthesis methods of nanoparticles and their applications in biomedical, regenerative medicine, food packaging, and environment, Discov. Appl. Sci. 6 (2024) 371. https://doi.org/10.1007/s42452-024-06040-8

[60]   M. Milovanovic, A. Arsenijevic, J. Milovanovic, T. Kanjevac, N. Arsenijevic, Nanoparticles in Antiviral Therapy, in: Antimicrob. Nanoarchitectonics, Elsevier, 2017: pp. 383–410. https://doi.org/10.1016/B978-0-323-52733-0.00014-8

[61]   M. Chakravarty, A. Vora, Nanotechnology-based antiviral therapeutics, Drug Deliv. Transl. Res. 11 (2021) 748–787. https://doi.org/10.1007/s13346-020-00818-0

[62]   A. Aygun, F. Gulbagca, E.E. Altuner, M. Bekmezci, T. Gur, H. Karimi-Maleh, F. Karimi, Y. Vasseghian, F. Sen, Highly Active PdPt Bimetallic Nanoparticles Synthesized By One Step Bioreduction Method: Characterizations, Anticancer, Antibacterial Activities and Evaluation of Their Catalytic Effect for Hydrogen Generation, Int. J. Hydrogen Energy 48 (2023) 6666–6679. https://doi.org/10.1016/j.ijhydene.2021.12.144

[63]   K. Nesrin, C. Yusuf, K. Ahmet, S.B. Ali, N.A. Muhammad, S. Suna, Ş. Fatih, Biogenic silver nanoparticles synthesized from Rhododendron ponticum and their antibacterial, antibiofilm and cytotoxic activities, J. Pharm. Biomed. Anal. 179 (2020) 112993. https://doi.org/10.1016/J.JPBA.2019.112993

[64]   N. Korkmaz, Y. Ceylan, P. Taslimi, A. Karadağ, A.S. Bülbül, F. Şen, Biogenic nano silver: Synthesis, characterization, antibacterial, antibiofilms, and enzymatic activity, Adv. Powder Technol. 31 (2020) 2942–2950. https://doi.org/10.1016/J.APT.2020.05.020

[65]   F. Gulbagca, A. Aygün, M. Gülcan, S. Ozdemir, S. Gonca, F. Şen, Green Synthesis of Palladium Nanoparticles: Preparation, Characterization, and Investigation of Antioxidant, Antimicrobial, Anticancer, and DNA Cleavage Activities, Appl. Organomet. Chem. 35 (2021). https://doi.org/10.1002/aoc.6272

[66]   S. Yasamineh, H.G. Kalajahi, P. Yasamineh, Y. Yazdani, O. Gholizadeh, R. Tabatabaie, H. Afkhami, F. Davodabadi, A.K. farkhad, D. Pahlevan, A. Firouzi-Amandi, K. Nejati-Koshki, M. Dadashpour, An overview on nanoparticle-based strategies to fight viral infections with a focus on COVID-19, J. Nanobiotechnology 2022 201 20 (2022) 1–26. https://doi.org/10.1186/S12951-022-01625-0

[67]   M.H. Mujahid, T.K. Upadhyay, F. Khan, P. Pandey, M.N. Park, A.B. Sharangi, M. Saeed, V.J. Upadhye, B. Kim, Metallic and metal oxide-derived nanohybrid as a tool for biomedical applications, Biomed. Pharmacother. 155 (2022) 113791. https://doi.org/10.1016/J.BIOPHA.2022.113791

[68]   K. Maduray, R. Parboosing, Metal Nanoparticles: a Promising Treatment for Viral and Arboviral Infections, Biol. Trace Elem. Res. 199 (2021) 3159–3176. https://doi.org/10.1007/s12011-020-02414-2

[69]   G. Tortella, O. Rubilar, P. Fincheira, J.C. Pieretti, P. Duran, I.M. Lourenço, A.B. Seabra, Bactericidal and Virucidal Activities of Biogenic Metal-Based Nanoparticles: Advances and Perspectives, Antibiot. 2021, Vol. 10, Page 783 10 (2021) 783. https://doi.org/10.3390/ANTIBIOTICS10070783

[70]   J. Taskinsoy, The Great Pandemic of the 21st Century: The Stolen Lives, SSRN Electron. J. (2020). https://doi.org/10.2139/ssrn.3689993

[71]   D. Yang, Application of Nanotechnology in the COVID-19 Pandemic, Int. J. Nanomedicine Volume 16 (2021) 623–649. https://doi.org/10.2147/IJN.S296383

[72]   L. Singh, H.G. Kruger, G.E.M. Maguire, T. Govender, R. Parboosing, The role of nanotechnology in the treatment of viral infections, Ther. Adv. Infect. Dis. 4 (2017) 105–131. https://doi.org/10.1177/2049936117713593

[73]   Y. Li, Z. Lin, M. Guo, Y. Xia, M. Zhao, C. Wang, T. Xu, T. Chen, B. Zhu, Inhibitory activity of selenium nanoparticles functionalized with oseltamivir on H1N1 influenza virus, Int. J. Nanomedicine Volume 12 (2017) 5733–5743. https://doi.org/10.2147/IJN.S140939

[74]   J.L. Speshock, R.C. Murdock, L.K. Braydich-Stolle, A.M. Schrand, S.M. Hussain, Interaction of silver nanoparticles with Tacaribe virus, J. Nanobiotechnology 8 (2010) 19. https://doi.org/10.1186/1477-3155-8-19

[75]   S. Gatadi, Y.V. Madhavi, S. Nanduri, Nanoparticle drug conjugates treating microbial and viral infections: A review, J. Mol. Struct. 1228 (2021) 129750. https://doi.org/10.1016/j.molstruc.2020.129750

[76]   H.H. Lara, L. Ixtepan-Turrent, E.N. Garza-Treviño, C. Rodriguez-Padilla, PVP-coated silver nanoparticles block the transmission of cell-free and cell-associated HIV-1 in human cervical culture, J. Nanobiotechnology 8 (2010) 15. https://doi.org/10.1186/1477-3155-8-15

[77]   X. Tian, J. Chen, X. Wang, Y. Xie, X. Zhang, D. Han, H. Fu, W. Yin, N. Wu, Global, regional,

and national HIV/AIDS disease burden levels and trends in 1990–2019: A systematic analysis for the global burden of disease 2019 study, Front. Public Heal. 11 (2023). https://doi.org/10.3389/fpubh.2023.1068664

[78]   S.S.A. Karim, G.J. Churchyard, Q.A. Karim, S.D. Lawn, HIV infection and tuberculosis in South Africa: an urgent need to escalate the public health response, Lancet 374 (2009) 921–933. https://doi.org/10.1016/S0140-6736(09)60916-8

[79]   E. Parker, M.A. Judge, E. Macete, T. Nhampossa, J. Dorward, D.C. Langa, C. De Schacht, A. Couto, P. Vaz, M. Vitoria, L. Molfino, R.T. Idowu, N. Bhatt, D. Naniche, P.N. Le Souëf, HIV infection in Eastern and Southern Africa: Highest burden, largest challenges, greatest potential, South. Afr. J. HIV Med. 22 (2021). https://doi.org/10.4102/sajhivmed.v22i1.1237

[80]   T. Mamo, E.A. Moseman, N. Kolishetti, C. Salvador-Morales, J. Shi, D.R. Kuritzkes, R. Langer, U. von Andrian, O.C. Farokhzad, Emerging Nanotechnology Approaches for HIV/AIDS Treatment and Prevention, Nanomedicine 5 (2010) 269–285. https://doi.org/10.2217/nnm.10.1

[81]   N.H. Khand, A.R. Solangi, S. Ameen, A. Fatima, J.A. Buledi, A. Mallah, S.Q. Memon, F. Sen, F. Karimi, Y. Orooji, A new electrochemical method for the detection of quercetin in onion, honey and green tea using Co3O4 modified GCE, J. Food Meas. Charact. 2021 154 15 (2021) 3720–3730. https://doi.org/10.1007/S11694-021-00956-0

[82]   A. Hojjati-Najafabadi, S. Salmanpour, F. Sen, P.N. Asrami, M. Mahdavian, M.A. Khalilzadeh, A Tramadol Drug Electrochemical Sensor Amplified by Biosynthesized Au Nanoparticle Using Mentha aquatic Extract and Ionic Liquid, Top. Catal. 65 (2022) 587–594. https://doi.org/10.1007/s11244-021-01498-x

[83]   A. Hojjati-Najafabadi, A. Aygun, R.N.E. Tiri, F. Gulbagca, M.I. Lounissaa, P. Feng, F. Karimi, F. Sen, Bacillus thuringiensis Based Ruthenium/Nickel Co-Doped Zinc as a Green Nanocatalyst: Enhanced Photocatalytic Activity, Mechanism, and Efficient H2Production from Sodium Borohydride Methanolysis, Ind. Eng. Chem. Res. 62 (2023) 4655–4664. https://doi.org/10.1021/ACS.IECR.2C03833

[84]   Y. Wu, E.E. Altuner, R.N. El Houda Tiri, M. Bekmezci, F. Gulbagca, A. Aygun, C. Xia, Q. Van Le, F. Sen, H. Karimi-Maleh, Hydrogen generation from methanolysis of sodium borohydride using waste coffee oil modified zinc oxide nanoparticles and their photocatalytic activities, Int. J. Hydrogen Energy 48 (2023) 6613–6623. https://doi.org/10.1016/J.IJHYDENE.2022.04.177

[85]   A. Rónavári, N. Igaz, D.I. Adamecz, B. Szerencsés, C. Molnar, Z. Kónya, I. Pfeiffer, M. Kiricsi, Green Silver and Gold Nanoparticles: Biological Synthesis Approaches and Potentials for Biomedical Applications, Molecules 26 (2021) 844. https://doi.org/10.3390/molecules26040844

[86]   S. Kumari, S. Raturi, S. Kulshrestha, K. Chauhan, S. Dhingra, K. András, K. Thu, R. Khargotra, T. Singh, A comprehensive review on various techniques used for synthesizing nanoparticles, J. Mater. Res. Technol. 27 (2023) 1739–1763. https://doi.org/10.1016/J.JMRT.2023.09.291

[87]   M. Mukhtar, M. Arshad, M. Ahmad, R.J. Pomerantz, B. Wigdahl, Z. Parveen, Antiviral potentials of medicinal plants, Virus Res. 131 (2008) 111–120. https://doi.org/10.1016/j.virusres.2007.09.008

[88]   M.F. Mahomoodally, Traditional Medicines in Africa: An Appraisal of Ten Potent African Medicinal Plants, Evidence-Based Complement. Altern. Med. 2013 (2013) 1–14. https://doi.org/10.1155/2013/617459

[89]   P. Thiviya, A. Gamage, D. Piumali, O. Merah, T. Madhujith, Apiaceae as an Important Source of Antioxidants and Their Applications, Cosmetics 8 (2021) 111. https://doi.org/10.3390/cosmetics8040111

[90]   P.B. Chouke, T. Shrirame, A.K. Potbhare, A. Mondal, A.R. Chaudhary, S. Mondal, S.R. Thakare, F. Nepovimova, M. Valis, K. Kuca, R. Sharma, R.G. Chaudhary, Bioinspired metal/metal oxide nanoparticles: A road map to potential applications, Mater. Today Adv. 16 (2022) 100314. https://doi.org/10.1016/j.mtadv.2022.100314

[91]   C.-Y. Hsu, A.M. Rheima, M.M. Kadhim, N.N. Ahmed, S.H. Mohammed, F.H. Abbas, Z.T. Abed,

Z.M. Mahdi, Z.S. Abbas, S.K. Hachim, F.K. Ali, Z.H. Mahmoud, E. Kianfar, An overview of nanoparticles in drug delivery: Properties and applications, South African J. Chem. Eng. 46 (2023) 233–270. https://doi.org/10.1016/j.sajce.2023.08.009

[92]   E. Sánchez-López, D. Gomes, G. Esteruelas, L. Bonilla, A.L. Lopez-Machado, R. Galindo, A. Cano, M. Espina, M. Ettcheto, A. Camins, A.M. Silva, A. Durazzo, A. Santini, M.L. Garcia, E.B. Souto, Metal-Based Nanoparticles as Antimicrobial Agents: An Overview, Nanomaterials 10 (2020) 292. https://doi.org/10.3390/NANO10020292

[93]   A. Luceri, R. Francese, D. Lembo, M. Ferraris, C. Balagna, Silver Nanoparticles: Review of Antiviral Properties, Mechanism of Action and Applications, Microorganisms 11 (2023) 629. https://doi.org/10.3390/microorganisms11030629

[94]   P. Allawadhi, V. Singh, A. Khurana, I. Khurana, S. Allwadhi, P. Kumar, A.K. Banothu, S. Thalugula, P.J. Barani, R.R. Naik, K.K. Bharani, Silver nanoparticle based multifunctional approach for combating COVID-19, Sensors Int. 2 (2021) 100101. https://doi.org/10.1016/j.sintl.2021.100101

[95]   S. Galdiero, A. Falanga, M. Vitiello, M. Cantisani, V. Marra, M. Galdiero, Silver Nanoparticles as Potential Antiviral Agents, Molecules 16 (2011) 8894–8918. https://doi.org/10.3390/molecules16108894

[96]   E.O. Mikhailova, Silver Nanoparticles: Mechanism of Action and Probable Bio-Application, J. Funct. Biomater. 11 (2020) 84. https://doi.org/10.3390/jfb11040084

[97]   R. Kumar, M. Nayak, G.C. Sahoo, K. Pandey, M.C. Sarkar, Y. Ansari, V.N.R. Das, R.K. Topno, Bhawna, M. Madhukar, P. Das, Iron oxide nanoparticles based antiviral activity of H1N1 influenza A virus, J. Infect. Chemother. 25 (2019) 325–329. https://doi.org/10.1016/j.jiac.2018.12.006

[98]   B. Aderibigbe, Metal-Based Nanoparticles for the Treatment of Infectious Diseases, Molecules 22 (2017) 1370. https://doi.org/10.3390/molecules22081370

[99]   P. Nath, M.A. Kabir, S.K. Doust, A. Ray, Diagnosis of Herpes Simplex Virus: Laboratory and Point-of-Care Techniques, Infect. Dis. Rep. 13 (2021) 518–539. https://doi.org/10.3390/idr13020049

[100]   A. Khurana, P. Allawadhi, V. Singh, I. Khurana, P. Yadav, K.B. Sathua, S. Allwadhi, A.K. Banothu, U. Navik, K.K. Bharani, Antimicrobial and anti-viral effects of selenium nanoparticles and selenoprotein based strategies: COVID-19 and beyond, J. Drug Deliv. Sci. Technol. 86 (2023) 104663. https://doi.org/10.1016/j.jddst.2023.104663

[101]   L.P. Arendsen, R. Thakar, A.H. Sultan, The Use of Copper as an Antimicrobial Agent in Health Care, Including Obstetrics and Gynecology, Clin. Microbiol. Rev. 32 (2019). https://doi.org/10.1128/CMR.00125-18

[102]   K. Imai, H. Ogawa, V.N. Bui, H. Inoue, J. Fukuda, M. Ohba, Y. Yamamoto, K. Nakamura, Inactivation of high and low pathogenic avian influenza virus H5 subtypes by copper ions incorporated in zeolite-textile materials, Antiviral Res. 93 (2012) 225–233. https://doi.org/10.1016/j.antiviral.2011.11.017

[103]   G. Borkow, H.H. Lara, C.Y. Covington, A. Nyamathi, J. Gabbay, Deactivation of Human Immunodeficiency Virus Type 1 in Medium by Copper Oxide-Containing Filters, Antimicrob. Agents Chemother. 52 (2008) 518–525. https://doi.org/10.1128/AAC.00899-07

[104]   N. Desai, D. Rana, S. Salave, R. Gupta, P. Patel, B. Karunakaran, A. Sharma, J. Giri, D. Benival, N. Kommineni, Chitosan: A Potential Biopolymer in Drug Delivery and Biomedical Applications, Pharmaceutics 15 (2023) 1313. https://doi.org/10.3390/pharmaceutics15041313

Biogenic Nanomaterials: Synthesis, Characterization, Applications, and Future Remarks   Materials Research Forum LLC
Materials Research Foundations 180 (2025)   https://doi.org/21741/9781644903759

Chapter 10

Investigation of the DNA Damage Inhibitory Effect of Biogenic Nanomaterials

N. Bazancir[1*], I. Meydan[1,] A. R. Kul[1], S. Kaptanoglu[1], S. Celikozlu[2], F. Sen[3*]

[1]Van Vocational School of Health Services, Van Yuzuncu Yil University, Van, Turkey.

[2]Department of Food Processing, Altıntaş Vocational School, Kütahya Dumlupınar University, Kutahya, Turkiye

[3]Sen Research Group, Department of Biochemistry, Kutahya Dumlupinar University, Kutahya, Turkiye.

nuranbazencir@yyu.edu.tr; fatih.sen@dpu.edu.tr

Abstract

DNA damage is an important biological process that disrupts cell and tissue function and is associated with cancer, aging and many other diseases. DNA damage can be caused by environmental factors (such as ultraviolet light, chemicals, and radiation) or cellular processes. Biological nanomaterials are being investigated as new therapeutics to help prevent DNA damage caused by interactions between biomolecules and nanoparticles. These nanomaterials can directly protect DNA from damage or stimulate DNA repair mechanisms. In this review, the results of studies investigating the effectiveness of biological nanomaterials in preventing DNA damage were examined.

Keywords

Biological Nanoparticles, DNA Damage, Inhibitory Effects, Green Synthesis

1.   Introduction

The study of bionanomaterials that protect against DNA damage is an important area of research at the intersection of biotechnology and nanotechnology. This type of application takes advantage of the effects of nanomaterials on DNA and helps evaluate nanomaterials for maintenance applications. Some bionanomaterials are naturally occurring substances produced or observed by living organisms. For example, nanomaterials produced by plant extracts, fungi and bacteria may have bioenergy potential. The effectiveness of these nanomaterials in preventing DNA damage may be due to factors such as their antioxidant properties and their ability to limit DNA repair [1].

As a research method, in vitro and in vivo experiments can be used to reduce the death rate of bionanomaterials due to DNA damage. In vitro experiments can use cell lines or isolated DNA samples to examine the effect of nanomaterials in preventing DNA damage. In contrast, in vivo experiments can more realistically assess the risk of death of nanomaterials and prevention of DNA damage in vivo. In general, studies can evaluate the properties of bionanomaterials such as antioxidant activity, DNA damage removal, free radical scavenging and DNA repair capacity. Given the benefits and potential or side effects of bio-nanomaterials in biological systems, this type of manipulation can be used to size and use DNA outer structures to take advantage of new electrical charge discharges. However, it should be noted that the potential of bio-nanomaterials to prevent DNA damage in these regions is not yet fully understood. Therefore, more research is needed on the sustainability and sustainability of the results of such outcomes. In this overview, various biomaterials are classified according to their structures, discussed in detail, and research examples regarding the use of these biomaterials in therapeutic applications are included, taking into account the current development of the literature [2].

Biotechnology is the use of living organisms or their products to create products or solve problems that are beneficial to humans (or their environment). Biological nanomaterials at the interface of biotechnology and nanotechnology, research in the fields of biology, biochemistry and genetics Today, new generation biomaterials developed in the field of medicine gain importance as an effective strategy in the prevention and solution of problems in diagnosis and treatment. The development of organisms at the biochemical, molecular and genetic levels produces diagnostic agents, chemicals for diagnostic and therapeutic research, vaccines, antibiotics, hormones, proteins and enzymes. At the same time, it is possible to produce mutations that can occur in vitro and in vivo in a short period of hundreds of thousands of years under normal conditions by using these bionanomaterials.

One of the new fields of science emerging today is nanotechnology. When we look at the development of nanotechnology, we can see that it is a new field in terms of scientific research. When you look at technology products, you will see that the shelves are covered with nanomaterials, which are the result of research in the field of nanotechnology. Thanks to these nanomaterial applications developed, many consumer products enter our lives. As a result, nanotechnology products are used in a wide variety of applications today [1].

The domestic and foreign contribution of nano products through technological development is very high. It has been revealed that the development phase of nanoproducts takes place thanks to national or international collaborations. These developments are considered as a measure of the state of university-industry cooperation. We believe that international cooperation guides development in the fields of technology, economy, culture and society. Thanks to the new technological developments called the kitchen of science, we encounter developments that lead to the production of nanomaterials with different properties. Today, research in nanotechnology has reached an astonishing scale. For example, nanomaterials with high surface/volume ratios and different properties have been obtained. Nanotechnology, whose importance is increasing day by day due to research, finds application in many fields such as catalysis, pharmaceutical industry, agriculture, optical materials, sensors, defense industry, space research, implants, textiles and energy storage [3].

Nanotechnology is a technology that emerges in the form of manipulation, separation, combination, loss or acquisition of new properties of existing structures at the atomic and molecular level [4]. Examining the applications in the field of nanotechnology reveals its effects in many areas such as production, health, environment, energy, defense, textile, medicine and agriculture [5–12]. For example, if we look at the development of the health sector, we can see that research on the prevention, diagnosis and treatment of diseases has led to the emergence of new fields of science. Today, research and development centers are of great importance for the rapid development and change of technology.

2.   What is DNA? What are DNA Structure and Functions?

Deoxyribonucleic acid (DNA) is a complex molecule that contains all the information that specifies the characteristics of an organism. DNA is found in the cell nucleus of all living things. But DNA does more than determine the structure and function of an organism. When an organism reproduces, some of its DNA is passed on to its offspring. This transfer of all or part of an organism's DNA helps maintain some degree of continuity from one generation to the next while allowing for subtle changes that add to the diversity of life [13].

2.1   What is DNA?

DNA carries the biological characteristics that distinguish species from one another. The characteristic features carried by DNA are transferred from adult organisms to their offspring through reproduction. In eukaryotic organisms, DNA is located in the nucleus of the cell. In order for the very long DNA molecule to fit into a very small cell nucleus, DNA molecules is packed. Chromosomes are forms of DNA packaged with histone proteins. The chromosome structure is seen in the cell nucleus only during mitosis or meiosis. At other times in the cell cycle, DNA is found untangled and distributed throughout the cell nucleus in the form of chromatic threads. The DNA in the cell nucleus is called "Nuclear DNA" [14].

The entire DNA contained in an organism is called the genome. All organisms have a small amount of DNA in the mitochondria, as well as in the cell nucleus. In sexual reproduction, an organism receives half of its nuclear DNA from the male parent and half from the female parent. However, mitochondrial DNA is only passed on to offspring from the female parent, because the egg cell is larger and has more cytoplasm than the sperm. DNA includes almost all parts of mathematics, chemistry, physics and biology [13].

2.2   DNA Structure and Function

The entire genome is replicated within the cells of the human body. The double-stranded DNA that makes up the genome contains about 3 billion base pairs. Human DNA consists of 23 pairs (46), divided into chromosomes. The number of genes encoding the functional protein is about 20-30,000. These genes are involved in embryogenesis, growth and development. They control reproduction and various metabolic functions. The number of functional genes in the genome makes up about 10% of the genome. Others have more than one similar copy within the genome (repeat order). These sequences result in thousands, possibly millions, of copies. Its functions have not yet been fully elucidated. Part of the gene may play a role not only in chromosome structure and function, but also in its expression [15].

DNA is a high molecular weight nucleic acid (NA) macromolecule (polynucleotide). A compound consisting of a base, a phosphate, and a sugar is called a nucleotide. Fundamentals of DNA macromolecules. It consists of a sugar (deoxyribose), four different bases (adenine, guanine, thymine, cytosine) and a phosphate group. The nucleotides are joined by 5' and 3' phosphodiester bonds to form polynucleotide macromolecules. Human genetic information is encoded by the sequence of four different bases within the DNA molecule. The structure of DNA determines how genetic information is transmitted from parent cells to daughter cells and how amino acids are arranged within proteins. A double-stranded DNA helix has two polynucleotide helices running in opposite directions. In these strands, adenine (A) thymine (T) and guanine (G) align with cytosine (C), and if the base order of one of these strands is known, the other strand also knows[15]. During cell division, the two strands separate and each strand that enters the daughter cell uses nucleotides from the medium to complete the pairing. During mitosis, daughter and parent cells have the same number of chromosomes. During meiosis, daughter cells are created using half of the parental chromosomes, so that one copy of each DNA is arranged. There is no information reduction in diploid organisms [16,17].

3.   DNA Damage

Changes in the molecular integrity of genetic material under the influence of external or internal factors are called "DNA damage". Genomes are exposed to various factors that cause DNA damage. Extrinsic sources include ultraviolet light from the sun, ionizing radiation from radon decay, aflatoxin from fungi, burnt tobacco, and many chemotherapeutic agents. Examples of endogenous sources are oxidative metabolism, spontaneous DNA changes, immunological diversity (D) and recombination mechanisms [13].

3.1   Factors Causing DNA Damage

a) Spontaneous or inherited gene mutations

b) Environmental factors: UV rays, ionizing radiation, electromagnetic waves, chemicals, alcohol, air pollution, malnutrition.

c) Factors resulting from natural cellular metabolism: Free radicals produced when mitochondria produce energy are referred to as inflammatory and detoxification processes. Cells respond to all this DNA damage in various metabolic ways. Severe DNA damage activates the cell's apoptotic pathway, leading to cell death. Cells can repair DNA damage through the 'DNA repair mechanism'. If DNA damage is not repaired during replication, mutations occur. As a result, various genetic diseases and cancers can occur. There are more than 100 genes responsible for the DNA repair mechanism, and the proteins encoded from these genes play a role in DNA repair. Every day, approximately 500,000 non-encoded or miscoded lesions occur in the DNA of all human cells [2].

DNA damage can ultimately change the structure of DNA, which in turn changes the genetic information that is passed on to other generations. Mild damage can be repaired primarily by the DNA repair system, but accumulation of moderate damage can lead to mutations and cancer. Severe damage can trigger 'cell death' by inducing apoptosis to protect the organism [13].

3.2   Types of DNA Damage

Various mutagenic agents can damage DNA, altering the DNA sequence. Oxidizing agents, alkylating agents, and high-energy electromagnetic radiation (such as UV and X-rays) are known mutagens. Different mutagens cause different damage to DNA. For example, UV light damages DNA by forming thymine dimers [18]. In contrast, free radicals and oxidants such as hydrogen peroxide can cause different types of damage. Such as base changes (especially guanosine) and double helix breaks [19]. All human cells are damaged by 500 bases of oxidant per day [20,21].

The most harmful of these oxidative damages are double chain breaks. Because it's hard to fix. They can cause point mutations, insertions and deletions, and chromosomal translocations in

DNA sequences [22]. Main type of damage deamination, depurination, alkylation, formation of T-T and T-C dimers, replication errors, double helix breaks (DSB) oxidative damage.

3.2.1 Deamination

During deamination, the amino groups of adenine (A) and cytosine (C) are converted to keto groups. It causes deamination of HNO_2 (nitrite), cytosine (C) => uracil (U) and adenine (A) => hypoxanthine. This is inconsistent with hypoxanthine cytosine formed by adenine deamination [2].

3.2.2 Depurination

Mammalian cells spontaneously lose about 10,000 purines over a 20 hour growth period at 37°C. Aflatoxins induce depurination (loss of purine bases), but depurination can also occur spontaneously. Lack of repair of depurin sequences can result in deletions. The presence of these mutations causes severe DNA loss during replication. A base cannot be added to the abasic region, or even if a base is added, the base becomes a mutated base [13].

3.2.3 Alkylation

Alkylation attaches alkyl groups such as methyl (CH_3-) and ethyl (CH_3-CH_2) to the amino and keto groups of nucleotides. Nitrosamines, ethyl methyl sulfonate and N-methyl-N1-nitrosoguanidine are the most important alkylating agents. The most important alkylation site is the oxygen at the 6th carbon atom of guanine [23]. O6-ethylguanine (or O6-methylguanine), formed by alkylation, pairs with thymine and acts as the base analog of adenine. This replaces the G:C base pairs with A:T base pairs when the DNA is paired. Many chemical mutagens cause base changes. These active ingredients are usually lower alkyls (such as methyl groups). It also consists of many mutagenic polycyclic moieties. O6-Methylguanine (mG) is formed in the presence of alkylating agents and is highly mutagenic. O6-methylguanine-DNA methyltransferase enzyme provides normal guanine formation by transferring CH_3 groups of mismethylated bases in DNA to cysteine residues. In this process, enzymes are also irreversibly inhibited and rendered dysfunctional. Therefore, in this repair, the number of enzymes is as important as their specificity [13].

3.2.4 Formation of T-T and T-C dimers

Absorption of UV light by nucleobases produces dimers (cyclobutane pyrimidine dimers), usually as a result of bond formation between strands of adjacent pyrimidine bases. DNA damage causes tanning and increased melanogenesis, which is responsible for 8% of all melanomas.

3.2.5 Replication Damages

An error caused by adding the wrong nucleotide during DNA replication. The frequency with which DNA polymerases make mistakes (misplacement of bases) influences the occurrence of spontaneous mutations. The most important factor affecting polymerase fidelity is the correction of 3'-5' exonuclease activity. This activity serves to remove bases accidentally added by polymerases and prevents mutation formation during replication [14].

3.2.6 Formation of Double Thread Fractures

They are produced by ionizing radiation, transposons, topoisomerases, endonucleases, mechanical stress on chromosomes, and single-strand breaks at single-stranded regions (as during replication or transcription). DSBs are the most dangerous type of DNA damage that occurs continuously throughout the life of the cell. DSBs pose a major threat to the genome as they can arise from both intrinsic and extrinsic factors and can lead to mutagenesis, oncogenic transformation or cell death [24].

3.2.7 DNA Damage Caused by Oxidative Stress

It is estimated that endogenous free radicals cause about 200,000 damage per day in every cell. Free radicals cause DNA attack, mutation and cell death. It reacts readily with hydroxyl radicals, bases and deoxyribose. Because hydrogen peroxide can easily cross membranes, it can enter the DNA of the cell nucleus, causing cellular dysfunction and even death. Therefore, DNA is a sensitive molecule [25]. ROS (reactive oxygen species) and RNS (reactive nitrogen species) cause little DNA damage in nature [16]. An endogenous reaction that causes DNA damage. These are oxidation, methylation, depurination and deamination. Reactive products such as nitric oxide or nitrogen dioxide (NO_2), peroxynitrite (ONOO-), nitrous oxide (N_2O_3) and nitric acid (HNO_3) show mutagenicity through nitrosation and deamination reactions. Different types of ROS cause different DNA damage. For example, O_2 and H_2O_2 do not react with bases, but OH radicals can bind to any of the four bases of DNA. It leads to the formation of reaction products [26]. Singlet oxygen specifically binds to guanine and causes damage [27]. Hydroxyl radicals react with purine bases at the C4, C5 and C8 positions to form C_4-OH, C_5-OH and C_8-OH purine radicals, respectively [28]. C4-OH and C5-OH purine radicals are dehydrated to form oxidized purine radicals. One electron oxidation and one electron reduction of the C_8-OH purine radical yields 8-hydroxypurine (7,8-dihydroxy-8-oxoprene) or formamidopyrimidine [28]. Both can occur in both aerobic and anaerobic environments. Reducing agents promote the formation of formamidpyrimidines, while an oxygen-containing environment seems more favorable for the formation of 8-OH-pyrimidines. 8-OH-guanine is a very common base damage product, so it is used as a damage indicator to measure oxidative DNA damage [29]. It is most commonly measured as 8-hydroxydeoxyguanosine (8-OH-dGua) nucleoside [22,23].

The interaction of the hydroxyl radical moieties with the sugar groups (5 carbons) of DNA is done by removing an H atom from one of the atoms [32]. Sugar radicals are produced by various reactions. In the anoxic system, the radical at the C4 center is cleaved, the DNA helix is broken, yielding an intact base and a modified sugar. Oxidation of Cl-centered radicals forms sugar lactones and liberates solid bases. In the absence of oxygen, base radicals join H atoms of adjacent sugar groups to form sugar radicals that lead to chain cleavage [33]. In the oxygen system, an oxygen molecule is added to a carbon-centered sugar radical to form a peroxyl radical. Most sugars are peroxyl radicals.

Their peculiarity is the breaking of carbon-carbon bonds and the formation of alkaline zones. The peroxyl radical in the C5 center is converted to an oxyl radical, resulting in DNA strand cleavage and cleavage, release of the stable base and exchange of sugar. Changes in DNA, the missing sugar group can either be released from the DNA helix or remain attached to the DNA via a phosphate bond [34]. Reaction of base radicals and sugar radicals. Differentially modified bases and sugars lead to uncontrolled sequencing, chain breaks and cross-linking of DNA and proteins. This type of damage, also called oxidative DNA damage, leads to mutagenesis, carcinogenesis and senescence [25,27].

4.   Nanotechnology

Nano, meaning dwarf, is technically defined as a unit of measure. The equivalent of this nanoscale, defined by the International System of Units, is expressed in nanometers. Putting these definitions together, we know that a nanometer (nm) equals one billionth of a meter. It is expressed as one billionth of the scientifically measured size [36]. Nanotechnology uses the nanoscale, which is billionths of a meter. Collect atoms and molecules separately and combine them with precision. This bonding process takes place at the nanoscale. The resulting nanoproducts appear as functionally designed and fabricated structures ranging in size from 1 to 100 nanometers. In other words, by changing the arrangement of atoms and molecules, they can be reduced to nano size and various physical, chemical and biological properties can be obtained. Efforts by scientists to find or improve existing ones have led to the technical study of atomic and molecular structures. For this purpose, first of all, the atomic arrangement of the structures found in nature is discussed [37].

Nanotechnology research is based on imitating nature's atomic arrangement principles. Nanostructures already exist in structures found in nature. The structures obtained with nanotechnology appear at the stage of fulfilling their function, whether they affect nature partially or not. Its small size presents some difficulties in its use and manufacture. Since these nanostructures are produced by physical and chemical means, they encounter undesirable problems such as temperature increase, loss of strength and friction. Today, we encounter applications that are rapidly spreading and occupying every moment and area of our lives. The purpose of these applications is to easily obtain new products that are lighter, more durable and more economical. When we look at the applications, we can see that it is one of the most important developments in the field of medicine. Thanks to the medical applications in the field of nanotechnology, it is expected to diagnose or treat many diseases that cannot be cured yet. As a result of new research, promising developments have been obtained. Further progress in these researches will enable the diagnosis and treatment of many diseases. This greatly increases the therapeutic dominance of the disease [38].

5.   Nanoparticles

Thanks to technical research in the field of nanotechnology, existing structures are reduced and more efficient structures are obtained. These highly efficient technologies result in well-crafted, durable, clean, safe and smart products. These new structures, called nanoscale materials, are divided into several classes such as nanocrystals, nanoparticles, nanotubes, nanowires and nanorods [39]. All these structures have various names and are called nanoparticles. Nanoparticles can be easily classified into carbon-based, metal-based and semiconductor-based. Nanoparticles have different properties and can be used in different ways. The functions and applications of these materials depend on the size and composition of the nanoparticles [40]. Therefore, polymer structures are widely used for nanoparticle production. Polymers exhibit unique properties when interacting, thus imparting different properties to nanoparticles during manufacture. Also, the toxicity of polymer-based nanoparticles is very low. This property alone is sufficient to make the polymer structure advantageous for nanoparticle production. The polymer structure allows mass production of the active substance and its in-house activity and use.

It can easily perform various tasks such as not being able to dissolve other substances during interaction and not be able to affect the structure of the substances used. In addition to these functions, they perform various functions and tasks in certain dimensions. This gives an incredible effect to the nanoparticles produced. The nanoparticles obtained by this interaction have a very high surface area/volume ratio. It appears as a differentiating element in the materials we use in our daily life. Nanoparticles used in healthcare provide targeted and more sensitive analysis in nuclear and radiological fields such as molecular imaging in drug discovery systems, targeted therapy, vaccine development, biomedical proteomics and genomics technology, magnetic resonance, ultrasound, fluorescence, computed tomography. In addition to these common and useful uses, it has the following molecular properties. It has been the subject of scientific research due to its possible toxic effects on the respiratory tract, blood, central nervous system, gastrointestinal system and skin [41].

5.1   Production of Nanoparticles

Nanoparticle production systems usually take the form of top-down and bottom-up production. These production methods use targeted physical, chemical and biological methods. Techniques such as hydrothermal/solvothermal, sol-gel, photochemical reduction, ultraviolet aerosol techniques, lithography, laser ablation and ultrasound are often used as examples of wet chemical processes in the production of metal nanoparticles [42].

5.2   Application Areas of Biological Nanoparticles

The application areas of biological nanotechnology are very wide and constantly evolving. As an interdisciplinary science, nanotechnology is gaining increasing importance in fields such as medicine and physics, chemistry, biology, computers, materials science and electronics. It is also used in the fields of cosmetics, agriculture and medicine.

6.   Green Nanotechnology

New techniques are needed because the traditional physical and chemical methods used for many years to produce nanoparticles have undesirable properties. This is the most environmentally friendly approach currently used in nanoparticle production. It is the ultimate plant/microbe mediated biosynthesis that can be produced by biological organisms such as plants and fungi in the presence of suitable substrates with desirable properties under environmental conditions. When plants and microorganisms are exposed to high concentrations of heavy metals, their morphology, cell membranes, enzymes and DNA are damaged and cell death occurs. Despite these adverse effects, organisms exposed to toxic concentrations of heavy metals have developed resistance mechanisms by enzymatically oxidizing heavy metals or reducing them to less toxic forms or covalent modifications [43].

This naturally occurring phenomenon has inspired scientists to use naturally occurring materials and biological structures to create perfectly engineered nanoscale materials. Green synthesis is one of the most environmentally friendly techniques in the synthesis of nanoparticles. Generally, organic compounds obtained from plants and microorganisms are used. These natural resources are generally readily available and abundant in nature, allowing for the biosynthesis of large quantities of nanoparticles [44]. Among the various natural materials used for nanoparticle synthesis, plant extracts seem to be the best candidates. Nanoparticles from plant extracts are more stable, vary in size and shape, and are produced more quickly compared to microbial production processes [45]. Knowledge gained from research has led to the development of the concept of green nanotechnology, which is environmentally friendly, less toxic and based on the production of nanoparticles from living cells. This concept is expressed in working techniques that help solve the problems caused by wastes, do not pose a risk to human health, and explore practical techniques in the field of nanotechnology [39]. Green plant extracts and microorganisms are widely used in green nanotechnology research. Many organisms have been used in this context, but examples of green plant extracts include *Aloe vera, Azadirachta indica, Camellia sinensis, Jatropha curcas,* and *Akalifa indica* [46].

6.1   Nanoparticle Synthesis from Plants

Nanoparticles synthesized from plants are said to be fast, stable and economical due to their easy availability and common properties. The ability of plant extracts to reduce metal ions was discovered in the early 20th century, but this natural mechanism is not fully understood. Silver nanoparticles have been produced from plants such as *Akhalifa indica, Allium sativum, Boswellia ovalifoliolata, Calotropis procera* and *Camellia sinensis.* It has been observed that silver nanoparticles obtained from plants containing abundant phytochemicals such as quinones and proteins are more stable [47].

It has been determined that silver nanoparticles (AgNP) synthesized from the extract of Salvia limbata, one of Iran's endemic plants, eliminate toxicity and protect the environment. Silver nanoparticle extracts in powder form, easily obtained by biological synthesis, are suitable for biomedical and medical research. Its production is environmentally friendly, practical, efficient and economical, and also has the potential to be a commercial product. The results of a related study showed that nanoparticles prepared with *Argyeria nervosa* seed extract have strong antagonistic effects against fungi and bacteria [46].

In another study, it was shown that silver nanoparticles produced from leaf extracts of *Ficus benghalensis* have environmentally friendly properties and that protein amino groups are important and play an important role in keeping the produced particles stable in solution [48].

In recent years, the ability of plants to reduce metal salts has been exploited. The harmony between the structural properties of silver nanoparticles and their antibacterial function and ability to reduce intracellular toxicity is unique to these particles. A study conducted in the dark using *Argyreia cymosa* outside of Rooibos tea showed that the antibacterial activity of silver nanoparticles synthesized using environmentally friendly green nanotechnology was determined by *E. coli, Pseudomonas aeruginosa, Proteus spp.* Metal ions in the medium were found to be stable during particle synthesis [48]. Silver nanoparticles obtained by green nanotechnology from plants such as *A. cymosa* are said to have significant potential in the field of medicine due to their strong antibacterial activities [48]. Bionanomaterials are involved in many roles such as biosynthesis of key enzymes, membrane stabilization, protection of DNA, ribosomes, production of proteins and carbohydrates [40,41]. In normal life, sustained DNA damage and delicate repair ensure stable DNA and long-term wellness [51]. A lot of research has been done to prevent DNA damage. Most research tries to take advantage of the antioxidant properties of plants, especially the structure of the plant. Again, many studies have shown that polyphenols in plant structures prevent DNA damage, are cheap and easy to use [52].

The study examined DNA damage in *H. persicum*, a plant belonging to the Umbelliferae family. As a result of the research, he reported that *H. persicum* plants prevent DNA damage [53].

In another study, E2 (estrogen) induced oxidative DNA uptake (covalent binding of carcinogenic compounds to DNA) was reduced from 83 to 97 in aqueous extracts of fennel, cumin, parsley and dill belonging to Umbelliferae concentrations. 0.4mg/ml. It has been shown that inhibition of DNA doping may be a result of scavenging of reactive oxygen species [54]. In another study, the plant *Diplotenia turcica* was used to synthesize CuNPs. In a study, the effect of CuNP/Dp on DNA damage was examined and it was reported that CuNP/Dp could prevent DNA damage at concentrations of 100-250 mg/day [55].

In a study, the protective effect of ethanol extracted from the seeds of thistle plant (thistle) from Diyarbakır province against DNA damage was investigated. Agarose gel electrophoresis was used in the study. In conclusion, it has been reported that *Silybum marianum* plants significantly reduce DNA damage [56].

Meydan et al. used agarose gel electrophoresis of rhubarb (Ukkun) plants native to Bahçesaray District, Van Province to examine the protective effect of plants against DNA damage by the addition of zinc nanoparticles (ZnO NP/Rr) [57].

A study using an aqueous grape extract to show the extent of cytotoxicity, antiproliferative and apoptotic trafficking was observed in situ. When *Vitis vinifera* cells were tested, green synthesized AgNPs exhibited anti-cancer effects. The results of this study suggest that AgNPs are a potential option for cancer therapy [58].

Sengupta et al. reported that in colorectal cancer induced by azoxymethane administration in male Sprague-Dawley rats that did not receive tomatoes, the expression levels of Cox-2 were significantly reduced in colorectal tissues of tomato-fed mice compared to cancer tissues, increased Cox-2 also increased PGE2 levels, thereby inhibiting the cell proliferation of tumors in these tissues. Shows a large increase [59].

In human prostate cancer, Zhang et al. reported that lycopene significantly reduced COX-2 expression in cancer tissue [60].

Talvas et al. reported that lycopene in diet tomatoes reduced p53 expression in human prostate cancer cells [61].

A study by Dias et al. reported that dietary lycopene in rats reduced p53 expression in colon cancer cells compared to controls [62].

Ferreira et al. studied the protective effect of melatonin against CP-induced DNA damage in the pineal gland. After intraperitoneal injection of 20 or 50 mg/kg of CP, melatonin treatment (1 mg/kg) was administered to the rats for 15 days. In this study, lucidum extract increased the expression level of Xpf gene in G. Rat bone marrow cells, while the group given melatonin increased the expression level of Xpf gene. Researchers suggest that melatonin can completely block chromosomal aberrations caused by CP, suggesting that melatonin can be used for more effective treatment of patients undergoing chemotherapy [63]. Increased free radical production, decreased antioxidant enzyme levels, and/or defective DNA repair machinery increase oxidative DNA damage. Therefore, they lead to the development of many degenerative diseases, especially cancer [55,56].

7.   Conclusion

The synthesis of biogenic nanomaterials is carried out by many methods, namely by using physical, chemical and biological or a combination of these. In these processes, situations arise where one method is superior or weaker than another method. Today, methods of producing samples that are inexpensive, do not cause any damage when released to nature, do not harm living organisms, are easily applicable and provide the desired properties are preferred.

New promising nanotechnological studies are being added every day. In this study, DNA damage and repair are mentioned. However, with advanced studies, developments in the field of nanotechnology will reach the highest levels in the field of biomaterials and medicine. In order to reveal these developments, researches and various tests are carried out in different branches of science. Since the effects of nanoparticles on the ecological system are not fully known, intensive studies are required to fill the gap in this field. In many studies, it has been shown that biomaterials protect DNA against damage or reduce DNA damage by activating these repair mechanisms. Studies in this area need to be increased. It is thought that this study will contribute to the development of alternatives.

References

[1]   N.A. Hanks, J.A. Caruso, P. Zhang, Assessing Pistia stratiotes for phytoremediation of silver nanoparticles and Ag(I) contaminated waters, J. Environ. Manage. 164 (2015) 41–45. https://doi.org/10.1016/j.jenvman.2015.08.026

[2]   J.E. Haber, DNA Repair: The Search for Homology, BioEssays 40 (2018). https://doi.org/10.1002/bies.201700229

[3]   R.G. Haverkamp, A.T. Marshall, The mechanism of metal nanoparticle formation in plants: limits on accumulation, J. Nanoparticle Res. 11 (2009) 1453–1463. https://doi.org/10.1007/s11051-008-9533-6

[4]   S. Malik, K. Muhammad, Y. Waheed, Nanotechnology: A Revolution in Modern Industry, Molecules 28 (2023) 661. https://doi.org/10.3390/molecules28020661

[5]   B. Şen, B. Demirkan, A. Savk, R. Kartop, M.S. Nas, M.H. Alma, S. Sürdem, F. Şen, High-performance graphite-supported ruthenium nanocatalyst for hydrogen evolution reaction, J. Mol. Liq. 268 (2018) 807–812. https://doi.org/10.1016/J.MOLLIQ.2018.07.117

[6]   B. Şen, N. Lolak, Ö. Paralı, M. Koca, A. Şavk, S. Akocak, F. Şen, Bimetallic PdRu/graphene oxide based Catalysts for one-pot three-component synthesis of 2-amino-4H-chromene derivatives, Nano-Structures & Nano-Objects 12 (2017) 33–40. https://doi.org/10.1016/J.NANOSO.2017.08.013

[7]   A. Hojjati-Najafabadi, A. Aygun, R.N.E. Tiri, F. Gulbagca, M.I. Lounissaa, P. Feng, F. Karimi, F. Sen, Bacillus thuringiensis Based Ruthenium/Nickel Co-Doped Zinc as a Green Nanocatalyst: Enhanced Photocatalytic Activity, Mechanism, and Efficient H2Production from Sodium Borohydride Methanolysis, Ind. Eng. Chem. Res. 62 (2023) 4655–4664. https://doi.org/10.1021/ACS.IECR.2C03833

[8]   E. Demir, B. Sen, F. Sen, Highly efficient Pt nanoparticles and f-MWCNT nanocomposites based counter electrodes for dye-sensitized solar cells, Nano-Structures & Nano-Objects 11 (2017) 39–45. https://doi.org/10.1016/J.NANOSO.2017.06.003

[9]   Y. Wu, E.E. Altuner, R.N. El Houda Tiri, M. Bekmezci, F. Gulbagca, A. Aygun, C. Xia, Q. Van Le, F. Sen, H. Karimi-Maleh, Hydrogen generation from methanolysis of sodium borohydride using waste coffee oil modified zinc oxide nanoparticles and their photocatalytic activities, Int. J. Hydrogen Energy 48 (2023) 6613–6623. https://doi.org/10.1016/J.IJHYDENE.2022.04.177

[10]   S. Günbatar, A. Aygun, Y. Karataş, M. Gülcan, F. Şen, Carbon-nanotube-based rhodium nanoparticles as highly-active catalyst for hydrolytic dehydrogenation of dimethylamineborane at room temperature, J. Colloid Interface Sci. 530 (2018) 321–327. https://doi.org/10.1016/J.JCIS.2018.06.100

[11]   F. Gulbagca, A. Aygün, M. Gülcan, S. Ozdemir, S. Gonca, F. Şen, Green Synthesis of Palladium Nanoparticles: Preparation, Characterization, and Investigation of Antioxidant, Antimicrobial, Anticancer, and DNA Cleavage Activities, Appl. Organomet. Chem. 35 (2021). https://doi.org/10.1002/aoc.6272

[12]   A. Hojjati-Najafabadi, S. Salmanpour, F. Sen, P.N. Asrami, M. Mahdavian, M.A. Khalilzadeh, A Tramadol Drug Electrochemical Sensor Amplified by Biosynthesized Au Nanoparticle Using Mentha aquatic Extract and Ionic Liquid, Top. Catal. 65 (2022) 587–594. https://doi.org/10.1007/s11244-021-01498-x

[13]   N. Chatterjee, G.C. Walker, Mechanisms of DNA damage, repair, and mutagenesis, Environ. Mol. Mutagen. 58 (2017) 235–263. https://doi.org/10.1002/em.22087

[14]   A.K. Basu, T. Nohmi, Chemically-Induced DNA Damage, Mutagenesis, and Cancer, Int. J. Mol. Sci. 19 (2018) 1767. https://doi.org/10.3390/ijms19061767

[15]   A. Travers, G. Muskhelishvili, DNA structure and function, FEBS J. 282 (2015) 2279–2295. https://doi.org/10.1111/febs.13307

[16]   B. Halliwell, O.I. Aruoma, DNA damage by oxygen-derived species. Its mechanism and measurement in mammalian systems, FEBS Lett. 281 (1991) 9–19. https://doi.org/10.1016/0014-5793(91)80347-6

[17]   J. Shendure, S. Balasubramanian, G.M. Church, W. Gilbert, J. Rogers, J.A. Schloss, R.H. Waterston, DNA sequencing at 40: past, present and future, Nature 550 (2017) 345–353. https://doi.org/10.1038/nature24286

[18]   T. Douki, A. Reynaud-Angelin, J. Cadet, E. Sage, Bipyrimidine Photoproducts Rather than Oxidative Lesions Are the Main Type of DNA Damage Involved in the Genotoxic Effect of Solar UVA Radiation, Biochemistry 42 (2003) 9221–9226. https://doi.org/10.1021/bi034593c

[19]   J. Cadet, T. Delatour, T. Douki, D. Gasparutto, J.-P. Pouget, J.-L. Ravanat, S. Sauvaigo, Hydroxyl radicals and DNA base damage, Mutat. Res. Mol. Mech. Mutagen. 424 (1999) 9–21. https://doi.org/10.1016/S0027-5107(99)00004-4

[20]   N.G. Dolinnaya, E.A. Kubareva, E.A. Romanova, R.M. Trikin, T.S. Oretskaya, Thymidine glycol: the effect on DNA molecular structure and enzymatic processing, Biochimie 95 (2013) 134–147. https://doi.org/10.1016/j.biochi.2012.09.008

[21]   D. Wang, Q. Liang, D. Tai, Y. Wang, H. Hao, Z. Liu, L. Huang, Association of urinary arsenic with the oxidative DNA damage marker 8-hydroxy-2 deoxyguanosine: A meta-analysis, Sci. Total Environ. 904 (2023) 166600. https://doi.org/10.1016/j.scitotenv.2023.166600

[22]   K. Valerie, L.F. Povirk, Regulation and mechanisms of mammalian double-strand break repair, Oncogene 22 (2003) 5792–5812. https://doi.org/10.1038/sj.onc.1206679

[23]   D.T. Beranek, Distribution of methyl and ethyl adducts following alkylation with monofunctional alkylating agents, Mutat. Res. Mol. Mech. Mutagen. 231 (1990) 11–30.

[24]   L.F. Agnez-Lima, J.T.A. Melo, A.E. Silva, A.H.S. Oliveira, A.R.S. Timoteo, K.M. Lima-Bessa, G.R. Martinez, M.H.G. Medeiros, P. Di Mascio, R.S. Galhardo, C.F.M. Menck, DNA damage by singlet oxygen and cellular protective mechanisms, Mutat. Res. Mutat. Res. 751 (2012) 15–28. https://doi.org/10.1016/j.mrrev.2011.12.005

[25]   A. Kore, B. Yang, B. Srinivasan, Recent Developments in the Synthesis of Substituted Purine Nucleosides and Nucleotides, Curr. Org. Chem. 18 (2014) 2072–2107. https://doi.org/10.2174/1385272819666140714174457

[26]   M.K. Türkdoğan, H. Ozbek, Z. Yener, I. Tuncer, I. Uygan, E. Ceylan, The role of Urtica dioica and Nigella sativa in the prevention of carbon tetrachloride-induced hepatotoxicity in rats, Phyther. Res. 17 (2003) 942–946. https://doi.org/10.1002/ptr.1266

[27]   E. van den Akker, J.T. Lutgerink, M.V.M. Lafleur, H. Joenje, J. Retèl, The formation of one-G

deletions as a consequence of singlet-oxygen-induced DNA dmage, Mutat. Res. Mol. Mech. Mutagen. 309 (1994) 45–52.

[28]   S. Steenken, Purine bases, nucleosides, and nucleotides: aqueous solution redox chemistry and transformation reactions of their radical cations and e- and OH adducts, Chem. Rev. 89 (1989) 503–520. https://doi.org/10.1021/cr00093a003.

[29]   M. Roginskaya, Y. Razskazovskiy, Oxidative DNA Damage and Repair: Mechanisms, Mutations, and Relation to Diseases, Antioxidants 12 (2023) 1623. https://doi.org/10.3390/antiox12081623

[30]   E. Horwood, B. Epe, DNA and free radicals, J Chichester 22 (1993) 41–65.

[31]   M. Dizdaroglu, Oxidative damage to DNA in mammalian chromatin, Mutat. Res. 275 (1992) 331–342. https://doi.org/10.1016/0921-8734(92)90036-O

[32]   H. WISEMAN, B. HALLIWELL, Damage to DNA by reactive oxygen and nitrogen species: role in inflammatory disease and progression to cancer, Biochem. J. 313 (1996) 17–29. https://doi.org/10.1042/bj3130017

[33]   L. Gros, M.K. Saparbaev, J. Laval, Enzymology of the repair of free radicals-induced DNA damage, Oncogene 21 (2002) 8905–8925. https://doi.org/10.1038/sj.onc.1206005

[34]   O.I. Aruoma, B. Halliwell, DNA and free radicals: Techniques, mechanisms and applications, (No Title) (1998).

[35]   J.R. Totter, Spontaneous cancer and its possible relationship to oxygen metabolism., Proc. Natl. Acad. Sci. 77 (1980) 1763–1767.

[36]   R. Nagraik, A. Sharma, D. Kumar, S. Mukherjee, F. Sen, A.P. Kumar, Amalgamation of biosensors and nanotechnology in disease diagnosis: Mini-review, Sensors Int. 2 (2021) 100089. https://doi.org/10.1016/J.SINTL.2021.100089

[37]   N. BAZANCİR, İ. MEYDAN, BÖLÜM 4, FARKLI BİLİM ALANLARINDA 49 (2022).

[38]   P. Taslimi, F. Türkan, A. Cetin, H. Burhan, M. Karaman, I. Bildirici, İ. Gulçin, F. Şen, Pyrazole[3,4-d] Pyridazine Derivatives: Molecular Docking and Explore of Acetylcholinesterase and Carbonic Anhydrase Enzymes Inhibitors as Anticholinergics Potentials, Bioorg. Chem. 92 (2019) 103213. https://doi.org/10.1016/j.bioorg.2019.103213

[39]   N. Bazancir, I. Meydan, Characterization of Zn nanoparticles of Platonus orientalis plant, investigation of DPPH radical extinquishing and antimicrobial activity, East. J. Med. 27 (2022).

[40]   R.G. Haverkamp, A.T. Marshall, D. van Agterveld, Pick your carats: nanoparticles of gold–silver–copper alloy produced in vivo, J. Nanoparticle Res. 9 (2007) 697–700. https://doi.org/10.1007/s11051-006-9198-y

[41]   C. Medina, M.J. Santos-Martinez, A. Radomski, O.I. Corrigan, M.W. Radomski, Nanoparticles: Pharmacological and Toxicological Significance, Br. J. Pharmacol. 150 (2007) 552–558. https://doi.org/10.1038/sj.bjp.0707130

[42]   N. Korkmaz, Y. Ceylan, P. Taslimi, A. Karadağ, A.S. Bülbül, F. Şen, Biogenic nano silver: Synthesis, characterization, antibacterial, antibiofilms, and enzymatic activity, Adv. Powder Technol. 31 (2020) 2942–2950. https://doi.org/10.1016/j.apt.2020.05.020

[43]   C.A.M. Bonilla, V. V Kouznetsov, "Green" Quantum Dots: Basics, Green Synthesis, and Nanotechnological Applications, Green Nanotechnology-Overview Furth. Prospect. 1 (2016) 174–192.

[44]   I. Meydan, H. Burhan, T. Gür, H. Seçkin, B. Tanhaei, F. Sen, Characterization of Rheum ribes with ZnO nanoparticle and its antidiabetic, antibacterial, DNA damage prevention and lipid peroxidation prevention activity of in vitro, Environ. Res. 204 (2022) 112363. https://doi.org/10.1016/J.ENVRES.2021.112363

[45]   O.U. Igwe, C.M. Ejiako, Bioconstruction of copper nanoparticle nitida and their ant, Res. J. Chem. Sci. 8 (2018) 10–15.

[46]   M. BEYKAYA, A. ÇAĞLAR, An Investigation on Synthesis of Silver-Nanoparticles (AgNP) and their Antimicrobial effectiveness by using Herbal Extracts, Afyon Kocatepe Univ. J. Sci. Eng. 16 (2016) 631–641. https://doi.org/10.5578/fmbd.34220

[47]   K. Nesrin, C. Yusuf, K. Ahmet, S.B. Ali, N.A. Muhammad, S. Suna, Ş. Fatih, Biogenic silver nanoparticles synthesized from Rhododendron ponticum and their antibacterial, antibiofilm and cytotoxic activities, J. Pharm. Biomed. Anal. 179 (2020) 112993. https://doi.org/10.1016/J.JPBA.2019.112993

[48]   M. Beykaya, A. Çağlar, Bitkisel özütler kullanılarak gümüş-nanopartikül (AgNP) sentezlenmesi ve antimikrobiyal etkinlikleri üzerine bir araştırma, Afyon Kocatepe Üniversitesi Fen ve Mühendislik Bilim. Derg. 16 (2016) 631–641

[49]   Q. Sun, J. Li, T. Le, Zinc oxide nanoparticle as a novel class of antifungal agents: current advances and future perspectives, J. Agric. Food Chem. 66 (2018) 11209–11220.

[50]   A.A. Malandrakis, N. Kavroulakis, C. V Chrysikopoulos, Use of copper, silver and zinc nanoparticles against foliar and soil-borne plant pathogens, Sci. Total Environ. 670 (2019) 292–299.

[51]   P.F. Yıldız, T.Ö. Odtü, U. Mogan, A. Merkezi, DNA Hasarı ve Tamirinde , Gıdaların Fonksiyonu İnflamasyon artışı Bağışıklık Sisteminde zayıflama, (2020).

[52]   İ. Meydan, G. Kizil, H. Demir, B. Ceken Toptanci, M. Kizil, In vitro DNA damage, protein oxidation protective activity and antioxidant potentials of almond fruit (Amygdalus trichamygdalus) parts (hull and drupe) using soxhlet ethanol extraction, Adv. Tradit. Med. 20 (2020) 571–579.

[53]   J. Asgarpanah, G. Dadashzadeh Mehrabani, M. Ahmadi, R. Ranjbar, M. Safi-Aldin Ardebily, Chemistry, pharmacology and medicinal properties of Heracleum persicum Desf. Ex Fischer: A review, J Med Plants Res 6 (2012) 1813–1820.

[54]   F. Aqil, J. Jeyabalan, R. Munagala, S. Ravoori, M. Vadhanam, D. Schultz, R. Gupta, Chemoprevention of Rat Mammary Carcinogenesis by Apiaceae Spices, Int. J. Mol. Sci. 18 (2017) 425. https://doi.org/10.3390/ijms18020425

[55]   H. Seçkin, Antimicrobial, antioxidant and Dna damage prevention effect of Nano-copper particles obtained from Diplotaenia turcica plant by green synthesis, Polish J. Environ. Stud. 30 (2021) 4187–4194.

[56]   A. Serçe, B.Ç. Toptancı, S.E. Tanrıkut, S. Altaş, G. Kızıl, S. Kızıl, M. Kızıl, Assessment of the antioxidant activity of Silybum marianum seed extract and its protective effect against DNA oxidation, protein damage and lipid peroxidation, Food Technol. Biotechnol. 54 (2016) 455–461.

[57]   İ. Meydan, H. Seçkin, Green synthesis, characterization, antimicrobial and antioxidant activities of zinc oxide nanoparticles using Helichrysum arenarium extract, Int. J. Agric. Environ. Food Sci. 5 (2021) 33–41. https://doi.org/10.31015/JAEFS.2021.1.5

[58]   G. Salman, S. Pehlivanoglu, C. Aydin Acar, S. Yesilot, Anticancer Effects of Vitis vinifera L. Mediated Biosynthesized Silver Nanoparticles and Cotreatment with 5 Fluorouracil on HT-29 Cell Line, Biol. Trace Elem. Res. 200 (2022) 3159–3170. https://doi.org/10.1007/s12011-021-02923-8

[59]   A. Sengupta, S. Ghosh, S. Das, Modulatory influence of garlic and tomato on cyclooxygenase-2 activity, cell proliferation and apoptosis during azoxymethane induced colon carcinogenesis in rat, Cancer Lett. 208 (2004) 127–136.

[60]   J.M. Chan, V. Weinberg, M.J. Magbanua, E. Sosa, J. Simko, K. Shinohara, S. Federman, M. Mattie, M. Hughes-Fulford, C. Haqq, Nutritional supplements, COX-2 and IGF-1 expression in men on active surveillance for prostate cancer, Cancer Causes Control 22 (2011) 141–150.

[61]   J. Talvas, C. Caris-Veyrat, L. Guy, M. Rambeau, B. Lyan, R. Minet-Quinard, J.-M.A. Lobaccaro, M.-P. Vasson, S. George, A. Mazur, Differential effects of lycopene consumed in tomato paste and lycopene in the form of a purified extract on target genes of cancer prostatic cells, Am. J. Clin. Nutr. 91 (2010) 1716–1724.

[62]   M.C. Dias, N.F.L. Vieiralves, M.I.F. V Gomes, D.M.F. Salvadori, M.A.M. Rodrigues, L.F. Barbisan, Effects of lycopene, synbiotic and their association on early biomarkers of rat colon

carcinogenesis, Food Chem. Toxicol. 48 (2010) 772–780.

[63] S.G. Ferreira, R.A. Peliciari-Garcia, S.A. Takahashi-Hyodo, A.C. Rodrigues, F.G. Amaral, C.M. Berra, S. Bordin, R. Curi, J. Cipolla-Neto, Effects of melatonin on DNA damage induced by cyclophosphamide in rats, Brazilian J. Med. Biol. Res. 46 (2013) 278–286.

[64] M.S. Cooke, M.D. Evans, M. Dizdaroglu, J. Lunec, Oxidative DNA damage: mechanisms, mutation, and disease, FASEB J. 17 (2003) 1195–1214.

[65] H. Seçkin, I. Meydan, Synthesis and characterization of *Sophora alopecuroides* L. green synthesized of Ag nanoparticles for the antioxidant, antimicrobial and DNA damage prevention activity, Brazilian J. Pharm. Sci. 58 (2022) e20992. https://doi.org/10.1590/S2175-97902022E20992

Biogenic Nanomaterials: Synthesis, Characterization, Applications, and Future Remarks   Materials Research Forum LLC
Materials Research Foundations 180 (2025)   https://doi.org/21741/9781644903759

Chapter 11

The Importance of Biogenetic Nanomaterial in Malaria

S. Yurekturk[1]*, S. Aydemir[2], H. Unal[1], D. Ikballı[3], F. Sen[3]*

[1]Van Health Services Vocational School, Van Yüzüncü Yıl University, Van, Turkiye

[2]Department of Parasitology, Faculty of Medicine, Van Yüzüncü Yıl University, Van, Turkiye

[3]Sen Research Group, Biochemistry Department, Faculty of Arts and Science, Dumlupınar University, Evliya Celebi Campus, 43100, Kutahya, Turkey

sehribanyurekturk@yyu.edu.tr; fatih.sen@dpu.edu.tr

Abstract

Malaria is an important protozoan infection that can cause morbidity and mortality in endemic areas. The first antimalarial drug to treat the disease was quinine. Over time, resistance to the drug developed, and different antimalarial drugs were produced. Due to some difficulties in treating the disease, researchers have stated that effective drug carrier nanomaterials should be developed instead of discovering new drugs. Nanomaterials can also be used to combat mosquitoes. The use of nanoparticles in the treatment of malaria, which constitutes an important public health problem, and in the fight against mosquitoes is promising for the eradication of the disease.

Keywords

Malaria, Nanotechnology, Nanomaterials, Antimalarial Drug Carriers, Malaria Eradication

1. Overview

The word "nano" is described as one billionth of a physical size. Nanotechnology is the branch of science that deals with the design, production, and application of materials, tools and systems with dimensions between 1 and 100 nanometers [1,2].

Nanotechnology has found wide application in many disciplines such as medicine, biology, physics, chemistry, food, agriculture, textile, electronics, materials science and environment [3–6]. Medicine is one of the most common uses of nanotechnology. Nano medicine is widely used in almost many fields for the treatment, diagnosis and control of illnesses [2,7].

Nanoparticles used in nanotechnology are defined as solid particles 1-100 nm in length in two or three dimensions. The different properties of nanoparticles allow the use of these particles in many fields [8]. Biomedical devices (computed tomography, magnetic resonance and ultrasound, etc.) are used in many health fields such as diagnosis, drug development, targeted therapy, vaccine development and disease prevention. The most widely studied nanoparticles are silver nanoparticles (AgNPs), gold nanoparticles (AuNP), carbon nanotubes (CNTs), iron oxide (FeO), zinc oxide nanoparticles, titanium oxide nanoparticles (TiO$_2$), and silica nanoparticles. AgNPs, one of the most preferred nanoparticles, can be used in water filtration systems, electronic devices, cancer treatment, textile fields, bio-detectors and dental biomaterials, as well as used as antimicrobial agents. AgNPs are known to eradicate or reduce the effects of infections caused by organisms including bacteria, viruses, parasites and fungi due to their high antimicrobial activation properties [9–12]. Nanoparticles are used in many different fields such as military, medicine, biomedicine, energy production, textile industry, food industry, dye removal, biosensors, electrochemical sensors, antibacterial and antifungal activity by taking advantage of their shape, size, easy functionalisation and environmentally friendly synthesis [13–28].

Nanoparticles are recognized as promising drug carriers known to be efficient in the treatment of many parasitic diseases through poor cellular permeability, non-specific distribution, low bioavailability and rapid elimination of parasitic agents from the body. In addition, nanoparticles have effects on the development of parasites localized in the host body. The effects of nanoparticles on the parasite include enhancing the function of the host's immune system against the parasite in the host body, causing changes in the morphology of parasites and causing cellular oxidative stress by attaching to the cell membrane of the infected cell [10,29,30].

Parasite-borne diseases have been causing major health problems in humans and animals throughout history. Parasites are thought to affect approximately four billion people around the world. Parasitic diseases are considered to be among the factors causing significant mortality and morbidity, especially in countries with low socio-economic status and inadequate environmental sanitation [31,32].

Parasitic diseases are considered to be an important health problem that still maintains its seriousness today, considering the damages they cause to the national economy such as labor force losses and treatment costs of diseases. Today, it is known that millions of people are affected by many parasitic diseases such as malaria, chagas disease, African trypanosomiasis, strongyloidiasis, cryptosporidiosis, amebiasis, giardiasis, leishmaniasis, ascariasis and schistosomiasis. According to World Health Organization data; in 2020 approximately 241 million people had malaria cases and 627 thousand people lost their lives due to malaria. It is also thought that the number of cases may increase in regions where malaria is endemic due to changing climatic conditions [11,31,33,34].

2. Malaria

Malaria is a common contagious disease all over the world, manifested by fever attacks, enlarged spleen, anemia, and accumulation of a special dye in the internal organs, manifested by the transmission of *Plasmodium* species (*Plasmodium malariae*, *Plasmodium vivax*, *Plasmodium falciparum*, *Plasmodium ovale* and *Plasmodium knowlesi*) by vector Anopheles mosquitoes [35].

Malaria is the most important protozoan-borne parasitic disease worldwide. The reproduction of Anophele mosquitoes is highest in tropical climates. Therefore, the prevalence is quite high in regions where the vector is dense. Malaria affects thousands of people every year and causes significant mortality and morbidity, especially in sub-Saharan African countries. Most deaths from malaria are caused by *P. falciparum* and are more common in children under 5 years of age [36,37].

Plasmodium sporozoites found in the salivary glands are released into the bloodstream and reach the liver within half an hour with the bite of the infected female anopheline. It infects hepatocytes in the liver parenchyma cell and reproduces asexually in this cell. Sporozoites of *P. ovale* and *P. vivax* species remain latent in the hepatic phase without division. This condition is called hypnozoite. Sporozoites in the hepatic phase cause relapses months or years after the first clinical symptoms appear. Sporozoites transform into merozoites within hepatocytes. The development of merozoites is the final life cycle that occurs in the liver parenchyma cells. After the rupture of hepatocytes, the released merozoites invade the erythrocytes. *Plasmodium* species feed, grow, and reproduce asexually within the erythrocyte. It carries out its nutrition by breaking down hemoglobin. Fragmented and undigested hemoglobins accumulate in the cytoplasm in the form of pigments. This structure, called the malaria pigment, plays an essential role in species differentiation of the parasite. Young trophozoites, expressed as signet rings in the erythrocyte, have a dot-like nucleus and annular cytoplasm. They show the appearance of young trophozoite, mature trophozoite (amoeboid), young schizont, mature schizont and merozoite in parasitic erythrocytes, respectively. Some of the new generation merozoites formed after a certain number of schizogonic cycles transform into male and female gametes (macrogametocytes and microgametocytes) in the erythrocyte they invade. The gametocyte stage is the final stage in humans. When sucking blood from a human, the female anopheles take female and male gametocytes from the erythrocytes, thus continuing the life cycle of the plasmodium species [35,38].

Plasmodium species known to cause disease in humans are *P. vivax*, *P. ovale*, *P. falciparum*, *P. malariae* and *P. knowlesi*. The disease is named according to the cause. Malaria caused by *P. vivax* is termed tercina or vivax malaria, malaria caused by *P. malariae* is called quartan malaria, and malaria caused by *P. falciparum* is called *P. falciparum* or tropical malaria [35].

2.1   *Plasmodium vivax*

It is the *Plasmodium* species that causes the highest incidence of malaria in humans worldwide. *P. vivax* is characterized by the ability of the disease to relapse with the activation of sporozoites (hypnozoites) that remain in a latent form in the liver. Vivax malaria is considered a benign and self-limiting infection. However, it is known to cause recurrent febrile seizures, severe anemia, respiratory distress and serious complications in pregnancy. Many African-origin populations also lack Duffy antigen, the red blood cell membrane protein required by *P. vivax* to invade erythrocytes. Because of this antigen deficiency, the risk of *P. vivax* in populations on this continent is very low. *P. vivax* shows selectivity on erythrocytes and infects young erythrocytes. Infected erythrocytes reach approximately 1.5-2 times the size of normal erythrocytes. Schüffner granules are seen in the erythrocyte. It completes its evolution in the erythrocyte in 48 hours. Fever attacks, characteristic of malaria, occur every 48 hours in this species [35,39].

2.2   *Plasmodium ovale*

Plasmodium ovale has a more limited geographic distribution throughout the world. Morphologically similar to *P. vivax*. The common features of these two strains are that they infect young erythrocytes, have a hypnotic phase, form Schüffner granules in the erythrocyte and complete their development in the erythrocyte in 48 hours. P. ovale-infected erythrocytes appear to grow little or no growth [35,40].

2.3   Plasmodium malariae

It was the first species identified as causing malaria. It infects aged erythrocytes. Infected erythrocytes have a normal appearance and do not grow. Ziemann granules are formed as malaria pigment. They complete their evolution in erythrocytes in 72 hours. Compared to other types of malaria, parasitemia is generally lower in these patients. Peripheral smear shows the classically described "band" form of the parasite. Clinically, fever attacks occurring every four days are characteristic features [35,41].

2.4   Plasmodium knowlesi

The zoonotic parasite *P. knowlesi*, is endemic in Southeast Asia. *P. knowlesi*, which causes malaria in macaque monkey species, also causes disease in humans. Compared to other plasmodium species, it completes erythrocyte evolution in 24 hours and completes erythrocyte evolution in the shortest time. Erythrocyte completes its evolution in 24 hours. Microscopically, *P. knowlesi* is difficult to distinguish from other plasmodium species. On microscopic examination, the ring forms of *P. knowlesi* resemble *P. falciparum*, while the trophozoite and schizont forms resemble *P. malariae*. Misdiagnosis of *P. knowlesi* as *P. malariae* can cause a delay in the administration of parenteral treatment, leading to a more severe course of the disease. The species distinction of the parasite can only be made by molecular methods [42].

2.5   Plasmodium falciparum

Plasmodium vivax is the most prevalent malaria species worldwide. However, P. falciparum causes the most clinically severe and fatal cases of malaria and is more common in Africa and tropical regions. Only gametocyte and signet ring form are seen in peripheral smear. Since it infects erythrocytes of all ages without showing selectivity, the level of parasitemia remains quite high. Malaria pigment called Maurer particles is formed in the erythrocyte. Erythrocyte phase can be completed in 48-72 hours or 24-36 hours [35].

3.   Malaria Clinic

Clinical signs of malaria vary according to the type and strains of malaria parasites, nutritional status and general health of the patient, genetic characteristics, immune system, age and pregnancy status. Typical clinical symptoms of the disease include chills, chills, fever and sweating. The period of these seizure types can be seen every 36-48-72 hours depending on the *Plasmodium* species. The other most important clinical manifestation is anemia. One of the most important factors causing anemia is that the parasite carries out most of its life cycle in the erythrocyte. In addition, destruction of uninfected erythrocytes as a result of hypersplenism in the spleen is also considered among the causes of anemia. The most clinically severe symptoms are seen in malaria caused by the *P. falciparum* species. Severe symptoms of *P. falciparum* malaria include metabolic acidosis, cerebral malaria, adult respiratory distress syndrome (ARDS), severe anemia due to hemolysis, coma, acute renal failure, acute tubular necrosis, seizures, and disseminated intravascular coagulation. Hypoglycemia and acidosis are among the most common metabolic complications. Cerebral malaria causes serious clinical symptoms in both adults and children. The course of the disease is particularly severe in children under five years of age, immunosuppressed persons and pregnant women [35,43].

In the liver stage, the first stage of development of *Plasmodium* species in the human body, the entry of merozoites into the peripheral circulation triggers the onset of clinical symptoms of the disease. In the erythrocyte stage, the merozoites inside the erythrocyte grow, develop and reproduce asexually. The increasing number of parasites fills and stretches the erythrocyte cell and causes it to rupture. Thus, free merozoites infect new erythrocytes and initiate the pathophysiology of the disease [43].

The clinical symptoms of the disease can be mild, severe and even fatal. The clinical manifestations of uncomplicated malaria are non-specific but similar to those of other infectious diseases. Non-specific clinical symptoms of the disease include fever, malaise, fatigue, nausea, vomiting, headache, and musculoskeletal pain. The severity of the disease depends on the immune status of the host and the *Plasmodium* species [35,43].

3.1   Plasmodium ovale and Plasmodium vivax

Although the clinical manifestations of *P. vivax* and *P. ovale* malaria are similar, *P. vivax* malaria has more severe clinical symptoms and a greater tendency to relapse [44].

Clinical symptoms may appear between 7 -10 days after infection. Initially, non-specific clinical symptoms such as headache, muscle aches, nausea and sometimes vomiting may be observed. If left untreated, within a few days the asexual cycles of the parasite's erythrocytic development become synchronized and malaria attacks occur. These seizures are repeated every 48 hours. Initially, the patient experiences classic malaria attacks with chills, followed by fever and sweating. In these types of malaria, seizures become milder and more irregular over time and may stop altogether. A few weeks after the infection occurs, the spleen enlarges and becomes enlarged. Rupture of the spleen is a rare but possible complication. Serious complications are rare but can lead to circulatory collapse, renal failure, bleeding, severe anemia, hemoglobinuria, jaundice, acute respiratory distress syndrome, coma, and death. Months or years after infection, sporozoites in the form of hypnozoites cause relapses [43,44].

3.2   Plasmodium malariae

Plasmodium malariae malaria has an incubation period of about 27 to 40 days, longer than *P. ovale* and *P. vivax* malaria. The first malaria attack and prodromal symptoms of this species are similar to *P. vivax* malaria. In this form, also called quartan malaria, malaria attacks occur every 72 hours. Laboratory findings may include a non-severe anemia and leukopenia. Proteinuria is a common symptom of *P. malariae* malaria and is associated with the development of nephrotic syndrome. It is known that the excess of antigens during the disease and the accumulation of circulating antigen-antibody complexes in the glomeruli cause renal failure. Glomerular damage associated with *P. malariae* infection can usually improve with treatment [43,44].

3.3   Plasmodium falciparum

Since *P. falciparum* infects erythrocytes of all ages, the parasitemia rate in the blood is quite high. Therefore, the high parasitemia rate in the blood causes the clinical symptoms of the disease to be severe. Untreated cases of this type of malaria, also called malignant tertiary malaria, can result in death. Non-specific symptoms of the disease appear 8-12 days after infection. Initially, weakness, fatigue, headache, and nausea are observed. As the parasite develops in the body, different organs and tissues show different clinical symptoms. The Schizogony stage of the parasite is seen not only in the circulating blood but also in the liver, spleen, and bone marrow. Infected erythrocytes adhere to healthy erythrocytes, capillary endothelium, and platelets, leading to the formation of rosettes and clumps that disrupt microcirculation. This leads to thrombocytopenia and anemia, which disrupts blood flow to other vital organs such as the kidneys. With the disruption of blood flow, tissue hypoxia may develop, and ischemia may occur. Parasite-infected erythrocytes cause ischemia due to blockages in these organs. In *P. falciparum*, malaria episodes can occur every 36-48 hours. Irregularities in the parasite development process in erythrocytes lead to irregular seizure intervals and consequently to variations in attack periods. Infection of more than 5% of erythrocytes in *P. falciparum* malaria has been associated with hypotension, hypoxia, hypoglycemia and metabolic acidosis. Severe malaria symptoms such as cerebral malaria, metabolic acidosis, adult respiratory distress syndrome (ARDS), severe anemia due to hemolysis, coma, tropical splenomegaly, acute renal failure and fever can be seen in *P. falciparum* malaria. Hypoglycemia and acidosis are the most common metabolic complications. Especially cerebral malaria is more common in *P. falciparum* malaria. The most serious complication of *P. falciparum* is cerebral malaria. In cerebral malaria, confusion and delirium may occur at the time of high fever. Severe headaches may develop, and the patient may rapidly fall into coma. Some patients may suddenly fall into coma without any symptoms. In Blackwater Fever, severe intravascular hemolysis, and acidic urine with high methemoglobin content are observed. It is also known to be associated with those who are hypersensitive to Blackwater Fever, quinine or artemisin derivatives [35,37,43,44].

3.4   Plasmodium knowlesi

It is the last *Plasmodium* species identified in humans. The clinical manifestations are similar to those of *P. falciparum* and *P. vivax* malaria. Since *P. knowlesi* has a similar appearance to *P. malariae*, it is very difficult to distinguish by microscopic examination. Molecular methods are preferred for definitive diagnosis [43,44].

4.   Diagnosis

Malaria is a serious public health problem today, especially in endemic areas. WHO estimates that malaria caused 241 million cases and 627,000 deaths worldwide in 2020. Although malaria is preventable and treatable, it is known to be endemic in 87 countries, mainly in tropical and subtropical regions. The majority of deaths are children and occur in sub-Saharan African countries. Rapid and accurate diagnosis is crucial to reduce the burden and impact of the disease, especially in children. Early diagnosis is one of the most important factors that reduce the mortality and morbidity of the disease [45,46].

The clinical signs and symptoms of malaria are non-specific. Since symptoms such as high fever, chills and headache seen in malaria are similar to the symptoms of other febrile diseases, clinical diagnosis of the disease can be difficult However, in endemic areas, individuals with these symptoms should be evaluated for malaria until proven otherwise [39,47].

Clinical symptoms are important in the diagnosis, but the definitive diagnosis is only possible with the diagnosis of the parasite. Microscopic, molecular, and serological methods are frequently used in the diagnosis of the disease. Microscopic examination, which is considered as the gold standard method, also examines thin and thick smear preparations made from peripheral blood. While the parasite is detected in thick smear preparations stained with Giemsa, the species of the parasite can be detected in thin smear preparations. The microscopic method is time consuming and requires experienced personnel. Only experienced microscopists can distinguish between clinically distinct malaria species. The fact that it cannot be applied in every environment due to the lack of basic laboratory infrastructure in some regions and the low amount of parasitemia in the examined sample are also considered as the advantages of this method. Rapid and effective diagnostics are needed for global control of malaria. Rapid diagnostic tests are easily performed without the need for experienced personnel and specialized equipment. RDTs are easy-to-administer, usually single-use tests that can provide results in a short time. The rapid diagnostic method is a method based on the detection of antigens obtained from the malaria-causing Plasmodium species. These rapid diagnostic tests (RDT) can provide results in a short time. With these rapid diagnostic tests (RDT), results can be obtained in a short time and there is no need for skilled microscopyists are the advantages of this method. Ambient conditions such as heat and humidity can significantly affect the sensitivity of these tests. RDTs diagnose the disease by detecting antigens produced during the erythrocytic cycle, such as histidine-rich protein 2 (HRP2), which is specific to *P. falciparum* or *P. vivax*, *Plasmodium* lactate dehydrogenase (pLDH) or aldolase, which are common to *Plasmodium* species. [35,37,44,47,48].

5.   Treatment

Quinine was the first antimalarial drug used in the treatment of malaria. Years later, it was determined that resistance to quinine developed, and chloroquine was produced. In the later period, resistance developed against proguanil, mefloquine, sulfadoxine-primethamine and atovaquone, which are antimalarial drugs produced. Since the use of newly produced antimalarial drugs alone in the treatment of malaria causes the development of resistance, WHO recommended the use of artemisinin-based combination therapies [34,49,50].

5.1   The Importance of Nanomaterial in Malaria

Nanotechnology can also be used to diagnose, treat, prevent, and control malaria. Nanotechnological approaches to disease diagnosis are based on specific *Plasmodium spp.* biomarkers. Biomarkers are indicators of substances, structures and biological processes that can be measured in the body to predict changes in disease and health status [46,51].

It is crucial to develop diagnostic methods that are rapid, sensitive, cost-effective, and applicable in any environment. Mostly RDTs are used to diagnose the disease, which is a rapid and sensitive method. Environmental factors such as heat and humidity can affect the sensitivity of this method. Detection of biomarkers is a key element of RDT. Six *Plasmodium* spp. biomarkers can be used in malaria diagnostics. These include histidine-rich protein II, aldolase, lactate dehydrogenase, hypoxanthine-guanine phosphoribosyl transferase, glutamate dehydrogenase and hemozoin [46].

Recently, nanomaterials such as Au NPs and magnetic NPs have been added to RDTs to investigate the specificity and sensitivity of the method. Magnetic nanoparticles (MNP) have been used in the development of an immune sensor for malaria diagnostics. In a study investigating MNP-enhanced diagnostics, covalent immobilization of a commercial monoclonal antibody against anti-HRP-2 to MNP was evaluated. A sandwich assay was successfully performed on magnetic nanoparticles using a second monoclonal antibody labeled with horseradish peroxidase (HRP) enzyme that binds to a different epitope of the target antigen for immunological reaction. This new method is said to be highly accurate, cost-effective, fast, and simple [52].

Gold nanoparticles (AuNP) are ideal NPs for biosensor analysis due to their properties such as high solubility in water, favorable morphology, size distribution. In recent years, fluorescence-based approaches using AuNP have been investigated for the generation of immunoassays. In a study, quantitative detection of P. falciparum heat shock protein 70 (Pf Hsp70) was performed using Au NP-based fluorescence immunoassay. This method was found to be a highly sensitive method for the detection of malaria antigen. Heat shock proteins are antigens recognized by the host immune system during infectious diseases. In addition, in the study, it was observed that antigen detection was successfully achieved by fluorescence immunoassay at 3% parasitemia in human blood culture samples infected with *Plasmodium* spp. and it was assumed that it could be detected in the presence of lower parasitemia [46,47].

In another study, a latex agglutination method using polystyrene NP conjugated with human polyclonal IgG antibodies specific for *P. falciparum* was developed and investigated. The study emphasized that it provides high sensitivity for *P. falciparum* detection. Further studies are needed to support the use of nanoparticle-enhanced methods in routine diagnosis [29,53].

There are some difficulties in the treatment of malaria. Poor adherence to treatment, complex life cycle of the parasite, mutation rate of the parasite, overall parasite burden, and efficacy of selected drugs are among these challenges. In particular, drug resistance is considered one of the most important causes of treatment failure. Some researchers stated that to overcome these disadvantages, effective drug carriers should be developed instead of discovering new drugs. It is thought that the success rate of treatment will increase by administering drugs to the area where drug interaction is desired through nano-structured drug carriers. With this system, it is aimed to have the least toxic effect and to accumulate the drug only in the desired area. Nanoparticles used in drug delivery are liposomes, nanostructured lipid carriers, solid lipid nanoparticles, polymeric nanoparticles, metallic nanoparticles and nanoemulsions. Studies have shown that several nanostructured systems such as polymeric nanoparticles, liposomes or dendrimers can increase the effectiveness of antimalarial treatments [38,50].

Liposomes are the most widely used nanoparticles for drug delivery. Due to their hydrophilic/hydrophobic structure, they have the ability to encapsulate water-soluble drugs in their core and phospholipid layer. Liposomes can be targeted to specific tissues in line with their unique properties or by binding of specific ligands to their surfaces. Thus, it can increase the pharmacokinetic effect of drugs in the desired target region. In addition, it can carry both hydrophilic and hydrophobic drugs, has a stable structure, is biodegradable and non-toxic, and can be administered by parenteral and cutaneous routes [38,50,54].

The nanoemulsion can encapsulate both water- and fat-soluble drugs. It is easy to prepare, both lipophilic and hydrophobic drugs are portable, have a long shelf life, can be used in oral, parenteral, and cutaneous administration routes. The disadvantages of nanoemulsions include the increased risk of toxicity due to the use of large amounts of surfactants [50].

Nanostructured lipid carriers are considered as an alternative to polymeric nanoparticles, emulsions, and liposomes. They are second generation lipid carriers created to reduce the complications of solid lipid nanoparticles. It shows properties similar to traditional drug carrier properties [50].

Solid lipid nanoparticles are composed of a solid lipid core containing surfactants no larger than 1,000 nm. Lipid-based formulations can reduce drug toxicity and increase bioavailability due to their unique physiological and biodegradable properties. Solid lipid nanoparticles can easily enter the cell because of their small nanoparticle size. The lipid entering the cell is decomposed by lysosomes and then released rapidly, they can directly affect the parasites inside the cell. By enhancing the therapeutic effect of antiparasitic drugs, nanomaterials are considered as a promising alternative approach in therapy [50,54].

It is known that silver, one of the most used nanoparticles, has been used in the treatment of wounds and burns since ancient times. Silver, one of the most widely used nanoparticles, has been used in the treatment of wounds and sores since ancient times. The antimicrobial effect of silver-containing compounds has long been investigated. Studies have shown that inorganic nanoparticles such as silver (gold, magnesium oxide, iron oxide, aluminum oxide, etc.) have antimalarial activity in malaria [37,47].

Studies have shown that inorganic nanoparticles such as silver (gold, magnesium oxide, iron oxide, aluminum oxide, etc.) have antimalarial activity [38,55]. The effects of the broad-spectrum antimicrobial activity of silver in malaria have been investigated. Studies have shown that silver nanoparticles inhibit the growth of *P. falciparum* in vivo and in vitro. In research investigating the effect of silver nanoparticles on liver damage caused by malaria, AgNPs were found to have antiplasmodial and hepatoprotective properties. Nanomaterials are considered as a promising alternative approach in treatment by enhancing the therapeutic effect of antiparasitic drugs [33,56,57].

Nanoparticles can be used as drug carriers in the treatment of disease, as well as in the fight against mosquitoes for disease control. Today, the lack of an effective vaccine against malaria and the difficulties in treatment reveal the importance of vector control. It is very important to break the mosquito-human transmission cycle for the eradication of malaria infection. Vector control plays an important role in malaria transmission and prevention. Insecticides widely used in the control of malaria are dichloro-diphenyl-trichloroethane (DDT), malathion, dieldrin, phenitrothion, Pirimiphos-methyl WP & EC, Pirimiphos-methyl CS, bendiocarb, propoxur WP, alpha-cypermethrin WP& SC, Bifenthrin WP, Cyfluthrin, deltamethrin, cyfluthrin, eto-fenprox and lambda-cyhalothrin. DDT, which is used in vector control, has been prohibited in many countries due to its toxicity and long-lasting effects on the environment. However, it is used in some countries where malaria is still a more serious public health problem. Chemical insecticides used for mosquito control can cause many serious problems, including toxic effects on humans, resistance development in mosquitoes, environmental pollution and toxicity in non-target organisms. Therefore, it is very important to develop safe and effective insecticides for vector control that do not harm the environment [33,35,58–61].

Insecticides developed with nano-biotechnology are considered as promising larvicide agents in vector control with their biologically compatible and non-toxic properties. In line with the studies, it was stated that plant-derived nanoparticle compounds should be used in the fight against mosquitoes with their larval and egg-killing, egg-laying deterrent, and growth inhibitory properties [50,58,61].

Metal nanoparticles synthesized from plants are known to be of great medical and veterinary importance. It has also been reported that these nanoparticles are effective ovicidal, larvicidal, pupicidal, adult mosquito killer and oviposition deterrents against mosquito species in the fight against malaria. While silver and gold nanoparticles are among the most frequently used products within the scope of green nanotechnology, it is known that copper, iron, lead, zinc oxide, titanium dioxide and carbon nanoparticles are used less frequently. For nanoparticles obtained from plant extracts, whole plants, plant tissue and fruits, plant extracts, seaweeds and microorganisms are used. This method, which is expressed as green nanotechnology, is more advantageous than other insecticide methods because it is cheap, does not require high pressure, temperature, and highly toxic use [62–64].

Mosquitoes have egg, larva, pupa, and adult stages in their life cycle. Therefore, in order to control the parasite, insecticides with high efficiency should be used at these stages. In the studies, successful results were obtained in the control of green nanoparticles and mosquitoes. In particular, there are many studies on the toxicity of nanoparticles in the larval and pupal stages of mosquitoes. However, there is limited information on its ovipositoric and oviposition deterrent properties, and further studies are needed in this area [65–67].

6. Conclusion

Malaria is a serious health problem especially in tropical and subtropical regions. There is currently no effective vaccine against malaria. The course of the disease is influenced by the complex life cycle of the parasite, the mutation rate of the parasite, the overall parasite burden and the effectiveness of the selected drugs. Since the drugs used in malaria treatment do not specifically target the pathogen, they may cause unwanted side effects. It is aimed to minimize unwanted side effects in treatment through nano-structured drug carriers that have the ability to bind only to the cells infected by the parasite. Issues related to drug solubility, biostability, toxicity and uncontrolled pharmacokinetics are thought to be overcome with drug carriers engineered into liposomes, polymers or dendrimers. There have been many studies reporting that the diagnosis and treatment of disease can be done with methods developed with nanoparticles. The methods developed to diagnose the disease must have high sensitivity and specificity. It is important to develop a safe and environmentally friendly method for the prophylaxis and treatment of the disease. The use of nanoparticles in the fight against mosquitoes, which are important in the diagnosis, treatment, and prevention of malaria, is promising for the eradication of the disease. With advancing technology, more studies are needed to develop and implement biocompatible systems that can benefit malaria diagnosis, control, and treatment.

Despite a large body of research on the synthesis of nanoparticulate plants for mosquito control, there is a gap between theoretical and practical applications. Eliminating the shortcomings in the practical application of these environmentally friendly methods will contribute to the eradication of the disease. In addition, mosquito-borne diseases such as Malaria, Yellow fever, Dengue, Zika and Chikungunya will be brought under control. With advancing technology, more studies are needed to develop and implement biocompatible systems that can benefit malaria diagnosis, control, and treatment.

References

[1]   C. Buzea, I.I. Pacheco, K. Robbie, Nanomaterials and Nanoparticles: Sources and Toxicity, Biointerphases 2 (2007) MR17–MR71. https://doi.org/10.1116/1.2815690

[2]   A. ÇÜÇEN, Y.T. ALTUNCI, Nanoteknolojik Yapı Malzemelerinin Mimaride Kullanım Olanaklarının Araştırılması, Tek. Bilim. Derg. 12 (2022) 17–23. https://doi.org/10.35354/tbed.984956

[3]   F. Şen, G. Gökağaç, Improving catalytic efficiency in the methanol oxidation reaction by inserting Ru in face-centered cubic Pt nanoparticles prepared by a new surfactant, tert-octanethiol, Energy and Fuels 22 (2008) 1858–1864. https://doi.org/10.1021/ef700575t

[4]   S. Günbatar, A. Aygun, Y. Karataş, M. Gülcan, F. Şen, Carbon Nanotube Based Rhodium Nanoparticles as Highly Active Catalyst for Hydrolytic Dehydrogenation of Dimethylamineborane at Room Temperature, J. Colloid Interface Sci. 530 (2018) 321–327.

https://doi.org/10.1016/j.jcis.2018.06.100

[5]   H. Goksu, Y. Yildiz, B. Çelik, M. Yazici, B. Kilbas, F. Sen, Eco-friendly hydrogenation of aromatic aldehyde compounds by tandem dehydrogenation of dimethylamine-borane in the presence of a reduced graphene oxide furnished platinum nanocatalyst, Catal. Sci. Technol. 6 (2016) 2318–2324. https://doi.org/10.1039/C5CY01462J

[6]   B. Şen, B. Demirkan, A. Savk, R. Kartop, M.S. Nas, M.H. Alma, S. Sürdem, F. Şen, High-performance graphite-supported ruthenium nanocatalyst for hydrogen evolution reaction, J. Mol. Liq. 268 (2018) 807–812. https://doi.org/10.1016/j.molliq.2018.07.117

[7]   M. Rai, A. Ingle, Role of nanotechnology in agriculture with special reference to management of insect pests, Appl. Microbiol. Biotechnol. 94 (2012) 287–293. https://doi.org/10.1007/s00253-012-3969-4

[8]   A. Şavk, K. Cellat, K. Arıkan, F. Tezcan, S.K. Gülbay, S. Kızıldağ, E.Ş. Işgın, F. Şen, Highly monodisperse Pd-Ni nanoparticles supported on rGO as a rapid, sensitive, reusable and selective enzyme-free glucose sensor, Sci. Rep. 9 (2019) 19228. https://doi.org/10.1038/s41598-019-55746-y

[9]   M.E. Vance, T. Kuiken, E.P. Vejerano, S.P. McGinnis, M.F. Hochella, D. Rejeski, M.S. Hull, Nanotechnology in the real world: Redeveloping the nanomaterial consumer products inventory, Beilstein J. Nanotechnol. 6 (2015) 1769–1780. https://doi.org/10.3762/bjnano.6.181

[10]   Z. Ferdous, A. Nemmar, Health Impact of Silver Nanoparticles: A Review of the Biodistribution and Toxicity Following Various Routes of Exposure, Int. J. Mol. Sci. 21 (2020) 2375. https://doi.org/10.3390/ijms21072375

[11]   C.A. Pimentel-Acosta, J. Ramírez-Salcedo, F.N. Morales-Serna, E.J. Fajer-Ávila, C. Chávez-Sánchez, H.H. Lara, A. García-Gasca, Molecular Effects of Silver Nanoparticles on Monogenean Parasites: Lessons from Caenorhabditis elegans, Int. J. Mol. Sci. 21 (2020) 5889. https://doi.org/10.3390/ijms21165889

[12]   Y. Yang, Z. Qin, W. Zeng, T. Yang, Y. Cao, C. Mei, Y. Kuang, Toxicity assessment of nanoparticles in various systems and organs, Nanotechnol. Rev. 6 (2017) 279–289. https://doi.org/10.1515/ntrev-2016-0047

[13]   R. Nagraik, A. Sharma, D. Kumar, S. Mukherjee, F. Sen, A.P. Kumar, Amalgamation of biosensors and nanotechnology in disease diagnosis: Mini-review, Sensors Int. 2 (2021) 100089. https://doi.org/10.1016/J.SINTL.2021.100089

[14]   A. Hojjati-Najafabadi, A. Aygun, R.N.E. Tiri, F. Gulbagca, M.I. Lounissaa, P. Feng, F. Karimi, F. Sen, Bacillus thuringiensis Based Ruthenium/Nickel Co-Doped Zinc as a Green Nanocatalyst: Enhanced Photocatalytic Activity, Mechanism, and Efficient H2Production from Sodium Borohydride Methanolysis, Ind. Eng. Chem. Res. 62 (2023) 4655–4664. https://doi.org/10.1021/ACS.IECR.2C03833

[15]   B. Sen, B. Demirkan, A. Şavk, S. Karahan Gülbay, F. Sen, Trimetallic PdRuNi nanocomposites decorated on graphene oxide: A superior catalyst for the hydrogen evolution reaction, Int. J. Hydrogen Energy (2018). https://doi.org/10.1016/j.ijhydene.2018.07.122

[16]   R. Ayranci, G. Baskaya, M. Guzel, S. Bozkurt, M. Ak, A. Savk, F. Sen, Enhanced optical and electrical properties of PEDOT via nanostructured carbon materials: A comparative investigation, Nano-Structures & Nano-Objects 11 (2017) 13–19. https://doi.org/10.1016/J.NANOSO.2017.05.008

[17]   K. Nesrin, C. Yusuf, K. Ahmet, S.B. Ali, N.A. Muhammad, S. Suna, Ş. Fatih, Biogenic silver nanoparticles synthesized from Rhododendron ponticum and their antibacterial and cytotoxic activities, J. Pharm. Biomed. Anal. 179 (2020) 112993. https://doi.org/10.1016/J.JPBA.2019.112993

[18]   N. Korkmaz, Y. Ceylan, P. Taslimi, A. Karadağ, A.S. Bülbül, F. Şen, Biogenic nano silver: Synthesis, characterization, antibacterial, antibiofilms, and enzymatic activity, Adv. Powder Technol. 31 (2020) 2942–2950. https://doi.org/10.1016/j.apt.2020.05.020

[19]   F. Şen, Ö. Demirbaş, M.H. Çalımlı, A. Aygün, M.H. Alma, M.S. Nas, The dye Removal from Aqueous Solution Using Polymer Composite Films, Appl. Water Sci. 8 (2018) 206.

https://doi.org/10.1007/s13201-018-0856-x

[20] Y. Wu, E.E. Altuner, R.N. El Houda Tiri, M. Bekmezci, F. Gulbagca, A. Aygun, C. Xia, Q. Van Le, F. Sen, H. Karimi-Maleh, Hydrogen generation from methanolysis of sodium borohydride using waste coffee oil modified zinc oxide nanoparticles and their photocatalytic activities, Int. J. Hydrogen Energy 48 (2023) 6613–6623. https://doi.org/10.1016/J.IJHYDENE.2022.04.177

[21] A. Hojjati-Najafabadi, S. Salmanpour, F. Sen, P.N. Asrami, M. Mahdavian, M.A. Khalilzadeh, A Tramadol Drug Electrochemical Sensor Amplified by Biosynthesized Au Nanoparticle Using Mentha aquatic Extract and Ionic Liquid, Top. Catal. 65 (2022) 587–594. https://doi.org/10.1007/s11244-021-01498-x

[22] N.H. Khand, A.R. Solangi, S. Ameen, A. Fatima, J.A. Buledi, A. Mallah, S.Q. Memon, F. Sen, F. Karimi, Y. Orooji, A new electrochemical method for the detection of quercetin in onion, honey and green tea using Co3O4 modified GCE, J. Food Meas. Charact. 2021 154 15 (2021) 3720–3730. https://doi.org/10.1007/S11694-021-00956-0

[23] A. Cherif, R. Nebbali, J.W. Sheffield, N. Doner, F. Sen, Numerical investigation of hydrogen production via autothermal reforming of steam and methane over Ni/Al2O3 and Pt/Al2O3 patterned catalytic layers, Int. J. Hydrogen Energy (2021). https://doi.org/10.1016/j.ijhydene.2021.04.032

[24] J.T. Abrahamson, B. Sempere, M.P. Walsh, J.M. Forman, F. Şen, S. Şen, S.G. Mahajan, G.L.C. Paulus, Q.H. Wang, W. Choi, M.S. Strano, Excess Thermopower and the Theory of Thermopower Waves, ACS Nano 7 (2013) 6533–6544. https://doi.org/10.1021/nn402411k

[25] F. Gulbagca, A. Aygün, M. Gülcan, S. Ozdemir, S. Gonca, F. Şen, Green Synthesis of Palladium Nanoparticles: Preparation, Characterization, and Investigation of Antioxidant, Antimicrobial, Anticancer, and DNA Cleavage Activities, Appl. Organomet. Chem. 35 (2021). https://doi.org/10.1002/aoc.6272

[26] M.B. Askari, P. Salarizadeh, A. Di Bartolomeo, F. Şen, Enhanced electrochemical performance of MnNi2O4/rGO nanocomposite as pseudocapacitor electrode material and methanol electro-oxidation catalyst, Nanotechnology 32 (2021). https://doi.org/10.1088/1361-6528/abfded

[27] R. Ayranci, G. Başkaya, M. Güzel, S. Bozkurt, F. Şen, M. Ak, Carbon Based Nanomaterials for High Performance Optoelectrochemical Systems, ChemistrySelect 2 (2017) 1548–1555. https://doi.org/10.1002/slct.201601632

[28] A. Aygun, F. Gulbagca, E.E. Altuner, M. Bekmezci, T. Gur, H. Karimi-Maleh, F. Karimi, Y. Vasseghian, F. Sen, Highly Active PdPt Bimetallic Nanoparticles Synthesized By One Step Bioreduction Method: Characterizations, Anticancer, Antibacterial Activities and Evaluation of Their Catalytic Effect for Hydrogen Generation, Int. J. Hydrogen Energy 48 (2023) 6666–6679. https://doi.org/10.1016/j.ijhydene.2021.12.144

[29] S. Abaza, Applications of nanomedicine in parasitic diseases, Parasitol. United J. 9 (2016) 1. https://doi.org/10.4103/1687-7942.192997

[30] M.S. Younis, E.A. el rahman Abououf, A.E. saeed Ali, S.M. Abd elhady, R.M. Wassef, In vitro Effect of Silver Nanoparticles on Blastocystis hominis, Int. J. Nanomedicine Volume 15 (2020) 8167–8173. https://doi.org/10.2147/IJN.S272532

[31] E. Yula, Ö. Deveci, M. İnci, A. Tekin, Intestinal parasites and report of etiological analysis in a state hospital, J. Clin. Exp. Investig. 2 (2011). https://doi.org/10.5799/ahinjs.01.2011.01.0213

[32] A. EKİCİ, Ş. YÜREKTÜRK, S. ELASAN, A.G. HALİDİ, S. KARAKUŞ, S. AYDEMİR, M. ŞAHİN, M. YASUL, H. YİLMAZ, HEALTH SERVICES VOCATIONAL SCHOOL STUDENTS' KNOWLEDGE LEVELS OF ON PARASITIC DISEASES, İnönü Üniversitesi Sağlık Hizmetleri Mesl. Yüksek Okulu Derg. 10 (2022) 1–11. https://doi.org/10.33715/inonusaglik.995026

[33] H. Barabadi, Z. Alizadeh, M.T. Rahimi, A. Barac, A.E. Maraolo, L.J. Robertson, A. Masjedi, F. Shahrivar, E. Ahmadpour, Nanobiotechnology as an emerging approach to combat malaria: A systematic

review, Nanomedicine Nanotechnology, Biol. Med. 18 (2019) 221–233. https://doi.org/10.1016/j.nano.2019.02.017

[34]   A. Abate, I. Bouyssou, S. Mabilotte, C. Doderer-Lang, L. Dembele, D. Menard, L. Golassa, Vivax malaria in Duffy-negative patients shows invariably low asexual parasitaemia: implication towards malaria control in Ethiopia, Malar. J. 21 (2022) 230. https://doi.org/10.1186/s12936-022-04250-2

[35]   B. DUNCAN, Basic Clinical Parasitology, Arch. Pediatr. Adolesc. Med. 137 (1983) 704. https://doi.org/10.1001/archpedi.1983.02140330086026

[36]   N. Nazari, Y. Hamzavi, M. Rezaei, P. Khoshbo, A brief review of malaria epidemiological trend in Kermanshah province, Iran, 1986–2014, J. Med. Life 15 (2022) 392–396. https://doi.org/10.25122/jml-2021-0374

[37]   N.M. Kafai, A.R. Odom John, Malaria in Children, Infect. Dis. Clin. North Am. 32 (2018) 189–200. https://doi.org/10.1016/j.idc.2017.10.008

[38]   L. Neves Borgheti-Cardoso, M. San Anselmo, E. Lantero, A. Lancelot, J.L. Serrano, S. Hernández-Ainsa, X. Fernàndez-Busquets, T. Sierra, Promising nanomaterials in the fight against malaria, J. Mater. Chem. B 8 (2020) 9428–9448. https://doi.org/10.1039/D0TB01398F

[39]   J. Popovici, C. Roesch, V. Rougeron, The enigmatic mechanisms by which Plasmodium vivax infects Duffy-negative individuals, PLOS Pathog. 16 (2020) e1008258. https://doi.org/10.1371/journal.ppat.1008258

[40]   I. Mueller, P.A. Zimmerman, J.C. Reeder, Plasmodium malariae and Plasmodium ovale – the 'bashful' malaria parasites, Trends Parasitol. 23 (2007) 278–283. https://doi.org/10.1016/j.pt.2007.04.009

[41]   W.E. Collins, G.M. Jeffery, Plasmodium malariae : Parasite and Disease, Clin. Microbiol. Rev. 20 (2007) 579–592. https://doi.org/10.1128/CMR.00027-07

[42]   B.E. Barber, M.J. Grigg, D.J. Cooper, D.A. van Schalkwyk, T. William, G.S. Rajahram, N.M. Anstey, Clinical management of Plasmodium knowlesi malaria, in: 2021: pp. 45–76. https://doi.org/10.1016/bs.apar.2021.08.004

[43]   A. Bartoloni, L. Zammarchi, CLINICAL ASPECTS OF UNCOMPLICATED AND SEVERE MALARIA, Mediterr. J. Hematol. Infect. Dis. 4 (2012) e2012026. https://doi.org/10.4084/mjhid.2012.026

[44]   E.S. Theel, B.S. Pritt, Parasites, Microbiol. Spectr. 4 (2016). https://doi.org/10.1128/microbiolspec.DMIH2-0013-2015

[45]   R. Varo, N. Balanza, A. Mayor, Q. Bassat, Diagnosis of clinical malaria in endemic settings, Expert Rev. Anti. Infect. Ther. 19 (2021) 79–92. https://doi.org/10.1080/14787210.2020.1807940

[46]   A. Guasch-Girbau, X. Fernàndez-Busquets, Review of the Current Landscape of the Potential of Nanotechnology for Future Malaria Diagnosis, Treatment, and Vaccination Strategies, Pharmaceutics 13 (2021) 2189. https://doi.org/10.3390/pharmaceutics13122189

[47]   B.S.S. Guirgis, C. Sá e Cunha, I. Gomes, M. Cavadas, I. Silva, G. Doria, G.L. Blatch, P. V. Baptista, E. Pereira, H.M.E. Azzazy, M.M. Mota, M. Prudêncio, R. Franco, Gold nanoparticle-based fluorescence immunoassay for malaria antigen detection, Anal. Bioanal. Chem. 402 (2012) 1019–1027. https://doi.org/10.1007/s00216-011-5489-y

[48]   J. Wu, Y. Peng, X. Liu, W. Li, S. Tang, Evaluation of wondfo rapid diagnostic kit (Pf-HRP2/PAN-pLDH) for diagnosis of malaria by using nano-gold immunochromatographic assay, Acta Parasitol. 59 (2014). https://doi.org/10.2478/s11686-014-0238-y

[49]   O. Salman, T. Erbaydar, Drug resistant malaria, TAF Prev. Med. Bull. 15 (2016) 368. https://doi.org/10.5455/pmb.1-1453411620

[50]   L. Gujjari, H. Kalani, S.K. Pindiprolu, B.P. Arakareddy, G. Yadagiri, Current challenges and nanotechnology-based pharmaceutical strategies for the treatment and control of malaria, Parasite Epidemiol. Control 17 (2022) e00244. https://doi.org/10.1016/j.parepi.2022.e00244

[51]   T. ULUDAĞ, Halk Sağlığı Bakış Açısı ile Biyoterörizm (Bioterrorism in Public Health Perspective), STED / Sürekli Tıp Eğitimi Derg. (2022). https://doi.org/10.17942/sted.904333

[52]   M. de Souza Castilho, T. Laube, H. Yamanaka, S. Alegret, M.I. Pividori, Magneto Immunoassays for Plasmodium falciparum Histidine-Rich Protein 2 Related to Malaria based on Magnetic Nanoparticles, Anal. Chem. 83 (2011) 5570–5577. https://doi.org/10.1021/ac200573s

[53]   R. Thiramanas, K. Jangpatarapongsa, U. Asawapirom, P. Tangboriboonrat, D. Polpanich, Sensitivity and specificity of PS/AA-modified nanoparticles used in malaria detection, Microb. Biotechnol. 6 (2013) 406–413. https://doi.org/10.1111/1751-7915.12021

[54]   Y. Sun, D. Chen, Y. Pan, W. Qu, H. Hao, X. Wang, Z. Liu, S. Xie, Nanoparticles for antiparasitic drug delivery, Drug Deliv. 26 (2019) 1206–1221. https://doi.org/10.1080/10717544.2019.1692968

[55]   B. Aderibigbe, Metal-Based Nanoparticles for the Treatment of Infectious Diseases, Molecules 22 (2017) 1370. https://doi.org/10.3390/molecules22081370

[56]   A. Mishra, N.K. Kaushik, M. Sardar, D. Sahal, Evaluation of antiplasmodial activity of green synthesized silver nanoparticles, Colloids Surfaces B Biointerfaces 111 (2013) 713–718. https://doi.org/10.1016/j.colsurfb.2013.06.036

[57]   K. Rahman, S.U. Khan, S. Fahad, M.X. Chang, A. Abbas, W.U. Khan, L. Rahman, Z.U. Haq, G. Nabi, D. Khan, Nano-biotechnology: a new approach to treat and prevent malaria, Int. J. Nanomedicine Volume 14 (2019) 1401–1410. https://doi.org/10.2147/IJN.S190692

[58]   K. Murugan, C. Panneerselvam, J. Subramaniam, M. Paulpandi, R. Rajaganesh, M. Vasanthakumaran, J. Madhavan, S.S. Shafi, M. Roni, J.S. Portilla-Pulido, S.C. Mendez, J.E. Duque, L. Wang, A.T. Aziz, B. Chandramohan, D. Dinesh, S. Piramanayagam, J.-S. Hwang, Synthesis of new series of quinoline derivatives with insecticidal effects on larval vectors of malaria and dengue diseases, Sci. Rep. 12 (2022) 4765. https://doi.org/10.1038/s41598-022-08397-5

[59]   K. MOTOHIRA, Y. IKENAKA, Y.B. YOHANNES, S.M.M. NAKAYAMA, V. WEPENER, N.J. SMIT, J.H.J. VAN VUREN, A.C. SOUSA, A.A. ENUNEKU, E.T. OGBOMIDA, M. ISHIZUKA, Dichlorodiphenyltrichloroethane (DDT) levels in rat livers collected from a malaria vector control region, J. Vet. Med. Sci. 81 (2019) 1575–1579. https://doi.org/10.1292/jvms.19-0168

[60]   M.O. Ndiath, Insecticides and Insecticide Resistance, in: 2019: pp. 287–304. https://doi.org/10.1007/978-1-4939-9550-9_18

[61]   V.N. Kalpana, K.M. Alarjani, V.D. Rajeswari, Enhancing malaria control using Lagenaria siceraria and its mediated zinc oxide nanoparticles against the vector Anopheles stephensi and its parasite Plasmodium falciparum, Sci. Rep. 10 (2020) 21568. https://doi.org/10.1038/s41598-020-77854-w

[62]   M. BEYKAYA, A. ÇAĞLAR, An Investigation on Synthesis of Silver-Nanoparticles (AgNP) and their Antimicrobial effectiveness by using Herbal Extracts, Afyon Kocatepe Univ. J. Sci. Eng. 16 (2016) 631–641. https://doi.org/10.5578/fmbd.34220

[63]   G. Benelli, Plant-mediated biosynthesis of nanoparticles as an emerging tool against mosquitoes of medical and veterinary importance: a review, Parasitol. Res. 115 (2016) 23–34. https://doi.org/10.1007/s00436-015-4800-9

[64]   V. Soni, P. Raizada, P. Singh, H.N. Cuong, R. S, A. Saini, R. V. Saini, Q. Van Le, A.K. Nadda, T.-T. Le, V.-H. Nguyen, Sustainable and green trends in using plant extracts for the synthesis of biogenic metal nanoparticles toward environmental and pharmaceutical advances: A review, Environ. Res. 202 (2021) 111622. https://doi.org/10.1016/j.envres.2021.111622

[65]   T. Santhoshkumar, A.A. Rahuman, G. Rajakumar, S. Marimuthu, A. Bagavan, C. Jayaseelan, A.A. Zahir, G. Elango, C. Kamaraj, Synthesis of silver nanoparticles using Nelumbo nucifera leaf extract and its larvicidal activity against malaria and filariasis vectors, Parasitol. Res. 108 (2011) 693–702. https://doi.org/10.1007/s00436-010-2115-4

[66]   K. Murugan, D. Dinesh, K. Kavithaa, M. Paulpandi, T. Ponraj, M.S. Alsalhi, S. Devanesan, J.

Subramaniam, R. Rajaganesh, H. Wei, S. Kumar, M. Nicoletti, G. Benelli, Hydrothermal synthesis of titanium dioxide nanoparticles: mosquitocidal potential and anticancer activity on human breast cancer cells (MCF-7), Parasitol. Res. 115 (2016) 1085–1096. https://doi.org/10.1007/s00436-015-4838-8

[67]   H. Fouad, L. Hongjie, D. Yanmei, Y. Baoting, A. El-Shakh, G. Abbas, M. Jianchu, Synthesis and characterization of silver nanoparticles using Bacillus amyloliquefaciens and Bacillus subtilis to control filarial vector Culex pipiens pallens and its antimicrobial activity, Artif. Cells, Nanomedicine, Biotechnol. 45 (2017) 1369–1378. https://doi.org/10.1080/21691401.2016.1241793

Biogenic Nanomaterials: Synthesis, Characterization, Applications, and Future Remarks   Materials Research Forum LLC
Materials Research Foundations 180 (2025)   https://doi.org/21741/9781644903759

Chapter 12

Biomimetic Nanomaterials and Their Applications in Medicine and Surgery

R. Orhun[1]*, M. H. Izzuddin[2], F. Sen[1]*

[1] Faculty of Health Sciences, Yuzuncu Yıl University, Van, Turkiye

[2]Sen Research Group, Department of Biochemistry, Kutahya Dumlupinar University, Kutahya, Turkiye

Email (reyhanorhun@yyu.edu.tr; fatih.sen@dpu.edu.tr)

Abstract

This book chapter discusses the applications of biogenic nanomaterials derived from natural sources in the fields of medicine, dentistry, and surgery. With the advancement of nanotechnology, biogenic nanomaterials have demonstrated a wide range of potential uses in medical and dental practices. This chapter underscores the future potential of biogenic nanomaterials derived from natural sources in the fields of medicine and dentistry, offering researchers, clinicians, and industrial practitioners' valuable insights into how biogenic nanomaterials can provide benefits in clinical applications.

Keywords

Biogenic Nanomaterials, Medicine, Dentistry, Surgery, Nanotechnology

1. Introduction

Imagine something small. Really, tiny—smaller than anything you can see under a microscope. Think of atoms and molecules, and now you are there; you are at the nanoscale.

Nature has used the art of biology at the nanoscale for thousands of years and impressed us with its perfect designs at the nanoscale. These harmonious relationships at the nanoscale have been the inspiration of science and provided discipline of conformity to the name of nanotechnology. Many cellular activities occur naturally at the nanoscale. For example, the diameter of hemoglobin, which carries oxygen in the body, is 5.5 nanometers. The diameter of a DNA strand is only 2 nanometers. These natural nano-features, such as the diameter of the oxygen molecules carried by hemoglobin, contain unique phenomena that inspire nanotechnology. Nanotechnology is a discipline that aims to study, understand and control this nano-sized world and has laid the foundations of the modern technological revolution.

Nanotechnology, defined as the understanding and control of matter at the nanoscale, with dimensions between approximately 1 and 100 nanometers, where unique phenomena enable new applications, has emerged as a "new technological revolution" in our lives.

Nanotechnology is a scientific discipline that creates new materials and systems by altering the physical, chemical, and biological properties of structures or components at the atomic level. Its main goal is to enhance the quality of life by efficiently producing materials at the nanoscale and achieving more effective use of limited resources. The desire to develop more durable, higher quality, more economical, longer-lasting, lighter, and more space-efficient devices has gained widespread acceptance in various aspects of life.

Research in this field aims to provide in-depth understanding of the design, production and application of these materials by examining the properties of materials at the nanoscale. Nanotechnology allows obtaining more durable and lightweight materials by manipulating and restructuring materials at nanoscales, creating structures such as nanotubes, nanowires and nanospheres [1–6]. Additionally, nanotechnology offers a wide range of applications, from drug delivery systems to solar cells, electromechanical systems to biosensors. This technology allows the development of new generation technological products and processes, based on the fact that changes made at the nanoscale level can have significant consequences at the macroscale level [7–10]. In this way, nanotechnology enables a significant transformation in industrial and scientific fields by providing solutions on important issues such as sustainability, energy efficiency and environmental impacts.

Nanotechnology brings many innovations using today's advanced technologies. Inspired by the natural world, nanotechnology has the potential to revolutionize the production of numerous products, ranging from the clothes we wear to the medications we take when we fall ill, from computer hard drives to even cleaning products, fundamentally altering the world we live in.

2.   Nanotechnology and Nanoscience

Nanotechnology is a scientific discipline that deals with the control and manipulation of structures at a very small scale. In the 21st century, significant advancements have been made in the use of nanotechnology in various fields, including medicine, biotechnology, computer technology, aviation, energy utilization, space exploration, materials, and manufacturing sectors [11].

The emergence of the concept of nanotechnology gained momentum after a speech by physicist Dr. Feynman in 1959, in which he discussed the possibility of arranging tools and devices at the molecular level [12].

The term "nano," which originates from the Greek word for dwarf, is used as a technical unit of measurement and represents one billionth of a physical quantity. A nanometer expresses a length that is one billionth of a meter, approximately representing the length obtained by aligning seven atoms side by side [13].

Nanoscience is a scientific discipline that focuses on understanding the differentiated physical and chemical properties of matter at the nano scale, which is one billionth of a meter (ranging from 0.1 nm to 100 nm). Although nanoscience has been an area of interest for chemists for centuries, recent research has shifted its focus to applied technology due to the discovery of new methods. The practical application of nanoscience is referred to as nanotechnology [14].

Nanotechnology is defined as the control, production, and engineering of matter at the atomic and molecular levels. The United States National Nanotechnology Initiative describes nanotechnology as the control of matter in the size range of 1 to 100 nm [5,6].

Nanotechnology deals with structures and components at the nanoscale, where the physical, chemical, and biological properties of materials and systems vary. The aim of nanoscience and nanotechnology is to produce materials and structures that can serve specific functions at the nanoscale, to create devices with defined properties and functions at the nano-level, and to integrate these devices into our daily lives [17].

Nanoscience and nanotechnology have permeated various aspects of our lives, starting from information technology and communication and extending to defense industries, aerospace and aviation technologies, and even molecular biology and genetic engineering [13].

At the core of nanotechnology lies the ability to work at the molecular level, to structure atom by atom, and to create comprehensive structures through new molecular arrangements [18].

Research on nanoparticles has shown significant progress as an important field of investigation in recent years. Nanotechnology, currently a sought-after and valuable area of research, is believed to initiate a new era for future generations, given the recognition of potential changes it can bring about [12].

By enabling the development of novel products in industries, information technology, healthcare, and many other fields, nanotechnology is expected to elevate the quality of life for societies.

3.   Nanomaterials

A nanomaterial is an object with at least one dimension at the nanoscale, measured in nanometers. Materials used in the field of nanotechnology and ranging in size from 1 to 100 nm are defined as nanoparticles, forming the foundation of nanotechnology and, consequently, nanomaterials.

Products obtained through nanoparticles are introduced into the realm of new technology materials. These structures, which are extremely small in size, exhibit various superior properties, such as mechanical alterations, changes in conductivity, alterations in surface characteristics, extended lifespan, and possessing a high surface-to-volume ratio [8,9].

Nanomaterials can be of two types:

1.   Naturally occurring unintentional nanomaterials (e.g., proteins, viruses produced during volcanic eruptions) or nanomaterials unintentionally generated through human activities (e.g., nanoparticles produced from diesel combustion).

2.   Intentionally manufactured nanomaterials, referring to nanomaterials produced through a defined manufacturing process [20].

What sets nanomaterials apart from materials with larger particles is not only the size of nanoparticles but also the distinct and superior properties they exhibit compared to voluminous materials. Nanomaterials possess different structural characteristics in terms of chemical reactions, energy absorption, and biological mobility [21].

When examining the advanced nanotechnologies used today, applications can be observed in various fields such as healthcare, pharmaceutical industry, textiles, electronics, automotive, food, and aerospace [22].

4.   Production of Nanoparticles

The first step towards the recent advancements in nanotechnology, which encompass the synthesis of nanostructured materials, is the production of nanoparticles [12].

4.1   Methods for Obtaining Nanostructures

Nanostructure can be obtained through two main approaches:

1.   Bottom-up approach: From atoms to molecules, and from molecules to materials. This approach involves the assembly of atoms in close proximity to form molecular structures.

2.   Top-down approach: From materials to molecules, and from molecules to atoms. This approach involves the mechanical or chemical breakdown of materials to disassemble their atoms and reorganize them.

Structures defined as nanoscale materials are categorized into various classes, including nanocrystals, nanoparticles, nanotubes, nanowires, and nanorods [23].

5.   Applications for Nanotechnology

In today's context, advancements in nanotechnology have gained significant popularity, continually opening up new areas for scientific research. With the increasing proliferation of nanotechnological applications, many consumers have begun to use nanotechnology-based products due to their distinct and unique properties [24].

Nanotechnology possesses a highly diverse range of potential applications, making it a multidisciplinary field. This characteristic allows the developments in nanotechnology to be applicable across various industries. In this regard, the application areas of nanotechnology in other scientific disciplines are outlined in Table 1 [17].

Table 1. Application Areas of Nanotechnology in Other Sciences

Electronics and Computer Technologies:	By relying on the principle of reducing the dimensions to the nanometer scale, as opposed to the conventional forms of electronic properties of materials, it has been observed that the processing power and capacities of electronic devices increase when these devices are obtained at the nanometer scale.
Aviation and Space Research:	High cost is considered the main challenge in this field. Nanoscale materials, being lighter, stronger, and more heat-resistant, have become a preferred choice for use in this field, leading to cost reduction.
Medical Research:	The application of nanotechnology in the medical field provides significant benefits for human health. The use of nanoparticles for the eradication or inhibition of cancerous tissues and the delivery of drugs to specific areas using nanorobots enable the destruction of only the affected cells, reducing the side effects of medication in the body. Diagnostic tools that move within the human body and implant coatings used in bone tissue injuries are among the applications of nanotechnology in medicine. In this context, the quality, safety, and efficacy of medical devices are enhanced through nanotechnology.

Material Science:	Due to the lightweight and enhanced strength of nanoscale materials, the production stage requires smaller quantities of materials and consumes less energy, leading to cost reduction.
Environment and Energy:	Nanotechnology applications facilitate the development of cleaner and sustainable production systems by reducing raw material and energy usage. Nanotechnology also plays a significant role in energy efficiency, storage, and generation.
Biotechnology and Agriculture:	Nanotechnology applications will contribute to the development of agricultural products. Novel drugs, fertilizers, disease-resistant plants and animals, introduced through nanotechnological studies, impact not only agriculture but also indirectly influence advancements in the healthcare sector.

Source: [17].

In his book titled "Nanotechnology and Its Applications," M.H. Fuleker discusses the significant roles of nanotechnology in various fields, which are listed as follows:

1.   Consumer Electronics and Information Technology

2.   Chemicals and Basic Materials

3.   Pharmaceuticals and Medical Products

4.   Household Cleaning Products

5.   Paints, Varnishes, and Coatings

6.   Chemistry (Personalized Catalysts)

7.   Information and Communication Technology (Nanoelectronics)

8.   Biomedical Applications (e.g., Lab-on-a-Chip, Biosensors, Medical Imaging, Prosthetics and Implants, Drug Delivery Devices)

9.   Environmental Remediation Technology

10.   Energy Capture and Storage Technology (e.g., Solar Cells, Batteries, Fuel Cells, Fuels, and Catalysts)

11.   Agriculture (e.g., Sensors, Seed Breeding)

12.   Food (Anti-bacterial powders, Pathogen, Contaminant Sensors, Environmental Monitors, Remote Sensing and Tracking Devices)

13.   Military Technology

14.   Textiles

15.   Lubricants, etc. [25].

6.   Nanotechnology and Health

The application of nanotechnologies in the medical sector is referred to as nanomedicine. As a subfield of nanotechnology, nanomedicine was defined by the European Science Foundation in 2004 as the use of nanoscale tools for disease prevention, diagnosis, treatment, and a better understanding of the complex pathophysiology of diseases [15].

In the field of medicine, nanotechnology employs nanoscale materials and nano-enabled techniques to diagnose, monitor, treat, and prevent diseases [20]. A living cell is a biological system that functions at the nano scale, and the goal of nanomedicine is to control and harness the functionality of this system to benefit the organism [26].

The development of nanomedicine involves the integration of nanotechnological methods with biotechnology. The application of nanotechnological methods on the foundation of knowledge accumulated in molecular biology has given rise to a new research area called nano-biotechnology. Nanomedicine utilizes nanotechnological methods that mimic the behavior of living systems [27].

Nanotechnology not only enables the production of materials that better integrate with the human body and are more durable but also facilitates the creation of structures that closely resemble or mimic biological molecules, leading to advancements in controlling biological activities. Nano fibers, for instance, can be used as artificial tissue scaffolds for cartilage, bone, blood vessels, heart, and nerve tissues within the organism [28].

Bio-nanotechnology is considered one of the most fascinating application areas of nanoscience by numerous experts. In recent years, nanotechnology has been intensively researched in various applications such as diagnosis, drug delivery, and molecular imaging, leading to remarkable results. There are numerous bio-nanotechnology products containing nanomaterials available in the global market [29].

Although the idea of nano-robots circulating within our blood vessels may seem like a dream at the moment, the advancements achieved so far indicate that this dream might become a reality in the future. The application of nanotechnology products as artificial organs and tissues in tissue engineering also shows promising developments [28].

Nanomedicine has the potential to contribute positively to healthcare services at different levels through new nanotechnology environments. The expected impacts of nanomedicine applications include:

1.   Detection of molecular changes responsible for disease pathogenesis.

2.   Disease diagnosis and imaging.

3.   Drug delivery and therapy.

4.   Multi-functional systems for combined therapeutic and diagnostic applications.

5.   Tools to report the in vivo efficacy of a therapeutic agent.

6.   Nano-scale technologies that accelerate scientific discoveries and fundamental research [30].

Nanotechnology applications have a wide range of potential contributions to the healthcare sector, including therapies, drug delivery systems, pharmaceuticals, new diagnostic tools, imaging agents and methods, implants, and structures used in tissue engineering. In the field of healthcare, nanotechnology has enabled the production of orthopedic prostheses, cardiovascular, neural, plastic and reconstructive, and dental implants, ophthalmic systems, catheters, insulin pumps, sutures, adhesives, and blood substitutes, among other surgical systems [20].

6.1   Use of Nanotechnology in Diagnostic Methods

The use of nanotechnology in diagnostic methods in medical science aims to produce molecular tools for disease prevention, early diagnosis, treatment, wound healing, utilizing the body's molecular information, and controlling cell functions to enhance and preserve health [31].

Current diagnostic and treatment methods in medicine may not always achieve sufficient performance, leading to challenges in diagnosing and treating certain diseases and injuries. To address this, nanotechnology applications have been developed in biomedical and medical fields, including smart drug carriers, medical imaging devices, biosensors, nanomachines (nano/biorobots), implants, and artificial tissues [32].

Accurate diagnosis is the first step in the effective treatment of a suspected disease in healthcare. Quick results are expected from diagnostic methods, while also requiring specificity, minimal error risk, accurate results, and high reliability. Nanomedicine aims to significantly improve the entire diagnostic process to meet these expectations [20].

Advancements in nanotechnology now allow the molecular-level monitoring, repair, construction, and control of human biological systems. Nanomedicine has been used in developing carriers that can facilitate easier delivery of drugs and vaccines into the body. Moreover, there have been significant developments in producing materials that better match the body and are more durable [16,22,23,24].

In MR (Magnetic Resonance) imaging, iron oxide nanoparticles are used as contrast agents to detect and label tumor tissues [32]. Thanks to the superparamagnetic properties of iron oxide nanoparticles, even very small tumor tissues in the body can be detected [35]. In this context, gold nanoparticles (AuNPs) are used in the diagnosis and treatment of certain cancer types due to their superior properties. For example, using gold nanoparticles as a bio-barcode test is one significant application, serving as an important diagnostic tool for prostate cancer. Studies have explored the use of gold nanoparticles (AuNPs) in cancer imaging, diagnosis, treatment, and drug delivery systems [23,24].

Nanotechnology-based systems have drawn attention in the diagnosis of Alzheimer's disease to improve central nervous system functions and imaging effectiveness of diseases. One of the diagnostic methods used to understand the pathophysiological condition of Alzheimer's disease involves Positron Emission Tomography (PET), which uses radio-labeled amyloid ligands [36].

7.   The Use of Nanomaterials in Drug Delivery and Treatment

Nanotechnology has been recognized as an important platform for nanoparticle-based drug delivery systems (NIDS), which have made significant advancements in drug delivery in recent years. These systems have been developed using nanotechnological approaches to enable targeted and effective drug delivery. Nanoparticles are typically structures with sizes ranging from 1 to 1,000 nm and can be made from various materials such as lipids, polymers, silica, proteins/peptides, oligonucleotides, and metals (such as gold, silver, and iron). This diversity allows different types of nanoparticles to provide unique chemical and physical properties for various purposes, such as carrying therapeutic agents, delivering intratumoral drugs, and tumor imaging.

Nanoparticle-based drug delivery systems offer several advantages compared to traditional drug formulations. Storing drugs within nanoparticles increases drug protection and stability, while preventing rapid distribution in the bloodstream, thereby promoting sustained drug effects. Moreover, the surface of nanoparticles can be easily modified to facilitate their attachment to targeted cells and tissues. This enables direct targeting of the drug to the desired site, reducing side effects and enhancing treatment effectiveness. Nanoparticles offer numerous other advantages, such as protecting drugs from premature degradation, enhancing drug absorption in selected tissues, controlling drug distribution and pharmacokinetics in tissues, facilitating intracellular passage, minimizing drug susceptibility to environmental factors, and reducing systemic toxicity [27,28].

Most traditional anticancer agents do not distinguish between cancerous and normal cells, leading to systemic toxicity and side effects. Multifunctional nanotechnological advancements aim to overcome these challenges. Nanoparticle applications minimize drug delivery to normal organs and tissues, thereby reducing systemic toxicity and enabling higher drug doses to be administered to cancer patients [38].

Nanoparticles utilize both physical and biological mechanisms to target cancerous tissues. Unlike normal tissues, solid tumor cells have leaky vascular systems. The weak structure of tumor vascularization provides an entry point for therapeutic nanoparticles circulating in the bloodstream. To penetrate the tumor, drug carriers need to be smaller than the size of pores in the vascular system [39].

Nanocarriers recognize and bind to target cells through ligand-receptor interactions. After the carriers are absorbed by the cells, the drug is released intracellularly. Nanotechnology offers unique opportunities for drug delivery systems and targeted treatment methods. Nano-sized carriers can enable drugs to reach the target site selectively and effectively, reducing side effects and achieving better treatment outcomes. The potential of nanotechnology in cancer treatment has garnered significant interest, particularly in the development of tumor-targeted drug carriers. The ability of nanoparticle drugs to accumulate in tumor regions through the Enhanced Permeability and Retention (EPR) effect suggests that they may be more effective and less toxic than free drugs. Studies have shown successful results with liposomal nanoparticle systems used in cancer treatment. Nanoparticle drug systems in cancer treatment offer significant advantages over traditional treatment methods. Approved liposomal nanoparticle drugs possess the ability to accumulate in tumor tissues through the EPR effect and passive targeting, reducing toxicity and enhancing treatment efficacy. However, the incorporation of active targeting strategies may further improve the specific delivery of these drug systems to cancer cells and enhance treatment outcomes. Future research on nanoparticle drugs will be crucial in advancing the efficacy and patient outcomes in cancer treatment[28,30].

The term "Theranostics," initially used by Funkhouser, defines a field that combines therapeutic methods with diagnostic imaging techniques. Theranostics is a novel area in nuclear medicine that integrates both diagnostic and therapeutic applications within a single system. With the rapid development of the Theranostics concept, loading diagnostic and therapeutic agents onto nanoparticles has given rise to the concept of "nanotheranostics." Nanotheranostics provide a suitable platform to monitor the pharmacokinetics and pharmacodynamics of drugs injected into the body. By combining molecular imaging with molecular therapy, nanotheranostics can be applied to various aspects of personalized treatment, such as early detection of diseases, disease staging, treatment selection, treatment planning, identification of side effects in the early stages of treatment, and planning for follow-up treatments [31]. With this approach, more effective treatments can be developed based on individual characteristics and responses of the disease, while side effects of drugs and treatments can be better monitored and treatment plans can be adapted more precisely.

Nanotechnology products can be deployed throughout the human body, including inside brain vessels, teeth, and other locations. Nanomolecules and drugs possess the remarkable ability to reach their targets without causing harm to healthy tissues. Nanorobots capable of circulating in the bloodstream can be utilized to remove lipids from the vascular lumen, allowing for intervention in cases of vascular blockages. Additionally, by strengthening the immune system with nanomolecules, infections can be treated rapidly. Furthermore, nanomolecules are employed in the early diagnosis and treatment of certain diseases such as cancer [32].

One of the most significant applications of nanotechnology lies in the treatment of neurodegenerative disorders. Alzheimer's disease, a progressive neurodegenerative condition characterized by cognitive impairments and memory problems, significantly affects people worldwide, and its impact is expected to increase in the coming years. Presently, an effective cure for Alzheimer's is lacking, and treatments are limited to symptomatic management. However, developments in nanotechnology offer a beacon of hope for the treatment of central nervous system diseases like Alzheimer's. Notably, the advancements in nanotechnology-based drug delivery systems enable the precise targeting of effective agents to specific tissues, facilitating the passage of drugs through the blood-brain barrier at desired concentrations and allowing for controlled and sustained release within the brain. Consequently, it is anticipated that nanotechnology-based drug delivery systems will pave the way for enhanced clinical responses and improved quality of life for Alzheimer's patients in the future [36].

These advancements also hold promise for the treatment of Parkinson's disease, the second most common neurodegenerative disorder after Alzheimer's, in its current therapies [31].

Nanovaccinology is a concept that has emerged in recent years due to the application of nanotechnology in the field of vaccine science. The field of nano-vaccines is rapidly advancing, thanks to the increasing use of nanotechnology devices and the provision of more information about the efficacy of polymeric drugs. Nanotechnology is progressively playing a more significant role in vaccine development, offering opportunities to enhance cellular and humoral immunity through nano-carrier-based delivery systems, thereby addressing challenges in vaccine production. The application of nanotechnology in immunomodulation has led to the synthesis of novel vaccine carriers, leading to rapid advancements in the field [42].

Nanoparticle therapies have shown significant impact in the treatment of anemia, stemming from the need to overcome the toxicity associated with the administration of free iron injections for iron replacement. In such applications, nanoparticle iron oxide colloids are employed as a therapeutic agent aimed at increasing iron concentration in the body [40].

In a study conducted by researchers in Finland, it has been observed that leading the Nano-Ear project and assisting individuals with hearing loss through implantable earphones has improved their hearing ability and helped overcome comprehension difficulties [32].

Another significant development in the field of Nanomedicine is the creation of "respirocytes," which are lab-engineered nanorobots that resemble red blood cells in structure and function. Respirocytes function similarly to erythrocytes and are capable of carrying oxygen. They can be likened to pressurized gas tanks, as they can carry oxygen 236 times more efficiently than erythrocytes of the same volume. Once introduced into the body, respirocytes are designed to release O2 and CO2 in a controlled manner. This technology, currently under development, is foreseen to offer significant therapeutic benefits in the treatment of various diseases once it enters clinical use [26].

8. Nanotechnology Applications in Dentistry

Nanotechnology and its applications in the fields of medicine and dentistry have a history of approximately thirty years. Although the progress of nanotechnology applications in dentistry has been relatively slower compared to its applications in medicine, nanoparticles are increasingly being used in dental applications. The use of nanoparticles in dentistry began with the addition of nanoparticle fillers to dental materials. Numerous dental materials now contain nanoparticles [43]. These advancements in dentistry have led to noteworthy research that translates the use of nanomaterials from theoretical foundations to clinical practice. Utilizing the advantageous physicochemical and biological properties of nanoparticles in dentistry, various oral diseases, including dental caries, periodontal diseases, pulp and periapical lesions, oral candidiasis, denture stomatitis, hyposalivation, and head, neck, and oral cancer, can be diagnosed, prevented, and treated. Nanoparticles can also enhance the mechanical and microbiological properties of dental prostheses and implants and can be utilized to increase the oral mucosal delivery of drugs. Currently, numerous nanomaterials are being employed in various sub-disciplines of dentistry [44].

Nowadays, the development of nanodentistry is intensively pursued through the combined use of nanomaterials, nanorobots, and biotechnologies, including tissue engineering. The goal of these endeavors is to achieve optimal oral health. To this end, nanotechnology is effectively employed in various aspects of dentistry, including tissue engineering, implantology, prosthetic dental treatment, and restorative dental treatment. One of the primary reasons for the preference of nanoparticles in dental materials is to enhance the optical properties of the material without compromising its mechanical strength and wear resistance. The aesthetic appearance of restorative materials is influenced by optical properties such as color, translucency, and smoothness [45].

Restorative dentistry is a significant field that focuses on the research and application of various materials to repair and restore damaged or impaired portions of teeth to regain their functionality. In recent years, there has been a notable evolution in restorative dentistry, particularly in materials used for dental color. Advances in nanotechnology have played a crucial role in the development of restorative materials [46].

Nanotechnology has played a significant role in the field of dental fillings. The use of nanoparticles instead of macro particles in traditional filling materials enhances the mechanical properties and durability of the filling material while also improving its aesthetic appearance. Furthermore, nanoparticle-filled dental materials are more compatible with tooth tissue, providing better responses to environmental stresses and ensuring the longevity of the filling [47].

Dental caries is one of the most common chronic diseases in oral health today, posing a significant problem in the field of dentistry. The development of preventive and restorative treatment methods in dentistry is of critical importance in effectively addressing this issue. Nanotechnology offers extensive potential for the improvement of preventive treatment methods in dentistry. The surface properties of materials produced in the nano-size range allow effective placement on tooth surfaces while also preventing bacterial plaque formation. The use of metal nanoparticles and antibacterial monomers enables dental composites and fillings to gain antibacterial activity, thus contributing to the prevention of dental caries [48].

Nanotechnology applications in dentistry can be mainly categorized under three headings: nano-diagnosis, nano-materials, and nano-robots (Table 2). In addition to these mentioned application areas, nanotechnology is widely utilized in various other aspects of dentistry. These include improving tooth durability and appearance, root canal disinfection, tooth replacement, light sensitization, surface disinfectants, stem cell imaging, and local anesthesia applications [43].

Table 2.   Nanotechnology Applications in Dentistry

Nano-Diagnosis	• Nano-pores • Nano-tubes • Quantum dots • Lab-on-a-chip (LOC) • Nano Electromechanical Systems (NEMS) • Oral fluid nano-sensor test • Optical nano-biosensors
Nano-Materials	• Nanocomposites • Prosthetic dental nanocomposites • Restorative materials • Bonding agents • Implant placement • Suture needles and materials • Nanocapsulation • Digital dental imaging • Large dental repair/nano tissue engineering • Nano glass ionomer restorative • Impression materials • Nano solutions • Nanomaterials for periodontal drug delivery • Bone graft materials • Plasma laser application for periodontology • Implants • Nano-vectors
Nano-Robots	• Dentin hypersensitivity • Local anesthesia • Dentif robots

Source: [43].

In recent years, advances in nanotechnology have offered new and promising approaches in dentistry.

9.   Nanotechnology and Surgery

With the advancements in nanotechnology, nanomaterials have been widely utilized in various fields owing to their exceptional properties, including specific surface area and high surface energy. Infected wounds represent a significant global health issue that requires effective antibacterial materials and dressings. In the treatment of infected wounds, micro/nano carbon materials such as activated carbon fibers, carbon nanotubes, graphene, carbon quantum dots, and carbon aerogels hold significant potential in offering novel alternatives.

In burn treatment, which requires special care for wound healing, the speed of recovery becomes even more critical due to the loss of the protective layer of the skin. In a study, burn wounds treated with silver nanoparticles showed a faster healing process compared to treatment with silver sulfadiazine. Additionally, better pain control, reduced risk of infection, and less scarring at the wound site were reported with the use of silver nanoparticles [49].

Nanomaterials used in wound healing play an active role in hemostasis, inflammation, proliferation, and antimicrobial inhibition. For a long time, metal nanoparticles such as silver, gold, and zinc have been among the materials used in dressings to accelerate wound healing [50].

In addition to the positive effects of nanoparticle use on wound healing, it has been indicated that they are also effective in the reconstruction of tissue defects and in preventing ischemia-reperfusion injury after limb replantation [41,42].

Nanomaterials with the ability to mimic cellular properties and used in nerve injuries have been shown to serve as temporary guides in promoting new nerve formation and creating a suitable environment for axon regeneration, as expressed by Aoron Tan et al.

Regarding the treatment of bone fractures, bacterial biofilm formation on titanium implants can lead to implant failure. However, it has been suggested that silver nanorobots can prevent plaque infections. In addition to bone fractures, nanotechnologically developed methods are available to enhance cartilage tissue regeneration in contact with titanium implants [29].

Nanotechnology is also utilized in the diagnosis and treatment of cardiovascular diseases. Particularly, chemical chemoreceptors are employed for the diagnosis and treatment of blocked arteries. This highlights the significance of nanotechnology applications in prolonging patient lifespans [53].

The future potential of nanotechnology in the biomedical field is vast. Nano-sized sensors, biosensors, and biomaterials enable the development of novel and effective approaches for various diagnostic and therapeutic applications in medicine. Furthermore, nanotechnology is expected to make significant contributions in areas such as regenerative medicine and tissue engineering.

10   Conclusion

The use of nanotechnology in the biomedical field can bring many innovations that can radically change diagnostic and treatment approaches in the field of medicine in the future. We can explain the expanded potential in this field as follows:

1. Nano-Sized Sensors and Biosensors: Nanotechnology allows the development of nano-sized sensors and biosensors. These sensors are capable of performing very sensitive biochemical analyses. For example, it allows rapid and sensitive detection of tumor markers, infections or

other pathological conditions. This allows for earlier diagnosis and more effective treatment options.

2. Nanotechnological Biomaterials: Nanotechnology contributes to the development of nanomaterials specifically designed for use in the biomedical field. For example, nanosized carriers enable more direct and effective delivery of drugs to the target. This increases treatment effectiveness while reducing side effects of medications.

3. Regenerative Medicine and Tissue Engineering: Nanotechnology creates a huge impact in the fields of regenerative medicine and tissue engineering. Nanosized carriers and artificial tissue matrices can help repair or even regenerate damaged tissues. This offers significant potential, especially in the treatment of injuries, organ failures and degenerative diseases.

4. Individualized Treatment: Nanotechnology allows the development of individualized treatment approaches. Treatment plans designed based on the genetic and biochemical characteristics of each patient make it possible to achieve more effective results. It can reduce side effects and provide more targeted treatment.

In summary, nanotechnology offers great potential in the biomedical field. Nanosized sensors, biosensors, biomaterials and tissue engineering applications herald a significant transformation in the fields of medical diagnosis and treatment. These technological advances can improve patients' quality of life and make healthcare more effective.

References

[1]   B. Şen, N. Lolak, Ö. Paralı, M. Koca, A. Şavk, S. Akocak, F. Şen, Bimetallic PdRu/graphene oxide based Catalysts for one-pot three-component synthesis of 2-amino-4H-chromene derivatives, Nano-Structures and Nano-Objects 12 (2017) 33–40. https://doi.org/10.1016/j.nanoso.2017.08.013

[2]   A. Hojjati-Najafabadi, A. Aygun, R.N.E. Tiri, F. Gulbagca, M.I. Lounissaa, P. Feng, F. Karimi, F. Sen, Bacillus thuringiensis Based Ruthenium/Nickel Co-Doped Zinc as a Green Nanocatalyst: Enhanced Photocatalytic Activity, Mechanism, and Efficient H2Production from Sodium Borohydride Methanolysis, Ind. Eng. Chem. Res. (2022). https://doi.org/10.1021/ACS.IECR.2C03833

[3]   E. Demir, B. Sen, F. Sen, Highly efficient Pt nanoparticles and f-MWCNT nanocomposites based counter electrodes for dye-sensitized solar cells, Nano-Structures & Nano-Objects 11 (2017) 39–45. https://doi.org/10.1016/J.NANOSO.2017.06.003

[4]   S. Eris, Z. Daşdelen, F. Sen, Investigation of electrocatalytic activity and stability of Pt@f-VC catalyst prepared by in-situ synthesis for Methanol electrooxidation, Int. J. Hydrogen Energy 43 (2018) 385–390. https://doi.org/10.1016/j.ijhydene.2017.11.063

[5]   B. Sen, E. Kuyuldar, A. Şavk, H. Calimli, S. Duman, F. Sen, Monodisperse ruthenium–copper alloy nanoparticles decorated on reduced graphene oxide for dehydrogenation of DMAB, Int. J. Hydrogen Energy 44 (2019) 10744–10751. https://doi.org/10.1016/j.ijhydene.2019.02.176

[6]   N. Korkmaz, Y. Ceylan, P. Taslimi, A. Karadağ, A.S. Bülbül, F. Şen, Biogenic nano silver: Synthesis, characterization, antibacterial, antibiofilms, and enzymatic activity, Adv. Powder Technol. 31 (2020) 2942–2950. https://doi.org/10.1016/j.apt.2020.05.020

[7]   E. Demir, A. Savk, B. Sen, F. Sen, A novel monodisperse metal nanoparticles anchored graphene oxide as Counter Electrode for Dye-Sensitized Solar Cells, Nano-Structures & Nano-Objects 12 (2017) 41–45. https://doi.org/10.1016/j.nanoso.2017.08.018

[8]   F. Şen, Ö. Demirbaş, M.H. Çalımlı, A. Aygün, M.H. Alma, M.S. Nas, The dye Removal from Aqueous Solution Using Polymer Composite Films, Appl. Water Sci. 8 (2018) 206. https://doi.org/10.1007/s13201-018-0856-x

[9]   A. Şavk, K. Cellat, K. Arıkan, F. Tezcan, S.K. Gülbay, S. Kızıldağ, E.Ş. Işgın, F. Şen, Highly monodisperse Pd-Ni nanoparticles supported on rGO as a rapid, sensitive, reusable and selective enzyme-free glucose sensor, Sci. Rep. 9 (2019) 19228. https://doi.org/10.1038/s41598-019-55746-y

[10]   Y. Yıldız, İ. Esirden, E. Erken, E. Demir, M. Kaya, F. Şen, Microwave (Mw)-assisted Synthesis of 5-Substituted 1H-Tetrazoles via [3+2] Cycloaddition Catalyzed by Mw-Pd/Co Nanoparticles Decorated on Multi-Walled Carbon Nanotubes, ChemistrySelect 1 (2016) 1695–1701. https://doi.org/10.1002/SLCT.201600265

[11]   İ. Meydan, H. Seçkin, Green synthesis, characterization, antimicrobial and antioxidant activities of zinc oxide nanoparticles using Helichrysum arenarium extract, Int. J. Agric. Environ. Food Sci. 5 (2021) 33–41. https://doi.org/10.31015/JAEFS.2021.1.5

[12]   İ. Yavuz, E.Ş. Yılmaz, Biyolojik Sistemli Nanopartiküller, Gazi Üniversitesi Fen Fakültesi Derg. 2 (2021) 93–108.

[13]   A. Öndürücü, E. Bilgin, Nanoteknoloji, Mühendis ve Makine 49 (2008) 9–20.

[14]   R. Darabi, F.E.D. Alown, A. Aygun, Q. Gu, F. Gulbagca, E.E. Altuner, H. Seckin, I. Meydan, G. Kaymak, F. Sen, H. Karimi-Maleh, Biogenic platinum-based bimetallic nanoparticles: Synthesis, characterization, antimicrobial activity and hydrogen evolution, Int. J. Hydrogen Energy 48 (2023) 21270–21284. https://doi.org/10.1016/J.IJHYDENE.2022.12.072

[15]   N. Öztürk, Hedeflendirilmiş Antikanser İlaç Yüklü Polimerik Nanotaşıyıcıların Formülasyonları ve Karakterizasyonları, (2019).

[16]   T. Gur, I. Meydan, H. Seckin, M. Bekmezci, F. Sen, Green synthesis, characterization and bioactivity of biogenic zinc oxide nanoparticles, Environ. Res. 204 (2022) 111897. https://doi.org/10.1016/J.ENVRES.2021.111897

[17]   H. Seçkin, İ. Meydan, Synthesis and Characterization of Veronica beccabunga Green Synthesized Silver Nanoparticles for The Antioxidant and Antimicrobial Activity, Turkish J. Agric. Res. 8 (2021) 49–55. https://doi.org/10.19159/TUTAD.805463

[18]   Z. TÜYLEK, Küçük Şeylerin Hikayesi: Nanomalzeme, Nevşehir Bilim ve Teknol. Derg. 5 (2016) 130–130. https://doi.org/10.17100/nevbiltek.284737

[19]   M.G. Lines, Nanomaterials for practical functional uses, J. Alloys Compd. 449 (2008) 242–245. https://doi.org/10.1016/J.JALLCOM.2006.02.082

[20]   L. Filipponi, D. Sutherland, Nanotechnologies: Principles, Applications, Implications and Hands-on Activities, European Comission, 2012. https://doi.org/10.2777/76945

[21]   Ö. Tarhan, V. Gökmen, Ş. Harsa, Nanoteknolojinin gıda bilim ve teknolojisi alanındaki uygulamaları, Gıda 35 (2010) 219–225.

[22]   Z. Tüylek, Nanoteknoloji Uygulamalarında Hayatımıza Yansımalar, Eurasian J. Biol. Chem. Sci. 4 (2021) 69–79. https://doi.org/10.46239/ejbcs.909023

[23]   A. Köroğlu, I. Kürkçüoğlu, E. Özkır, M. Ateş, THE NANOTECHNOLOGY CONCEPT IN DENTISTRY, Süleyman Demirel Üniversitesi Sağlık Bilim. Derg. 5 (n.d.) 77–80.

[24]   N.A. Hanks, J.A. Caruso, P. Zhang, Assessing Pistia stratiotes for phytoremediation of silver nanoparticles and Ag (I) contaminated waters, J. Environ. Manage. 164 (2015) 41–45.

[25]   M.H. Meshkatalsadat, J. Safaei-Ghomi, S. Moharramipour, M. Nasseri, Chemical characterization of volatile components of Tagetes minuta L. cultivated in south west of Iran by nano scale injection, Dig. J. Nanomater. Biostructures 5 (2010) 101–106.

[26]   F. Şenel, Nanotıp, Bilim ve Tek. 497 (2009) 79–83.

[27]   A.Z. Şengil, Teknolojik değişim süreci: Nanoteknoloji ve nanotıp, İstanbul ve Ankara'da Aile Hekim. Uygulamalarının Başladığı Bugünlerde Bu Konuları Dergimize Taşımak 84 (2010).

[28]   G. Suepueren, Z.E. Kanat, A. Cay, T. Kırcı, T. Gueluemser, I. Tarakçıoğlu, Nano fibres (Part 2),

Biogenic Nanomaterials: Synthesis, Characterization, Applications, and Future Remarks
Materials Research Foundations 180 (2025)

Materials Research Forum LLC
https://doi.org/21741/9781644903759

Text. Appar. 17 (2007) 83–89.

[29]   İ. Demirhan, SAĞLIK & BİLİM 2022: Nanotıp, Efe Academy Publications, 2022.

[30]   O. FAROKHZAD, R. LANGER, Nanomedicine: Developing smarter therapeutic and diagnostic modalities☆, Adv. Drug Deliv. Rev. 58 (2006) 1456–1459. https://doi.org/10.1016/j.addr.2006.09.011

[31]   S. Syed, A. Zubair, M. Frieri, Immune Response to Nanomaterials: Implications for Medicine and Literature Review, Curr. Allergy Asthma Rep. 13 (2013) 50–57. https://doi.org/10.1007/s11882-012-0302-3

[32]   Z. Tüylek, Nanotıp ve yeni tedavi yöntemleri, Avrasya Sağlık Bilim. Derg. 4 (2021) 121–131.

[33]   H. DEMİRTAŞ, C. ŞENGEL TÜRK, GOLD NANOPARTICULES AND USES IN CANCER, Ankara Univ. Eczac. Fak. Derg. (2021) 70–95. https://doi.org/10.33483/jfpau.773430

[34]   F. Li, Y. Li, J. Feng, Y. Dong, P. Wang, L. Chen, Z. Chen, H. Liu, Q. Wei, Ultrasensitive amperometric immunosensor for PSA detection based on Cu2O@CeO2-Au nanocomposites as integrated triple signal amplification strategy, Biosens. Bioelectron. 87 (2017) 630–637. https://doi.org/10.1016/j.bios.2016.09.018

[35]   C.E. Neumaier, G. Baio, S. Ferrini, G. Corte, A. Daga, MR and Iron Magnetic Nanoparticles. Imaging Opportunities in Preclinical and Translational Research, Tumori J. 94 (2008) 226–233. https://doi.org/10.1177/030089160809400215

[36]   D. Nadir; GÜN, Alzheimer hastalığı tedavisinde nano boyutlu ilaç taşıyıcı sistemlerin kullanımı, Ankara Üniversitesi Eczac. Fakültesi Derg. 40 (2016) 54–73. https://doi.org/10.1501/Eczfak_0000000579

[37]   N. Taneja, A. Alam, R.S. Patnaik, T. Taneja, S. Gupta, S.M. K, Understanding Nanotechnology in the Treatment of Oral Cancer: A Comprehensive Review, Crit. Rev. Ther. Drug Carr. Syst. 38 (2021) 1–48. https://doi.org/10.1615/CritRevTherDrugCarrierSyst.2021036437

[38]   İ.M. ALKAÇ, B. VURAL, Nanomaterial Based Treatment Strategies in Triple Negative Breast Cancer: Traditional Review, Turkiye Klin. J. Med. Sci. 41 (2021) 491–500. https://doi.org/10.5336/medsci.2021-84439

[39]   K.B. Sutradhar, M.L. Amin, Nanotechnology in Cancer Drug Delivery and Selective Targeting, ISRN Nanotechnol. 2014 (2014) 1–12. https://doi.org/10.1155/2014/939378

[40]   A.C. Anselmo, S. Mitragotri, Nanoparticles in the clinic, Bioeng. Transl. Med. 1 (2016) 10–29. https://doi.org/10.1002/btm2.10003

[41]   M. EKİNCİ, D. İLEM-ÖZDEMİR, NANOTHERANOSTICS, Ankara Univ. Eczac. Fak. Derg. (2021) 131–155. https://doi.org/10.33483/jfpau.717067

[42]   E. DÖNMEZ, H.T. YÜKSEL DOLGUN, Ş. KIRKAN, Nanopartiküler Aşılar, J. Anatol. Environ. Anim. Sci. 6 (2021) 578–584. https://doi.org/10.35229/jaes.970713

[43]   Y. Dağlıoğlu, M.C. Yavuz, Dişhekimliğinde nanoteknoloji ve uygulamaları Nanotechnology in dentistry and their applications, Ege Univ. Fac. Dent. J. 41 (2020) 149–160.

[44]   G. Moraes, C. Zambom, W.L. Siqueira, Nanoparticles in Dentistry: A Comprehensive Review, Pharmaceuticals 14 (2021) 752. https://doi.org/10.3390/ph14080752

[45]   F. KOSHİ, E. CENGİZ, F. ER, N. ULUSOY, RESTORATİF DİŞ HEKİMLİĞİNDE NANOTEKNOLOJİ, Atatürk Üniversitesi Diş Hekim. Fakültesi Derg. 25 (2016). https://doi.org/10.17567/dfd.36138

[46]   P. Oyar, DİŞ HEKİMLİĞİNDE KULLANILAN NANOPARTİKÜLLER, KULLANIM ALANLARI VE BİYOUYUMLULUK, Atatürk Üniversitesi Diş Hekim. Fakültesi Derg. 24 (2015). https://doi.org/10.17567/dfd.00381

[47]   Z. Khurshid, M. Zafar, S. Qasim, S. Shahab, M. Naseem, A. AbuReqaiba, Advances in Nanotechnology for Restorative Dentistry, Materials (Basel). 8 (2015) 717–731. https://doi.org/10.3390/ma8020717

[48]   S. Ekrikaya, S. Demirbuğa, Nanotechnology Use of for Antibacterial and Remineralizing Effect in Management of Dental Caries, J. Ege Univ. Sch. Dent. 44 (2023) 77–85. https://doi.org/10.5505/eudfd.2023.15046

[49]   A. Tan, R. Chawla, N. G, S. Mahdibeiraghdar, R. Jeyaraj, J. Rajadas, M.R. Hamblin, A.M. Seifalian, Nanotechnology and regenerative therapeutics in plastic surgery: The next frontier, J. Plast. Reconstr. Aesthetic Surg. 69 (2016) 1–13. https://doi.org/10.1016/j.bjps.2015.08.028

[50]   Y. Liu, Q. Li, H. Zhang, S. Yu, L. Zhang, Y. Yang, Research progress on the use of micro/nano carbon materials for antibacterial dressings, New Carbon Mater. 35 (2020) 323–335. https://doi.org/10.1016/S1872-5805(20)60492-9

[51]   K. Amin, R. Moscalu, A. Imere, R. Murphy, S. Barr, Y. Tan, R. Wong, P. Sorooshian, F. Zhang, J. Stone, J. Fildes, A. Reid, J. Wong, The Future Application of Nanomedicine and Biomimicry in Plastic and Reconstructive Surgery, Nanomedicine 14 (2019) 2679–2696. https://doi.org/10.2217/nnm-2019-0119

[52]   M. Kiraz, S. Cevik, A. Demirel, Y.E. Gergin, O. Ozdemir, Nanoteknoloji ve nanonöroşirürji, Türk Nöroşirürji Derg. (2018).

[53]   S.K. Sahoo, S. Parveen, J.J. Panda, 775 The present and future of nanotechnology in human health care, Nanomedicine in Cancer (2017) 775–806

Biogenic Nanomaterials: Synthesis, Characterization, Applications, and Future Remarks   Materials Research Forum LLC
Materials Research Foundations 180 (2025)   https://doi.org/21741/9781644903759

Chapter 13

Effect of Biogenic Nanomaterials on Pregnancy

G. Sargin[a*], M. Akin[b,c], F. Sen[b*]

[a]Van Yuzuncu Yıl University Vocational School of Health Services, Van, Turkiye

[b]Sen Research Group, Department of Biochemistry, Dumlupinar University, Kutahya, Turkiye

[c]Department of Materials Science & Engineering, Faculty of Engineering, Dumlupinar University, Kutahya, Turkiye

gulumsargin@yyu.edu.tr; fatihsen1980@gmail.com

Abstract

Nanoparticles, used in many areas including product structure, technology and human health, have the potential to prevent pregnancy-related complications and cross the placental barrier. The current research on the effects of nanoparticles on the fetus is inadequate, with findings that are both positive and negative. Further in vivo and in vitro studies are required to gain a full understanding of the impact of nanoparticles on the developing fetus. This chapter provides a comprehensive overview of the impact of nanoparticles on pregnancy, emphasising the necessity for further research to fully comprehend the risks and benefits associated with their utilisation.

Keyword

Nanoparticles and Pregnancy, Placental Barrier, Foetal Development, Nanoparticle Toxicity, In Vivo and In Vitro Studies

1. Introduction

Today, nanotechnology, a new branch of science, is developing rapidly and is an industrial activity area used in a wide variety of fields with the development and production of nanoparticles [1–3]. At this point; energy [4,5], production, health [6], medicine, environment [7], defence and many other areas . Considering that many more of these materials, which offer a widespread area of use, will be available for sale in the near future, it is critical to assess the risks that these new technological products will pose after their use, especially in Turkey. Therefore, the effects of NPs on individuals (from cellular to systemic effects) and the level of interaction with the environment are the subject of research. While the unique properties of nanoparticles (NPs) generate excitement about their potential applications to benefit various aspects of our lives [8,9]. they also raise concerns about the toxicological effects of their potential impact on human health. They may pose public health concerns for risk groups such as pregnant women and developing fetuses, as the body may be more sensitive and vulnerable to external influences. However, the possibility of developing safe and effective NP-containing drug formulations in the nanomedical field at the point of application holds great promise for humanity in revolutionizing treatment during pregnancy. At this point, with regard to the use of NPs during pregnancy, care should be taken to ensure that the health of the mother and fetus is not affected and comprehensive studies should be carried out on the effects of nanoparticles ingested by the mother that may be transported to the placenta and the effects that may occur after placement in the placenta. The importance of research on the placenta has increased, especially given its many essential functions in ensuring a healthy pregnancy and delivery [10].

Nanoparticles have a very common usage area because they have various properties [11–14]. Nanoparticles used in health-related fields are used for many different purposes such as biomedical technologies to provide more specific analysis, molecular imaging in many radiological fields such as computerized tomography, magnetic resonance, fluorescence and ultrasound, targeted therapy, drug development systems, vaccine development [9,15–17]. With the increase in the use of these common and beneficial areas, it has become inevitable to include the molecular properties of nanoparticles in studies that need to be scientifically examined for possible toxic effects on the respiratory system, nervous system, blood, gastrointestinal system and skin [18,19].

Humanity's reproductive goal requires healthy women to be able to conceive quickly, and mothers to be able to experience no adverse events during pregnancy or overcome any complications that may arise. According to world statistics, more than 100 million births occur within a year [20]. In terms of the incidence of maternal or fetal complications as a result of these deliveries, more than 10% are affected. The most common fetal complications are intrauterine growth retardation (IUGR), pregnancy failure and early delivery. In addition, the incidence of maternal complications such as preeclampsia and gestational diabetes mellitus (GDM) has also increased [21,22]. In this case, complications during pregnancy can cause significant maternal and fetal effects. However, although it causes neonatal morbidity and mortality, there is no method to prevent these complications in pregnant women or a drug to treat them. Nanomedicine is developing new platforms that enable the use of drugs in targeted therapies during pregnancy. The first trimester of pregnancy is generally the most sensitive period. This is because the fetus is already forming and taking shape. With the systematic formation of organs taking place at this time, the fetus is highly susceptible to teratogenic influences from outside, such as drugs [23–25]. The placenta is completely exposed to the mother's blood during the first months of pregnancy. Therefore, any medication used by the mother can reach the embryo by entering the bloodstream or by transporting chemicals to the placenta. With this in mind, it is necessary to review and discuss the progress and scope of surface functional nanoparticles that can be used for targeted treatment of pregnancy complications. One of the things that needs to be known is the anatomy of the developing placenta during pregnancy, which is the basis for developing placental targeted therapy, and then to revisit what is known about the mechanisms of nanoparticle transplacental transport pathways. Studies have generally shown that nanoparticles with functionalized surface areas have recently been studied to target the uterus and placenta [26].

The placenta anatomy, which is the sensitive structure of the developing fetus, is the most important biological barrier that attracts more attention in terms of health and an organ that separates the two circulatory systems of the mother and fetus. During pregnancy, which is a physiological process, the placenta is formed in the uterus very close to the uterine wall. It is an important organ for maternal-fetal physiological exchange. Nanoparticles administered

Biogenic Nanomaterials: Synthesis, Characterization, Applications, and Future Remarks   Materials Research Forum LLC
Materials Research Foundations 180 (2025)   https://doi.org/21741/9781644903759

maternally determine the degree of exposure of the fetus in the placenta. It varies greatly between different species in terms of placental shape and structure [27]. The placenta is similar to the blood-brain barrier in many ways. This is because it acts as a protective shield during the fetal period and does not allow for the indiscriminate exchange of substances between the mother's blood and the nourished fetüs [28]. Since the human placenta contains disc-shaped structures, it is discoidal and is in contact with maternal blood through the villi, resulting in a distinct hemochorial placenta. It is filled with both fetal and maternal blood, as seen in Figure 1. The chorionic plate and chorionic villi located on the placenta consist of maternal tissue such as the decidua basalis of the uterus. The decidua septa, originating from the decidua basalis, separate the placenta into separate functional units known as cotyledons [29,30]. The vessels branching from the surface of the placenta divide and enter a network covered with a thick layer of cells. As a result, villous tree structures are formed. On the mother's side, these villous tree structures turn into lobes called "Cotyledons". This barrier contains an outer trophoblast containing numerous capillaries, endothelial cells, multinucleated syncytiotrophoblast, and cytotrophoblast [30]. It is also the layer that determines the speed of substances that want to cross the placenta [31,32]. However, since the development of these barriers is not fully formed until the tenth week of pregnancy, maternal blood reaches the fetus directly without any obstacles, and it is a dangerous process for the fetus. Substances in maternal blood can pass to the fetus via diffusion from the extracellular fluid. In this period when the development of the placental barrier continues (between the 10th and 12th weeks of pregnancy), the drugs and similar chemicals used are substances with small molecular weight, lipophilic, low protein binding and low ionization degree. Therefore, it easily crosses the placental barrier [33].

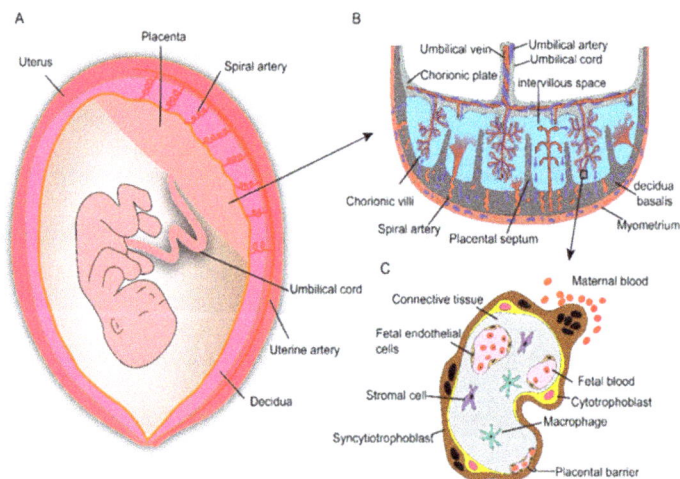

Figure 1. Schematic description of the structures that make up the anatomy of the placenta. (A) It is a significant barrier that adheres to the uterine wall and where maternal-fetal physiological exchange takes place. (B) Anatomical structure of the human placenta. Oxygen and nutrient-rich blood is taken to the fetus with the help of the umbilical vein located in the umbilical cord, and

waste products from the fetus pass through the placenta through two different umbilical arteries. Through the regenerated spiral arteries, the intervillous space is filled with maternal blood. (C) The placental barrier and its principal cell types are composed of syncytiotrophoblasts, cytotrophoblasts, and fetal endothelial cells [26]. Copyright © 2019 MDPI.

1.1   Transplacental transport mechanisms of nanoparticles

As nanoparticles can reach the placenta, the purpose of specifically controlling them is clear. However, our knowledge on how nanoparticles pass through the placental barrier to the fetus is very limited. In order for nanoparticles to reach the placenta, they must pass through two different structures: (a) the layer formed by the syncytiotrophoblast and cytotrophoblast, and (b) the endothelial cells that form the fetal capillaries within the villi [34–36].

The structure separating the fetal circulation from the intervillous space is, of course, the maternal-fetal barrier. The placental area surface, which forms a fairly large surface area with large microvilli and syncytiotrophoblast, allows for significant endocytic uptake with optimum diffusion and exchange [37]. A multinucleated layer called syncytiotrophoblast, which covers the villi and intervillous space on the placenta, interacts primarily with the maternal blood. Immediately below the syncytiotrophoblast layer is the cytotrophoblast layer. When we divide pregnancy into three trimesters, these layers are thicker in the first trimester and thinner in the second and third trimesters [38]. After these layers there is also a trophoblastic structure called the basal lamina, connective tissue, which develops from extraembryonic mesoderm, and the foetal vascular endothelium [37]. Figure 2 shows the various transport routes for the passage of material across the placenta.

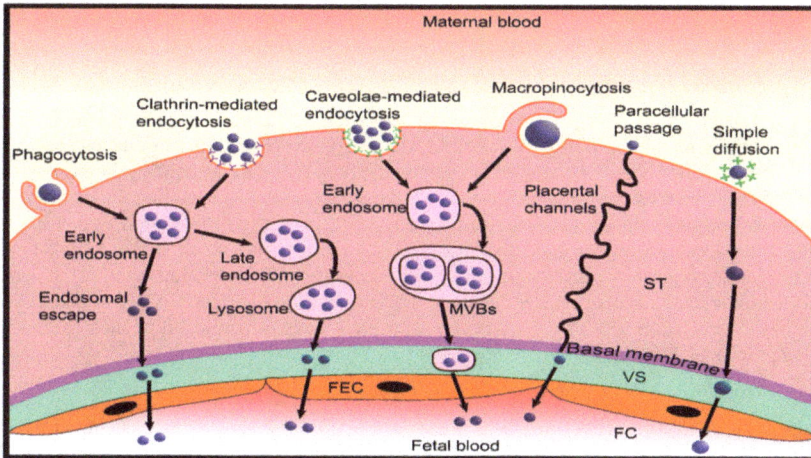

Figure 2. *Diagram describing the transport mechanism of nanoparticles to the placenta [26]. Copyright © 2019 MDPI.*

Substances such as nanoparticles can often use a number of pathways to reach the placenta and cross the barrier. Among these, paracellular and transcellular routes are most commonly used.

Small-sized nanoparticles can penetrate the syncytiotrophoblast (ST) through placental ducts and thus penetrate the villous stroma (VS). They can then diffuse into the lumen of fetal capillaries (FC) via fetal endothelial cells (FEC). It forms the transcellular transition from endocytosis and exocytosis. Nanoparticles can enter syncytiotrophoblasts via phagocytosis, endocytosis, endocytosis and macropinocytosis. Subsequent secretion associated with multivesicular bodies (MVBs) via the endosomal escape pathway, lysosomal secretion and exocytosis may follow. Nanoparticles cross FEC and reach fetal blood. Cationic nanoparticles can directly assemble by affecting the negatively charged trophoblast cell membrane and direct movement by direct diffusion towards the basement membrane, thus diffusing through the FEC into the fetal blood [26].

Studies in diverse animal models have shown that during gestation physiological adaptations occur for the formation of another living creature. Due to these adaptations, the immune system is lower than its normal level during pregnancy, resulting in a greater susceptibility to foreign bodies such as microorganisms that may come from outside [39]. As an example, some studies have proven intrauterine transmission in COVID-19 disease. Percentage of the patients infectious with COVID-19 show thrombotic events of the placenta, resulting in pulmonary inflammation, susceptibility to bacterial inoculation, suppression of the immune system as well as changes in the pulmonary microbiome due to COVID-19, resulting in death during pregnancy [40,41]. However, the most common complications for the fetus are early termination of pregnancy and fetal loss, abnormal intrauterine growth and premature birth [42,43]. Although NP has been used in different pathologies and in the assessment of some metabolic defects, the lack of research on NP during pregnancy and its effects on the fetus is frightening. Biogenic nanomaterials may be used to sensitively control drug application during pregnancy and may provide a new way to treat pathological outcomes in pregnancy. However, the smallest risk of side effects in the fetus or mother should not be underestimated [26].

1.2   Placental Permeability

Many NP species have the ability to cross some biological barriers and have been found to have negative effects on important organs like brain, kidney and liver. The impact of nanomaterials on reproductive toxicity has only recently attracted attention. NPs can accumulate in these tissues after the blood crosses the testicular barrier and the epithelial barrier covering the placenta and reproductive organs. NP accumulation can adversely affect Leydig cells, Sertoli cells and germ cells, as well as sperm quality, quantity, curvature and motility. As a result, it causes dysfunction of reproductive organs by disrupting primary and secondary follicle development [44]. It can be said that untargeted nanoparticles accumulate ineffectively in organs, especially in organs with high blood circulation and also in structures with high vascular permeability such as tumours during pregnancy [45].

1.3   Toxicity of NPs in Embryonic Development

The toxicity of NPs may vary according to a number of factors. These may include the amount ingested, shape, size, weight, material and how much surface area is covered. The entry of NPs into the body of a pregnant woman can be by any route, e.g. inhalation, vascular injection, ingestion or skin penetration, and many maternal toxic distress reactions such as inflammation, ROS, apoptosis and endocrine dyscrasia can occur. In the period of pregnancy, NPs usually penetrate the placenta by passive diffusion or endocytosis and may reach the fetus. The toxic effects of NPs on the fetus may result in abnormal embryonic development and foetal death,

Biogenic Nanomaterials: Synthesis, Characterization, Applications, and Future Remarks   Materials Research Forum LLC
Materials Research Foundations 180 (2025)   https://doi.org/21741/9781644903759

foetal inflammation, genotoxicity, cytotoxicity, low birth weight, reproductive system failure, nerve impairment and immune system triggering. Figure 3 illustrates the toxic effects that may occur on the fetus.

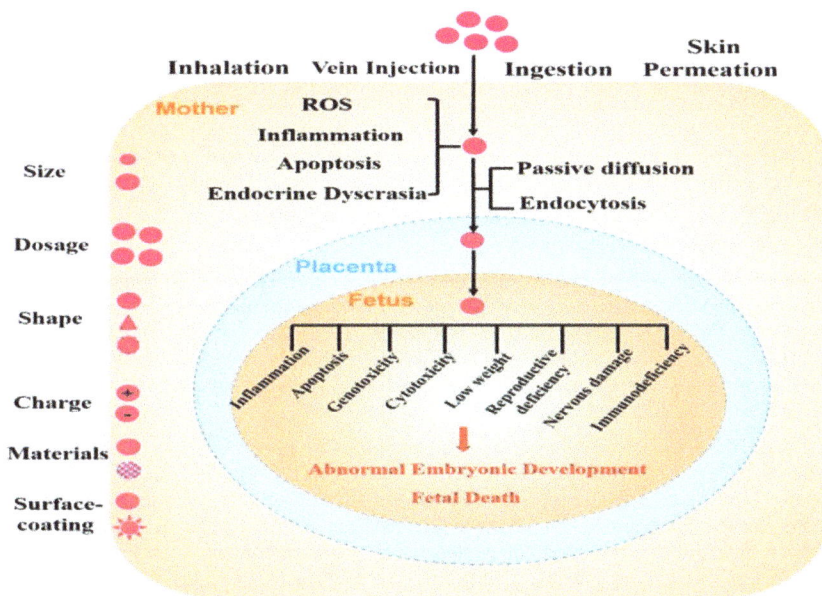

Figure 3. Potential toxic effects on the fetus [46]. Copyright © 2017 Oncotarget.

Studies have shown that placental pores or ducts with diameters ranging from 15 to 25 nm exist, starting from the basal trophoblastic surface and extending to the syncytiotrophoblasts under normal intravascular pressure [47]. In addition, there are studies reporting that flexural ducts in the placenta are continuously present and passed from the fetus to the mother [48–50]. Therefore, placental channels can cross the placenta by paracellular transmission (also known as passive diffusion) of nanoparticles below 25 nm. The human placental delivery system can represent an ethically validated role model and can be used to study and determine the transplacental transport of nanoparticles. Studies using this model have shown that silica particles and polystyrene nanoparticles of certain sizes (ranging between 25-50 nm) can be transferred to the palsent through passive diffusion, one of the transportation method [51]. There is an active pathway known as transcellular transport, which involves both endocytosis and exocytosis and involves intricate linked vesicular systems. Through this transcellular transport, nanoparticle transport via the plasma membrane can have more steps [52,53]. However, as the size of the nanoparticles increases, more is taken up by phagocytosis [54]. As the size of nanoparticles decreases, their passage to the placenta is via pinocytosis [55]. This is because there are clathrin-coated areas between placental syncytiotrophoblasts and microvilli. The vesicles that macropinocytosis has in large numbers are found in excess in stasis. This facilitates the uptake of nanoparticles by trophoblasts via pinocytosis. In short, although the transport of nanoparticles through the placental barrier is via pinocytosis and exocytosis, the transport of placental trophoblasts to the capillary endothelium, small liposomes and nanoparticles (<60 nm) is via macropinocytosis [56].

1.4   Uses of Surface Functional Nanoparticles Affecting Placenta

It is known that there are many research groups worldwide established to develop placenta-focused nanoparticles to be used to prevent or treat complications during pregnancy. In placental trophoblasts, the epidermal growth factor receptor (EGFR) appears to be highly elevated compared to normal human tissues. Based on this situation, doxorubicin-loaded nanocells have been developed to treat ectopic pregnancy by creating EGFR coated with specific antibodies on the surface of trophoblasts [57]. In vitro studies have shown that EGFR-targeted nanocells are recruited by human placental explants and as a result, placental choriocarcinoma cell death is greatly increased [58–60].

1.4.1   Tumor Targeting Peptides

Liposomes loaded with tumour-targeting peptides have been successfully found to reach and target the endothelium of placental trophoblasts and spiral arteries in the uterus. In a mouse model with limited fetal growth, when liposomes loaded with growth factor 2 years were administered, it was reported that it significantly increased fetal and placental growth and a transition of growth factor transfer to the fetus. In vivo experiments have shown that there can be a significant increase in fetal and placental weights [61].

Nanodrugs for various releases are the principle and the amount of nano medicine research results is rapidly increasing along with clinical developments [62]. However, more studies are needed. The translocation of nanoparticles in the placenta and its consequences need to be comprehensively understood. This leads to the need for the development of new nanoparticle platforms that will reduce fetal effects while treating complications that occur during pregnancy. Nevertheless, these developments, nanomedicine targeted therapy for pregnancy complications show great promise. While we study how nanoparticle transfer occurs throughout the placenta, it is necessary to consider a number of mechanisms. For example, polystyrene nanoparticles about 50 nm in size reach the placenta through diffusion and the transtrophoblastic duct system, while those 120 nm in size can pass the placenta through vesicles (those coated with caveolin) [51]. Therefore, crossing the placental barrier, nanoparticle transfer depends on particle properties and functionalisation [63].

During the gestation period, the use of nanomedicine is only with perinatal therapeutics. In the goal-directed treatment of pregnancy complications, surface-functionalised nanoparticles can be effective materials [26]. Nanomaterials used for this purpose are prepared in such a way that they readily pass through to access the fetal bloodstream. Maximising the dose given to the mother in order to increase the beneficial effect in the fetal area will prevent contamination and settlement in the maternal compartment. The desired clinical effect can be achieved with one of the factors affecting nanoparticle transfer [63].

Table 1. *Properties of nanomaterials that can be used in the treatment of complications that may occur during pregnancy.*

Characteristic	Benefit
Goal setting ability	Allows targeting of mother, placenta or fetus while trying experimental drugs, reducing the risk to the fetus.
Bioavailability of drugs	A very low drug concentration will limit adverse side effects to the mother or fetus by preventing the administration of high doses of drugs with the same biological effect.
Prevention of the onset and deterioration of the administered drug through maternal clearance mechanisms	Encapsulation of drugs in nanoparticles prevents first-pass metabolism through the liver and limits uptake by non-target cells. Facilitates the passage of drugs into the placenta that otherwise pass through the maternal circulation
Emulsification of stable or non-soluble therapeutic substances	It enables the use of advanced treatment methods such as siRNA in placental treatments.
Nanoparticle features	The delivery of nanoparticles to the gestating placenta could be modified to take advantage of existing pathways and mechanisms to enable the design of a systemically administered drug designed primarily to be received via a specific route.
Huge carrying capacity with large surface area	Development of more effective drugs to eliminate harmful side effects on the mother or fetus by reducing the doses of drugs used

The increasing industrial uses and consumption of nanomaterials in products that are quite common require a thorough examination of their safety. However, in addition to the advantages of the widespread introduction of nanomaterials into our lives, a significant disadvantage is their genotoxic potential. Genotoxicity, chemical or physical agents genetic it can define coding as a condition that can occur with it affecting. Genotoxic events may occur with permanent changes in the genetic material within the cell, such as reversible (repairable temporary damage) or irreversible (mutations). A mutated germ cell is transferred to the generation and can result in genetic deformations. For example, a mutation in a critical gene in somatic cells can develop cancerous cells. Therefore, a certain pathway and scope for genotoxicity testing during or before pregnancy is required [64,65].

Genotoxicity tests are usually done in vitro based on a single cell type. However, in the archives made for nanoparticles, such studies are in vivo form. According to the results obtained, DNA-induced changes are usually associated with an infection [66,67]. However, in recent studies, an intense effort has been made to produce more complex 3D structures by preparing in vitro models similar to in vivo cultures made by combining multiple cell types. Using 2D models to evaluate genotoxicity in humans is thought to provide more reliable data in toxicology studies than the 3D model. Newly developed algorithms are used for many of these models. For example, models of skin, liver and lung tissue [65,68]. In vitro methods to address NM-induced toxicity have advantages over in vivo approaches, such as simplicity, cost-effectiveness, and short time required for research. They can also help to elucidate the mechanisms of action of NMs on cells and provide a basis for assessing the potential risks of exposure [69].

The toxicity of nanomaterials is considered a significant drawback as a genotoxic effect, such as DNA damage, can lead to mutation and potentially to the development of emerging cancers and birth defects. DNA damage can be a consequence of cytotoxicity, a condition that can occur through direct interaction of nanoparticles with DNA or indirectly through the effect of induction of oxidative stress. Therefore, it is important to evaluate genotoxicity at concentrations of non-cytotoxic nanomaterials. Given the exposure to the use of multiple nanomaterials, their toxic effects need to be clarified and the mechanisms taking place in their toxicity need to be elucidated. There is a need for screening methods that aim to accurately demonstrate and evaluate toxicity. This is because the introduction of these high-throughput methods for toxicity testing of nanomaterials allows multiple materials to be evaluated at different concentrations and in different cell types, reducing the impact of inter-experimental variation. The data obtained with significant time and cost savings provide us with valuable information [69,70].

During pregnancy, until embryological and fetal development is fully formed, it is necessary to know how drug or nanoparticle transfer through the placenta is transported. Determining the safety of a particular nanoparticle requires extensive testing using combinations of cell and cell lines, perfused tissue and placental transfer modeling in maternal and placental tissues. Research to evaluate the negative and toxic effects on the fetus requires carefully designed animal studies in vivo [28].

Nanoparticles that leak into the bloodstream and can reach the placenta, mothers and fetuses form a vulnerable population [85]. Intrauterine exposure to nanoparticles not only affects fetal development but also reveals that the etiology of some diseases that occur throughout life may be of fetal origin and may adversely affect health in the later stages. Several epidemiological studies have shown that prenatal exposure to (ultra) fine particles is associated with different adverse health outcomes, including an increased risk of preterm birth (<37 weeks' gestation), retarded development of newborns (<2500 g) and cardiovascular problems, respiratory disorders and neurodevelopmental abnormalities [86–88]. Taking into account the above mentioned, studies in the form of in vivo, in vitro/ex vivo can better determine the levels and breakdown of placenta, nanoparticles, evaluations can be performed more accurately. In addition, the studies to be carried out in different pregnancy periods are important in understanding the toxicity of nanoparticles. It is conceivable that the development of toxicity from nanoparticles can be caused by dose and frequency [89].

Table 2. Potential risks of nanoparticles during reproduction [71–73].

Mechanism of damage	Explanation	Examples
Direct or fetal oxidative damage to the placenta	Direct passage of nanoparticles into placental or foetal tissues may result in the induction of ROS and inflammation.	Exposure and interaction of the fetus with Titanium dioxide can lead to the destruction of genes, dysregulation of neurotransmitters and brain development [74]. Injection of amine-functionalised MWCNTs into pregnant mice caused DNA defect in the fetal organ, whereas the use of antioxidants diminished these effects [75]. As a result of oxidised MWCNTs crossing the placental barrier and reaching the fetus, stenosis in the vessels and decrease in their density were observed [76].
DNA damage and Genotoxicity	Serious DNA damage such as DNA lesions and mutations may occur.	Placental passage of gold nanoparticles caused dimension-dependent epigenetic impacts [77]. If the mother received diesel nanoparticles, it caused DNA destruction in the offspring [78]. DNA damage was increased in the liver of the offspring of the mother who received cobalt and chromium nanoparticles [79].
Fetal exposure after placental transfer	Nanoparticles accumulate in foetal organs after reaching the fetus.	Silica and titanium dioxide nanoparticles were detected in the foetal brain and liver, and the mothers had smaller uteruses and fetuses [80].
Teratogenicity	Exposure to nanoparticles or drugs may cause teratogenicity.	Studies on pregnant women have reported that carbon nanotubes cause fetal malformations and placental injury [81–84].

In animal model studies, no differences were observed in any physiological events in the placenta when exposed to nanoparticles. However, increased toxicity and DNA damage were observed in neonatal blood and liver [90]. This highlights the need for further research and careful evaluation in trophoblast or placental explant tissue to determine which nanomaterials are safe to use in order to minimise risks and avoid any effects on the human placenta, both directly and indirectly. It can be emphasised once again that nanoparticles ingested in early pregnancy may have significant developmental implications in any harm to vulnerable developing gametes or embryos [46,72].

As the use of nanomaterials has become widespread, the positive effects of nanomaterials on health as well as their negative effects have begun to cause concern in humans. Animal experiments confirm the biological effects of nanomaterials and that nanoparticles crossing the placental barrier can cause neurotoxicity in offspring. However, the difficulty of understanding and interpreting the effects of nanoparticles on pregnant animals can create controversy. It has been shown that silica and titanium dioxide nanoparticles with a diameter of 70 nm and 35 nm given to pregnant mice can cause adverse conditions such as pregnancy complications after intravenous injection. As a result of the study, small silica and titanium dioxide nanoparticles were found in the tissues of the fetus, placenta, liver, and brain, and these women had smaller uterus and fetus than normal, as shown in Table 2 [80].

Cell toxicity and genetic toxicity tests of metal nanoparticles used in the health field are increasing. Various nanoparticles, including metals, have been shown to have chemotherapy and genotoxic potential in some tests for humans. In most of the studies, an increase in DNA damage was found in parallel with the increase in nanoparticle concentration [91]. Due to their small size, metal nanoparticles can directly enter the cell and nucleus, cause free radical formation in the cell, and bind to DNA and cause genetic damage [92]. Nano titanium dioxide was administered subcutaneously in pregnant rats and it was observed that it passed to the offspring. Monitoring of the offspring revealed brain damage and reproductive problems in male offspring. Brain damage was observed in fish exposed to nanoparticles for 48 hours. If these structures can be taken up by cells, they can enter the food chain through bacteria. It has been reported that nanoparticles of silver, gold, chromium, cobalt, copper oxide, zinc oxide, aluminum oxide and titanium dioxide, whose effects were investigated in different cell cultures by various cell toxicity and genetic toxicity tests, have chemotherapy and genotoxic potentials [93]. Generally, studies of in vivo genetic toxicity tests of metal nanoparticles, especially in mammals, are very limited. In order to establish a connection between the physicochemical properties of nanoparticles and their toxic effects, the uptake and accumulation of nanoparticles in living systems and cells should be well known [94]. For this reason, it is very important to carry out in vivo studies on different species of organisms, covering different doses and exposure times.

2.   Conclusion

The fact that the fetus has completed its development in the womb and is ready for birth does not mean that its defense against internal and external factors is strong enough. This occurs because of the rapid development and alters metabolism during pregnancy, which makes the fetus more susceptible to external toxic substances compared to most adults. As nanomaterials play a greater part of everyday practice, whether occupational or environmental, the risk of exposure to them during pregnancy is increasing. Although unintentional exposure or taking more than the required amount may cause inconvenient situations, we should accurately state the benefits that arise when nanomaterials are developed and used for diagnostic or therapeutic purposes during pregnancy. There is a great need for research to be conducted and analysed to determine the usefulness of nanomaterials. We need more research to determine whether the NP will reach the placenta and, if so, the interaction with the mother and the positive or negative effects on the fetus.

Some studies have observed a greater total blood volume during pregnancy as a result of the use of nanomaterials, increased blood flow in the kidneys, decreased intestinal transit and differences in the degree of binding of proteins in the blood and various important physiological events. However, further specific studies to compare the pharmacokinetic findings obtained from a non-pregnant population with a pregnant population will further enlighten us.

The development of targeted therapies for reproductive system organs using nanoparticles is an important development. In addition to safe treatments for factors affecting women's health and reproductive capacity, it has also taken its place in having a healthy pregnancy or preventing serious pregnancy complications. However, our knowledge about the passage of nanoparticles from the placental barrier to the fetus is quite restricted and it is stated that this is a new area that needs further research.

References

[1]   F. Ameen, H. Karimi-Maleh, R. Darabi, M. Akin, A. Ayati, S. Ayyildiz, M. Bekmezci, R. Bayat, F. Sen, Synthesis and characterization of activated carbon supported bimetallic Pd based nanoparticles and their sensor and antibacterial investigation, Environ. Res. 221 (2023) 115287. https://doi.org/10.1016/J.ENVRES.2023.115287

[2]   R. Bayat, M. Akin, B. Yilmaz, M. Bekmezci, M. Bayrakci, F. Sen, Biogenic platinum based nanoparticles: Synthesis, characterization and their applications for cell cytotoxic, antibacterial effect, and direct alcohol fuel cells, Chem. Eng. J. Adv. 14 (2023) 100471. https://doi.org/10.1016/j.ceja.2023.100471

[3]   R. Nagraik, A. Sharma, D. Kumar, S. Mukherjee, F. Sen, A.P. Kumar, Amalgamation of biosensors and nanotechnology in disease diagnosis: Mini-review, Sensors Int. 2 (2021) 100089. https://doi.org/10.1016/J.SINTL.2021.100089

[4]   M. Akin, M. Bekmezci, R. Bayat, I. Isik, F. Sen, Ultralight covalent organic frame graphene aerogels modified platinum-magnetite nanostructure for direct methanol fuel cell, Fuel. 357 (2024) 129771. https://doi.org/10.1016/J.FUEL.2023.129771

[5]   M. Akin, M. Bekmezci, R. Bayat, F.S. Kuegou, I. Isik, G. Kaya, F. Sen, Synthesis of platinum-nickel nanoparticles supported by carbon and titanium oxide structures for efficient and enhanced formic acid oxidation, Fuel. 373 (2024) 132258. https://doi.org/10.1016/J.FUEL.2024.132258

[6]   N. Korkmaz, Y. Ceylan, P. Taslimi, A. Karadağ, A.S. Bülbül, F. Şen, Biogenic nano silver: Synthesis, characterization, antibacterial, antibiofilms, and enzymatic activity, Adv. Powder Technol. 31 (2020) 2942–2950. https://doi.org/10.1016/j.apt.2020.05.020

[7]   M. Bekmezci, M. Akin, G.N. Gules, R. Bayat, F. Sen, Innovative chelation strategies for facile synthesis of bimetallic nanomaterials with remarkable photocatalytic and biochemical activities, Next Res. 1 (2024) 100001. https://doi.org/10.1016/J.NEXRES.2024.100001

[8]   A. Aygun, F. Gulbagca, E.E. Altuner, M. Bekmezci, T. Gur, H. Karimi-Maleh, F. Karimi, Y. Vasseghian, F. Sen, Highly active PdPt bimetallic nanoparticles synthesized by one-step bioreduction method: Characterizations, anticancer, antibacterial activities and evaluation of their catalytic effect for hydrogen generation, Int. J. Hydrogen Energy. (2022). https://doi.org/10.1016/J.IJHYDENE.2021.12.144

[9]   F. Gulbagca, A. Aygün, M. Gülcan, S. Ozdemir, S. Gonca, F. Şen, Green synthesis of palladium nanoparticles: Preparation, characterization, and investigation of antioxidant, antimicrobial, anticancer, and DNA cleavage activities, Appl. Organomet. Chem. 35 (2021) e6272. https://doi.org/10.1002/AOC.6272

[10]   L. Aengenheister, R.R. Favaro, D.M. Morales-Prieto, L.A. Furer, M. Gruber, C. Wadsack, U.R. Markert, T. Buerki-Thurnherr, Research on nanoparticles in human perfused placenta: State of the art and

perspectives, Placenta. 104 (2021) 199–207. https://doi.org/10.1016/J.PLACENTA.2020.12.014

[11]   E. Demir, B. Sen, F. Sen, Highly efficient Pt nanoparticles and f-MWCNT nanocomposites based counter electrodes for dye-sensitized solar cells, Nano-Structures & Nano-Objects. 11 (2017) 39–45. https://doi.org/10.1016/J.NANOSO.2017.06.003

[12]   T. Gur, I. Meydan, H. Seckin, M. Bekmezci, F. Sen, Green synthesis, characterization and bioactivity of biogenic zinc oxide nanoparticles, Environ. Res. 204 (2022) 111897. https://doi.org/10.1016/J.ENVRES.2021.111897

[13]   A. Hojjati-Najafabadi, S. Salmanpour, F. Sen, P.N. Asrami, M. Mahdavian, M.A. Khalilzadeh, A Tramadol Drug Electrochemical Sensor Amplified by Biosynthesized Au Nanoparticle Using Mentha aquatic Extract and Ionic Liquid, Top. Catal. 65 (2022) 587–594. https://doi.org/10.1007/S11244-021-01498-X/TABLES/2

[14]   A. Cherif, R. Nebbali, J.W. Sheffield, N. Doner, F. Sen, Numerical investigation of hydrogen production via autothermal reforming of steam and methane over Ni/Al2O3 and Pt/Al2O3 patterned catalytic layers, Int. J. Hydrogen Energy. 46 (2021) 37521–37532. https://doi.org/10.1016/J.IJHYDENE.2021.04.032

[15]   M. Akin, M. Bekmezci, R. Bayat, Z.K. Coguplugil, F. Sen, F. Karimi, H. Karimi-Maleh, Mobile device integrated graphene oxide quantum dots based electrochemical biosensor design for detection of miR-141 as a pancreatic cancer biomarker, Electrochim. Acta. 435 (2022) 141390. https://doi.org/10.1016/J.ELECTACTA.2022.141390

[16]   R. Darabi, H. Karimi-Maleh, M. Akin, K. Arikan, Z. Zhang, R. Bayat, M. Bekmezci, F. Sen, Simultaneous determination of ascorbic acid, dopamine, and uric acid with a highly selective and sensitive reduced graphene oxide/polypyrrole-platinum nanocomposite modified electrochemical sensor, Electrochim. Acta. 457 (2023) 142402. https://doi.org/10.1016/J.ELECTACTA.2023.142402

[17]   C. Demir, A. Aygun, M.K. Gunduz, B.Y. Altınok, T. Karahan, I. Meydan, E. Halvaci, R.N.E. Tiri, F. Sen, Production of plant-based ZnO NPs by green synthesis; anticancer activities and photodegradation of methylene red dye under sunlight, Biomass Convers. Biorefinery 2024. (2024) 1–16. https://doi.org/10.1007/S13399-024-06172-2

[18]   G. Medina-Pérez, F. Fernández-Luqueño, E. Vazquez-Nuñez, F. López-Valdez, J. Prieto-Mendez, A. Madariaga-Navarrete, M. Miranda-Arámbula, Remediating Polluted Soils UsingNanotechnologies: Environmental Benefitsand Risks, Polish J. Environ. Stud. 28 (2019) 1013–1030. https://doi.org/10.15244/PJOES/87099

[19]   P. Taslimi, F. Türkan, A. Cetin, H. Burhan, M. Karaman, I. Bildirici, İ. Gulçin, F. Şen, Pyrazole[3,4-d]pyridazine derivatives: Molecular docking and explore of acetylcholinesterase and carbonic anhydrase enzymes inhibitors as anticholinergics potentials, Bioorg. Chem. 92 (2019) 103213. https://doi.org/10.1016/J.BIOORG.2019.103213

[20]   H. Wang, C.A. Liddell, M.M. Coates, M.D. Mooney, C.E. Levitz, A.E. Schumacher, H. Apfel, M. Iannarone, B. Phillips, K.T. Lofgren, L. Sandar, R.E. Dorrington, I. Rakovac, T.A. Jacobs, X. Liang, M. Zhou, J. Zhu, G. Yang, Y. Wang, S. Liu, Y. Li, A.A. Ozgoren, S.F. Abera, I. Abubakar, T. Achoki, A. Adelekan, Z. Ademi, Z.A. Alemu, P.J. Allen, M.A. AlMazroa, E. Alvarez, A.A. Amankwaa, A.T. Amare, W. Ammar, P. Anwari, S.A. Cunningham, M.M. Asad, R. Assadi, A. Banerjee, S. Basu, N. Bedi, T. Bekele, M.L. Bell, Z. Bhutta, J.D. Blore, B.B. Basara, S. Boufous, N. Breitborde, N.G. Bruce, L.N. Bui, J.R. Carapetis, R. Cárdenas, D.O. Carpenter, V. Caso, R.E. Castro, F. Catalá-Lopéz, A. Cavlin, X. Che, P.P.C. Chiang, R. Chowdhury, C.A. Christophi, T.W. Chuang, M. Cirillo, I. Da Costa Leite, K.J. Courville, L. Dandona, R. Dandona, A. Davis, A. Dayama, K. Deribe, S.D. Dharmaratne, M.K. Dherani, U. Dilmen, E.L. Ding, K.M. Edmond, S.P. Ermakov, F. Farzadfar, S.M. Fereshtehnejad, D.O. Fijabi, N. Foigt, M.H. Forouzanfar, A.C. Garcia, J.M. Geleijnse, B.D. Gessner, K. Goginashvili, P. Gona, A. Goto, H.N. Gouda, M.A. Green, K.F. Greenwell, H.C. Gugnani, R. Gupta, R.R. Hamadeh, M. Hammami, H.L. Harb, S. Hay, M.T. Hedayati, H.D. Hosgood, D.G. Hoy, B.T. Idrisov, F. Islami, S. Ismayilova, V. Jha, G.

Jiang, J.B. Jonas, K. Juel, E.K. Kabagambe, D.S. Kazi, A.P. Kengne, M. Kereselidze, Y.S. Khader, S.E.A.H. Khalifa, Y.H. Khang, D. Kim, Y. Kinfu, J.M. Kinge, Y. Kokubo, S. Kosen, B.K. Defo, G.A. Kumar, K. Kumar, R.B. Kumar, T. Lai, Q. Lan, A. Larsson, J.T. Lee, M. Leinsalu, S.S. Lim, S.E. Lipshultz, G. Logroscino, P.A. Lotufo, R. Lunevicius, R.A. Lyons, S. Ma, A.A. Mahdi, M.B. Marzan, M.T. Mashal, T.T. Mazorodze, J.J. McGrath, Z.A. Memish, W. Mendoza, G.A. Mensah, A. Meretoja, T.R. Miller, E.J. Mills, K.A. Mohammad, A.H. Mokdad, L. Monasta, M. Montico, A.R. Moore, J. Moschandreas, W.T. Msemburi, U.O. Mueller, M.M. Muszynska, M. Naghavi, K.S. Naidoo, K.M.V. Narayan, C. Nejjari, M. Ng, J. De Dieu Ngirabega, M.J. Nieuwenhuijsen, L. Nyakarahuka, T. Ohkubo, S.B. Omer, A.J. Paternina Caicedo, V. Pillay-Van Wyk, D. Pope, F. Pourmalek, D. Prabhakaran, S.U.R. Rahman, S.M. Rana, R.Q. Reilly, D. Rojas-Rueda, L. Ronfani, L. Rushton, M.Y. Saeedi, J.A. Salomon, U. Sampson, I.S. Santos, M. Sawhney, J.C. Schmidt, M. Shakh-Nazarova, J. She, S. Sheikhbahaei, K. Shibuya, H.H. Shin, K. Shishani, I. Shiue, I.D. Sigfusdottir, J.A. Singh, V. Skirbekk, K. Sliwa, S.S. Soshnikov, L.A. Sposato, V.K. Stathopoulou, K. Stroumpoulis, K.M. Tabb, R.T. Talongwa, C.M. Teixeira, A.S. Terkawi, A.J. Thomson, A.L. Thorne-Lyman, H. Toyoshima, Z.T. Dimbuene, P. Uwaliraye, S.B. Uzun, T.J. Vasankari, A.M.N. Vasconcelos, V.V. Vlassov, S.E. Vollset, S. Waller, X. Wan, S. Weichenthal, E. Weiderpass, R.G. Weintraub, R. Westerman, J.D. Wilkinson, H.C. Williams, Y.C. Yang, G.K. Yentur, P. Yip, N. Yonemoto, M. Younis, C. Yu, K.Y. Jin, M. El Sayed Zaki, S. Zhu, T. Vos, A.D. Lopez, C.J.L. Murray, Global, regional, and national levels of neonatal, infant, and under-5 mortality during 1990-2013: A systematic analysis for the Global Burden of Disease Study 2013, Lancet. 384 (2014) 957–979. https://doi.org/10.1016/S0140-6736(14)60497-9

[21]   M.A. Rodger, M.T. Betancourt, P. Clark, P.G. Lindqvist, D. Dizon-Townson, J. Said, U. Seligsohn, M. Carrier, O. Salomon, I.A. Greer, The association of factor V leiden and prothrombin gene mutation and placenta-mediated pregnancy complications: a systematic review and meta-analysis of prospective cohort studies, PLoS Med. 7 (2010). https://doi.org/10.1371/JOURNAL.PMED.1000292

[22]   K.S. Khan, D. Wojdyla, L. Say, A.M. Gülmezoglu, P.F. Van Look, WHO analysis of causes of maternal death: a systematic review, Lancet. 367 (2006) 1066–1074. https://doi.org/10.1016/S0140-6736(06)68397-9

[23]   R.A.H. Kinch, Diethylstilbestrol in pregnancy: an update, Can. Med. Assoc. J. 127 (1982) 812. https://pmc.ncbi.nlm.nih.gov/articles/PMC1862229/ (accessed October 22, 2024).

[24]   J.S. Choi, G. Koren, I. Nulman, Pregnancy and isotretinoin therapy, C. Can. Med. Assoc. J. 185 (2013) 411. https://doi.org/10.1503/CMAJ.120729

[25]   N. Vargesson, Thalidomide-induced teratogenesis: History and mechanisms, Birth Defects Res. Part C Embryo Today Rev. 105 (2015) 140–156. https://doi.org/10.1002/BDRC.21096

[26]   B. Zhang, R. Liang, M. Zheng, L. Cai, X. Fan, Surface-Functionalized Nanoparticles as Efficient Tools in Targeted Therapy of Pregnancy Complications, Int. J. Mol. Sci. . 20 (2019) 3642. https://doi.org/10.3390/IJMS20153642

[27]   E.M. Van Der Aa, J.H.J. Copius Peereboom-Stegeman, J. Noordhoek, F.W.J. Gribnau, F.G.M. Russel, Mechanisms of drug transfer across the human placenta, Pharm. World Sci. 20 (1998) 139–148. https://doi.org/10.1023/A:1008656928861

[28]   N. Pritchard, T. Kaitu'U-Lino, L. Harris, S. Tong, N. Hannan, Nanoparticles in pregnancy: the next frontier in reproductive therapeutics, Hum. Reprod. Update. 27 (2021) 280–304. https://doi.org/10.1093/HUMUPD/DMAA049

[29]   C. Gundacker, J. Neesen, E. Straka, I. Ellinger, H. Dolznig, M. Hengstschläger, Genetics of the human placenta: implications for toxicokinetics, Arch. Toxicol. 2016 9011. 90 (2016) 2563–2581. https://doi.org/10.1007/S00204-016-1816-6

[30]   M. Ceckova-Novotna, P. Pavek, F. Staud, P-glycoprotein in the placenta: Expression, localization, regulation and function, Reprod. Toxicol. 22 (2006) 400–410. https://doi.org/10.1016/J.REPROTOX.2006.01.007

[31]   A.C. Enders, T.N. Blankenship, Comparative placental structure, Adv. Drug Deliv. Rev. 38 (1999) 3–15. https://doi.org/10.1016/S0169-409X(99)00003-4

[32]   D. Evseenko, J.W. Paxton, J.A. Keelan, Active transport across the human placenta: impact on drug efficacy and toxicity, Expert Opin. Drug Metab. Toxicol. 2 (2006) 51–69. https://doi.org/10.1517/17425255.2.1.51

[33]   K.L. Audus, Controlling drug delivery across the placenta, Eur. J. Pharm. Sci. 8 (1999) 161–165. https://doi.org/10.1016/S0928-0987(99)00031-7

[34]   J.A. Keelan, J.W. Leong, D. Ho, K.S. Iyer, Therapeutic and Safety Considerations of Nanoparticle-Mediated Drug Delivery in Pregnancy, Nanomedicine. 10 (2015) 2229–2247. https://doi.org/10.2217/NNM.15.48

[35]   V. Menezes, A. Malek, J. A. Keelan, Nanoparticulate Drug Delivery in Pregnancy: Placental Passage and Fetal Exposure, Curr. Pharm. Biotechnol. 12 (2011) 731–742.

[36]   M. Saunders, Transplacental transport of nanomaterials, Wiley Interdiscip. Rev. Nanomedicine Nanobiotechnology. 1 (2009) 671–684. https://doi.org/10.1002/WNAN.53

[37]   R.A. Polin, W.W. Fox, S.H. Abman, Fetal and Neonatal Physiology: Fifth Edition, Fetal Neonatal Physiol. Fifth Ed. (2011) 1–2038. https://books.google.com/books/about/Fetal_and_Neonatal_Physiology_E_Book.html?id=38X8Kc_YM MUC (accessed October 22, 2024).

[38]   C.P. Sibley, P. Brownbill, J.D. Glazier, S.L. Greenwood, Knowledge needed about the exchange physiology of the placenta, Placenta. 64 (2018) S9–S15. https://doi.org/10.1016/J.PLACENTA.2018.01.006

[39]   A.P. Kourtis, J.S. Read, D.J. Jamieson, Pregnancy and Infection, N. Engl. J. Med. 370 (2014) 2211–2218. https://doi.org/10.1056/NEJMRA1213566

[40]   H. Chen, J. Guo, C. Wang, F. Luo, X. Yu, W. Zhang, J. Li, D. Zhao, D. Xu, Q. Gong, J. Liao, H. Yang, W. Hou, Y. Zhang, Clinical characteristics and intrauterine vertical transmission potential of COVID-19 infection in nine pregnant women: a retrospective review of medical records, Lancet. 395 (2020) 809–815. https://doi.org/10.1016/S0140-6736(20)30360-3

[41]   J.J. Mulvey, C.M. Magro, L.X. Ma, G.J. Nuovo, R.N. Baergen, Analysis of complement deposition and viral RNA in placentas of COVID-19 patients, Ann. Diagn. Pathol. 46 (2020) 151530. https://doi.org/10.1016/J.ANNDIAGPATH.2020.151530

[42]   W. Guan, Z. Ni, Y. Hu, W. Liang, C. Ou, J. He, L. Liu, H. Shan, C. Lei, D.S.C. Hui, B. Du, L. Li, G. Zeng, K.-Y. Yuen, R. Chen, C. Tang, T. Wang, P. Chen, J. Xiang, S. Li, J. Wang, Z. Liang, Y. Peng, L. Wei, Y. Liu, Y. Hu, P. Peng, J. Wang, J. Liu, Z. Chen, G. Li, Z. Zheng, S. Qiu, J. Luo, C. Ye, S. Zhu, N. Zhong, Clinical Characteristics of Coronavirus Disease 2019 in China, N. Engl. J. Med. 382 (2020) 1708–1720. https://doi.org/10.1056/NEJMOA2002032

[43]   J. Yang, R. D'Souza, A. Kharrat, D.B. Fell, J.W. Snelgrove, K.E. Murphy, P.S. Shah, Coronavirus disease 2019 pandemic and pregnancy and neonatal outcomes in general population: A living systematic review and meta-analysis , Acta Obstet. Gynecol. Scand. 101 (2022) 7–24. https://doi.org/10.1111/AOGS.14277

[44]   R. Wang, B. Song, J. Wu, Y. Zhang, A. Chen, L. Shao, Potential adverse effects of nanoparticles on the reproductive system, Int. J. Nanomedicine. 13 (2018) 8487–8506. https://doi.org/10.2147/IJN.S170723

[45]   S.T. Jahan, S.M.A. Sadat, M. Walliser, A. Haddadi, Targeted Therapeutic Nanoparticles: An Immense Promise to Fight against Cancer, J. Drug Deliv. 2017 (2017) 1–24. https://doi.org/10.1155/2017/9090325

[46]   C.C. Hou, J.Q. Zhu, Nanoparticles and female reproductive system: how do nanoparticles affect oogenesis and embryonic development, Oncotarget. 8 (2017) 109799.

https://doi.org/10.18632/ONCOTARGET.19087

[47]   S. Kertschanska, G. Kosanke, P. Kaufmann, Pressure dependence of so-called transtrophoblastic channels during fetal perfusion of human placental villi, Microsc. Res. Tech. 38 (1997) 52–62. https://doi.org/10.1002/(SICI)1097-0029(19970701/15)38:1/2

[48]   A.C. Enders, T.N. Blankenship, K.C. Lantz, S.S. Enders, Morphological variation in the interhemal areas of chorioallantoic placentae: A review, Placenta. 19 (1998) 1–19. https://doi.org/10.1016/S0143-4004(98)80030-1

[49]   C. Bosco, C. Buffet, M.A. Bello, R. Rodrigo, M. Gutierrez, G. García, Placentation in the degu (Octodon degus): analogies with extrasubplacental trophoblast and human extravillous trophoblast, Comp. Biochem. Physiol. A. Mol. Integr. Physiol. 146 (2007) 475–485. https://doi.org/10.1016/J.CBPA.2005.12.013

[50]   S. Kertschanska, B. Štulcová, P. Kaufmann, J. Štulc, Distensible transtrophoblastic channels in the rat placenta, Placenta. 21 (2000) 670–677. https://doi.org/10.1053/PLAC.2000.0558

[51]   P. Wick, A. Malek, P. Manser, D. Meili, X. Maeder-Althaus, L. Diener, P.A. Diener, A. Zisch, H.F. Krug, U. Von Mandach, Barrier capacity of human placenta for nanosized materials, Environ. Health Perspect. 118 (2010) 432–436. https://doi.org/10.1289/EHP.0901200

[52]   R. Sakhtianchi, R.F. Minchin, K.B. Lee, A.M. Alkilany, V. Serpooshan, M. Mahmoudi, Exocytosis of nanoparticles from cells: Role in cellular retention and toxicity, Adv. Colloid Interface Sci. 201–202 (2013) 18–29. https://doi.org/10.1016/J.CIS.2013.10.013

[53]   N. Tetro, S. Moushaev, M. Rubinchik-Stern, S. Eyal, The Placental Barrier: the Gate and the Fate in Drug Distribution, Pharm. Res. 35 (2018). https://doi.org/10.1007/S11095-017-2286-0

[54]   Y. Liu, A. Ibricevic, J.A. Cohen, J.L. Cohen, S.P. Gunsten, J.M.J. Fréchet, M.J. Walter, M.J. Welch, S.L. Brody, Impact of hydrogel nanoparticle size and functionalization on in vivo behavior for lung imaging and therapeutics, Mol. Pharm. 6 (2009) 1891–1902. https://doi.org/10.1021/MP900215P

[55]   S.D. Conner, S.L. Schmid, Regulated portals of entry into the cell, Nat. . 422 (2003) 37–44. https://doi.org/10.1038/nature01451

[56]   C.D. Ockleford, A. Whyte, Differeniated regions of human placental cell surface associated with exchange of materials between maternal and foetal blood: coated vesicles, J. Cell Sci. 25 (1977) 293–312. https://doi.org/10.1242/JCS.25.1.293

[57]   T.J. Kaitu'u-Lino, S. Pattison, L. Ye, L. Tuohey, P. Sluka, J. MacDiarmid, H. Brahmbhatt, T. Johns, A.W. Horne, J. Brown, S. Tong, Targeted Nanoparticle Delivery of Doxorubicin Into Placental Tissues to Treat Ectopic Pregnancies, Endocrinology. 154 (2013) 911–919. https://doi.org/10.1210/EN.2012-1832

[58]   S.J. Kang, H.Y. Jeong, M.W. Kim, I.H. Jeong, M.J. Choi, Y.M. You, C.S. Im, I.H. Song, T.S. Lee, Y.S. Park, Anti-EGFR lipid micellar nanoparticles co-encapsulating quantum dots and paclitaxel for tumor-targeted theranosis, Nanoscale. 10 (2018) 19338–19350. https://doi.org/10.1039/C8NR05099F

[59]   J. Huang, W. Huang, Z. Zhang, X. Lin, H. Lin, L. Peng, T. Chen, Highly Uniform Synthesis of Selenium Nanoparticles with EGFR Targeting and Tumor Microenvironment-Responsive Ability for Simultaneous Diagnosis and Therapy of Nasopharyngeal Carcinoma, ACS Appl. Mater. Interfaces. 11 (2019) 11177–11193. https://doi.org/10.1021/ACSAMI.8B22678

[60]   N. Groysbeck, A. Stoessel, M. Donzeau, E.C. Da Silva, M. Lehmann, J.M. Strub, S. Cianferani, K. Dembélé, G. Zuber, Synthesis and biological evaluation of 2.4 nm thiolate-protected gold nanoparticles conjugated to Cetuximab for targeting glioblastoma cancer cells via the EGFR, Nanotechnology. 30 (2019) 184005. https://doi.org/10.1088/1361-6528/AAFF0A

[61]   F. Beards, L.E. Jones, J. Charnock, K. Forbes, L.K. Harris, Placental Homing Peptide-microRNA Inhibitor Conjugates for Targeted Enhancement of Intrinsic Placental Growth Signaling, Theranostics. 7 (2017) 2940. https://doi.org/10.7150/THNO.18845

[62]   S. Hua, M.B.C. de Matos, J.M. Metselaar, G. Storm, Current trends and challenges in the clinical translation of nanoparticulate nanomedicines: Pathways for translational development and commercialization, Front. Pharmacol. 9 (2018) 403086. https://doi.org/10.3389/FPHAR.2018.00790

[63]   C. Muoth, L. Aengenheister, M. Kucki, P. Wick, T. Buerki-Thurnherr, Nanoparticle transport across the placental barrier: pushing the field forward!, Nanomedicine . 11 (2016) 941–957. https://doi.org/10.2217/NNM-2015-0012

[64]   D.W. Galbraith, M. Emily, K. Calvert, G. Histology, L. Microscopy, F. Janat, A. Azqueta, M. Dusinska, The use of the comet assay for the evaluation of the genotoxicity of nanomaterials, Front. Genet. 6 (2015) 239. https://doi.org/10.3389/FGENE.2015.00239

[65]   S.H. Doak, B. Manshian, G.J.S. Jenkins, N. Singh, In vitro genotoxicity testing strategy for nanomaterials and the adaptation of current OECD guidelines, Mutat. Res. Toxicol. Environ. Mutagen. 745 (2012) 104–111. https://doi.org/10.1016/J.MRGENTOX.2011.09.013

[66]   S.J. Evans, M.J.D. Clift, N. Singh, J. De Oliveira Mallia, M. Burgum, J.W. Wills, T.S. Wilkinson, G.J.S. Jenkins, S.H. Doak, Critical review of the current and future challenges associated with advanced in vitro systems towards the study of nanoparticle (secondary) genotoxicity, Mutagenesis. 32 (2017) 233–241. https://doi.org/10.1093/MUTAGE/GEW054

[67]   S. Pfuhler, T.R. Downs, A.J. Allemang, Y. Shan, M.E. Crosby, Weak silica nanomaterial-induced genotoxicity can be explained by indirect DNA damage as shown by the OGG1-modified comet assay and genomic analysis, Mutagenesis. 32 (2017) 5–12. https://doi.org/10.1093/MUTAGE/GEW064

[68]   Y. Kohl, E. Rundén-Pran, E. Mariussen, M. Hesler, N. El Yamani, E.M. Longhin, M. Dusinska, Genotoxicity of Nanomaterials: Advanced In Vitro Models and High Throughput Methods for Human Hazard Assessment—A Review, Nanomater. . 10 (2020) 1911. https://doi.org/10.3390/NANO10101911

[69]   A.R. Collins, B. Annangi, L. Rubio, R. Marcos, M. Dorn, C. Merker, I. Estrela-Lopis, M.R. Cimpan, M. Ibrahim, E. Cimpan, M. Ostermann, A. Sauter, N. El Yamani, S. Shaposhnikov, S. Chevillard, V. Paget, R. Grall, J. Delic, F.G. de-Cerio, B. Suarez-Merino, V. Fessard, K.N. Hogeveen, L.M. Fjellsbø, E.R. Pran, T. Brzicova, J. Topinka, M.J. Silva, P.E. Leite, A.R. Ribeiro, J.M. Granjeiro, R. Grafström, A. Prina-Mello, M. Dusinska, High throughput toxicity screening and intracellular detection of nanomaterials, Wiley Interdiscip. Rev. Nanomedicine Nanobiotechnology. 9 (2017) e1413. https://doi.org/10.1002/WNAN.1413

[70]   A. Kumar, A. Dhawan, Genotoxic and carcinogenic potential of engineered nanoparticles: An update, Arch. Toxicol. 87 (2013) 1883–1900. https://doi.org/10.1007/S00204-013-1128-Z

[71]   L. Campagnolo, M. Massimiani, A. Magrini, A. Camaioni, A. Pietroiusti, Physico-Chemical Properties Mediating Reproductive and Developmental Toxicity of Engineered Nanomaterials, Curr. Med. Chem. 19 (2012) 4488–4494. https://doi.org/10.2174/092986712803251566

[72]   J. Das, Y.J. Choi, H. Song, J.H. Kim, Potential toxicity of engineered nanoparticles in mammalian germ cells and developing embryos: treatment strategies and anticipated applications of nanoparticles in gene delivery, Hum. Reprod. Update. 22 (2016) 588–619. https://doi.org/10.1093/HUMUPD/DMW020

[73]   Y. Zhang, J. Wu, X. Feng, R. Wang, A. Chen, L. Shao, Current understanding of the toxicological risk posed to the fetus following maternal exposure to nanoparticles, Expert Opin. Drug Metab. Toxicol. 13 (2017) 1251–1263. https://doi.org/10.1080/17425255.2018.1397131

[74]   M. Shimizu, H. Tainaka, T. Oba, K. Mizuo, M. Umezawa, K. Takeda, Maternal exposure to nanoparticulate titanium dioxide during the prenatal period alters gene expression related to brain development in the mouse, Part. Fibre Toxicol. 6 (2009). https://doi.org/10.1186/1743-8977-6-20

[75]   X. Huang, F. Zhang, X. Sun, K.Y. Choi, G. Niu, G. Zhang, J. Guo, S. Lee, X. Chen, The genotype-dependent influence of functionalized multiwalled carbon nanotubes on fetal development, Biomaterials. 35 (2014) 856–865. https://doi.org/10.1016/J.BIOMATERIALS.2013.10.027

[76]   W. Qi, J. Bi, X. Zhang, J. Wang, J. Wang, P. Liu, Z. Li, W. Wu, Damaging Effects of Multi-

walled Carbon Nanotubes on Pregnant Mice with Different Pregnancy Times, Sci. Reports . 4 (2014) 1–13. https://doi.org/10.1038/srep04352

[77]   R. Balansky, M. Longobardi, G. Ganchev, M. Iltcheva, N. Nedyalkov, P. Atanasov, R. Toshkova, S. De Flora, A. Izzotti, Transplacental clastogenic and epigenetic effects of gold nanoparticles in mice, Mutat. Res. Mol. Mech. Mutagen. 751–752 (2013) 42–48. https://doi.org/10.1016/J.MRFMMM.2013.08.006

[78]   R. Reliene, A. Hlavacova, B. Mahadevan, W.M. Baird, R.H. Schiestl, Diesel exhaust particles cause increased levels of DNA deletions after transplacental exposure in mice, Mutat. Res. 570 (2005) 245–252. https://doi.org/10.1016/J.MRFMMM.2004.11.010

[79]   A. Sood, S. Salih, D. Roh, L. Lacharme-Lora, M. Parry, B. Hardiman, R. Keehan, R. Grummer, E. Winterhager, P.J. Gokhale, P.W. Andrews, C. Abbott, K. Forbes, M. Westwood, J.D. Aplin, E. Ingham, I. Papageorgiou, M. Berry, J. Liu, A.D. Dick, R.J. Garland, N. Williams, R. Singh, A.K. Simon, M. Lewis, J. Ham, L. Roger, D.M. Baird, L.A. Crompton, M.A. Caldwell, H. Swalwell, M. Birch-Machin, G. Lopez-Castejon, A. Randall, H. Lin, M.S. Suleiman, W.H. Evans, R. Newson, C.P. Case, Signalling of DNA damage and cytokines across cell barriers exposed to nanoparticles depends on barrier thickness, Nat. Nanotechnol. . 6 (2011) 824–833. https://doi.org/10.1038/nnano.2011.188

[80]   K. Yamashita, Y. Yoshioka, K. Higashisaka, K. Mimura, Y. Morishita, M. Nozaki, T. Yoshida, T. Ogura, H. Nabeshi, K. Nagano, Y. Abe, H. Kamada, Y. Monobe, T. Imazawa, H. Aoshima, K. Shishido, Y. Kawai, T. Mayumi, S.I. Tsunoda, N. Itoh, T. Yoshikawa, I. Yanagihara, S. Saito, Y. Tsutsumi, Silica and titanium dioxide nanoparticles cause pregnancy complications in mice, Nat. Nanotechnol. 6 (2011) 321–328. https://doi.org/10.1038/nnano.2011.41

[81]   A. Pietroiusti, M. Massimiani, I. Fenoglio, M. Colonna, F. Valentini, G. Palleschi, A. Camaioni, A. Magrini, G. Siracusa, A. Bergamaschi, A. Sgambato, L. Campagnolo, Low doses of pristine and oxidized single-wall carbon nanotubes affect mammalian embryonic development, ACS Nano. 5 (2011) 4624–4633. https://doi.org/10.1021/NN200372G

[82]   T. Fujitani, K.I. Ohyama, A. Hirose, T. Nishimura, D. Nakae, A. Ogata, Teratogenicity of multi-wall carbon nanotube (MWCNT) in ICR mice, J. Toxicol. Sci. 37 (2012) 81–89. https://doi.org/10.2131/JTS.37.81

[83]   J.A. Keelan, Nanoparticles versus the placenta, Nat. Nanotechnol. . 6 (2011) 263–264. https://doi.org/10.1038/nnano.2011.65

[84]   M. Ema, K.S. Hougaard, A. Kishimoto, K. Honda, Reproductive and developmental toxicity of carbon-based nanomaterials: A literature review, Nanotoxicology. 10 (2016) 391–412. https://doi.org/10.3109/17435390.2015.1073811

[85]   H. Bové, E. Bongaerts, E. Slenders, E.M. Bijnens, N.D. Saenen, W. Gyselaers, P. Van Eyken, M. Plusquin, M.B.J. Roeffaers, M. Ameloot, T.S. Nawrot, Ambient black carbon particles reach the fetal side of human placenta, Nat. Commun. . 10 (2019) 1–7. https://doi.org/10.1038/s41467-019-11654-3

[86]   L.J. Luyten, N.D. Saenen, B.G. Janssen, K. Vrijens, M. Plusquin, H.A. Roels, F. Debacq-Chainiaux, T.S. Nawrot, Air pollution and the fetal origin of disease: A systematic review of the molecular signatures of air pollution exposure in human placenta, Environ. Res. 166 (2018) 310–323. https://doi.org/10.1016/J.ENVRES.2018.03.025

[87]   P. Dadvand, J. Parker, M.L. Bell, M. Bonzini, M. Brauer, L.A. Darrow, U. Gehring, S. V. Glinianaia, N. Gouveia, E.H. Ha, J.H. Leem, E.H. van den Hooven, B. Jalaludin, B.M. Jesdale, J. Lepeule, R. Morello-Frosch, G.G. Morgan, A.C. Pesatori, F.H. Pierik, T. Pless-Mulloli, D.Q. Rich, S. Sathyanarayana, J. Seo, R. Slama, M. Strickland, L. Tamburic, D. Wartenberg, M.J. Nieuwenhuijsen, T.J. Woodruff, Maternal exposure to particulate air pollution and term birth weight: A multi-country evaluation of effect and heterogeneity, Environ. Health Perspect. 121 (2013) 367–373. https://doi.org/10.1289/EHP.1205575

[88]   X. Li, S. Huang, A. Jiao, X. Yang, J. Yun, Y. Wang, X. Xue, Y. Chu, F. Liu, Y. Liu, M. Ren, X.

Chen, N. Li, Y. Lu, Z. Mao, L. Tian, H. Xiang, Association between ambient fine particulate matter and preterm birth or term low birth weight: An updated systematic review and meta-analysis, Environ. Pollut. . 227 (2017) 596–605. https://doi.org/10.1016/J.ENVPOL.2017.03.055

[89]   M. Semmler-Behnke, J. Lipka, A. Wenk, S. Hirn, M. Schäffler, F. Tian, G. Schmid, G. Oberdörster, W.G. Kreyling, Size dependent translocation and fetal accumulation of gold nanoparticles from maternal blood in the rat, Part. Fibre Toxicol. 11 (2014) 1–12. https://doi.org/10.1186/S12989-014-0033-9/TABLES/3

[90]   R.D. Brohi, L. Wang, H.S. Talpur, D. Wu, F.A. Khan, D. Bhattarai, Z.U. Rehman, F. Farmanullah, L.J. Huo, Toxicity of nanoparticles on the reproductive system in animal models: A review, Front. Pharmacol. 8 (2017) 260180. https://doi.org/10.3389/FPHAR.2017.00606

[91]   H. Xie, M.M. Mason, J.P. Wise, Genotoxicity of metal nanoparticles, Rev. Environ. Health. 26 (2011) 251–268. https://doi.org/10.1515/REVEH.2011.033

[92]   P.R.-N. Perceptions, undefined 2006, The biological effects of nanoparticles, Nanotechnol. Perceptions. 2 (2006) 283–298. https://nano-ntp.com/index.php/nano/article/download/305/214 (accessed October 22, 2024).

[93]   K. Donaldson, C.A. Poland, R.P.F. Schins, Possible genotoxic mechanisms of nanoparticles: Criteria for improved test strategies, Nanotoxicology. 4 (2010) 414–420. https://doi.org/10.3109/17435390.2010.482751

[94]   K. Nesrin, C. Yusuf, K. Ahmet, S.B. Ali, N.A. Muhammad, S. Suna, Ş. Fatih, Biogenic silver nanoparticles synthesized from Rhododendron ponticum and their antibacterial, antibiofilm and cytotoxic activities, J. Pharm. Biomed. Anal. 179 (2020) 112993. https://doi.org/10.1016/J.JPBA.2019.112993

Biogenic Nanomaterials: Synthesis, Characterization, Applications, and Future Remarks   Materials Research Forum LLC
Materials Research Foundations 180 (2025)   https://doi.org/21741/9781644903759

Chapter 14

Use of Biogenic Nanomaterials in Controlled Drug Release Systems

N. Akman[a*], M. Bekmezci[b,c], F. Sen[b*]

[a]Faculty of Health Sciences Van Yuzuncu Yil University, Van, Turkiye,

[b]Sen Research Group, Department of Biochemistry, Dumlupinar University, Kutahya, Türkiye

[c]Department of Materials Science & Engineering, Faculty of Engineering, Dumlupinar University, Kutahya, Turkiye

nurakman@yyu.edu.tr; fatihsen1980@gmail.com

Abstract

By removing the drawbacks of conventional medications, such as their poor selectivity, limited bioavailability, and adverse effects, nanomaterials enable targeted drug delivery to the intended location. Since biogenic nanomaterials don't contain any harmful compounds, they are an ecologically acceptable alternative. They are derived from biological sources including fungus, bacteria, and plants. Although nanotechnology makes it possible to create novel therapeutic approaches in biomedicine and pharmacology, nanoparticles are also employed in targeted drug delivery systems to boost therapeutic effectiveness. With controlled drug delivery systems, this chapter offers a significant viewpoint on delivering medications at low dosages to the intended location.

Keywords

Nanotechnology, Biogenic, Nanomaterials, Drug Delivery, Controlled Delivery

1. Introduction

In the last fifty years, nanoparticles have been used in many areas due to their sensor [1–3], energy [4–6], and antibacterial/anticancer [7] properties [8–17]. However, in the fight against infectious and non-infectious diseases, the use of nanomaterials has increased the effectiveness of the use of nanomaterials in delivering drugs to the target site due to the limited properties of traditional drugs such as specificity, biodiversity, biocompatibility, biodegradability, and parenteral drug administration routes. In recent years, carbohydrate-based polymers have been used as a very interesting nanomaterial in drug delivery systems due to their properties such as biocompatible, natural, environmentally friendly, low cost, and improved encapsulation efficiency [18]. The first nanoparticles now approved by the FDA included lipid systems such as liposomes and micelles. Inorganic nanoparticles, including gold and other magnetic nanoparticles, can be found inside these liposomes and micelles, making them useful for therapeutic purposes, imaging, and drug delivery [19].

Nanomaterials stay in the blood circulation for a long time, allowing each of the co-administered drugs to circulate at a specific dose, reducing the drug fluctuation in the plasma and the side effects of drugs [20]. With nanocarriers, the potential therapeutic efficacy of drugs can be increased, systemic side effects can be reduced, and patient compliance can be improved by reducing the dose and frequency of administration.

2.   Biogenic Nanomaterials

In the last decade, scientists focused on nanotechnological methods to diagnose and treat cancer due to the increase in the prevalence of cancer. Nanotechnology covers different multidisciplinary fields such as medicine, biomaterial, and electronics [21]. New advances in biology and medicine have resulted from the field of nanotechnology's explosive growth. Nano-based drugs provide a great opportunity for the development of the health sector by developing more effective treatment methods for diseases with high mortality rates. Nanotechnology-based applications enable the synthesis, maneuvering, and design of nanostructures produced from nanoparticles or nanomaterials by nanotechnological techniques [22].

Organic, inorganic, and carbon-based nanoparticles with lengths ranging from 1 to 100 nanometers are common classifications for nanomaterials [5]. In drug delivery systems that aim to distribute the active components of drugs, organic nanomaterials such as ferritin, liposomes, dendrimers, and micelles are frequently used [21]. Due to the absence of any toxic substances in the synthesis, the synthesis process can be carried out under ambient temperature and pressure, and it is environmentally friendly, many researchers have focused on biogenic nanomaterials. Synthesis of biogenic nanomaterials is synthesized using biological tools such as fungi, bacteria, lichens, actinominesets, algae, and plants [23]. The enzymes secreted by these biological agents we have mentioned have been shown to act as agents that provide stability to the nanoparticles and reduce the metal salts of proteins to nanoparticles, making them compatible with various biological applications [23]. The antibacterial properties of silver and gold nanoparticles have been demonstrated in the treatment of bacterial and fungal infections. Despite this, biogenic nanomaterials have been the subject of extensive research with the aim of improving the therapeutic and pharmacological effects of anticancer medications in particular. Biogenic nanoparticles act as nano transporters in the targeted distribution of drugs, and the nanoparticle is involved in blood circulation before the easy passage of the added drug into the epithelial cell barrier reaches the target site and affects the drug delivery mechanism by the drug mandatory [24].

2.1   Use of Biogenic Nanomaterials in Controlled Drug Release Systems

The drug distribution system is defined as a method that expresses the absorption and distribution of the drug through various carriers after the drug is delivered to the appropriate site to ensure that the drug is delivered to the desired tissue, organ, cell, and subcellular structures. The primary objective of drug delivery systems is to enhance the pharmacological actions of medicinal medications and address various issues, including drug aggregation, restricted solubility, low bioavailability, poor biodistribution, declining selectivity, and the emergence of new adverse effects [25]. With the development of drug distribution systems, the complexity of diseases, the toxicity that develops in multi-drug treatments, and the obstacles that limit the distribution of drugs are tried to be removed [26].

The non-specific biodistribution, unregulated distribution, and uncontrolled high dosages inherent to these substances result in systemic adverse effects when distributed via traditional drug distribution systems [27]. Traditional drugs cannot cross the blood-brain barrier due to the tightness of glial cells in the endothelial layer of the blood-brain barrier. Especially in brain tumors, the drug must pass to the target area. As a result of combining conventional drugs with nanocarriers, nanocarriers have been reported to have an effect by crossing the blood-brain barrier with receptor-mediated endocytosis, reaching the targeted tumor cell [24]. Nanoparticles are structures that control the size of the nanoparticle in drug delivery systems, its surface area properties, its effect on drug distribution, release time, and specific location so that the drug can have maximum effect [28]. Nanoparticles used in drug dispersion systems can be broken down into biological structures, biocompatibility, time-release, etc, it is based on the mechanical properties that are designed to be made and the production process is easy, and generally in drug distribution systems, nanoparticles can protect the drug against degradation, be better absorbed, modifying pharmacokinetics and diffusion by epithelial pathway, etc, by changing the distribution of the drug, they increase its penetration into the cell and increase its therapeutic effectiveness [29].

Conventional drug applications are characterized by limited drug efficacy, poor biodistribution, and poor selectivity. These limitations and disadvantages can be addressed with controlled drug delivery systems. In controlled drug distribution systems, the drug is transported to the area where it will be affected, and the effect of the drug on vital tissues and possible side effects are minimized. Controlled drug distribution systems prevent rapid degradation or rapid clearance of the drug and increase the concentration of the drug in the target tissue, allowing the drug to be used at low doses. Controlled drug distribution systems are of great importance, especially when there is a difference between the concentration and dose of the drug that causes toxicity [30].

These new distribution systems increase the efficacy and safety of the drug by controlling the amount, timing, and in situ release of the drug by bypassing the barriers to reach the therapeutic target [31].

During the years 2015-2018, significant progress has been made in the drug distribution system in areas such as pharmacology, materials, and biomedicine. In recent years, nanoparticles as carriers have shown great potential in drug delivery systems [25]. The clinical use of nanoparticles aroused great interest in the 1960s with the discovery of liposomes. Nanotechnology has helped make great discoveries in pharmacological studies and drug research, especially in oncology, in the diagnosis, imaging, and treatment of various diseases. Today, more than 30 types of nanoparticles have been approved for clinical use. Interest in nanoparticles in drug delivery has greatly increased, with the United States Food and Drug Administration (FDA) approving lipid nanoparticle-based carriers for mRNA vaccines against COVID-19 [32]. For example, protein-based nanoparticles have the advantage of traditional drugs in drug distribution systems that they do not show toxic properties compared to conventional drugs and are easy to produce non-antigenic. The plasma protein albumin-based nanoparticles allow a significant amount of the drug to bind to the particle-matrix, due to the different drug binding sites of the albumin molecule, and at the same time, its size is quite small, being biogenic, it is among the advantages of emulsion, desolvation, coacervation under normal conditions and both easy adaptation and easy preparation of patients [29].

Nanoparticles aim to increase the uptake of poorly soluble drugs in drug delivery systems, to target drugs to a certain region, and to increase drug bioavailability, increasing the targeting of drugs to the disease site and increasing therapeutic efficacy (Figure 1) [33].

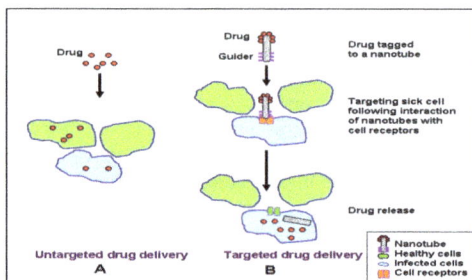

Figure 1. *A: Targeting of free drug molecules in drug distribution systems. B: Reaching the target of the drug molecule combined with the nanoparticle Reprinted (adapted) with permission from reference [33]. Copyright 2007, Journal of Occupational Medicine and Toxicology/Springer Nature.*

The utilisation of nanotechnology in the creation of novel pharmaceutical products may prove to be a crucial step in the development of new therapeutic agents. Using nanotechnology;

• The distribution of water-soluble drugs can be expanded

• Distribution of drugs to a specific tissue or cell can be improved

• Drugs can be passed through tight epithelial and endothelial barriers.

• Large macromolecule drugs can be transferred to the targeted area.

• In two or more drug treatments, drugs can be transferred to the target area.

• The real-time in vivo efficacy of a therapeutic agent can be determined [34].

The ability of biogenic nanoparticles to combine with bioactive drugs and compounds with biofuel leads to an increase in the therapeutic efficacy of conventional drugs, leading to a synergistic effect, and an increase in the effectiveness of conventional drugs. Combining the antibiotics Amikacin, Canamycin, Ampicillin, Cephotaxime, and Teracyclin with *Urtica dioica* silver nanoparticles *Escherichia coli, Staphylococcus aureus, Staphylococcus epidermidis, Bacillus cereus*, etc, showing a synergistic effect against *Salmonella typhimurium* led to an increase in the therapeutic efficacy of these drugs, again the combination of silver nanoparticles with tetracycline, which significantly increased wound healing generated in albino mice, while preventing infection of the wound [35].

For example, using nanotechnology, Pandey and Khuller developed nanoparticles by nano-encapsulation to improve oral use of injectable antibiotics such as streptomycin in drug delivery systems [36]. Elechiguerra et al. They explained that metal nanoparticles can interact with viruses. They reported that 1-10 nm silver nanoparticles interacted with HIV-1 virus in a size-dependent manner, and silver nanoparticles prevented HIV-1 virus from binding to the host cell [37]. It has been reported that gold nanoparticles are used in cell culture imaging techniques to display functional cells as a single molecule [38]. It has been reported that nanoborates are involved in both the diagnosis and treatment of cancer disease [19]. For example, nanomedicine-based nanosystems have more advantages in infectious diseases than conventional drugs used. These advantages are defined as increased solubility of the drug, its bioavailability, improved epithelial permeability, prolonging the half-life of the drug, reaching the targeted tissue of the drug, and minimizing its toxic effects [39]. Pourseif et al., (2023), controlled drug distribution in the treatment of wound infections and continuous release of chitosan and niozoma combination for 72 hours, which they developed for antibacterial purposes, the strong antibacterial activity and cytotoxic effect in human fibroblasts is the basis for practices that will help reduce the use of excess antibiotics, and this combination is not just for the treatment of wound infections, but rather for the treatment of human fibroblasts, they also reported that it could be used in infections such as dental implants and urinary probes [39]. In the last decade, the effect of nanoparticles on the development and application of new delivery systems has been investigated to extend the time of stay in the ocular of the coli in eye diseases and increase mucoadecision and permeability through the cornea [40]. In the treatment of central nervous system disorders such as Alzheimer's, Parkinson's, Multiple Sclerosis, brain tumors, traumatic brain injuries, and stroke, the blood-brain barrier is a major obstacle in the transition of the therapeutic drug to the brain [41]. Rivastigmine used to treat Alzheimer's is more effective in a preclinical trial compared to its free form when conjugated with the liposomal carrier of sodium tauroclate [42]. For the treatment of Parkinson's disease, dopamine-dependent liposomes were implanted in the striatum of an experimental Parkinson-formed rat, in which liposome-linked dopamine is released continuously for 25 days, and 40 days of high dopamine levels he was identified as staying [43]. Micelles, nanoemulsions, nanosuspensions, liposomes, nanoparticles, dendrimers, niosomes, etc, new drug delivery systems based on nanotechnology, such as nano wafers, are more effective than conventional color, and these distributions of drugs increase the duration of the drug's stay on the ocular surface, while facilitating transcorneal permeability [44].

Nanoparticles used as carriers in drug delivery systems generally consist of different biodegradable materials with a minimum diameter of less than 100 nm or natural synthetic, lipid, and metals [33]. Organic, inorganic, metallic, and polymeric nanoparticles, including dendrimers, micelles, and liposomes, are often considered in designing target-specific drug delivery systems (Figure 2) [45]. It is aimed to increase the effectiveness of drugs with low solubility and less absorption by combining them with dendrimers, micelles, liposomes, and organic, inorganic, and metallic nanoparticles [46]. The size, shape, and other inherent biophysical and chemical characteristics of these nanoparticles influence the degree to which they bind to different drug delivery vehicle molecules. To illustrate, polymeric nanoparticles are optimal for the efficient distribution of medication, given their diameter range of 10 to 1000 nm [47].

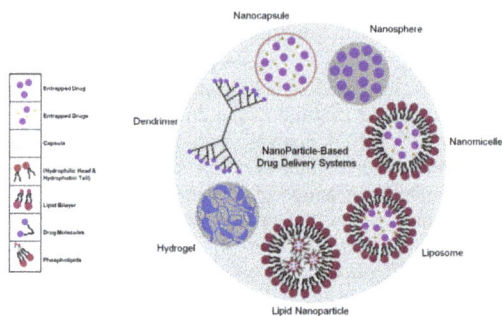

Figure 2. *The most commonly used nanoparticles in drug delivery systems. Reprinted (adapted) with permission from reference [45]. Copyright 2022, Journal of Translational Medicine /Springer Nature.*

2.2   Use of Dendrimer Nanoparticles in Drug Delivery Systems

The specific structures of dendrimers provide flexibility to therapeutic drugs either by covalent conjugation or by electrostatic adsorption [48]. Dendrimer nanoparticles are employed in a variety of biological and pharmacological applications, as well as in the development of dendrimer-based products (such as Vivagel) and drug carriers (including covalent conjugates and non-covalent encapsulation of medicines), as evidenced by numerous biochemical studies. Dendrimers offer a significant advantage over alternative drug delivery methods due to a number of factors, including their high ligand and functionality, high structural and chemical density, interaction with several environmental groups, and modifiable molecular size and structure. These benefits facilitate the utilisation of dendrimers in the development of pharmaceutical formulations and medical devices in accordance with their intended purpose [49]. Due to the physical and chemical properties of dendrimers, it is known that they increase the biodistribution and pharmacokinetic efficiency of drugs. In addition, amphiphilic dendrimers with lipid dendrimers conjugate can form complexes in the nanosize range, transmitting RNA through endocytosis intracellular uptake to the cytoplasm, thus providing nanotechnology-based nucleic acid distribution [50]. Combining and applying curcumin with dendrimary nanoparticles in treating glioblastoma tumor cells increased the anticancer effect of curcumin [51]. Dendrimer nanocarriers increase the penetration of the drug into the cornea in ophthalmic drug applications, resulting in the continuous release of the drug, leading to an increase in the effectiveness of the drug [52]. Sarin et al combined doxorubicin-conjugated gadolinium (Gd)-diethyltriaminepentaacetic acid (DTPA) chelated G5 PAMAM dendrimer in the treatment of brain tumors [53]. It has been reported that Gd-D5 doxorubicin between 7-10 nm crosses the blood-brain barrier, provides sufficient concentration of doxorubicin in brain tumor cells, and is more effective than free doxorubicin in shrinking brain tumor cells [54]. Dendrimer-based nanocarriers will greatly contribute to the treatment of neurological diseases with the advantages of drugs crossing the blood-brain barrier, increasing treatment efficacy, and decreasing side effects [55]. In a study conducted by Mbatha et al., (2021), gold nanoparticles were combined with folic acid, and this nanocomplex was applied with PAMAM-generation-5 dendrimers. The optimal binding rate of this nanoparticle showed that the bound mRNA was effectively protected from nucleases [56].

2.3 Use of Micellar Nanoparticles in Drug Delivery Systems

The micellar nanoparticle is a groundbreaking nanotechnology-based formulation of transdermal therapeutics developed in the mid-1990s. Micellar-based nanoparticle emulsions are nanoparticles that allow effective and rapid passage of drugs into the systemic circulation through topical application [57].

Micelles increase the water solubility of hydrophobic drugs, including anticancer drugs. This is a significant challenge in the drug delivery system, as the increased solubility of a hydrophobic drug in water increases the movement of the drug into the tumor cell. In short, micelles provide more effective administration of drugs, and complete transport of the drug to the targeted area through the bloodstream, which consists mostly of water [58].

It is known as an effective drug carrier, especially with its polymeric micelles solubility, target-selectiveness, inhibition of P-glycoprotein, complete integration with drugs, and localization in subcellular structures [59]. In the late or early 1990's Prof. Developed by Kataoka. Polymeric micelles are carriers that are about 100 nm in size, increasing the solubility of many drugs that are not good in solubility [60]. Under environmental factors such as the stimulant sensitivity of micelles, redox reactions, pH, temperature, light, enzyme activity, their stability, chemical structure, and physiochemical properties can easily change. In addition, micellar versions of the drugs increased absorption in all drug administration pathways (parenteral, nasal, oral, and ocular) and reduced side effects of the drugs [61]. These properties allow micelles to release the drug form by responding to stimuli, thereby reducing the side effects of antineoplastic drugs on different tissues, particularly by targeting the tumor cell [62]. It is expected that polymeric micelles will increase the accumulation of anticancer drugs in the tumor tissue by using the method of enhanced permeability and retention of anticancer drugs and enable the drugs to pass into the nucleus of the tumor cell through physical capture, through chemical conjugation or high stability. To ensure that the micelles do not pass through the normal vessel wall, its size is adjusted in the range of 20-100 nm, and the incidence of side effects of drugs is expected to decrease [63]. For example, raloxifene is a selective estrogen receptor modulator belonging to the pentathiophene class, in the commercial sale it is in tablet form. But about 60% of the oral dose of raloxifene is absorbed, leading to a decrease in the therapeutic efficacy of the drug buddha for this purpose the transdermal form of raloxifene has been developed by complexing with micellar nanoparticles, and the therapeutic efficacy of the drug has increased [57].

2.4 Use of Liposome Nanoparticles in Drug Delivery Systems

Liposomes; genes, vaccines, imaging techniques, and antifungal and cytotoxic drugs are located in the most clinically used nanometric system used to deliver agents to the target site. Nanoparticle liposome nanoparticles were first investigated as drug carriers in drug delivery. Liposomes consist of a single or multiple encapsulated lipid layer [64] and the sizes of liposomes used clinically are 50-300 nm in diameter [65]. Liposomes were first described as a closed bilayer phospholipid system in 1965, and a few years later it was reported that they could be used in drug delivery systems due to their ability to bind to liposomes [66].

The field of lipid investigation has recently been included in over 600 clinical trials, including those pertaining to the utilization of medication delivery devices. In order to reduce the rapid clearance of the drug and allow it sufficient time to circulate through the bloodstream to reach target tissues and cells, liposomal nanoparticles enabled particle surface hydration technology (with polyethylene glycol) was employed to gain a deeper understanding of the biodistribution and pharmacokinetics of lipid-drug particles [65].

The application of liposome nanoparticles in anticancer drugs to treat different cancer indications is widely used in the clinic. Due to their vascular leakage and lymphatic drainage properties, densely layered liposomes can passively collect in the solid tumor layer. Nanometer-sized liposomes called Doxil, Caelyx, and Myocet that contain doxorubicin are used to treat multiple myeloma, Kaposi's sarcoma, and ovarian cancer [67]. DepoCyt (multivesicular liposomes containing cytarabine) has been approved for the treatment of cancer due to its sustained release properties [68]. New generation liposomes were developed in a controlled manner by triggering specific drug distribution sensitive to internal and external environmental stimuli (such as pH, temperature, and enzymes). ThermoDox is combined with a temperature-sensitive nanometer-sized liposome and is now used in clinical trials for the treatment of hyperthermia in oncology [69].

2.5   Use of Polymers Nanoparticles in Drug Delivery Systems

Biologists, engineers, physicians, and chemists have led to significant biotechnological advances in tissue engineering, starting in the 1960s, in the drug delivery systems, biomaterials, and medical device development, of polymers for over fifty years, major researchers such as Roseman, Folkman, Peppas, Heller, Speiser, and Ringsdorf have discovered that polymers can be found in macroscopic drug stores, in suture materials, they have led us to witness an evolution by the effective use in implants and by increasing the effectiveness of the drug in drug delivery systems [70]. Soluble polymers can be hydrolyzed in vivo environments and are classified according to whether the sources from which they are obtained are natural. Natural polymers such as hydroxy acids, chitosan, and polyanhydrates are widely used. Synthetic polymers are used in fewer drug distribution systems than polymers derived from natural sources. In particular, the ability to increase oral use of drugs in the chitosan drug distribution system has been used in wound healing, and antibacterial and anti-inflammatory effects in wounds [71]. For example, hyaluronic acid-based polymers have been used as biological conjugates in drug delivery systems. These biological conjugates improve the bio-solubility, pharmacokinetics, and clearance of drugs [72]. Macro-based polymers were originally developed to control the time of drug release in medical devices and later became effective in more controlled dispersion through the diffusion of drugs. Ocusert is the first polymer-drug used in the effective clinic in the drug delivery system. This implant was placed behind the eyelid, allowing the distribution of pilocarpine at a constant rate over a week and showing fewer side effects compared to traditionally used eye drops [73]. In diseases of the central nervous system, it is difficult to transfer drugs to the central nervous system due to the blood-brain barrier. Poly-butyl cyanoacrylate has been tried as the first polymer-based nanoparticle to facilitate the passage of drugs into the central nervous system. In this study, butyl cyanoacrylate nanoparticles were administered intravenously by making them complex with a margin opioid activity. The dalargin-butylcyanoacrylate complex has been reported to have a strong analgesic effect [74]. The poly-lactic acid polymer formed by breviscapine (flavonoid) penetrated the blood-brain barrier in a size-dependent manner, allowing larger particles to pass into the brain. Chitosan is a naturally occurring, biodegradable, biocompatible polysaccharide with higher nanoparticle-forming ability [75]. A significant amount of estradiol was found in the central nervous system by intranasal delivery of estradiol-laden chitosan nanoparticles [76]. The property of the nanoparticle, which will improve the effectiveness or passage of the drug in the central nervous system, is non-toxic, stable in the blood, non-immunogenic, can avoid the reticuloendothelial system, can stay in the circulation for a long time, can, it must have properties that can be negatively or neutrally charged [77].

2.6   Use of Organic and Inorganic Nanoparticles in Drug Delivery Systems

Organic nanoparticles are prepared from a variety of materials, including polymers and lipids. Compared to organic materials, inorganic nanoparticles have a more nontoxic, more hydrophilic, more biocompatible, and more stable structure. Inorganic nanoparticles increase their functionality by combining with various specific ligands to increase their affinity for the molecule or target cell. Inorganic nanoparticles protect drugs against degradation as well as controlled drug release profiles and can reduce the frequency and dose of drug administration, thus providing a significant reduction in the toxicity of drugs, especially cancer drugs [78].

The most important feature that distinguishes inorganic nanoparticles from conventional polymer and lipid nanoparticles is that they show strong theranostic effects [79]. Generally, inorganic nanoparticles are large in a magnetic field made of metals such as gold silver, and magnetic nanoparticles, namely Ni, Co, Fe, Fe_3O_4, and FePt. It is prepared from superparamagnetic with magnetic moments, quantum dots, and fluorescent nanoparticles such as SiO_2 [80].

The advantage of using inorganic nanoparticles in drug delivery systems as stimulant-sensitive is due to their easy synthesis and functionalization, their highly tuneable optical properties, and their excellent biocompatibility. In drug delivery systems, nanoparticles are useful for providing sustained or instantaneous release of antimicrobials during infection [81]. Due to its shape and size-dependent adjustable properties, metallic nanoparticles have become the central focus for biomedical applications as well as strong antimicrobial effects, copper, silver, titanium, zinc, etc, metallic nanoparticles such as iron are used against microorganisms with antimicrobial resistance due to their strong antimicrobial effects. For example, copper nanoparticles from *Sida acuta* showed antibacterial activity against *Staphylococcus aureus*, *Proteus vulgaris*, and *Escherichia coli*, and linkomycin, vancomycin, etc, the use of traditional antibiotics such as penicillin G and rifampicin has been reported to increase the antibacterial activity of these antibiotics [82]. Copper nanoparticles have been reported to cause rapid wound epithelialization after anal rectum surgery in humans, reducing inflammation in the wound site [35]. It has also been reported that copper nanoparticles have been used as antimicrobial coating fields in medical devices to prevent bacterial contamination. Zinc oxide nanoparticles increase the antibacterial activity of more than 25 antibiotics used in infections caused by *Staphylococcus aureus* and *Escherichia coli*, penicillins, aminoglycosides, cephalosporins, glycopeptides, macrolides, tetracyclines, fluoroquinolones, fluoroquinolones, the drug, such as Linkozamides, has been reported to cause an increase in antibacterial activity [83].

Furthermore, a comparative study between biological and chemical nanoparticles reported that biological nanoparticles showed higher antimicrobial effects than chemically synthesized nanoparticles. Mukherjee et al., (2012) in a study that biogenically synthesized nickel nanoparticles were more biocompatible than chemically synthesized nickel nanoparticles and specifically *Proteus vulgaris*, *S. aureusa*, *K. pneumonia*, *etc.* they found that it had more antibacterial effects against *V. choleraea* [84]. Macrophages as a result of oxidative stress in diseases such as infection, inflammation, injury, IL-1, IL-1, and TNF-Preformed steroids and anti-inflammatory drugs used in increasing levels of inflammatory cytokines, such as steroids and anti-inflammatory drugs have been reported. With the addition of biogenic nanoparticles to anti-inflammatory drugs, there has been an increase in both the anti-inflammatory effect and a decrease in the incidence of undesirable side effects. In one study, silver nanoparticles produced from *Selaginella myosurus* showed strong anti-inflammatory effects in vivo and invitro conditions [85], similarly, silver nanoparticles have been reported to relieve inflammation by inhibiting histamine, serotonin, quinine, prostaglandins, and cyclooxygenase, which are highly effective in experimental rat posterior paw edema [86].

The use of silver nanoparticles, which are inorganic nanoparticles, is wide and diverse. Due to their antimicrobial properties, silver nanoparticles are suitable for use in biomedicine as well as an antiseptic in various production and consumption products [87,88]. Although the use of silver nanoparticles is limited due to their cytotoxic effects, they are used in the treatment of various diseases, including ocular neovascular diseases, multi-drug resistant tumors, and immunological and inflammatory diseases [89]. Silver nanoparticles are used in bone prostheses, as an antiseptic for antibacterial action in surgical materials, in dental implants, and in dressings for antibacterial effects in wounds [42]. As one of the leading causes of morbidity and mortality in the world, fungal infections, especially those caused by multi-drug-resistant strains, come from. The low therapeutic effects of traditionally used antifungal drugs, the presence of nephrotoxic and hepatotoxic effects, and the high level of side effects such as myelotoxicity limit their use in treatment. In a study conducted for this purpose, biogenic silver nanoparticles were synthesized using the aqueous shell extract of *Punica granatum*. This composition obtained from *Candida tropikis*, showed strong antifungal action against *Candida albicans* and *Candida glabrata* strains and was reported to exhibit synergistic activity against *Candida glabrata,* which is multidrug-resistant with itraconazole, which is used as an antifungal [90].

Gold nanoparticles are used in various biomedical fields due to their good biocompatibility, small size, low toxicity, easy surface modifications, and their ability to take part in controlled drug distribution [91]. Gold nanoparticles are amino acids, proteins/enzymes, etc, the fact that DNA and conjugate species can easily conjugate to large biomolecules without altering their biological properties and that their biocompatibility is high and nontoxic, gives them a great advantage over other metal-containing nanoparticles [24].

Because gold nanoparticles interact with bases such as thiols and their biocompatibility is high, they have been particularly effective in malignancies in the female genital system, epithelial ovarian cancer, metastasis of ovarian cancer, and in preventing progression of metastasis. It has also been observed to inhibit vascular endothelial growth factor, stimulating multiple myeloma on cell profiler [23].

Iron oxide nanoparticles have been used as contrast agents in hyperthermia, magnetically targeted drug delivery, and magnetic resonance imaging. Iron oxide can be used as a contrast agent to visualize nanoparticle deposition in cancerous tissue by magnetic resonance imaging [92]. Iron oxide nanoparticles exhibit indirect or direct antitumor activity through non-toxic wavelength radiation by reducing the production of reactive oxygen species that lead to oxidative stress. This antitumor activity is caused by the covalent bonding of iron oxide to the tumor cell [93]. The United States has approved the use of iron oxide nanoparticles in the treatment of prostate cancer, radiotherapy, and chemotherapy, as well as in the treatment of brain tumors [94].

Inorganic nanoparticles are being tested in the preclinical phase stages for their use in the treatment of diseases such as pneumonia, acquired immunodeficiency syndrome (AIDS), tuberculosis, malaria, and diabetes, as well as being used in the treatment of cancer disease [95].

3.   Conclusion

Nanocarriers in drug delivery systems have been developed to enhance the therapeutic and pharmacological effects of conventional drugs. Nanotechnology-based drug delivery systems ensure the safe transport of nanocarriers and drug molecules to the target site, increase the pharmacological and therapeutic properties of traditional drugs, protect against degradation of drugs, dissolve water-insoluble drugs, and provide solubility, reducing the possible side effects of traditional drugs has been seen as a promising strategy for clinical use of drugs, with an increase in their transport to the target area and their increase in their properties such as controlled release. The sizes and large surface areas of the nanoparticles increase the bioavailability of drugs, allowing the accumulation of drugs in target tissues and cells, increasing the passage of drugs to areas where the passage of drugs such as the central nervous system is difficult, and, in cancer treatment, it increases the penetration of drugs into cancer cells, leading to an increase in the therapeutic effectiveness of drugs and a decrease in their toxic effects, increasing the effectiveness of treatment. As a result, with the development of nanotechnology, easier transport of the drug to the target site in drug delivery systems, the therapeutic effectiveness of the drug is increased and the hand of both patients and physicians will be strengthened by bringing solutions to problems such as the failure to continue the treatment of patients due to the side effects of the drug. More extensive studies on the inclusion of nanoparticles effectively in medical treatments are needed.

References

[1]   F. Karimi, N. Zare, M. Bekmezci, M. Akin, R. Bayat, B. Seyitoglu, K. Arikan, I. Isik, F. Sen, Enzyme-free glucose detection via scalable and economical fabrication of nickel-polyvinylpyrrolidone-modified multi-walled carbon nanotubes, Electrochim. Acta 496 (2024) 144341. https://doi.org/10.1016/J.ELECTACTA.2024.144341

[2]   M. Bekmezci, R. Bayat, V. Erduran, F. Sen, Biofunctionalization of functionalized nanomaterials for electrochemical sensors, Funct. Nanomater. Electrochem. Sensors Princ. Fabr. Methods, Appl. (2022) 55–69. https://doi.org/10.1016/B978-0-12-823788-5.00003-X

[3]   M. Bekmezci, R. Bayat, M. Akin, Z.K. Coguplugil, F. Sen, Modified screen-printed electrochemical biosensor design compatible with mobile phones for detection of miR-141 used to pancreatic cancer biomarker, Carbon Lett. 33 (2023) 1863–1873. https://doi.org/10.1007/s42823-023-00545-9

[4]   M. Akin, M. Bekmezci, R. Bayat, I. Isik, F. Sen, Ultralight Covalent Organic Frame Graphene Aerogels Modified Platinum Magnetite Nanostructure for Direct Methanol Fuel Cell, Fuel 357 (2024) 129771. https://doi.org/10.1016/j.fuel.2023.129771

[5]   M. Bekmezci, G.N. Gules, R. Bayat, F. Sen, Modification of multi-walled carbon nanotubes with platinum–osmium to develop stable catalysts for direct methanol fuel cells, Anal. Methods 15 (2023) 1223–1229. https://doi.org/10.1039/D2AY02002E

[6]   M. Akin, M. Bekmezci, R. Bayat, F.S. Kuegou, I. Isik, G. Kaya, F. Sen, Synthesis of platinum-nickel nanoparticles supported by carbon and titanium oxide structures for efficient and enhanced formic acid oxidation, Fuel 373 (2024) 132258. https://doi.org/10.1016/J.FUEL.2024.132258

[7]   M. Bekmezci, M. Akin, G.N. Gules, R. Bayat, F. Sen, Innovative chelation strategies for facile

synthesis of bimetallic nanomaterials with remarkable photocatalytic and biochemical activities, Next Res. 1 (2024) 100001. https://doi.org/10.1016/j.nexres.2024.100001

[8]   A. Şavk, K. Cellat, K. Arıkan, F. Tezcan, S.K. Gülbay, S. Kızıldağ, E.Ş. Işgın, F. Şen, Highly monodisperse Pd-Ni nanoparticles supported on rGO as a rapid, sensitive, reusable and selective enzyme-free glucose sensor, Sci. Rep. 9 (2019) 19228. https://doi.org/10.1038/s41598-019-55746-y

[9]   N.H. Khand, A.R. Solangi, S. Ameen, A. Fatima, J.A. Buledi, A. Mallah, S.Q. Memon, F. Sen, F. Karimi, Y. Orooji, A new electrochemical method for the detection of quercetin in onion, honey and green tea using Co3O4 modified GCE, J. Food Meas. Charact. 2021 154 15 (2021) 3720–3730. https://doi.org/10.1007/S11694-021-00956-0

[10]   F. Gulbagca, A. Aygün, M. Gülcan, S. Ozdemir, S. Gonca, F. Şen, Green Synthesis of Palladium Nanoparticles: Preparation, Characterization, and Investigation of Antioxidant, Antimicrobial, Anticancer, and DNA Cleavage Activities, Appl. Organomet. Chem. 35 (2021). https://doi.org/10.1002/aoc.6272

[11]   A. Hojjati-Najafabadi, S. Salmanpour, F. Sen, P.N. Asrami, M. Mahdavian, M.A. Khalilzadeh, A Tramadol Drug Electrochemical Sensor Amplified by Biosynthesized Au Nanoparticle Using Mentha aquatic Extract and Ionic Liquid, Top. Catal. 65 (2022) 587–594. https://doi.org/10.1007/s11244-021-01498-x

[12]   A. Aygun, F. Gulbagca, E.E. Altuner, M. Bekmezci, T. Gur, H. Karimi-Maleh, F. Karimi, Y. Vasseghian, F. Sen, Highly Active PdPt Bimetallic Nanoparticles Synthesized By One Step Bioreduction Method: Characterizations, Anticancer, Antibacterial Activities and Evaluation of Their Catalytic Effect for Hydrogen Generation, Int. J. Hydrogen Energy 48 (2023) 6666–6679. https://doi.org/10.1016/j.ijhydene.2021.12.144

[13]   K. Nesrin, C. Yusuf, K. Ahmet, S.B. Ali, N.A. Muhammad, S. Suna, Ş. Fatih, Biogenic silver nanoparticles synthesized from Rhododendron ponticum and their antibacterial, antibiofilm and cytotoxic activities, J. Pharm. Biomed. Anal. 179 (2020) 112993. https://doi.org/10.1016/J.JPBA.2019.112993

[14]   N. Korkmaz, Y. Ceylan, P. Taslimi, A. Karadağ, A.S. Bülbül, F. Şen, Biogenic nano silver: Synthesis, characterization, antibacterial, antibiofilms, and enzymatic activity, Adv. Powder Technol. 31 (2020) 2942–2950. https://doi.org/10.1016/J.APT.2020.05.020

[15]   R. Nagraik, A. Sharma, D. Kumar, S. Mukherjee, F. Sen, A.P. Kumar, Amalgamation of biosensors and nanotechnology in disease diagnosis: Mini-review, Sensors Int. 2 (2021) 100089. https://doi.org/10.1016/J.SINTL.2021.100089

[16]   Y. Wu, E.E. Altuner, R.N. El Houda Tiri, M. Bekmezci, F. Gulbagca, A. Aygun, C. Xia, Q. Van Le, F. Sen, H. Karimi-Maleh, Hydrogen generation from methanolysis of sodium borohydride using waste coffee oil modified zinc oxide nanoparticles and their photocatalytic activities, Int. J. Hydrogen Energy 48 (2023) 6613–6623. https://doi.org/10.1016/J.IJHYDENE.2022.04.177

[17]   F. Şen, Ö. Demirbaş, M.H. Çalımlı, A. Aygün, M.H. Alma, M.S. Nas, The dye Removal from Aqueous Solution Using Polymer Composite Films, Appl. Water Sci. 8 (2018) 206. https://doi.org/10.1007/s13201-018-0856-x

[18]   B.S. Sivamaruthi, P. kumar Nallasamy, N. Suganthy, P. Kesika, C. Chaiyasut, Pharmaceutical and biomedical applications of starch-based drug delivery system: A review, J. Drug Deliv. Sci. Technol. 77 (2022). https://doi.org/10.1016/J.JDDST.2022.103890

[19]   J.K. Patra, G. Das, L.F. Fraceto, E.V.R. Campos, M.D.P. Rodriguez-Torres, L.S. Acosta-Torres, L.A. Diaz-Torres, R. Grillo, M.K. Swamy, S. Sharma, S. Habtemariam, H.S. Shin, Nano based drug delivery systems: recent developments and future prospects, J. Nanobiotechnology 2018 161 16 (2018) 1–33. https://doi.org/10.1186/S12951-018-0392-8

[20]   M.M. de Villiers, P. Aramwit, G.S. Kwon, eds., Nanotechnology in Drug Delivery, (2009). https://doi.org/10.1007/978-0-387-77667-5

[21]   G. Kah, R. Chandran, H. Abrahamse, Biogenic Silver Nanoparticles for Targeted Cancer Therapy and Enhancing Photodynamic Therapy, Cells 12 (2023). https://doi.org/10.3390/CELLS12152012

[22]   A.C. Aydin, S. Yesilot, C. Aydin, Silver Nanoparticles; A New Hope In Cancer Therapy?, East J Med 24 (2019) 111–116. https://doi.org/10.5505/ejm.2019.66487

[23]   M. Rai, A.P. Ingle, J. Trzcińska-Wencel, M. Wypij, S. Bonde, A. Yadav, G. Kratošová, P. Golińska, Biogenic Silver Nanoparticles: What We Know and What Do We Need to Know?, Nanomaterials 11 (2021) 2901. https://doi.org/10.3390/NANO11112901

[24]   R.M. Tripathi, S.J. Chung, Biogenic nanomaterials: Synthesis, characterization, growth mechanism, and biomedical applications, J. Microbiol. Methods 157 (2019) 65–80. https://doi.org/10.1016/J.MIMET.2018.12.008

[25]   C. Li, J. Wang, Y. Wang, H. Gao, G. Wei, Y. Huang, H. Yu, Y. Gan, Y. Wang, L. Mei, H. Chen, H. Hu, Z. Zhang, Y. Jin, Recent progress in drug delivery, Acta Pharm. Sin. B 9 (2019) 1145–1162. https://doi.org/10.1016/J.APSB.2019.08.003

[26]   Y. Herdiana, N. Wathoni, S. Shamsuddin, M. Muchtaridi, Scale-up polymeric-based nanoparticles drug delivery systems: Development and challenges, OpenNano 7 (2022) 100048. https://doi.org/10.1016/J.ONANO.2022.100048

[27]   F. Laffleur, V. Keckeis, Advances in drug delivery systems: Work in progress still needed?, Int. J. Pharm. 590 (2020). https://doi.org/10.1016/J.IJPHARM.2020.119912

[28]   Z. Liu, Y. Jiao, Y. Wang, C. Zhou, Z. Zhang, Polysaccharides-based nanoparticles as drug delivery systems, Adv. Drug Deliv. Rev. 60 (2008) 1650–1662. https://doi.org/10.1016/J.ADDR.2008.09.001

[29]   A.O. Elzoghby, W.M. Samy, N.A. Elgindy, Albumin-based nanoparticles as potential controlled release drug delivery systems, J. Control. Release 157 (2012) 168–182. https://doi.org/10.1016/j.jconrel.2011.07.031

[30]   A.Z. Wilczewska, K. Niemirowicz, K.H. Markiewicz, H. Car, Nanoparticles as Drug Delivery Systems, Pharmacol. Reports 64 (2012) 1020–1037. https://doi.org/10.1016/S1734-1140(12)70901-5

[31]   B. Begines, T. Ortiz, M. Pérez-Aranda, G. Martínez, M. Merinero, F. Argüelles-Arias, A. Alcudia, Polymeric Nanoparticles for Drug Delivery: Recent Developments and Future Prospects, Nanomater. (Basel, Switzerland) 10 (2020) 1–41. https://doi.org/10.3390/NANO10071403.

[32]   S. Asad, A.C. Jacobsen, A. Teleki, Inorganic nanoparticles for oral drug delivery: opportunities, barriers, and future perspectives, Curr. Opin. Chem. Eng. 38 (2022) 100869. https://doi.org/10.1016/J.COCHE.2022.100869

[33]   S.S. Suri, H. Fenniri, B. Singh, Nanotechnology-based drug delivery systems, J. Occup. Med. Toxicol. 2 (2007) 16. https://doi.org/10.1186/1745-6673-2-16

[34]   O.C. Farokhzad, R. Langer, Impact of nanotechnology on drug delivery, ACS Nano (2009). https://doi.org/10.1021/nn900002m.

[35]   C. Tyavambiza, P. Dube, M. Goboza, S. Meyer, A.M. Madiehe, M. Meyer, Wound Healing Activities and Potential of Selected African Medicinal Plants and Their Synthesized Biogenic Nanoparticles, Plants (Basel, Switzerland) 10 (2021). https://doi.org/10.3390/PLANTS10122635

[36]   R. Pandey, G.K. Khuller, Nanoparticle-based oral drug delivery system for an injectable antibiotic - streptomycin. Evaluation in a murine tuberculosis model, Chemotherapy 53 (2007) 437–441. https://doi.org/10.1159/000110009

[37]   J.L. Elechiguerra, J.L. Burt, J.R. Morones, A. Camacho-Bragado, X. Gao, H.H. Lara, M.J. Yacaman, Interaction of silver nanoparticles with HIV-1, J. Nanobiotechnology 3 (2005) 6. https://doi.org/10.1186/1477-3155-3-6

[38]   G. Peleg, A. Lewis, M. Linial, L.M. Loew, Nonlinear optical measurement of membrane potential around single molecules at selected cellular sites, Proc. Natl. Acad. Sci. U. S. A. 96 (1999) 6700–6704.

https://doi.org/10.1073/PNAS.96.12.6700

[39]   T. Pourseif, R. Ghafelehbashi, M. Abdihaji, N. Radan, E. Kaffash, M. Heydari, M. Naseroleslami, N. Mousavi-Niri, I. Akbarzadeh, Q. Ren, Chitosan -based nanoniosome for potential wound healing applications: Synergy of controlled drug release and antibacterial activity, Int. J. Biol. Macromol. 230 (2023). https://doi.org/10.1016/J.IJBIOMAC.2023.123185

[40]   M. Mofidfar, B. Abdi, S. Ahadian, E. Mostafavi, T.A. Desai, F. Abbasi, Y. Sun, E.E. Manche, C.N. Ta, C.W. Flowers, Drug delivery to the anterior segment of the eye: A review of current and future treatment strategies, Int. J. Pharm. 607 (2021). https://doi.org/10.1016/J.IJPHARM.2021.120924

[41]   C. Ekhator, M.Q. Qureshi, A.W. Zuberi, M. Hussain, N. Sangroula, S. Yerra, M. Devi, M.A. Naseem, S.B. Bellegarde, P.R. Pendyala, Advances and Opportunities in Nanoparticle Drug Delivery for Central Nervous System Disorders: A Review of Current Advances., Cureus 15 (2023) e44302. https://doi.org/10.7759/cureus.44302

[42]   N.B. Mutlu, Z. Değim, Ş. Yilmaz, D. Eiz, A. Nacar, New perspective for the treatment of Alzheimer diseases: liposomal rivastigmine formulations, Drug Dev. Ind. Pharm. 37 (2011) 775–789. https://doi.org/10.3109/03639045.2010.541262

[43]   M.J. During, A. Freese, A.Y. Deutch, P.G. Kibat, B.A. Sabel, R. Langer, R.H. Roth, Biochemical and behavioral recovery in a rodent model of Parkinson's disease following stereotactic implantation of dopamine-containing liposomes, Exp. Neurol. 115 (1992) 193–199. https://doi.org/10.1016/0014-4886(92)90053-S

[44]   A.L. Onugwu, C.S. Nwagwu, O.S. Onugwu, A.C. Echezona, C.P. Agbo, S.A. Ihim, P. Emeh, P.O. Nnamani, A.A. Attama, V. V. Khutoryanskiy, Nanotechnology based drug delivery systems for the treatment of anterior segment eye diseases, J. Control. Release 354 (2023) 465–488. https://doi.org/10.1016/J.JCONREL.2023.01.018

[45]   F. Karamali, S. Behtaj, S. Babaei-Abraki, H. Hadady, A. Atefi, S. Savoj, S. Soroushzadeh, S. Najafian, M.H. Nasr Esfahani, H. Klassen, Potential therapeutic strategies for photoreceptor degeneration: the path to restore vision, J. Transl. Med. 20 (2022) 572. https://doi.org/10.1186/s12967-022-03738-4

[46]   A. Zeeshan, M.• Farhan, A. Siddiqui, Nanomedicine and drug delivery: a mini review, Int. Nano Lett. 2014 41 4 (2014) 1–7. https://doi.org/10.1007/S40089-014-0094-7

[47]   B.V. Bonifácio, P.B. da Silva, M.A. dos S. Ramos, K.M.S. Negri, T.M. Bauab, MarlusChorilli, Nanotechnology-based drug delivery systems and herbal medicines : a review, Int. J. Nanomedicine (2014).

[48]   X. Du, B. Shi, J. Liang, J. Bi, S. Dai, S.Z. Qiao, Developing functionalized dendrimer-like silica nanoparticles with hierarchical pores as advanced delivery nanocarriers, Adv. Mater. 25 (2013) 5981–5985. https://doi.org/10.1002/ADMA.201302189

[49]   A. Santos, F. Veiga, A. Figueiras, Dendrimers as Pharmaceutical Excipients: Synthesis, Properties, Toxicity and Biomedical Applications, Materials (Basel). 13 (2019) 65. https://doi.org/10.3390/MA13010065

[50]   A. Aurelia Chis, C. Dobrea, C. Morgovan, A.M. Arseniu, L.L. Rus, A. Butuca, A.M. Juncan, M. Totan, A.L. Vonica-Tincu, G. Cormos, A.C. Muntean, M.L. Muresan, F.G. Gligor, A. Frum, Applications and Limitations of Dendrimers in Biomedicine, Molecules 25 (2020) 3982. https://doi.org/10.3390/MOLECULES25173982

[51]   J. Gallien, B. Srinageshwar, K. Gallo, G. Holtgrefe, S. Koneru, P.S. Otero, C.A. Bueno, J. Mosher, A. Roh, D.S. Kohtz, D. Swanson, A. Sharma, G. Dunbar, J. Rossignol, Curcumin Loaded Dendrimers Specifically Reduce Viability of Glioblastoma Cell Lines., Molecules 26 (2021). https://doi.org/10.3390/molecules26196050

[52]   S. Mignani, S. El Kazzouli, M. Bousmina, J.P. Majoral, Expand classical drug administration ways by emerging routes using dendrimer drug delivery systems: a concise overview, Adv. Drug Deliv.

Rev. 65 (2013) 1316–1330. https://doi.org/10.1016/J.ADDR.2013.01.001

[53]   H. Sarin, A.S. Kanevsky, H. Wu, A.A. Sousa, C.M. Wilson, M.A. Aronova, G.L. Griffiths, R.D. Leapman, H.Q. Vo, Physiologic upper limit of pore size in the blood-tumor barrier of malignant solid tumors, J. Transl. Med. 7 (2009) 51. https://doi.org/10.1186/1479-5876-7-51

[54]   J. Zhao, B. Zhang, S. Shen, J. Chen, Q. Zhang, X. Jiang, Z. Pang, CREKA peptide-conjugated dendrimer nanoparticles for glioblastoma multiforme delivery, J. Colloid Interface Sci. 450 (2015) 396–403. https://doi.org/10.1016/J.JCIS.2015.03.019

[55]   Y. Zhu, C. Liu, Z. Pang, Dendrimer-Based Drug Delivery Systems for Brain Targeting, Biomolecules 9 (2019). https://doi.org/10.3390/BIOM9120790

[56]   L.S. Mbatha, F. Maiyo, A. Daniels, M. Singh, Dendrimer-Coated Gold Nanoparticles for Efficient Folate-Targeted mRNA Delivery In Vitro, Pharmaceutics 13 (2021). https://doi.org/10.3390/PHARMACEUTICS13060900

[57]   R. Lee, D. Shenoy, R. Sheel, Micellar Nanoparticles: Applications for Topical and Passive Transdermal Drug Delivery, Handb. Non-Invasive Drug Deliv. Syst. (2010) 37–58.

[58]   Z.L. Tyrrell, Y. Shen, M. Radosz, Fabrication of micellar nanoparticles for drug delivery through the self-assembly of block copolymers, Prog. Polym. Sci. 35 (2010) 1128–1143. https://doi.org/10.1016/J.PROGPOLYMSCI.2010.06.003

[59]   J. Gong, M. Chen, Y. Zheng, S. Wang, Y. Wang, Polymeric Micelles Drug Delivery System in Oncology, J. Control. Release 159 (2012) 312–323. https://doi.org/10.1016/j.jconrel.2011.12.012

[60]   D. Hwang, J.D. Ramsey, A. V. Kabanov, Polymeric micelles for the delivery of poorly soluble drugs: From nanoformulation to clinical approval, Adv. Drug Deliv. Rev. 156 (2020) 80–118. https://doi.org/10.1016/j.addr.2020.09.009

[61]   A. Guzmán Rodríguez, M. Sablón Carrazana, C. Rodríguez Tanty, M.J.A. Malessy, G. Fuentes, L.J. Cruz, Smart Polymeric Micelles for Anticancer Hydrophobic Drugs, Cancers (Basel). 15 (2022) 4. https://doi.org/10.3390/cancers15010004

[62]   A. Cazacu, E.-L. Ursu, I. Negut, B. Bita, Polymeric Micellar Systems—A Special Emphasis on "Smart" Drug Delivery, Pharm. 2023, Vol. 15, Page 976 15 (2023) 976. https://doi.org/10.3390/PHARMACEUTICS15030976

[63]   Y. Matsumura, Polymeric Micellar Delivery Systems in Oncology, Jpn. J. Clin. Oncol. 38 (2008) 793–802. https://doi.org/10.1093/JJCO/HYN116

[64]   R. Rahnama, E. Shafiei, M.R. Jamali, Preconcentration of copper using 1,5-diphenyl carbazide as the complexing agent via dispersive liquid-liquid microextraction and determination by flame atomic absorption spectrometry, J. Chem. (2013). https://doi.org/10.1155/2013/962365

[65]   J.C. Kraft, J.P. Freeling, Z. Wang, R.J.Y. Ho, Emerging research and clinical development trends of liposome and lipid nanoparticle drug delivery systems, J. Pharm. Sci. 103 (2014) 29–52. https://doi.org/10.1002/JPS.23773

[66]   T.M. Allen, P.R. Cullis, Liposomal drug delivery systems: from concept to clinical applications, Adv. Drug Deliv. Rev. 65 (2013) 36–48. https://doi.org/10.1016/J.ADDR.2012.09.037

[67]   S. Verma, S. Dent, B.J.W. Chow, D. Rayson, T. Safra, Metastatic breast cancer: the role of pegylated liposomal doxorubicin after conventional anthracyclines, Cancer Treat. Rev. 34 (2008) 391–406. https://doi.org/10.1016/J.CTRV.2008.01.008

[68]   S. Mantripragada, A lipid based depot (DepoFoam® technology) for sustained release drug delivery, Prog. Lipid Res. 41 (2002) 392–406. https://doi.org/10.1016/S0163-7827(02)00004-8

[69]   C.D. Landon, Nanoscale Drug Delivery and Hyperthermia: The Materials Design and Preclinical and Clinical Testing of Low Temperature-Sensitive Liposomes Used in Combination with Mild Hyperthermia in the Treatment of Local Cancer, Open Nanomed. J. 3 (2011) 24–37. https://doi.org/10.2174/1875933501103010038

[70]   N. Kamaly, B. Yameen, J. Wu, O.C. Farokhzad, Degradable Controlled-Release Polymers and Polymeric Nanoparticles: Mechanisms of Controlling Drug Release, Chem. Rev. 116 (2016) 2602. https://doi.org/10.1021/ACS.CHEMREV.5B00346

[71]   L. Casettari, L. Illum, Chitosan in nasal delivery systems for therapeutic drugs, J. Control. Release 190 (2014) 189–200. https://doi.org/10.1016/J.JCONREL.2014.05.003

[72]   G. Huang, H. Huang, Application of hyaluronic acid as carriers in drug delivery, Drug Deliv. 25 (2018) 766–772. https://doi.org/10.1080/10717544.2018.1450910

[73]   C.J. Kearney, D.J. Mooney, Macroscale delivery systems for molecular and cellular payloads, Nat. Mater. 12 (2013) 1004–1017. https://doi.org/10.1038/NMAT3758

[74]   J. Kreuter, R.N. Alyautdin, D.A. Kharkevich, A.A. Ivanov, Passage of peptides through the blood-brain barrier with colloidal polymer particles (nanoparticles), Brain Res. 674 (1995) 171–174. https://doi.org/10.1016/0006-8993(95)00023-J

[75]   K. Nagpal, S.K. Singh, D.N. Mishra, Chitosan nanoparticles: a promising system in novel drug delivery, Chem. Pharm. Bull. (Tokyo). 58 (2010) 1423–1430. https://doi.org/10.1248/CPB.58.1423

[76]   X. Wang, N. Chi, X. Tang, Preparation of estradiol chitosan nanoparticles for improving nasal absorption and brain targeting, Eur. J. Pharm. Biopharm. 70 (2008) 735–740. https://doi.org/10.1016/J.EJPB.2008.07.005

[77]   T. Patel, J. Zhou, J.M. Piepmeier, W.M. Saltzman, Polymeric Nanoparticles for Drug Delivery to the Central Nervous System, Adv. Drug Deliv. Rev. 64 (2011) 701. https://doi.org/10.1016/J.ADDR.2011.12.006

[78]   W. Paul, C.P. Sharma, Inorganic nanoparticles for targeted drug delivery, Biointegration Med. Implant Mater. (2019) 333–373. https://doi.org/10.1016/B978-0-08-102680-9.00013-5

[79]   P. Ma, H. Xiao, C. Li, Y. Dai, Z. Cheng, Z. Hou, J. Lin, Inorganic nanocarriers for platinum drug delivery, Mater. Today 18 (2015) 554–564. https://doi.org/10.1016/J.MATTOD.2015.05.017

[80]   A. Mittal, I. Roy, S. Gandhi, Magnetic Nanoparticles: An Overview for Biomedical Applications, Magnetochemistry 8 (2022) 107. https://doi.org/10.3390/magnetochemistry8090107

[81]   S.C. T Moorcroft, D.G. Jayne, S.D. Evans, Z. Yuin Ong, S.C. T Moorcroft, S.D. Evans, Z.Y. Ong, D.G. Jayne, Stimuli-Responsive Release of Antimicrobials Using Hybrid Inorganic Nanoparticle-Associated Drug-Delivery Systems, Macromol. Biosci. 18 (2018) 1800207. https://doi.org/10.1002/MABI.201800207

[82]   P. Singh, A. Garg, S. Pandit, V.R.S.S. Mokkapati, I. Mijakovic, Antimicrobial Effects of Biogenic Nanoparticles, Nanomater. (Basel, Switzerland) 8 (2018). https://doi.org/10.3390/NANO8121009

[83]   C. Ashajyothi, K.H. Harish, N. Dubey, R.K. Chandrakanth, Antibiofilm activity of biogenic copper and zinc oxide nanoparticles-antimicrobials collegiate against multiple drug resistant bacteria: a nanoscale approach, J. Nanostructure Chem. 6 (2016) 329–341. https://doi.org/10.1007/S40097-016-0205-2

[84]   A. Mukherjee, D. Sarkar, S. Sasmal, A Review of Green Synthesis of Metal Nanoparticles Using Algae, Front. Microbiol. 12 (2021) 2152. https://doi.org/10.3389/FMICB.2021.693899/BIBTEX

[85]   S. Patil, R. Chandrasekaran, Biogenic nanoparticles: a comprehensive perspective in synthesis, characterization, application and its challenges, J. Genet. Eng. Biotechnol. 2020 181 18 (2020) 1–23. https://doi.org/10.1186/S43141-020-00081-3

[86]   P.B.E. Kedi, F.E. Meva, L. Kotsedi, E.L. Nguemfo, C.B. Zangueu, A.A. Ntoumba, H.E.A. Mohamed, A.B. Dongmo, M. Maaza, Eco-friendly synthesis, characterization, in vitro and in vivo anti-inflammatory activity of silver nanoparticle-mediated Selaginella myosurus aqueous extract, Int. J. Nanomedicine 13 (2018) 8537–8548. https://doi.org/10.2147/IJN.S174530

[87]   Z. Huang, X. Jiang, D. Guo, N. Gu, Controllable synthesis and biomedical applications of silver

nanomaterials, J. Nanosci. Nanotechnol. 11 (2011) 9395–9408. https://doi.org/10.1166/JNN.2011.5317

[88]   C. Marambio-Jones, E.M.V. Hoek, A review of the antibacterial effects of silver nanomaterials and potential implications for human health and the environment, J. Nanoparticle Res. 2010 125 12 (2010) 1531–1551. https://doi.org/10.1007/S11051-010-9900-Y

[89]   Y. Ghosn, M.H. Kamareddine, A. Tawk, C. Elia, A. El Mahmoud, K. Terro, N. El Harake, B. El-Baba, J. Makdessi, S. Farhat, Inorganic Nanoparticles as Drug Delivery Systems and Their Potential Role in the Treatment of Chronic Myelogenous Leukaemia, Technol. Cancer Res. Treat. 18 (2019). https://doi.org/10.1177/1533033819853241

[90]   M.T. Yassin, A.A.F. Mostafa, A.A. Al-askar, F.O. Al-otibi, Synergistic Antifungal Efficiency of Biogenic Silver Nanoparticles with Itraconazole against Multidrug-Resistant Candidal Strains, Cryst. 2022, Vol. 12, Page 816 12 (2022) 816. https://doi.org/10.3390/CRYST12060816

[91]   P. Podsiadlo, V.A. Sinani, J.H. Bahng, N.W.S. Kam, J. Lee, N.A. Kotov, Gold nanoparticles enhance the anti-leukemia action of a 6-mercaptopurine chemotherapeutic agent, Langmuir 24 (2008) 568–574. https://doi.org/10.1021/LA702782K

[92]   D.C. Ferreira Soares, S.C. Domingues, D.B. Viana, M.L. Tebaldi, Polymer-hybrid nanoparticles: Current advances in biomedical applications, Biomed. Pharmacother. 131 (2020). https://doi.org/10.1016/J.BIOPHA.2020.110695

[93]   P.V. Rao, D. Nallappan, K. Madhavi, S. Rahman, L. Jun Wei, S.H. Gan, Phytochemicals and Biogenic Metallic Nanoparticles as Anticancer Agents, Oxid. Med. Cell. Longev. (2016). https://doi.org/10.1155/2016/3685671

[94]   M. Vinardell, M. Mitjans, Antitumor Activities of Metal Oxide Nanoparticles, Nanomaterials 5 (2015) 1004–1021. https://doi.org/10.3390/nano5021004

[95]   S. Bharti, G. Kaur, S. Jain, S. Gupta, S.K. Tripathi, Characteristics and mechanism associated with drug conjugated inorganic nanoparticles, J. Drug Target. 27 (2019) 813–829. https://doi.org/10.1080/1061186X.2018.1561888

Chapter 15

Use of Biogenic Nanomaterial as a Chemotherapeutic Material

A.F Komuroglu[1*], R. Bayat[2,3], F. Sen[2*]

[1]Department of Biochemistry and Clinical Biochemistry, School of Medicine, Yuzuncu Yil University, Van, Turkiye

[2]Sen Research Group, Department of Biochemistry, Dumlupinar University, Kutahya, Turkiye

[3]Department of Materials Science & Engineering, Faculty of Engineering, Dumlupinar University, Kutahya, Turkiye

aukomuroglu@yyu.edu.tr; fatih.sen@dpu.edu.tr

Abstract

Cancer is one of the most common health problems worldwide and is responsible for countless deaths. Nanomedicine plays an important role in the development of effective treatment strategies for cancer. Scientists are now exploring novel approaches to create techniques for the detection and treatment of cancer, using advancements in technology. Biologically mediated nanoparticles provide the advantages of being cost-effective, eco-friendly, secure, and compatible with living organisms. Biogenic nanoparticles are believed to exert their anticancer effects by inducing the generation of reactive oxygen species inside cellular compartments, hence triggering the activation of autophagic, apoptotic, and necrotic pathways resulting in cell death.

Keywords

Cancer, Biogenic Nanoparticles, Green Synthesis, Nanotechnology, Nanomedicine

1. Introduction

Nanotechnology is a branch of science that aims to design, develop, produce and characterize matter 1-100 nm in size by examining it at atomic and molecular level [1–4]. Nanotechnology emerges as a rapidly developing field, as it has a wide range of applications in many different branches of science and technology [5–8]. Nanotechnology is important in the development of sustainable technologies to protect human well-being and the environment[9,10]. Rapid advances in nanotechnology have been with the development of new synthesis protocols and characterization techniques [11]. Nanoparticles, which have been chemically and physically synthesized for many years, are known as traditional nanoparticle synthesis methods. The use of stabilizing chemicals (polyvinylpyrrolidone, sodium dodecylbenzyl sulfate, etc.) and reducing agents (methoxypolyethylene glycol, sodium borohydride, etc.) is required in order to synthesize nanoparticles and maintain their stability using traditional methods [12–20].

With these traditional methods, nanoparticles is preferred more in laboratory studies because it provides low costs to synthesize. However, there are some limitations in continuing the synthesis of nanoparticles with biotechnological applications and traditional methods [12,18,21–25]. These limitations are the limited synthesis of nanoparticles, weak morphological properties, and the chemicals used in the synthesis and preservation of its stability are among the most harmful to the ecosystem [12,26–28]. In order to avoid these limitations and to use nanoparticles in wider areas, interest in biological resources has increased tremendously in recent years [29].

Currently, it is feasible to produce nanoparticles by the use of ecologically sustainable green chemical methods. Actinomycetes, bacteria, fungus, viruses, yeasts, and other plant components have been used to create biogenic nanoparticles in the merging of nanotechnology with biology [27]. Green nanotechnology refers to the growing area of nanoscience and nanotechnology that focuses on the use of nanoscale materials and structures for biological purposes (usually 1-100 nm) [30,31]. Green nanotechnology nanoparticle synthesis methods provide new possibilities for synthesizing nanoparticles using natural reducing and stabilizing agents [32]. Microorganisms possess reductase enzymes inside their structures, enabling them to effectively store and detoxify heavy metals. Due to these properties, it is possible to reduce metal salts to metal nanoparticles without the use of any chemicals [29]. Biogenic nanoparticles provide a cost-effective, eco-friendly, and biocompatible alternative to chemical and physical methods, eliminating the need for energy and harmful substances (Fig. 1) [32,33].

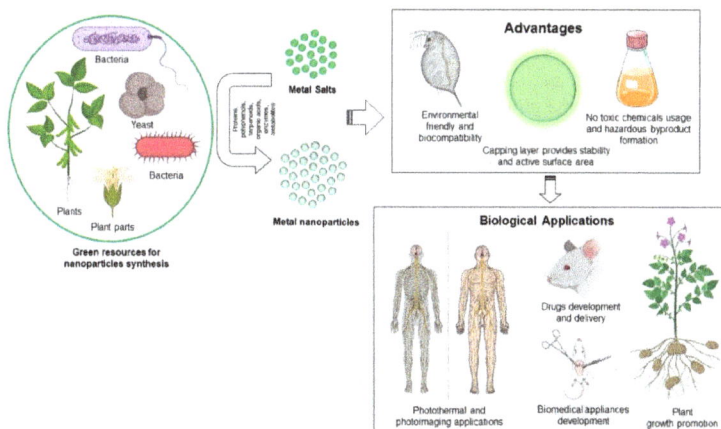

Figure 1. Green synthesis of metallic nanoparticle, their advantages and biological applications. Reprinted with permission from reference [33]. Copyright 2024. Nanomaterials.

2.   Biological Resources Used in Green Nanotechnology

Green synthesis aims to produce environmentally friendly nanoparticles with an approach that is not harmful to human health. In this context, roots, fruits, leaves and flowers of plants can be used to synthesize nanoparticles, while algae, bacteria or fungi can also be preferred [34–37]. The biological resources used in green nanotechnology are indicated in Fig. 2.

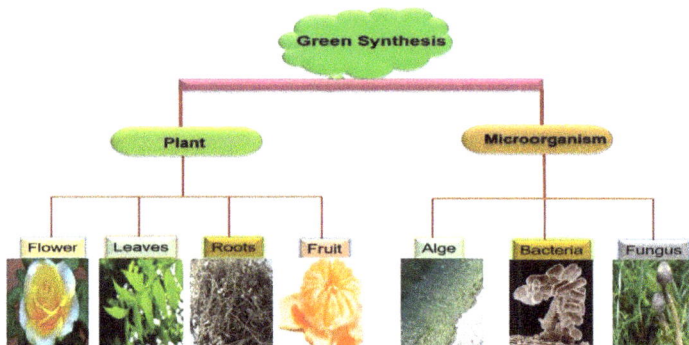

Figure 2. Biological resources utilized in green nanotechnology. Reprinted with permission from reference [34]. Copyright 2025. Micro and Nano Technologies.

2.1   Herbal Synthesis of Biogenic Nanoparticles

Plants have antioxidants and secondary metabolites, and the information is increasing day by day. Biomolecules in plants protect the cellular components from oxidative damage and can also reduce metal ions to nanoparticles. Using plants biogenic nanoparticles can be synthesized from many parts of the plant like plant tissues, leaves, stems, roots, fruits, bark and flower essence [38].

In the field of green nanotechnology, which is popular today, herbal extracts are widely used in nanoparticle synthesis. The reason why green synthesis is preferred is the ability of plants to convert solar energy into chemical energy, large biomass potential, easy accessibility, naturalness and reliability [39,40]. The fact that plants are easily accessible and have various biomolecules or metabolites plays a major role in the synthesis of biogenic nanoparticles [32,41–44].

Various herbs aloe vera (Aloe barbadensis Miller), clover (Medicago sativa), Lemon (Citrus limon), oat (Avena sativa), Tulsi (Osimum sanctum), Coriander (Coriandrum sativum), Neem (Azadirachta indica), Mustard (Brassica juncea) and lemongrass (Cymbopogon flexuosus) were used to synthesize silver nanoparticles and gold nanoparticles [45]. A research demonstrated that the use of Coffea arabica seed extract for the green manufacture of silver nanoparticles exhibited both antibacterial properties and induced a shift in $AgNO_3$ concentration [46]. Umashankari et al. (2012) reported that green synthesis of silver nanoparticles using extract of mangrove leaves is both more impressive and friendly to the ecosystem[47]. Satyanavi et al. (2012) reported that silver nanoparticles obtained by green synthesis from the Suaeda monoica plant have anticancer activity [48]. Another research using the aqueous extract of Calotropis procera flowers showed the creation of cubic and rectangular nanoparticles[49].

Biogenic nanoparticles synthesized from plants may open up new opportunities in the treatment of many types of cancer due to their biocompatibility, effectiveness and low cost. Although there has been a great increase in research on the cytotoxic effects of biogenic nanoparticles synthesized from plants in recent years, they are still far from clinical trials due to insufficient in vivo data. Additionally, further studies are needed to analyze the role of phytoconstituents in the formation of nanoparticles [3,50].

2.2 Bacterial Synthesis of Biogenic Nanoparticles

Research in the field of bacterial manufacture of nanoparticles, particularly metal nanoparticles, mostly focuses on prokaryotic cells [51,52]. Bacteria are a good choice in nanoparticle studies because of the ability of prokaryotic cells to adapt to difficult conditions. At the same time, bacteria are among the reasons for choosing them in nanoparticle synthesis because they can multiply rapidly, can be grown at affordable costs, can be easily manipulated, and temperature, oxygen, and incubation times can be easily controlled under growth conditions [21,53,54]. According to He et al., alterations in the growth media impact the synthesis of nanoparticles with varying shapes and sizes throughout the incubation period [55]. The bacteria that have been extensively researched for their ability to synthesize metal nanoparticles include *Pseudomonas, Klebsiella, Bacillus, Escherichia, Enterobacter, Aeromonas, Lactobacillus, Corynebacterium, Brevibacterium, Rhodobacter, Sargassum, Trichoderma, and Streptomyces* species [29]. Although there is an intense interest in gold and silver in bacterial nanoparticle synthesis, there are also studies on nanoparticle production with cadmium, palladium, titanium metals [12,21,29].

Johnston et al. (2013) reported that *Delftia acidovorans* bacteria were effective in the synthesis of pure gold nanoparticles, according to a study they conducted. According to reports, the *Delftia* bacteria is implicated in the synthesis of a short non-ribosomal peptide that is responsible for the manufacturing of gold nanoparticles. Johnston et al., (2013) reported that gold ions are associated with the resistance mechanism of *Delftia acidovorans* in the production of delftibactin in gold nanoparticle synthesis. Gold nanoparticles attached to delftibactin are nanoparticles that do not have any toxic effects for cells [56].

2.3 Synthesis of Biogenic Nanoparticles by Fungi

The utilization of fungus in the production of metal nanoparticles has been well acknowledged due to its notable benefits in comparison to the utilization of bacteria. The fact that the reproduction processes of fungi are easier than bacteria, their economic feasibility and their micelles that allow an increase in surface area are the most important features of their use as nanoparticles [57]. Mukherjee et al. found that fungus produce more protein than bacteria. According to them this helps in nanoparticle synthesis especially in case of silver nanoparticles [58]. The most commonly used fungal species in studies are *Candida albicans, Pichia jadinii, Yarrowia lipoytica, Hansenula anomala*. The ability of fungi to secrete bioactive metabolites with antimicrobial activity has also increased studies on endophytic fungi [59,60]. The use of *Fusarium oxysporum* fungus has been explored in many research for synthesis of metal nanoparticles especially silver nanoparticles. According to him in the process of making 5-15 nm gold nanoparticles these nanoparticles are coated with fungus to make them stable. This is done by the proteins released by the fungus[61].

2.4   Synthesis of Biogenic Nanoparticles by Algae

Algae are well known to accumulate heavy metal ions in large quantities and can convert them into more manageable forms [62,63]. Because of their unique properties, algae are considered as good organisms for bio-nanomaterials production. Algae extracts contain carbohydrates, proteins, minerals, oils, lipids, polyunsaturated fatty acids, bioactive compounds (antioxidants: tocopherols, polyphenols) and colors (carotenoids). Specifically, algae are used in the production of silver and gold nanoparticles by synthesis [64]. In extracellular and intracellular nanoparticle synthesis, cyanobacteria and microalgae species *Euglena intermedia, Euglena gracilis, Botryococcus braunii, Amphora sp., Chlorealla sp., Anabaena sp., Chlamydomonas sp., Synechococcus sp., Synechocystis sp., Spirulina platensis* [65].

3.   The Chemotherapeutic Use of Biogenic Nanomaterials

Cancer is a group of diseases characterised by uncontrolled cell growth that forms tumours. These tumours can be localised or metastasise to other organs and grow new tumours [66]. Cancer is a complex condition caused by a combination of environmental and genetic factors [67]. A cancerous condition is characterized by uncontrolled cell division followed by invasion of healthy cells and tissues [68]. This process comprises a broad range of up-regulation or down-regulation of cell adhesion receptors or receptors associated to cell motility, genes involved in tumor suppression, and various other cellular genes and proteins that affect cell growth and death [68].

Cancer is a significant contributor to mortality rates in both developed and developing nations around the globe [69]. Approximately 12.5% of all fatalities are attributed to cancer, with almost 70% of these cancer-related deaths occurring in low- and middle-income nations. Approximately 20% of individuals get cancer before reaching the age of 75. However, it is believed that 10% of individuals in this age range succumb to cancer. The yearly expenditure on cancer treatment in the United States is projected to surpass $150 billion [68]. Cancer ranks as the second most prevalent cause of mortality worldwide, with an annual occurrence of 14 million new cases and resulting in 8 million fatalities. The 2014 global cancer report states that in 2012, there were 1.5 million deaths from lung cancer, 745,000 deaths from liver cancer, 723,000 deaths from stomach cancer, 694,000 deaths from colon cancer, 521,000 deaths from breast cancer, and 400,000 deaths from esophageal cancer, among other forms of cancer [70]. It has been reported that 13.2 million people will die of cancer each year [71]. The global cancer burden is expected to reach approximately 28.4 million cases by 2040 [4].

Conventional antineoplastic treatments for cancer include chemotherapy, radiation, targeted therapy, immunotherapy, stem cell transplantation, cancer vaccines, surgery, and radiosurgery. These medicines are often used in combination and may lead to adverse effects. However, since traditionally used antineoplastic treatments have high fatigue, weakness, diarrhea, nerve damage, hair loss, oral aphthae and toxicity, new applications in cancer treatment have come to the fore [68,72–74]. Usually, the first choice in cancer treatment is chemotherapy. Chemotherapy molecules used for this purpose are widely used as primary targets to destroy cancer cells. Since these treatments are not target specific, both healthy and cancerous cells are destroyed by these chemotherapeutic molecules [4,74]. Hormone and immune therapies used in cancer treatment may cause serious side effects in patients, such as damage to normal cells and organs [75]. In light of the progress made in identifying and treating the problem at an early stage, it is crucial to explore alternative therapies, tools, and medications to effectively combat this ailment. Hence, it is essential to create novel therapeutic molecules that possess biocompatibility and cost-effectiveness.

Recent research has shown favorable outcomes for the use of biogenic nanoparticles in the treatment of cancer. In the 1900s, researchers were interested in reduction processes involved in the formation of metal particles from plant extract ions [3,76]. However, the precise mechanism by which these particles exhibit anticancer properties remained unclear. However, in recent years, the cost-effectiveness of plant extracts and their anticancer properties by taking part in the biological reduction reactions of plant extracts and metal salts at room temperature have been emphasized. It has been reported that nanoparticles (especially silver and gold) involved in green synthesis have antineoplastic activity in leukemia cancer type [77,78].

Nanoparticle production is thought to be straightforward, secure, economical, and eco-friendly by using biological principles [79]. Nanomedicine has revolutionized the fields of biomedical engineering, pharmacology, chemistry, and biology by specifically targeting advancements in cancer diagnosis, research, and management [80]. The main objective of cancer therapy is to enhance patient longevity and enhance their quality of life. The efficacy of a therapeutic agent in cancer therapy is determined by its ability to selectively inhibit and eliminate tumor cells while minimizing damage to adjacent healthy tissue. Nanoparticles are engineered to enhance the targeting and uptake of drugs, hence improving the effectiveness of therapy and minimizing the occurrence of significant side effects often associated with existing cancer therapies [81]. Nanotechnology is becoming more pervasive in everyday activities and has the potential to revolutionize disease detection and treatment in a variety of ways [80].

3.1   *Biogenic Nanoparticles in Cancer Therapy*

Several nanoparticles have been specifically engineered for the purpose of diagnosing or treating cancer. The prominent anti-cancer nanoparticles are silver (Ag), gold (Au), zinc (Zn), and copper (Cu). The nanoparticles (NPs) have the ability to potentially fight cancer by generating reactive oxygen species inside the cells, which triggers the activation of autophagy, apoptosis, and cell death pathways [82].

Metal nanoparticles have become a valuable tool in cancer treatment due to their enhanced ability to silence genes, target specific cells, and transport drugs more effectively in tumor cells. In addition to these therapeutic benefits, MNP has also been used as a diagnostic tool for cancer cells. Metal nanoparticles possess a diminutive size that enables them to effectively convey active chemicals to tumor cells [83].

Due to the efficacy and safety of nanoparticles, USA FDA, Myocet™ (Perrigo, Dublin, Ireland), DaunoXome® (Gilead Sciences, Foster City, CA, USA), Doxil® (Johnson & Johnson, New Brunswick, NJ, USA) and Abraxane have recently been used. ® (Celgene, Summit, NJ, USA) has approved anticancer drugs based on nanotechnology [84].

The standard therapeutic options for cancer have often lacked the ability to differentiate between healthy cells and malignant cells. As a result, individuals undergoing these therapies experience various adverse effects and systemic toxicities [83]. Tumors often include a diverse group of cells, making it difficult to effectively target all cell types with medications. Some treatments are unable to penetrate tumor cells adequately, which may lead to the development of resistance to many drugs. Metal nanoparticles possess remarkable characteristics that make them an appealing platform for researchers to provide a less intrusive alternative, with the potential to enhance the well-being of a significant number of people globally [83].

NPs interact with the biological system due to their significant surface-to-volume ratios, which allow atoms at the cellular level to readily begin diverse reactions. The distinctive structure of NPs affects their incorporation or penetration into cells. The charge residing on the surface of the NPs influences the duration and speed at which they circulate in the bloodstream, as well as their rates of absorption and displacement. Cationic nanoparticles, in contrast to anionic nanoparticles, have the ability to impair the integrity of the plasma membrane, hinder the construction of organelles, and disrupt normal cellular activity [85].

3.2   Anticancer Activities of Plant-Derived Biogenic Nanoparticles

Plants are known to possess a diverse range of antioxidants and secondary metabolites, as documented. The biomolecules in plants collaborate to safeguard cellular components against oxidative harm, resulting in the reduction of metal ions into nanoparticles. Biogenic nanoparticles may be generated from several plant components, including plant tissues, leaves, stems, roots, fruits, bark, and flower pulp [34,38].

Plants are beneficial for NP synthesis due to their abundance of diverse biomolecules or metabolites, which are both safe and easily accessible, and contribute to the stability and reduction of NPs. Plant-derived therapeutic medications have surfaced in contemporary medicine as a potential tool in the management of cancer (Fig. 3) [82]. Recently, there has been significant research on the development of efficient methods for creating metallic (NPs) that possess anti-cancer characteristics. These studies have been conducted both in laboratory settings (in vitro) and in living organisms (in vivo) [86].

The processes by which biogenic nanoparticles combat cancer have been thoroughly investigated, revealing that the restricted functional groups of these nanoparticles have a direct or indirect role in enhancing anticancer efficacy, lowering toxicity, and boosting bioavailability and absorption [87]. The variation in the phytochemical composition of the biological material used in the manufacture of distinct nanoparticles leads to variations in their anticancer capabilities [88].

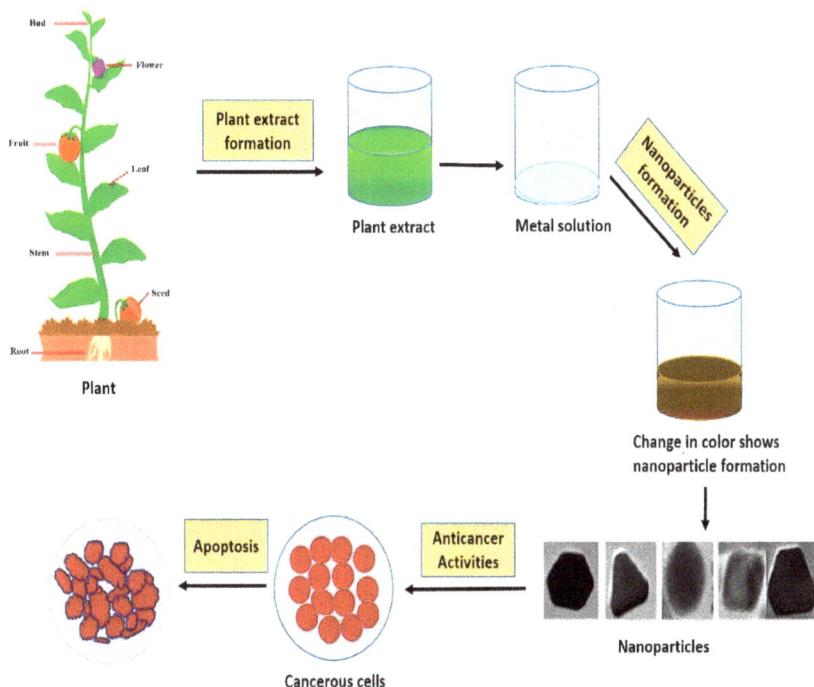

Figure 3. *Plant-mediated synthesis of metallic np's and their anticancer activity. Reprinted with permission from reference [82]. Copyright 2024. Cancers.*

The size distribution and shape of nanoparticles during their production are influenced not only by biological sources but also by experimental variables such as pH, temperature, and concentration of metal salts. Nanoparticles composed of metal demonstrate varying levels of toxicity depending on their size and form [89].

Zangeh et al. (2020) used Spinacia oleracea leaf, popularly known as spinach, which is widely used in the synthesis of silver nanoparticles in an acute myeloid leukemia model and compared it with the anticancer drug doxorubicin. Silver nanoparticles coated with *Spinacia oleracea* provided anticancer effect only in cancer cells, while doxorubicin showed anticancer properties against normal cells. Spinacia oleracea extracts include over 13 distinct flavonoids that exhibit anticancer and antioxidant properties. When combined with silver nanoparticles as a reducing coating agent, a synergistic effect has been reported [90].

AgNPs produced using biogenic means shown significant anticancer activity against the cervical cancer cell lines HeLa and Siha. AgNPs of hexagonal and triangular shapes, ranging in size from 2-18 nm, exhibited significant inhibitory effects on the Siha cancer cell line [91].

Zangeh et al. (2020) used *Hibiscus sabdariffa* to make gold nanoparticles. The antitumor activity of these gold nanoparticles was tested in an acute myeloid leukemia model. The abundance of polyphenolic compounds in unprocessed and uncontaminated parts of Hibiscus sabdariffa is key to its anticancer effect. These compounds are selective cytotoxic, induce autophagy not apoptosis and anti-metastatic, so good against certain type of cancer [90]. According to their study, Liu et al. (2021) said that the anticancer activity of gold nanoparticles made from *Allium sativum* L. leaf extract is due to free radical reduction and high antioxidant capacity by reducing tumor volume [92].

In vitro study of silver and gold nanoparticles biogenically synthesised from *Acalypha indica* Linn leaf extract on MDA-MB-231 human breast cancer cells. Nanoparticles from plant extract and its ability to induce apoptosis was confirmed by caspase-3 activation and DNA fragmentation [93]. Another lab study found biogenic silver nanoparticles (AgNPs) from *Sesbania grandiflora* leaves had cytotoxic effect on MCF-7 human breast cancer cells. MTT, acridine orange/ethidium bromide (AO-EB), Hochest and comet assay were used to determine the cytotoxicity. AgNPs were found to cause cellular damage including cell membrane disruption, oxidative stress and apoptosis [79].

The anticancer activity of gold nanoparticle biogenically synthesised from *Muntingia calabura* was studied in Hep2 line. AuNP from *Muntingia calabura* was found to be cytotoxic to Hep2 cell line and inhibit cell growth by inducing apoptosis. Further studies showed that apoptosis in Hep2 cells by AuNp is associated with cell membrane disruption, nuclear morphology alteration and cell cycle arrest in G2 phase [94].

Inorganic nanoparticles interact with intracellular macromolecules like cells, proteins and DNA. Internalization of nanoparticles by cells results in production of reactive oxygen species which in turn induces oxidative stress. Nanoparticles can cross the nuclear membrane and can interact with DNA directly or indirectly. However the mechanism of this interaction is unknown [79]. The toxicity of nanoparticles can be influenced by several parameters like dose, duration of exposure and particle size. According to reports physiologically produced AgNPs induce cell damage in Hep-2 cell line by generating reactive oxygen species (ROS) [79].

ZnO nanoparticle synthesized from water extract of *Monsonia burkena* plant has growth inhibitory effect on A549 lung cancer cell lines[95].

AuNPs synthesized using *Lonicera japonica* flowers have been shown to inhibit cell viability by 95% in cervical cancer cells. It has been stated that alkaloids and phenolic compounds found in the plant are a factor contributing to nanoparticle formation and anticancer activity[32].

3.3   Anticancer Activities of Algae-Derived Biogenic Nanoparticles

Algae, which is used in many areas such as feed, food, medicines, skin care and fertilizer, is now also used to synthesize green nanoparticles. Algae, which are rich in bioactive substances, are increasingly trending towards the use of biogenic nanoparticle synthesis due to their easy production and rapid growth. Depending on the area of the nanoparticles, algae-derived nanoparticles can be intracellular or extracellular. Algae are also known as bio nanofactories because they are stable, easy to handle and do not require cell maintenance. Phytochemicals used in the synthesis of metallic nanoparticles are a significant source of phytochemicals found in algae, which are naturally occurring plants[96].

Anti-cancer effect was investigated in AuNPs lung cancer cell line (A549) synthesized by biogenic method using red seaweed *Champia parvula*. It has been stated that AuNPs administration inhibits cell growth in cell cancer cell lines. In addition, it has been shown that AuNPs application changes cell morphology and decreases cell viability. Research has shown that nanoscale gold particles are capable of entering cells and subcellular structures, such as the nucleus. This enables the delivery of therapeutic substances into cells and enhances their ability to be absorbed by cells, leading to increased toxicity [80]. In another study, the anticancer effect of biogenic. AuNPs synthesized using the red seaweed *Hypnea valantiae* in lung cancer cells (A549) was investigated. It has been shown that AuNP exhibits potent cytotoxic activity against lung cancer cells and controls cell migration and colony formation[80].

Ramaswamy et al. (2016) found that the combination of copper sulfate and *Sargassum polycystum*, a kind of brown seaweed, together with anticancer medicines, had a notable impact on breast cancer cells[97]. Additionally, copper nanoparticles containing *Prosopis cineraria* had anticancer properties against MCF-7 and HeLa cancer cells in humans prior research has shown that copper nanoparticles produced using environmentally friendly methods including *Nerium oleander, Magnolia kobus,* and *Eclipta prostrata* have significant anticancer properties against the human HepG2 cancer cell line [98].

The anticancer activity of ZnO NPs produced via *Anabaena cylindrica* was investigated in the melanoma cell line (B16F10). Anticancer activity was evaluated using MTT, Acridine Orange EtBt assays. The synthesized ZnO NPs have been shown to have high efficiency in killing cancer cells with low dose and less toxicity [99].

3.4 Anticancer Activities of Bacteria-Derived Biogenic Nanoparticles

Selenium nanoparticles synthesized from many bacteria such as *Escherichia coli, Ralstonia eutropha, Enterobacter cloacae, Burkholderia fungorum, Stenotrophomonas maltophilia, Lactobacillus acidophilus, Bacillus megaterium, Bacillus subtilis* have been reported to have cancer cell selectivity and very low anticancer activity on normal cells [100].

Studies have shown that silver nanoparticles produced from *Ganoderma enigmaticum* and *Trametes ljubarsky* white rot fungus have cytotoxic properties in human lung cancer cells when tested in laboratory conditions [93].

The antitumor activity of selenium nanoparticle synthesized from *Lactobacillus casei* ATCC 393 strain was investigated in the CT26 syngenic collateral cancer model. It has been reported that selenium nanoparticles derived from *L.casei* show cancer-specific antiproliferative activity in vitro. Additionally, selenium nanoparticles synthesized from *L.casei* were found to induce apoptosis and increase reactive oxygen levels in cancer cells[101].

4. Anticancer Mechanism of Action of Biogenic Nanoparticles

Biogenic nanoparticles can intervene at the cellular and intracellular levels. These interventions include physical, chemical and biological factors. Due to their physicochemical properties, NPs are easily taken up by cancer cells and interact with different components of the cell (proteins, lipids, nucleic acids, membranes and organelles), causing serious changes in the cell. Molecular and cellular changes caused by NPs can either cause inhibitory effects (mainly in terms of anti-cancer properties such as tumor invasion, growth and cell viability) or stimulatory effects such as the apoptotic pathway and cell death. On the other hand, the anticancer potential of biogenic nanoparticles may be related to the properties of biomolecules or phytocompounds found in their structures [102].

Tumor cells undergo cell death by necrosis and apoptosis, which may be quantitatively differentiated based on their appearance. The nuclear contents and cytoplasm of necrotic cells are seen to seep out from the cells, while the nuclei of apoptotic cells have a reduced size and tightly packed chromatin [103]. Metallic nanoparticles have cytotoxic capacity against various cancer cell lines such as necrosis, stimulation of signaling pathways, lysosomal damage, caspase-mediated signal transduction and apoptosis (Figure 4) [70].

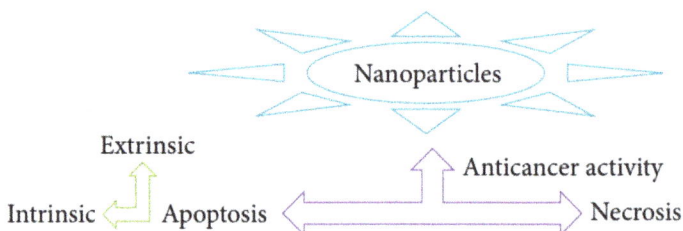

Figure 4. *A simplified diagram of anticancer activities triggered by nanoparticles in tumor cells. Reprinted with permission from reference [70]. Copyright 2024. Oxidative Medicine and Cellular.*

AgNPs synthesized from *Albizia adianthifolia* significantly induce apoptosis and necrosis in human lung carcinoma cells [104]. Apoptosis may be categorized into two distinct pathways: intrinsic and extrinsic. Mitochondria have a crucial function in intrinsic apoptosis by causing the mitochondrial membrane to lose its electrical potential via the opening of a channel called the mitochondrial permeability transition pore. This mechanism ultimately leads to a decrease in ATP levels and is triggered by intrinsic apoptosis. The extrinsic apoptotic pathway is facilitated by the CD95 death receptor, which enlists the adaptor protein fas-associated death domain. The adaptor protein FADD interacts with and stimulates caspase-8 by forming a death-inducing signalling complex [70].

Necrosis arises from the rupture of cellular and nuclear membranes due to severe physiological circumstances. The rupture of the cellular membrane distinguishes necrosis from apoptosis [70].

5.   Conclusion and Future Prospects

Despite significant progress in cancer therapy, cancer remains a prominent cause of mortality globally. Given the absence of a successful cancer treatment, the conventional pharmaceuticals used in cancer therapy have significant side effects and harm healthy cells in addition to the intended target cells. Consequently, we are compelled to pursue new investigations in order to enhance the effectiveness of anticancer treatments. Scientists are endeavoring to create new ways in order to improve diagnosis and therapy with enhanced efficiency, specificity, and toxicity. Recently, NP biological and green synthesis has attracted great interest in biomedicine and various nanoparticles have been used frequently in the development of cancer diagnosis and treatment. Nanoparticles are believed to possess anticancer characteristics by inducing the generation of reactive oxygen species inside cellular compartments and triggering the activation of autophagy, apoptotic, and necrotic death pathways. In many studies mentioned above, it has been reported that biogenic materials are used in cancer treatment and have a cytotoxic effect on cancer cells. At this point, the biocompatibility of nanoparticles obtained by green synthesis suggests that they will be effective in cancer treatment with their toxic effects only on the target cell and strong antioxidant effects. Metal nanoparticles provide significant promise for addressing the challenges associated with traditional cancer therapy. Therefore, scientists are trying to develop new approaches in search of finding better diagnosis and treatment with maximum efficacy, specificity and less toxicity. Nevertheless, the majority of research on biogenic nanoparticles has been conducted in laboratory settings, namely in vitro, with little availability of in vivo data. Hence, it is crucial to conduct extensive and extensive research on biogenic nanoparticles in live organisms to build upon the discoveries obtained in laboratory settings and to advance the development of anticancer medications while also uncovering the processes by which biogenic nanoparticles function.

References

[1]   T. Gur, I. Meydan, H. Seckin, M. Bekmezci, F. Sen, Green synthesis, characterization and bioactivity of biogenic zinc oxide nanoparticles, Environ. Res. 204 (2022) 111897. https://doi.org/10.1016/J.ENVRES.2021.111897

[2]   D. Kulkarni, R. Sherkar, C. Shirsathe, R. Sonwane, N. Varpe, S. Shelke, M.P. More, S.R. Pardeshi, G. Dhaneshwar, V. Junnuthula, S. Dyawanapelly, Biofabrication of nanoparticles: sources, synthesis, and biomedical applications, Front. Bioeng. Biotechnol. 11 (2023). https://doi.org/10.3389/fbioe.2023.1159193

[3]   V.K. Chaturvedi, B. Sharma, A.D. Tripathi, D.P. Yadav, K.R. Singh, J. Singh, R.P. Singh, Biosynthesized nanoparticles: a novel approach for cancer therapeutics, Front. Med. Technol. 5 (2023). https://doi.org/10.3389/fmedt.2023.1236107

[4]   G. Kah, R. Chandran, H. Abrahamse, Biogenic Silver Nanoparticles for Targeted Cancer Therapy and Enhancing Photodynamic Therapy, Cells 12 (2023). https://doi.org/10.3390/CELLS12152012

[5]   N. Bazancir, I. Meydan, Characterization of Zn nanoparticles of Platonus orientalis plant, investigation of DPPH radical extinquishing and antimicrobial activity, East. J. Med. 27 (2022).

[6]   M. Bekmezci, D.B. Subasi, R. Bayat, M. Akin, Z.K. Coguplugil, F. Sen, Synthesis of a functionalized carbon supported platinum–iridium nanoparticle catalyst by the rapid chemical reduction method for the anodic reaction of direct methanol fuel cells, New J. Chem. 46 (2022) 21591–21598. https://doi.org/10.1039/D2NJ03209K

[7]   K. Arikan, H. Burhan, R. Bayat, F. Sen, Glucose nano biosensor with non-enzymatic excellent sensitivity prepared with nickel–cobalt nanocomposites on f-MWCNT, Chemosphere (2021) 132720. https://doi.org/10.1016/J.CHEMOSPHERE.2021.132720

[8]   F. Karimi, M. Akin, R. Bayat, M. Bekmezci, R. Darabi, E. Aghapour, F. Sen, Application of Quasihexagonal Pt@PdS2-MWCNT catalyst with High Electrochemical Performance for Electro-Oxidation of Methanol, 2-Propanol, and Glycerol Alcohols For Fuel Cells, Mol. Catal. 536 (2023) 112874. https://doi.org/10.1016/j.mcat.2022.112874

[9]   N. Bazancir, İ. Meydan, Geriatrik Beslenme, in: Beslenme ve Sağlık, 2023.

[10]   R. Bayat, M. Bekmezci, M. Akin, I. Isik, F. Sen, Nitric Oxide Detection Using a Corona Phase Molecular Recognition Site on Chiral Single-Walled Carbon Nanotubes, ACS Appl. Bio Mater. 6 (2023) 4828–4835. https://doi.org/10.1021/acsabm.3c00573

[11]   M. Rai, A. Gade, A. Yadav, Biogenic Nanoparticles: An Introduction to What They Are, How They Are Synthesized and Their Applications, in: Met. Nanoparticles Microbiol., Springer Berlin

Heidelberg, Berlin, Heidelberg, 2011: pp. 1–14. https://doi.org/10.1007/978-3-642-18312-6_1

[12]   K.N. Thakkar, S.S. Mhatre, R.Y. Parikh, Biological synthesis of metallic nanoparticles, Nanomedicine Nanotechnology, Biol. Med. 6 (2010) 257–262. https://doi.org/10.1016/j.nano.2009.07.002

[13]   R. Bayat, E. Halvaci, T. Kozak, M. Bekmezci, F. Sen, Electrocatalytic performance of Shape-Controlled synthesized PdNi-rGO nano Cube on sodium borohydride and methanol electrooxidation, Fuel 366 (2024) 131248. https://doi.org/10.1016/j.fuel.2024.131248

[14]   P. Taslimi, F. Türkan, A. Cetin, H. Burhan, M. Karaman, I. Bildirici, İ. Gulçin, F. Şen, Pyrazole[3,4-d] Pyridazine Derivatives: Molecular Docking and Explore of Acetylcholinesterase and Carbonic Anhydrase Enzymes Inhibitors as Anticholinergics Potentials, Bioorg. Chem. 92 (2019) 103213. https://doi.org/10.1016/j.bioorg.2019.103213

[15]   R. Ulus, Y. Yıldız, S. Eriş, B. Aday, F. Şen, M. Kaya, Functionalized Multi-Walled Carbon Nanotubes (f-MWCNT) as Highly Efficient and Reusable Heterogeneous Catalysts for the Synthesis of Acridinedione Derivatives, ChemistrySelect 1 (2016) 3861–3865. https://doi.org/10.1002/SLCT.201600719

[16]   S. Ertan, F. Şen, S. Şen, G. Gökağaç, Platinum nanocatalysts prepared with different surfactants for C1-C3 alcohol oxidations and their surface morphologies by AFM, J. Nanoparticle Res. 14 (2012) 1–12. https://doi.org/10.1007/S11051-012-0922-5

[17]   M.B. Askari, P. Salarizadeh, A. Di Bartolomeo, F. Şen, Enhanced electrochemical performance of MnNi2O4/rGO nanocomposite as pseudocapacitor electrode material and methanol electro-oxidation catalyst, Nanotechnology 32 (2021). https://doi.org/10.1088/1361-6528/abfded

[18]   F. Şen, G. Gökağaç, S. Şen, High performance Pt nanoparticles prepared by new surfactants for C1 to C3 alcohol oxidation reactions, J. Nanoparticle Res. 15 (2013) 1979. https://doi.org/10.1007/s11051-013-1979-5

[19]   Y. Yıldız, İ. Esirden, E. Erken, E. Demir, M. Kaya, F. Şen, Microwave (Mw)-assisted Synthesis of 5-Substituted 1 H -Tetrazoles via [3+2] Cycloaddition Catalyzed by Mw-Pd/Co Nanoparticles Decorated on Multi-Walled Carbon Nanotubes, ChemistrySelect 1 (2016) 1695–1701. https://doi.org/10.1002/slct.201600265

[20]   S. Eris, Z. Daşdelen, F. Sen, Investigation of electrocatalytic activity and stability of Pt@f-VC catalyst prepared by in-situ synthesis for Methanol electrooxidation, Int. J. Hydrogen Energy 43 (2018) 385–390. https://doi.org/10.1016/j.ijhydene.2017.11.063

[21]   N. Pantidos, Biological Synthesis of Metallic Nanoparticles by Bacteria, Fungi and Plants, J. Nanomed. Nanotechnol. 05 (2014). https://doi.org/10.4172/2157-7439.1000233

[22]   M. Shah, D. Fawcett, S. Sharma, S. Tripathy, G. Poinern, Green Synthesis of Metallic Nanoparticles via Biological Entities, Materials (Basel). 8 (2015) 7278–7308. https://doi.org/10.3390/ma8115377

[23]   N.H. Khand, A.R. Solangi, S. Ameen, A. Fatima, J.A. Buledi, A. Mallah, S.Q. Memon, F. Sen, F. Karimi, Y. Orooji, A new electrochemical method for the detection of quercetin in onion, honey and green tea using Co3O4 modified GCE, J. Food Meas. Charact. 2021 154 15 (2021) 3720–3730. https://doi.org/10.1007/S11694-021-00956-0

[24]   H. Goksu, Y. Yildiz, B. Çelik, M. Yazici, B. Kilbas, F. Sen, Eco-friendly hydrogenation of aromatic aldehyde compounds by tandem dehydrogenation of dimethylamine-borane in the presence of a reduced graphene oxide furnished platinum nanocatalyst, Catal. Sci. Technol. 6 (2016) 2318–2324. https://doi.org/10.1039/C5CY01462J

[25]   K. Nesrin, C. Yusuf, K. Ahmet, S.B. Ali, N.A. Muhammad, S. Suna, Ş. Fatih, Biogenic silver nanoparticles synthesized from Rhododendron ponticum and their antibacterial, antibiofilm and cytotoxic activities, J. Pharm. Biomed. Anal. 179 (2020) 112993. https://doi.org/10.1016/J.JPBA.2019.112993

[26]   S.S. Birla, V. V. Tiwari, A.K. Gade, A.P. Ingle, A.P. Yadav, M.K. Rai, Fabrication of silver nanoparticles by Phoma glomerata and its combined effect against Escherichia coli, Pseudomonas aeruginosa and Staphylococcus aureus, Lett. Appl. Microbiol. 48 (2009) 173–179. https://doi.org/10.1111/J.1472-765X.2008.02510.X

[27]   M. Rai, A. Yadav, A. Gade, Silver nanoparticles as a new generation of antimicrobials, Biotechnol. Adv. 27 (2009) 76–83. https://doi.org/10.1016/j.biotechadv.2008.09.002

[28]   A. Hojjati-Najafabadi, S. Salmanpour, F. Sen, P.N. Asrami, M. Mahdavian, M.A. Khalilzadeh, A Tramadol Drug Electrochemical Sensor Amplified by Biosynthesized Au Nanoparticle Using Mentha aquatic Extract and Ionic Liquid, Top. Catal. (2021). https://doi.org/10.1007/s11244-021-01498-x

[29]   P. Singh, Y.-J. Kim, D. Zhang, D.-C. Yang, Biological Synthesis of Nanoparticles from Plants and Microorganisms, Trends Biotechnol. 34 (2016) 588–599. https://doi.org/10.1016/j.tibtech.2016.02.006

[30]   A. Saxena, R.M. Tripathi, R.P. Singh, Biological synthesis of silver nanoparticles by using onion (Allium cepa) extract and their antibacterial activity, Dig J Nanomater Bios 5 (2010) 427–432.

[31]   M. V Yezhelyev, X. Gao, Y. Xing, A. Al-Hajj, S. Nie, R.M. O'Regan, Emerging use of nanoparticles in diagnosis and treatment of breast cancer, Lancet Oncol. 7 (2006) 657–667. https://doi.org/10.1016/S1470-2045(06)70793-8

[32]   S. Patil, R. Chandrasekaran, Biogenic nanoparticles: a comprehensive perspective in synthesis, characterization, application and its challenges, J. Genet. Eng. Biotechnol. 2020 181 18 (2020) 1–23. https://doi.org/10.1186/S43141-020-00081-3

[33]   P. Singh, A. Garg, S. Pandit, V.R.S.S. Mokkapati, I. Mijakovic, Antimicrobial Effects of Biogenic Nanoparticles, Nanomater. (Basel, Switzerland) 8 (2018). https://doi.org/10.3390/NANO8121009

[34]   B. Bharti, R. Kumar, H. Kumar, H. Li, X. Zha, F. Ouyang, Advanced applications and current status of green nanotechnology in the environmental industry, in: Green Funct. Nanomater. Environ. Appl., Elsevier, 2022: pp. 303–340. https://doi.org/10.1016/B978-0-12-823137-1.00012-9

[35]   F. Gol, A. Aygun, A. Seyrankaya, T. Gur, C. Yenikaya, F. Sen, Green synthesis and characterization of Camellia sinensis mediated silver nanoparticles for antibacterial ceramic applications (vol 250, 123037, 2020), Mater. Chem. Phys. 270 (2021).

[36]   B. Sahin, A. Aygun, H. Gunduz, K. Sahin, E. Demir, S. Akocak, F. Sen, Cytotoxic effects of platinum nanoparticles obtained from pomegranate extract by the green synthesis method on the MCF-7 cell line, COLLOIDS AND SURFACES B-BIOINTERFACES 204 (2021).

[37]   S. Wahab, A. Salman, Z. Khan, S. Khan, C. Krishnaraj, S.-I. Yun, Metallic Nanoparticles: A Promising Arsenal against Antimicrobial Resistance—Unraveling Mechanisms and Enhancing Medication Efficacy, Int. J. Mol. Sci. 24 (2023) 14897. https://doi.org/10.3390/ijms241914897

[38]   A. Rana, K. Yadav, S. Jagadevan, A comprehensive review on green synthesis of nature-inspired metal nanoparticles: Mechanism, application and toxicity, J. Clean. Prod. 272 (2020) 122880. https://doi.org/10.1016/J.JCLEPRO.2020.122880

[39]   M. Nasrollahzadeh, S.M. Sajadi, Z. Issaabadi, M. Sajjadi, Biological Sources Used in Green Nanotechnology, in: 2019: pp. 81–111. https://doi.org/10.1016/B978-0-12-813586-0.00003-1

[40]   P. Velusamy, G.V. Kumar, V. Jeyanthi, J. Das, R. Pachaiappan, Bio-Inspired Green Nanoparticles: Synthesis, Mechanism, and Antibacterial Application, Toxicol. Res. 32 (2016) 95–102. https://doi.org/10.5487/TR.2016.32.2.095

[41]   R. Bayat, M. Akin, B. Yilmaz, M. Bekmezci, M. Bayrakci, F. Sen, Biogenic platinum based nanoparticles: Synthesis, characterization and their applications for cell cytotoxic, antibacterial effect, and direct alcohol fuel cells, Chem. Eng. J. Adv. 14 (2023) 100471. https://doi.org/10.1016/j.ceja.2023.100471

[42]   D. Sharma, S. Kanchi, K. Bisetty, Biogenic synthesis of nanoparticles: A review, Arab. J. Chem. (2019). https://doi.org/10.1016/j.arabjc.2015.11.002

[43]   A. Aygun, F. Gülbagca, L.Y. Ozer, B. Ustaoglu, Y.C. Altunoglu, M.C. Baloglu, M.N. Atalar, M.H. Alma, F. Sen, Biogenic platinum nanoparticles using black cumin seed and their potential usage as antimicrobial and anticancer agent, J. Pharm. Biomed. Anal. (2019) 112961. https://doi.org/10.1016/j.jpba.2019.112961

[44]   A. Aygün, S. Özdemir, M. Gülcan, K. Cellat, F. Şen, Synthesis and Characterization of Reishi Mushroom-mediated Green Synthesis of Silver Nanoparticles for the Biochemical Applications, J. Pharm. Biomed. Anal. (2019) 112970. https://doi.org/10.1016/j.jpba.2019.112970

[45]   P. Singh, S. Pandit, V.R.S.S. Mokkapati, A. Garg, V. Ravikumar, I. Mijakovic, Gold nanoparticles in diagnostics and therapeutics for human cancer, Int. J. Mol. Sci. 19 (2018). https://doi.org/10.3390/ijms19071979

[46]   T.M. Pereira, V.L.P. Polez, M.H. Sousa, L.P. Silva, Modulating physical, chemical, and biological properties of silver nanoparticles obtained by green synthesis using different parts of the tree Handroanthus heptaphyllus (Vell.) Mattos, Colloid Interface Sci. Commun. 34 (2020) 100224. https://doi.org/10.1016/j.colcom.2019.100224

[47]   J. Umashankari, D. Inbakandan, T.T. Ajithkumar, T. Balasubramanian, Mangrove plant, Rhizophora mucronata (Lamk, 1804) mediated one pot green synthesis of silver nanoparticles and its antibacterial activity against aquatic pathogens, Aquat. Biosyst. 8 (2012) 11. https://doi.org/10.1186/2046-9063-8-11

[48]   K. Satyavani, S. Gurudeeban, T. Ramanathan, T. Balasubramanian, Toxicity Study of Silver Nanoparticles Synthesized from Suaeda monoica on Hep-2 Cell Line., Avicenna J. Med. Biotechnol. 4 (2012) 35–9.

[49]   S.A. Babu, H.G. Prabu, Synthesis of AgNPs using the extract of Calotropis procera flower at room temperature, Mater. Lett. 65 (2011) 1675–1677. https://doi.org/10.1016/j.matlet.2011.02.071

[50]   R.A. El-Fitiany, A. AlBlooshi, A. Samadi, M.A. Khasawneh, Biogenic synthesis and physicochemical characterization of metal nanoparticles based on Calotropis procera as promising sustainable materials against skin cancer, Sci. Rep. 14 (2024) 25154. https://doi.org/10.1038/s41598-024-76422-w

[51]   A. Arora, E. Lashani, R.J. Turner, Bacterial synthesis of metal nanoparticles as antimicrobials, Microb. Biotechnol. 17 (2024). https://doi.org/10.1111/1751-7915.14549

[52]   D. Lahiri, M. Nag, H.I. Sheikh, T. Sarkar, H.A. Edinur, S. Pati, R.R. Ray, Microbiologically-Synthesized Nanoparticles and Their Role in Silencing the Biofilm Signaling Cascade, Front. Microbiol. 12 (2021). https://doi.org/10.3389/fmicb.2021.636588

[53]   A.L. Campaña, A. Saragliadis, P. Mikheenko, D. Linke, Insights into the bacterial synthesis of metal nanoparticles, Front. Nanotechnol. 5 (2023) 1216921. https://doi.org/10.3389/FNANO.2023.1216921

[54]   A. Carrapiço, M.R. Martins, A.T. Caldeira, J. Mirão, L. Dias, Biosynthesis of Metal and Metal Oxide Nanoparticles Using Microbial Cultures: Mechanisms, Antimicrobial Activity and Applications to Cultural Heritage, Microorganisms 11 (2023) 378. https://doi.org/10.3390/microorganisms11020378

[55]   S. He, Z. Guo, Y. Zhang, S. Zhang, J. Wang, N. Gu, Biosynthesis of gold nanoparticles using the bacteria Rhodopseudomonas capsulata, Mater. Lett. 61 (2007) 3984–3987. https://doi.org/10.1016/j.matlet.2007.01.018

[56]   C.W. Johnston, M.A. Wyatt, X. Li, A. Ibrahim, J. Shuster, G. Southam, N.A. Magarvey, Gold biomineralization by a metallophore from a gold-associated microbe, Nat. Chem. Biol. 9 (2013) 241–243. https://doi.org/10.1038/nchembio.1179

[57]   S. Ummartyotin, N. Bunnak, J. Juntaro, M. Sain, H. Manuspiya, Synthesis of colloidal silver

nanoparticles for printed electronics, Comptes Rendus. Chim. 15 (2012) 539–544. https://doi.org/10.1016/j.crci.2012.03.006

[58]   P. Mukherjee, A. Ahmad, D. Mandal, S. Senapati, S.R. Sainkar, M.I. Khan, R. Parishcha, P. V. Ajaykumar, M. Alam, R. Kumar, M. Sastry, Fungus-Mediated Synthesis of Silver Nanoparticles and Their Immobilization in the Mycelial Matrix: A Novel Biological Approach to Nanoparticle Synthesis, Nano Lett. 1 (2001) 515–519. https://doi.org/10.1021/nl0155274

[59]   X. Zhang, Y. Qu, W. Shen, J. Wang, H. Li, Z. Zhang, S. Li, J. Zhou, Biogenic synthesis of gold nanoparticles by yeast Magnusiomyces ingens LH-F1 for catalytic reduction of nitrophenols, Colloids Surfaces A Physicochem. Eng. Asp. 497 (2016) 280–285. https://doi.org/10.1016/j.colsurfa.2016.02.033

[60]   A.S. Vijayanandan, R.M. Balakrishnan, Biosynthesis of cobalt oxide nanoparticles using endophytic fungus Aspergillus nidulans, J. Environ. Manage. 218 (2018) 442–450. https://doi.org/10.1016/j.jenvman.2018.04.032

[61]   A. Ahmad, P. Mukherjee, S. Senapati, D. Mandal, M.I. Khan, R. Kumar, M. Sastry, Extracellular biosynthesis of silver nanoparticles using the fungus Fusarium oxysporum, Colloids Surfaces B Biointerfaces 28 (2003) 313–318. https://doi.org/10.1016/S0927-7765(02)00174-1

[62]   E.-S. Salama, H.-S. Roh, S. Dev, M.A. Khan, R.A.I. Abou-Shanab, S.W. Chang, B.-H. Jeon, Algae as a green technology for heavy metals removal from various wastewater, World J. Microbiol. Biotechnol. 35 (2019) 75. https://doi.org/10.1007/s11274-019-2648-3

[63]   S.Y. Cheng, P.-L. Show, B.F. Lau, J.-S. Chang, T.C. Ling, New Prospects for Modified Algae in Heavy Metal Adsorption, Trends Biotechnol. 37 (2019) 1255–1268. https://doi.org/10.1016/j.tibtech.2019.04.007

[64]   P. Khanna, A. Kaur, D. Goyal, Algae-based metallic nanoparticles: Synthesis, characterization and applications, J. Microbiol. Methods 163 (2019) 105656. https://doi.org/10.1016/j.mimet.2019.105656

[65]   S.A. Dahoumane, M. Mechouet, F.J. Alvarez, S.N. Agathos, C. Jeffryes, Microalgae: An outstanding tool in nanotechnology, Bionatura 1 (2016). https://doi.org/10.21931/RB/2016.01.04.7

[66]   X. Li, H. Xu, Z.-S. Chen, G. Chen, Biosynthesis of Nanoparticles by Microorganisms and Their Applications, J. Nanomater. 2011 (2011) 1–16. https://doi.org/10.1155/2011/270974

[67]   M. Buttacavoli, N.N. Albanese, G. Di Cara, R. Alduina, C. Faleri, M. Gallo, G. Pizzolanti, G. Gallo, S. Feo, F. Baldi, P. Cancemi, Anticancer activity of biogenerated silver nanoparticles: an integrated proteomic investigation, Oncotarget 9 (2018) 9685–9705. https://doi.org/10.18632/oncotarget.23859

[68]   M. Ovais, A.T. Khalil, A. Raza, M.A. Khan, I. Ahmad, N.U. Islam, M. Saravanan, M.F. Ubaid, M. Ali, Z.K. Shinwari, Green Synthesis of Silver Nanoparticles Via Plant Extracts: Beginning a New Era in Cancer Theranostics, Nanomedicine 11 (2016) 3157–3177. https://doi.org/10.2217/nnm-2016-0279

[69]   H. Barabadi, M.A. Mahjoub, B. Tajani, A. Ahmadi, Y. Junejo, M. Saravanan, Emerging Theranostic Biogenic Silver Nanomaterials for Breast Cancer: A Systematic Review, J. Clust. Sci. 30 (2019) 259–279. https://doi.org/10.1007/s10876-018-01491-7

[70]   P.V. Rao, D. Nallappan, K. Madhavi, S. Rahman, L. Jun Wei, S.H. Gan, Phytochemicals and Biogenic Metallic Nanoparticles as Anticancer Agents, Oxid. Med. Cell. Longev. (2016). https://doi.org/10.1155/2016/3685671

[71]   V. Nayak, K.R. Singh, R. Verma, M.D. Pandey, J. Singh, R. Pratap Singh, Recent advancements of biogenic iron nanoparticles in cancer theranostics, Mater. Lett. 313 (2022) 131769. https://doi.org/10.1016/j.matlet.2022.131769

[72]   Y. Cai, B. Karmakar, H.S. AlSalem, A.F. El-kott, M.Z. Bani-Fwaz, S. Negm, A.A.A. Oyouni, O. Al-Amer, G.E.-S. Batiha, Oak gum mediated green synthesis of silver nanoparticles under ultrasonic conditions: Characterization and evaluation of its antioxidant and anti-lung cancer effects, Arab. J. Chem. 15 (2022) 103848. https://doi.org/10.1016/j.arabjc.2022.103848

[73]   M. Arruebo, N. Vilaboa, B. Sáez-Gutierrez, J. Lambea, A. Tres, M. Valladares, Á. González-

Fernández, Assessment of the Evolution of Cancer Treatment Therapies, Cancers (Basel). 3 (2011) 3279–3330. https://doi.org/10.3390/cancers3033279

[74]   B. Liu, H. Zhou, L. Tan, K.T.H. Siu, X.-Y. Guan, Exploring treatment options in cancer: tumor treatment strategies, Signal Transduct. Target. Ther. 9 (2024) 175. https://doi.org/10.1038/s41392-024-01856-7

[75]   H.J. Han, C. Ekweremadu, N. Patel, Advanced drug delivery system with nanomaterials for personalised medicine to treat breast cancer, J. Drug Deliv. Sci. Technol. 52 (2019) 1051–1060. https://doi.org/10.1016/j.jddst.2019.05.024

[76]   Z. Liu, Z. Zhang, X. Du, Y. Liu, A. Alarfaj, A. Hirad, S.A. Ansari, Z. Zhang, Novel green synthesis of silver nanoparticles mediated by Curcumae kwangsiensis for anti-lung cancer activities: a preclinical trial study, Arch. Med. Sci. 19 (2023) 1463–1471. https://doi.org/10.5114/aoms/134059

[77]   E. Mostafavi, A. Zarepour, H. Barabadi, A. Zarrabi, L.B. Truong, D. Medina-Cruz, Antineoplastic activity of biogenic silver and gold nanoparticles to combat leukemia: Beginning a new era in cancer theragnostic, Biotechnol. Reports 34 (2022) e00714. https://doi.org/10.1016/j.btre.2022.e00714

[78]   D.T. Debela, S.G.Y. Muzazu, K.D. Heraro, M.T. Ndalama, B.W. Mesele, D.C. Haile, S.K. Kitui, T. Manyazewal, New approaches and procedures for cancer treatment: Current perspectives, SAGE Open Med. 9 (2021). https://doi.org/10.1177/20503121211034366

[79]   M. Jeyaraj, G. Sathishkumar, G. Sivanandhan, D. MubarakAli, M. Rajesh, R. Arun, G. Kapildev, M. Manickavasagam, N. Thajuddin, K. Premkumar, A. Ganapathi, Biogenic silver nanoparticles for cancer treatment: An experimental report, Colloids Surfaces B Biointerfaces 106 (2013) 86–92. https://doi.org/10.1016/j.colsurfb.2013.01.027

[80]   S. Viswanathan, T. Palaniyandi, P. Kannaki, R. Shanmugam, G. Baskar, A.M. Rahaman, L.T.D. Paul, B.K. Rajendran, A. Sivaji, Biogenic synthesis of gold nanoparticles using red seaweed Champia parvula and its anti-oxidant and anticarcinogenic activity on lung cancer, Part. Sci. Technol. 41 (2023) 241–249. https://doi.org/10.1080/02726351.2022.2074926

[81]   R.B.S.M.N. Mydin, W.N. Rahman, R.M. Lazim, A. Mohd Gazzali, N.H.M. Azlan, S. Moshawih, Targeted Therapeutic Nanoparticles for Cancer and Other Human Diseases, in: Nanotechnol. Appl. Energy, Drug Food, Springer International Publishing, Cham, 2019: pp. 187–207. https://doi.org/10.1007/978-3-319-99602-8_8

[82]   A. Andleeb, A. Andleeb, S. Asghar, G. Zaman, M. Tariq, A. Mehmood, M. Nadeem, C. Hano, J.M. Lorenzo, B.H. Abbasi, A Systematic Review of Biosynthesized Metallic Nanoparticles as a Promising Anti-Cancer-Strategy, Cancers (Basel). 13 (2021) 2818. https://doi.org/10.3390/cancers13112818

[83]   S. Jain, N. Saxena, M.K. Sharma, S. Chatterjee, Metal nanoparticles and medicinal plants: Present status and future prospects in cancer therapy, Mater. Today Proc. 31 (2020) 662–673. https://doi.org/10.1016/j.matpr.2020.06.602

[84]   K.T. Nguyen, Targeted Nanoparticles for Cancer Therapy: Promises and Challenges, J. Nanomed. Nanotechnol. 02 (2011). https://doi.org/10.4172/2157-7439.1000103e

[85]   E. Fröhlich, The role of surface charge in cellular uptake and cytotoxicity of medical nanoparticles, Int. J. Nanomedicine (2012) 5577. https://doi.org/10.2147/IJN.S36111

[86]   D. Nath, P. Banerjee, Green nanotechnology – A new hope for medical biology, Environ. Toxicol. Pharmacol. 36 (2013) 997–1014. https://doi.org/10.1016/j.etap.2013.09.002

[87]   V. Sharma, P. Singh, A.K. Pandey, A. Dhawan, Induction of oxidative stress, DNA damage and apoptosis in mouse liver after sub-acute oral exposure to zinc oxide nanoparticles, Mutat. Res. Toxicol. Environ. Mutagen. 745 (2012) 84–91. https://doi.org/10.1016/j.mrgentox.2011.12.009

[88]   X. Qu, P.J.J. Alvarez, Q. Li, Applications of nanotechnology in water and wastewater treatment, Water Res. 47 (2013) 3931–3946. https://doi.org/10.1016/j.watres.2012.09.058

[89]   R.H. Taha, Green synthesis of silver and gold nanoparticles and their potential applications as therapeutics in cancer therapy; a review, Inorg. Chem. Commun. 143 (2022) 109610. https://doi.org/10.1016/j.inoche.2022.109610

[90]   M.M. Zangeneh, A. Zangeneh, Novel green synthesis of *Hibiscus sabdariffa* flower extract conjugated gold nanoparticles with excellent anti-acute myeloid leukemia effect in comparison to daunorubicin in a leukemic rodent model, Appl. Organomet. Chem. 34 (2020). https://doi.org/10.1002/aoc.5271

[91]   A. Mishra, S.J. Mehdi, M. Irshad, A. Ali, M. Sardar, M. Moshahid, A. Rizvi, Effect of Biologically Synthesized Silver Nanoparticles on Human Cancer Cells, Sci. Adv. Mater. 4 (2012) 1200–1206. https://doi.org/10.1166/sam.2012.1414

[92]   Q. Liu, F. Wu, Y. Chen, S.T. Alrashood, S.A. Alharbi, Anti-human colon cancer properties of a novel chemotherapeutic supplement formulated by gold nanoparticles containing Allium sativum L. leaf aqueous extract and investigation of its cytotoxicity and antioxidant activities, Arab. J. Chem. 14 (2021) 103039. https://doi.org/10.1016/j.arabjc.2021.103039

[93]   C. Krishnaraj, P. Muthukumaran, R. Ramachandran, M.D. Balakumaran, P.T. Kalaichelvan, Acalypha indica Linn: Biogenic synthesis of silver and gold nanoparticles and their cytotoxic effects against MDA-MB-231, human breast cancer cells, Biotechnol. Reports 4 (2014) 42–49. https://doi.org/10.1016/j.btre.2014.08.002

[94]   P.S. Kumar, M.V. Jeyalatha, J. Malathi, S. Ignacimuthu, Anticancer effects of one-pot synthesized biogenic gold nanoparticles (Mc-AuNps) against laryngeal carcinoma, J. Drug Deliv. Sci. Technol. 44 (2018) 118–128. https://doi.org/10.1016/j.jddst.2017.12.008

[95]   L.K. Ruddaraju, S.V.N. Pammi, P.N.V.K. Pallela, V.S. Padavala, V.R.M. Kolapalli, Antibiotic potentiation and anti-cancer competence through bio-mediated ZnO nanoparticles, Mater. Sci. Eng. C 103 (2019) 109756. https://doi.org/10.1016/j.msec.2019.109756

[96]   S. Sampath, Y. Madhavan, M. Muralidharan, V. Sunderam, A.V. Lawrance, S. Muthupandian, A review on algal mediated synthesis of metal and metal oxide nanoparticles and their emerging biomedical potential, J. Biotechnol. 360 (2022) 92–109. https://doi.org/10.1016/j.jbiotec.2022.10.009

[97]   S.V.P. Ramaswamy, S. Narendhran, R. Sıvaraj, Potentiating effect of ecofriendly synthesis of copper oxide nanoparticles using brown alga: antimicrobial and anticancer activities, Bull. Mater. Sci. 39 (2016) 361–364. https://doi.org/10.1007/s12034-016-1173-3

[98]   L. Dou, X. Zhang, M.M. Zangeneh, Y. Zhang, Efficient biogenesis of Cu2O nanoparticles using extract of Camellia sinensis leaf: Evaluation of catalytic, cytotoxicity, antioxidant, and anti-human ovarian cancer properties, Bioorg. Chem. 106 (2021) 104468. https://doi.org/10.1016/j.bioorg.2020.104468

[99]   P. Bhattacharya, K. Chatterjee, S. Swarnakar, S. Banerjee, Green Synthesis of Zinc Oxide Nanoparticles via Algal Route and its Action on Cancerous Cells and Pathogenic Microbes, Adv. Nano Res. 3 (2020) 15–27. https://doi.org/10.21467/anr.3.1.15-27

[100]   A. Ullah, J. Mu, F. Wang, M.W.H. Chan, X. Yin, Y. Liao, Z.A. Mirani, S. Sebt-e-Hassan, S. Aslam, M. Naveed, M.N. Khan, Z. Khatoon, M.R. Kazmi, Biogenic Selenium Nanoparticles and Their Anticancer Effects Pertaining to Probiotic Bacteria—A Review, Antioxidants 11 (2022) 1916. https://doi.org/10.3390/antiox11101916

[101]   K. Spyridopoulou, E. Tryfonopoulou, G. Aindelis, P. Ypsilantis, C. Sarafidis, O. Kalogirou, K. Chlichlia, Biogenic selenium nanoparticles produced by Lactobacillus casei ATCC 393 inhibit colon cancer cell growth in vitro and in vivo, Nanoscale Adv. 3 (2021) 2516–2528. https://doi.org/10.1039/D0NA00984A

[102]   I. Karmous, A. Pandey, K. Ben Haj, A. Chaoui, Efficiency of the Green Synthesized Nanoparticles as New Tools in Cancer Therapy: Insights on Plant-Based Bioengineered Nanoparticles, Biophysical Properties, and Anticancer Roles, Biol. Trace Elem. Res. 196 (2020) 330–342.

https://doi.org/10.1007/s12011-019-01895-0

[103]   C. Cheng, A.E. Porter, K. Muller, K. Koziol, J.N. Skepper, P. Midgley, M. Welland, Imaging carbon nanoparticles and related cytotoxicity, J. Phys. Conf. Ser. 151 (2009) 012030. https://doi.org/10.1088/1742-6596/151/1/012030

[104]   R. Govender, A. Phulukdaree, R.M. Gengan, K. Anand, A.A. Chuturgoon, Silver nanoparticles of Albizia adianthifolia: the induction of apoptosis in human lung carcinoma cell line, J. Nanobiotechnology 11 (2013) 5. https://doi.org/10.1186/1477-3155-11-5

Chapter 16

Other Applications of Biogenic Nanomaterials: Removal of Toxic Agents from Wastewater Using Biogenic Nanomaterials

F. Mutlag[1*,] H. Elaibi[1,2], R. Mahious[3], S. Celikozlu[4], F. Sen[1*]

[1]Department of Biochemistry, Sen Research Group, Kutahya Dumlupinar University, 43000, Kutahya, Turkiye

[2]Department of Pharmacy. Al Safwa University College, Karbala, 56001, Iraq

[3]Department of Biology, Kutahya Dumlupinar University, 43000, Kutahya, Turkiye

farah.mutlag@ogr.dpu.edu.tr;fatih.sen@dpu.edu.tr

Abstract

Biogenic nanomaterials (BNMs) provide a sustainable and effective method for water cleanup by decomposing organic and inorganic contaminants, eliminating heavy metals, and disinfecting harmful microbes. Water pollution from pesticides, heavy metals, and pathogens threatens human health and socio-economic progress. Biological nanomaterials (BNMs), sourced from biological origins, offer an environmentally sustainable solution for wastewater treatment, targeting pollution from industrial discharges, agricultural runoff, and urban garbage. Notwithstanding their potential, additional research is necessary to enhance large-scale applications. This article examines recent developments in BNMs for water purification, emphasizing their processes, efficacy, and future potential in environmental sustainability.

Keywords

Biogenic Nanomaterials, Pollution Removal, Wastewater Treatment, Heavy Metal Removal, Nanomaterial Applications.

1.   Introduction

Chemicals, minerals, and energy are in high demand due to rising urbanization and industrial output, among other things [1]. One inevitable effect of environmental degradation is the discharge of harmful waste into the environment caused by different human activities [2,3] The creation of environmentally friendly, cost-effective, and efficient materials and processes to lessen or eradicate contaminants and diseases in water and wastewater is a highly sought-after goal [4]. The ever-growing presence of harmful microbes and organic pollutants has elevated wastewater management to the forefront of international attention. Because these toxins are already in the world, we need to find ways to reduce their danger to people and the environment without compromising safety [5,6].

Biological sources such as bacteria, fungi, algae, and plant extracts are utilized to produce biogenic nanoparticles, representing an innovative and eco-friendly method for the elimination of hazardous substances from wastewater [7][8][9][10]. The unique chemical characteristics and vast surface area of these materials permit the efficient absorption and destruction of contaminants [11].

Nanotechnology has emerged as a realistic and novel technique to wastewater treatment in recent years, providing improved answers to traditional methods. One major, environmentally beneficial, and quite successful approach is the use of biogenic nanomaterials [12–14]. Biogenic nanomaterials are created by a complicated system of biological components, including plants, bacteria, and enzymes [15,16]. These nanoparticles are extremely effective at capturing and solubilizing dangerous materials. It has various remarkable physicochemical features, including selective adsorption, and a large surface area [17] . Recent advances in biotechnology and nanotechnology have made it easier to remove toxic metals, herbicides, medications, and colors from wastewater, increasing the potential utility of these goods. This approach eliminates a variety of contaminants through redox interactions, catalysis, and the adsorption properties of biogenic nanoparticles [18].

Heavy metals such as lead, mercury, cadmium, and arsenic are retained by biogenic nanomaterials through strong interactions, thereby preventing their detrimental effects on human health and water quality. Organic contaminants, such as pesticides, pharmaceuticals, and dyes, are converted into non-toxic molecules through the use of biogenic nanomaterials and catalytic and enzymatic processes [19,20].

Biogenic nanomaterials also eliminate bacteria, viruses, and protozoa by breaching their cell walls or generating reactive oxygen species that induce microbial inactivation, thereby fulfilling another critical function[21]. Biogenic nanoparticles are suitable for large-scale water treatment applications, in contrast to conventional chemical treatments, due to their sustainable generation and absence of all adverse effects [22].

Though they seem very promising, further research is required to make them more dependable and helpful in daily life [23]. The scientific community wants everyone to have access to clean water and seeks for steady development. This is achieved by bettering the long-term usage of these small particles in wastewater cleansing.

1.   Synthesis of Biogenic Nanomaterials

Biogenic nanomaterials are created by living organisms like bacteria, fungus, algae, and plants. They have special chemical and physical features [24]. Since this green synthesis method replaces dangerous chemicals and energy-intensive procedures, it is preferred for its various benefits: durability, cost-effectiveness, and minimum environmental impact [25].

This technique is environmentally favorable and produces sustainable nanomaterials without the use of hazardous chemicals or a significant amount of energy. Bacteria and fungi are employed to generate nanomaterials through the processes of biomineralization and biological reduction [26]. Microorganisms aid metal ions transform into stable nanoparticles by producing enzymes and proteins. Depending on the metabolic pathways involved, this action may occur within or outside of microbial cells [27].

Some bacteria and fungi are extremely efficient at producing metallic and metal oxide nanoparticles due to their metal-resistant systems, which allow them to collect and convert metal ions into nanostructures [28].

Natural solvents and stabilizers derived from plants, such as leaves, roots, flowers, and seeds, are employed in plant-based synthesis. Nanoparticles rely on unique plant compounds such as terpenoids, alkaloids, flavonoids, and tannins [29]. These biomolecules aid in the reduction of metal ions and the formation of nanoparticles when combined with liquids containing metal ions. One main advantage of this approach is the fast, cheap, shape- and size-adjustable synthesis of stable nanoparticles [30].

By means of bioaccumulation and bioreduction systems, algal-mediated processes involving both unicellular and multicellular algae can produce nanomaterials. Pigments, polysaccharides, and proteins made by algae help nanoparticles to be reduced and stabilized [31]. This method enables the synthesis of metal and metal oxide nanoparticles, which have potential applications in biological and environmental sciences. Large-scale production finds algae perfect because of their power to absorb heavy metals and their adaptability in numerous conditions [32].

Enzyme-assisted synthesis uses certain enzymes to help nanomaterials to be created. Enzymes derived from plants or microbes among other biocatalysts help to produce nanostructures under control. This method enables exact synthesis of nanoparticles by facilitating control over their size, crystalline, and form. Nitrate reductase, hydrogenase, and oxidoreductases are enzymes that facilitate the synthesis of stable nanoparticles through the reduction of metal ions [33,34].

Several variables, such as pH, temperature, reaction duration, and precursor concentration, influence the biogenic nanomaterial production process. Reaching ideal results with these criteria guarantees the synthesis of nanomaterials displaying the necessary physicochemical characteristics for different uses [27,28]. Apart from reducing negative environmental effects, this green manufacturing method improves the biocompatibility of nanomaterials, therefore generating interesting prospects for their use in several fields, including energy storage, environmental restoration, and medicine [37], Table 1 shows the main approaches for the manufacture of biogenic nanoparticles.

2.1 Plant-Mediated Synthesis

Among numerous biomolecules, plants may independently synthesis alkaloids, flavonoids, terpenoids, and phenolic chemicals [38,39]. Plant extracts can be produced by crushing or boiling plant parts including leaves, flowers, seeds, and roots [40]. Metal ions, including silver (Ag), gold (Au), and zinc oxide (ZnO), are converted into nanoparticles using phytochemicals included in plant-mediated synthesis agents such as *Eichhornia crassipes* (water hyacinth), *Azadirachta indica* (Neem), *Ocimum sanctum* (holy basil) extract, and *Moringa oleifera* [41,42].

2.2 Microbial Synthesis

The manufacture of nanomaterials significantly utilizes microorganisms, including bacteria, fungi, and algae. Metal ions can be spontaneously transformed into nanoparticles via metabolic pathways or enzymatic activity [43]. Microbial organisms emit biomolecules, including proteins, enzymes, and secondary metabolites, which function as reducing and stabilizing agents [44] .

Bacteria, including *Escherichia coli* and *Pseudomonas aeruginosa*, synthesize silver and gold nanoparticles. Fungi, such as *Aspergillus niger* and its related species, generate rather stable nanoparticles by the action of extracellular enzymes [45]. Microalgae utilize photosynthesis to produce highly stable nanoparticles, facilitating the nanoparticle manufacturing process. The biological source utilized influences the effectiveness of nanoparticles in pollutant removal, hence impacting the morphology, stability, and reactivity of the resultant nanomaterials [46].

2.3   Enzyme-Mediated Synthesis

By reducing metal ions, enzymes, whether intracellular or extracellular, can effectively stabilize the nanoparticles that are produced. In addition to expediting the reduction of metal ions, enzymes facilitate the production of precise nanoparticles. The formation of gold and silver nanoparticles is facilitated by the enzyme nitrate reductase, which is present in numerous bacteria and plants [47,48].

2.4   Biomineralization

Biomineralization is the biological process via which organisms precipitate minerals and synthesize nanomaterials [49]. This technique emulates natural phenomena, including the formation of shells by mollusks and the development of bones in vertebrates. Magnetotactic bacteria synthesize magnetic iron oxide nanoparticles (magnetite) within their cells [50].

2.5   Hybrid or Synergistic Methods

Sometimes to boost the efficiency of a process, biogenic nanomaterials are created by combining plant extracts, microbial systems, and enzymes. Example multi-metal nanoparticles by combining plant extracts with microbial fermenting [51].

Table 1. A comparison of different methods for making biogenic nanomaterials.

Synthesis Method	Mechanism	Examples of Nanoparticles	Key Advantages	Key Limitations	References
Microbial Synthesis	Microorganisms use metabolic or enzymatic processes to reduce metal ions into nanoparticles.	Silver (*Escherichia coli*) Gold (*Pseudomonas aeruginosa*) Iron oxide (*Aspergillus niger*)	Eco-friendly Scalable Good shape and size control	Time-intensive Sensitive to growth conditions	[41,42]
Plant-Based Synthesis	Bioactive compounds (e.g., alkaloids, phenolics) in plant extracts act as reducing and capping	Silver (*Azadirachta indica*) Gold (*Ocimum sanctum*) Zinc oxide (*Moringa oleifera*)	Rapid Renewable raw materials No toxic chemicals	Variability in plant composition Requires optimization	[43,44]

	agents.				
Enzyme-Mediated Synthesis	Enzymes catalyze the reduction of metal ions and stabilize nanoparticles.	Gold (via nitrate reductase) Silver (via reductase enzymes)	High specificity Controlled particle morphology	Costly enzyme production Limited scalability	[45,46]
Biomineralization	Organisms naturally precipitate metal ions into nanomaterials through biomineralization pathways.	Magnetic iron oxide (Magnetotactic bacteria)	Highly sustainable Precise particle control	Limited material types Requires specialized organisms	[47,48]
Hybrid Approaches	Combines microbial, plant, or enzymatic methods to enhance synthesis efficiency and nanoparticle properties.	Multi-metal nanoparticles using Pseudomonas with plant extracts	Versatile Customizable Enhanced yields	Complex optimization Higher production costs	[60,61]

Table 1 delineates five distinct strategies for the production of biogenic nanomaterials, encompassing their mechanisms, examples, advantages, and disadvantages. Microbial and plant-based technologies are ecologically sustainable and scalable; nevertheless, enzyme-mediated and biomineralization methods offer precision but have restricted applicability. Hybrid techniques use the benefits of both approaches, resulting in heightened complexity and expense.

3. Mechanisms of Toxic Agent Removal

Biogenic nanoparticles are great in eliminating harmful substances from wastewater because of their distinctive physical and chemical characteristics which include a big surface area, the capacity to be modified for specialized needs, and enhanced reactivity [62].

By means of physical, chemical, and biological interactions between pollutants and biogenic nanoparticles, several techniques assist to remove harmful substances from wastewater. Adsorption, catalytic degradation, ion exchange, precipitation, redox reactions, and microbial inactivation are essential processes in water purification [63,64].

Adsorption is among the most efficient processes for water filtration. Biogenic nanomaterials have high interactions with contaminants owing to the multitude of functional groups on their surface, including hydroxyl, carboxyl, and amine groups [18]. Heavy metals like lead, cadmium, arsenic, and mercury adhere to nanomaterial surfaces by electrostatic attraction, van der Waals forces, or chelation. Hydrophobic interactions and π-π stacking augment the affinity of organic contaminants for nanoparticles, hence promoting their elimination. This encompasses pigments, agrochemicals, and pharmaceuticals [65].

The mitigation of the toxicity of complex compounds relies on a fundamental process known as catalytic degradation. Biogenic nanoparticles serve as effective oxidation and reduction catalysts for the removal of organic contaminants, including pesticides, industrial dyes, and pharmaceutical residues. Illuminating biological metal oxides and other photocatalytic nanomaterials generates reactive oxygen species. These microbes oxidize contaminants to produce benign chemicals such as water and carbon dioxide [66]. Biogenic nanoparticles are biomolecules that promote the biological oxidation of specific deleterious chemicals. Furthermore, these biomolecules facilitate the breakdown of compounds by enzymatic action [67,68].

Ion exchange is the process of substituting less hazardous ions produced by nanomaterials with more toxic metal ions in wastewater [69]. This method is especially efficient in eliminating heavy metals such as lead and arsenic by replacing them with sodium, potassium, or calcium ions. Biogenic nanomaterials can enhance pollution removal efficiency and ion exchange capacity by modifying their surface charge and chemical composition [70].

A fundamental method by which biogenic nanomaterials enable the sedimentation of insoluble metal complexes from water is precipitation. This happens when nanoparticles interact with dissolved metal ions to produce solid precipitates that might be easily removed with sedimentation or filtration. Living components such proteins, polysaccharides, and enzymes included in biogenic nanomaterials help to promote nucleation and crystal formation, therefore enhancing the process.[71,72]

Redox reactions are the movement of electrons between hazardous contaminants and biogenic nanomaterials that cause their change into less harmful shapes [73]. Heavy metals, such as mercury and chromium, can exist in multiple oxidation states; yet, certain states are more perilous than others [74]. Biogenic nanomaterials promote the reduction of these metals to less toxic forms via electron donation. This technique is advantageous for detoxifying hexavalent chromium, which is transformed into the less toxic trivalent chromium by nanomaterial-mediated reduction.[75]

One of the key methods for eliminating pathogens from wastewater is microbial inactivation [76]. Strong antibacterial effects of biogenic nanomaterials are achieved by means of rupture of microbial cell walls, generation of reactive oxygen species, and interference with cellular metabolism among several channels [77]. Silver, copper, and zinc oxide metal-based nanomaterials release ions that enter microbial cells and target essential macromolecules, therefore killing the cells [78]. During bacterial respiration, certain nanomaterials disrupt electron transport systems, therefore reducing energy generation and resulting in microbial inactivity.

Often working jointly, these systems help biogenic nanomaterials for wastewater treatment to be more generally effective. Large-scale water purification users find these materials perfect since they can adsorb, decompose, and neutralize dangerous compounds with low environmental impact [79,80], as seen in Table 2.

3.1   Adsorption

Biogenic nanomaterials possess a substantial surface area, elevated porosity, and functional groups such as hydroxyl and carboxyl, endowing them with exceptional adsorption properties. Due to their properties, they can physically or chemically bind contaminants [81]. For example, nanoparticles of iron oxide or zinc oxide generated during biosynthesis can sequester heavy metals including arsenic, cadmium, and lead [82]. Electrostatic interactions, hydrogen bonds, and van der Waals forces are the interactions that bind chemical dyes and organic toxins. This procedure is comprehensible and effective for a diverse range of pollutants [83].

3.2   Catalysis

3.2.1   Photocatalysis

Upon exposure to sunshine or UV radiation, nanoparticles like TiO_2 and ZnO generate reactive oxygen species that facilitate the degradation of organic pollutants and pigments [84].

3.2.2   Redox Reactions

Biologically active nanomaterials can repair contaminants via redox reactions by either donating or receiving electrons. Nanoscale zero-valent iron (nZVI) can convert hexavalent chromium (Cr(VI)) into the less hazardous trivalent form (Cr(III)) [85,86]. This technique is particularly efficacious for the purification of wastewater containing metals or persistent organic pollutants, which present a considerable challenge. Additionally, metallic BNMs augment electron transfer pathways, transforming poisonous chemicals into less damaging variants [87].

3.2.3   Fenton-Like Reactions

Hazardous organic chemicals in wastewater are broken down by reactive oxygen species (ROS) generated by iron-based nanoparticles. Industrial effluents often contain phenolic compounds; nevertheless, biogenic iron oxide nanoparticles can facilitate their degradation [88].

3.3   Antibacterial Action

Numerous biogenic nanoparticles are highly effective in treating wastewater, particularly in eliminating dangerous microbes due to their antibacterial properties [7,8]. Rationale: The infiltration of silver, copper, and zinc oxide nanoparticles into bacterial membranes impairs enzyme performance and produces reactive oxygen species (ROS), resulting in bacterial death [89]. Silver nanoparticles produced from plant extracts exhibit potential against *E. coli* and *Pseudomonas aeruginosa*. This aids in eradicating potentially hazardous bacteria from wastewater, hence ensuring safety for subsequent use or disposal [90].

3.4   Precipitation

Biogenic nanomaterials augment chemical precipitation by converting heavy metals into insoluble forms that can be readily eliminated [91–93]. Iron oxide nanoparticles transform chromium and arsenic into insoluble hydroxides, facilitating their precipitation [94]. This technique effectively eliminates heavy metals from aquatic environments, thereby diminishing both toxicity and mobility [95].

3.5   Ion Exchange

Certain biogenic nanoparticles can sequester harmful ions and liberate beneficial ones via a mechanism known as ion exchange with contaminants in wastewater [96]. Biogenically generated nanomaterials derived from zeolite can replace hazardous heavy metal ions such as mercury or lead [97]. The significance of this method lies in its efficacy in eliminating toxic metals from water while generating minimal secondary contamination [98].

Table 2. Key mechanisms of toxic agent removal using biogenic nanomaterials.

Mechanism	Description	Examples	Key Applications	References
Adsorption	Nanomaterials bind contaminants via surface interactions.	Heavy metals (Pb, Cd), dyes, organic pollutants.	Simple, widely applicable for pollutants.	[99,100]
Catalysis	Nanoparticles facilitate the breakdown of toxic substances through redox, photocatalysis, or Fenton-like reactions.	Iron oxide (Fenton), zinc oxide (photocatalysis), gold (redox).	Degradation of pesticides, dyes, pharmaceuticals.	[101]
Ion Exchange	Replaces toxic ions with benign ones.	Zeolite nanomaterials replace Hg or Pb with Na or K ions.	Effective for heavy metal removal.	[102,103]
Precipitation	Converts toxic agents into insoluble forms for easy removal.	Iron oxide converts arsenic or chromium into insoluble hydroxides.	Heavy metal detoxification.	[104,105]
Antimicrobial Activity	Disrupts microbial membranes, enzyme activity, or generates ROS to kill pathogens.	Silver, zinc oxide, and copper oxide nanoparticles.	Pathogen removal in wastewater.	[92,93,94]

Table 2 shows Biogenic nanoparticles can eliminate deleterious compounds from wastewater by multiple mechanisms, such as adsorption, catalysis, ion exchange, precipitation, antimicrobial activity, and redox reactions. These processes aim to eliminate heavy metals, organic pollutants, and pathogens to purify water effectively and sustainably.

4.   Applications of Biogenic Nanomaterials in Wastewater Toxic Agent Removal

Biogenic nanoparticles are fundamental for the effective elimination of heavy metals, organic contaminants, pathogens, and excess nutrients in wastewater treatment [109]. They guarantee safe, pure water and provide a long-lasting remedy for water pollution because of their adaptability, economy, and environmental sustainability [110].

4.1   Heavy Metal Adsorption

Biogenic nanoparticles have developed as an effective way for eliminating dangerous substances, particularly heavy metals, from wastewater. Heavy metals like lead, mercury, cadmium, arsenic, and chromium are poisonous and persistent, posing substantial environmental and health dangers. Biogenic nanomaterials, which are created naturally by bacteria, fungi, plants, and algae, provide an environmentally benign alternative with strong metal adsorption and removal capabilities [111,112].

These nanostructures can remove heavy metals from water in a variety of ways. Biogenic nanoparticles have a large surface area and functional groups, which allow them to attract and hold onto metal ions. This process is called adsorption, and it makes the metal ions immobile [113]. Certain biogenic nanomaterials can replace hazardous metal ions with safer substitutes in order to regulate the elimination process. Some nanomaterials make it easier to remove dangerous metals by lowering their toxicity or making them insoluble, which helps with chemical reduction and precipitation, two processes that are important for this purpose [112]. Heavy metals are less dangerous and less mobile in water environments because they bioaccumulate in microbial nanomaterials that are formed by bacteria and fungi [114]

Water treatment requires a diverse array of biogenic nanoparticles. Metallic nanoparticles, particularly zero-valent iron, are effective at converting arsenic and chromium into less harmful forms, while silver nanoparticles are highly good at binding with lead and mercury [111,115]. Metal oxide nanoparticles, such as magnetite, have the advantage of magnetic separation following contaminant adsorption, but titanium dioxide causes metal breakdown when exposed to light. Carbon-based nanomaterials with large surface areas, such as carbon dots and graphene oxide, allow for superior metal trapping. Furthermore, the capacity to bind heavy metals makes microbial nanomaterials generated from fungi and bacteria useful for bioremediation [113,116].

Biogenic nanoparticles provide numerous advantages over present treatment techniques. They are environmentally friendly because they are made from natural materials and contain no hazardous chemical residues. Their high adsorption capacity allows for the absorption of small amounts of heavy metals, resulting in cleaner water. They are more cost effective than produced nanoparticles and biodegradable, lowering the risk of secondary pollution. Certain materials can sometimes be more sustainable than others because of their capacity to be recycled and reused [117,118].

4.2   Dye Removal

Major environmental problems brought on by industrial operations such printing, textile manufacture, and medicines include the elimination of colors from effluent. In this sense, biogenic nanomaterials offer great potential. Because they are so highly poisonous, non-biodegradable, and resistant to standard wastewater treatment systems, synthetic colors contaminate water and compromise aquatic ecosystems and human health [119,120]. Biogenic nanomaterials, which are derived from natural sources such as bacteria, fungus, plants, and algae, provide an effective and environmentally favorable method for removing color from dye sources.

There are numerous methods by which biogenic nanoparticles can eliminate pigments. The adsorption process is one of the most effective methods, as the nanomaterials' high surface area and functional groups bond dye molecules, preventing further contamination and effectively trapping them. Some biogenic nanoparticles use photocatalysis, which involves absorption of light energy by nanomaterials such as zinc oxide and titanium dioxide breaks down complicated dye molecules into less damaging counterparts Redox processes are also vital since biogenic nanoparticles chemically change the structure of dye molecules such that they either disintegrate or precipitate. Biosorption is sometimes utilized by microbial nanomaterials, wherein bacteria or fungus-derived nanoparticles sequester pigment molecules, rendering them immobile through biological interactions.

A range of biogenic nanoparticles is utilized for the removal of colors from wastewater. Silver and gold are metal nanoparticles that demonstrate significant adsorption capacity and catalytic activity, facilitating the degradation of dye molecules [121].

4.3   Organic Pollutant Degradation

Biogenic nanoparticles have evolved as a sustainable and effective way to break down organic pollutants in wastewater, therefore addressing the mounting environmental issues associated to industrial and agricultural activities. Organic pollutants such as pesticides, drugs, petroleum hydrocarbons, and industrial solvents pose severe hazards to aquatic ecosystems and human health because of their toxicity and endurance. Long-lasting pollution happens because standard wastewater treatment methods can't completely break down these harmful substances. Biogenic nanoparticles are an eco-friendly and effective way to break down harmful organic materials into safer substances. They come from natural sources like plants, bacteria, mushrooms, and algae [4,122].

Biogenic nanoparticles can breakdown organic pollutants in many ways. In photocatalysis, one of the more effective techniques, nanomaterials such zinc oxide and titanium dioxide absorb light energy and generate reactive species that break down complicated organic compounds into less harmful, simpler forms. Adsorption helps organic pollutants to be trapped on the surface of nanoparticles, therefore allowing their slow decomposition. Redox reactions especially help to eliminate contaminants by chemically altering their structures, reducing their toxicity, and promoting biodegradation [15,123,124].

Another crucial method is enzymatic catalysis, in which a range of organic contaminants can be broken down by microbial nanomaterials produced from fungus and bacteria. Small natural particles are used to clean wastewater by breaking down harmful organic materials. Some metal nanoparticles that can successfully break down harmful substances include zero-valent iron, gold, and silver. [125,126].

Metal oxide nanoparticles, including magnetite, zinc oxide, and titanium dioxide, are frequently employed to degrade persistent organic pollution due to their photocatalytic and adsorption qualities [127]. Carbon-based nanoparticles like graphene oxide and carbon dots have a large surface area that helps move electrons and absorb substances, which can assist in breaking down harmful toxins [128,129]. Microbial nanomaterials, created by bacteria, fungi, and algae, provide an eco-friendly way to clean up organic pollutants by using natural biological processes.

4.4   Pathogen Inactivation

The high efficiency of biogenic nanoparticles in inactivating bacteria in wastewater addresses a major environmental and public health issue. Bacteria, viruses, and fungi are among the harmful microorganisms commonly found in wastewater, and if left untreated, they can spread infectious diseases. Conventional disinfection techniques including UV treatment and chlorination could have negative effects including inadequate pathogen elimination and the generation of toxic byproducts [130]. Derived from natural sources including bacteria, fungi, algae, and plants, biogenic nanomaterials provide an efficient and environmentally benign approach of eradicating diseases from contaminated water sources [131].

Biogenic nanoparticles could deactivate many different types of infections using several approaches. When nanoparticles and microbial cells interact closely, the cell membrane may be physically damaged, allowing significant components to flow out and finally causes cell death [132]. Reactive oxygen species produced by certain nanomaterials can damage microbial structures and prevent their completion of biological roles. Released by nanoparticles like copper and silver, metal ions change cellular metabolism and enzyme function, therefore neutralizing infections [133]. Adsorption is required since it immobilizes and traps pathogens on the surface of the nanomaterial so preventing their development and dissemination. Natural antibacterial drugs generated by specific microbial nanoparticles also help to inactivate pathogens [134].

Different biogenic nanoparticles help to deactivate microorganisms in wastewater. Metal nanoparticles are effective antibacterial agents because of the way they interact with the metabolic activity and membranes of germs. Metal oxide nanoparticles, including zinc oxide, titanium dioxide, and magnetite, create reactive oxygen species that can kill microbes and viruses [135]. Carbon nanomaterials, such as graphene oxide and carbon nanoparticles, have vast surface areas, which help them collect and deactivate microorganisms. Microbial nanomaterials are made up of bacteria, fungi, and algae, and they contain bioactive compounds that kill bacteria naturally [136].

5.   Advantages and Disadvantages of Bio-Nanomaterials (BNMs)

5.1   Advantages

The synthesis of biogenic nanomaterials utilizes renewable biological resources, including plants, microbes, and enzymes. This mitigates the adverse effects of nanomaterial production on the environment by decreasing the utilization of non-renewable chemical reagent [137].

Biogenic synthesis is generally conducted under benign settings, unlike conventional nanomaterial synthesis methods that frequently require elevated temperatures and toxic chemicals. This results in reduced environmental harm and energy usage [138].

Biogenic nanomaterials have elevated surface reactivity owing to their substantial active surface area in relation to volume, facilitating pollutant elimination [139]. Consequently, they may enhance their capacity to absorb or eliminate hazardous compounds, so expediting the removal of contaminants from water and air, as seen in Table 3.

5.2   Disadvantages

Properties may vary due to biological factors and environmental effects. The protracted duration of biologic synthesis renders it inefficient for large-scale production ; Certain BNMs may ultimately cease functioning or become ineffective under specific conditions [140,141].

Table 3. Advantages and disadvantages of bio-nanomaterials (BNMs)

Advantages	Disadvantages	References
Environmental Sustainability: Derived from renewable biological resources, reducing environmental impact.	Variability in Properties: Properties can vary depending on biological sources and environmental conditions.	[142,143]
Low Toxicity and Eco-Friendliness: Non-toxic and safe for use in sensitive applications like water treatment and medicine.	Scalability Issues: Scaling up the synthesis process for industrial use can be challenging.	[144,145]
Cost-Effectiveness: Requires less energy and fewer chemicals compared to traditional chemical synthesis methods.	Longer Synthesis Time: Biogenic synthesis is slower, making it less efficient for large-scale production.	[146,147]
High Surface Area and Reactivity: Large surface area enhances reactivity, improving their effectiveness in various applications.	Limited Control Over Nanoparticle Morphology: Achieving precise control over size, shape, and surface properties can be difficult.	[148,149]
Versatility in Applications: Can be used in various fields such as water purification, medicine, agriculture, and electronics.	Stability and Shelf-Life Issues: Some BNMs may degrade or lose efficacy over time under certain environmental conditions.	[150,151]

Table 3 shows useful biogenic nanomaterials (BNMs) are low-toxic, reasonably priced, environmentally beneficial ones. These features help them in environmental and medical therapy. Property variability, scale, long synthesis periods, and nanoparticle morphological control have to be resolved before industrial application.

6.   Future Directions of Biogenic Nanomaterials (BNMs)

Biogenic nanoparticles possess significant promise across various industries. Improving synthesis methods will be a priority. Researchers are seeking methods to enhance production dependability and scalability. Enhancing biological resources and cultivation conditions will facilitate the large-scale production of high-quality, consistent biogenic nanomaterials. Modifying nanomaterials is essential. Biogenic nanomaterials can be altered in dimensions, morphology, and surface characteristics to enhance efficacy and broaden their uses. These advancements may facilitate environmental remediation, sensing technology, and individualized therapy. Biogenic nanoparticles enhance green technologies; nonetheless, sustainability continues to pose difficulty. These materials enhance waste management and mitigate industrial environmental effects. They address global environmental challenges by offering a more sustainable nanomaterial alternative derived from renewable resources.

7. Conclusion

Due to the increasing pollution levels caused by agricultural and industrial operations, it is critical to remove dangerous substances from wastewater. These nanoparticles can clean water by eliminating harmful contaminants and heavy metals. They not only reduce garbage accumulation and encourage a circular economy, but they also help to valorize waste by transforming agricultural or industrial leftovers into functional nanoparticles. Despite the abundance of accessible solutions, challenges exist. Investigations are undertaken to increase manufacturing capacity while maintaining quality and consistency. Well-defined regulatory frameworks are becoming increasingly important in protecting human health and the environment from nanoparticle exposure. This study looks into the use of BNMs, and the procedures used to remove organic contaminants, decolorize things, and sequester heavy metals. We analyze potential environmental implications, realistic scalability, and any further problems. Biogenic nanomaterials are gaining traction as a potential long-term solution to these issues.

References

[1]   V. Saxena, Water Quality, Air Pollution, and Climate Change: Investigating the Environmental Impacts of Industrialization and Urbanization, Water, Air, Soil Pollut. 236 (2025) 1–40. https://doi.org/10.1007/s11270-024-07702-4

[2]   G.I. Edo, L.O. Itoje-akpokiniovo, P. Obasohan, V.O. Ikpekoro, P.O. Samuel, A.N. Jikah, L.C. Nosu, H.A. Ekokotu, U. Ugbune, E.E.A. Oghroro, Impact of environmental pollution from human activities on water, air quality and climate change, Ecol. Front. (2024). https://doi.org/10.1016/j.ecofro.2024.02.014

[3]   K.D. Patil, J. De, V.K. Patil, M.M. Kulkarni, Environmental Effects and Threats of Waste: Understanding Threats and Challenges to Ecosystem, Health, and Sustainability and Mitigation Strategies, in: From Waste to Wealth, Springer, 2024: pp. 37–69. https://doi.org/10.1007/978-981-99-7552-5_3

[4]   H. Sable, V. Kumar, V. Singh, S. Rustagi, S. Chahal, V. Chaudhary, Strategically engineering advanced nanomaterials for heavy-metal remediation from wastewater, Coord. Chem. Rev. 518 (2024) 216079. https://doi.org/10.1016/j.ccr.2024.216079

[5]   S. V Rathod, P. Saras, S.M. Gondaliya, Environmental pollution: threats and challenges for management, in: Eco-Restoration Polluted Environ., CRC Press, 2025: pp. 1–34. https://doi.org/10.1201/9781003423393-1

[6]   M. Awasthi, K. Vaibhav, A.K. Choudhary, A.K. Gautam, A. Chandra, E-Waste Management: An Essential Deed to Safeguard Future, Glob. Waste Manag. (2025) 85–114. https://doi.org/10.1002/9781394318414.ch5

[7]   A. Aygun, F. Gulbagca, E.E. Altuner, M. Bekmezci, T. Gur, H. Karimi-Maleh, F. Karimi, Y. Vasseghian, F. Sen, Highly Active PdPt Bimetallic Nanoparticles Synthesized By One Step Bioreduction Method: Characterizations, Anticancer, Antibacterial Activities and Evaluation of Their Catalytic Effect for Hydrogen Generation, Int. J. Hydrogen Energy 48 (2023) 6666–6679. https://doi.org/10.1016/j.ijhydene.2021.12.144

[8]   F. Gulbagca, A. Aygün, M. Gülcan, S. Ozdemir, S. Gonca, F. Şen, Green synthesis of palladium nanoparticles: Preparation, characterization, and investigation of antioxidant, antimicrobial, anticancer,

and DNA cleavage activities, Appl. Organomet. Chem. 35 (2021) e6272.
https://doi.org/10.1002/AOC.6272

[9]   A. Hojjati-Najafabadi, A. Aygun, R.N.E. Tiri, F. Gulbagca, M.I. Lounissaa, P. Feng, F. Karimi, F.
Sen, Bacillus thuringiensis Based Ruthenium/Nickel Co-Doped Zinc as a Green Nanocatalyst: Enhanced
Photocatalytic Activity, Mechanism, and Efficient H2Production from Sodium Borohydride
Methanolysis, Ind. Eng. Chem. Res. 62 (2023) 4655–4664. https://doi.org/10.1021/ACS.IECR.2C03833

[10]   K. Nesrin, C. Yusuf, K. Ahmet, S.B. Ali, N.A. Muhammad, S. Suna, Ş. Fatih, Biogenic silver
nanoparticles synthesized from Rhododendron ponticum and their antibacterial, antibiofilm and cytotoxic
activities, J. Pharm. Biomed. Anal. 179 (2020) 112993. https://doi.org/10.1016/J.JPBA.2019.112993

[11]   P. Rath, L.K. Bhardwaj, P. Yadav, A.K. Bhardwaj, A Synthesis of Biogenic Nanoparticles (NPs)
for the Treatment of Wastewater and Its Application: A Review, Biog. Wastes-Enabled Nanomater.
Synth. Appl. Environ. Sustain. (2024) 127–148. https://doi.org/10.1007/978-3-031-59083-2_5

[12]   G.S. El-Sayyad, D. Elfadil, M.A. Mosleh, Y.A. Hasanien, A. Mostafa, R.S. Abdelkader, N.
Refaey, E.M. Elkafoury, G. Eshaq, E.A. Abdelrahman, Eco-friendly strategies for biological synthesis of
green nanoparticles with promising applications, Bionanoscience 14 (2024) 3617–3659.
https://doi.org/10.1007/s12668-024-01494-x

[13]   A. Rasool, S. Sri, M. Zulfajri, F.S.H. Krismastuti, Nature inspired nanomaterials, advancements
in green synthesis for biological sustainability, Inorg. Chem. Commun. (2024) 112954.
https://doi.org/10.1016/j.inoche.2024.112954

[14]   Y. Liang, H. Demir, Y. Wu, A. Aygun, R.N. Elhouda Tiri, T. Gur, Y. Yuan, C. Xia, C. Demir, F.
Sen, Y. Vasseghian, Facile synthesis of biogenic palladium nanoparticles using biomass strategy and
application as photocatalyst degradation for textile dye pollutants and their in-vitro antimicrobial activity,
Chemosphere 306 (2022) 135518. https://doi.org/10.1016/J.CHEMOSPHERE.2022.135518

[15]   A. Hojjati-Najafabadi, S. Salmanpour, F. Sen, P.N. Asrami, M. Mahdavian, M.A. Khalilzadeh, A
Tramadol Drug Electrochemical Sensor Amplified by Biosynthesized Au Nanoparticle Using Mentha
aquatic Extract and Ionic Liquid, Top. Catal. 65 (2022) 587–594. https://doi.org/10.1007/s11244-021-
01498-x

[16]   A. Aygün, S. Özdemir, M. Gülcan, K. Cellat, F. Şen, Synthesis and Characterization of Reishi
Mushroom Mediated Green Synthesis of Silver Nanoparticles for the Biochemical Applications, J. Pharm.
Biomed. Anal. 178 (2020) 112970. https://doi.org/10.1016/j.jpba.2019.112970

[17]   P. V Mane, R.M. Rego, P.L. Yap, D. Losic, M.D. Kurkuri, Unveiling cutting-edge advances in
high surface area porous materials for the efficient removal of toxic metal ions from water, Prog. Mater.
Sci. (2024) 101314. https://doi.org/10.1016/j.pmatsci.2024.101314

[18]   S.E. Sanni, B.A. Oni, E.E. Okoro, S. Pandya, Recent advances in the use of biogenic
nanomaterials and photocatalysts for wastewater treatment: challenges and future prospects, Front.
Nanotechnol. 6 (2024) 1469309. https://doi.org/10.3389/fnano.2024.1469309

[19]   Y. Wu, E.E. Altuner, R.N. El Houda Tiri, M. Bekmezci, F. Gulbagca, A. Aygun, C. Xia, Q. Van
Le, F. Sen, H. Karimi-Maleh, Hydrogen generation from methanolysis of sodium borohydride using
waste coffee oil modified zinc oxide nanoparticles and their photocatalytic activities, Int. J. Hydrogen
Energy 48 (2023) 6613–6623. https://doi.org/10.1016/J.IJHYDENE.2022.04.177

[20]   A.O. Iyiola, M.O. Ipinmoroti, O.O. Akingba, J.S. Ewutanure, S.B. Setufe, J. Bilikoni, E. Ofori-
Boateng, O.M. Wangboje, Organic chemical pollutants within water systems and sustainable management
strategies, in: Water Cris. Sustain. Manag. Glob. South, Springer, 2024: pp. 211–251.
https://doi.org/10.1007/978-981-97-4966-9_7

[21]   H.N. Onyeaka, O.F. Nwabor, Food preservation and safety of natural products, Academic Press,
2022. https://doi.org/10.1016/B978-0-323-85700-0.00005-8

[22]   L.E. Mofokeng, E. Makhado, P. Ndungu, A comprehensive review on biogenic metallic, metal

oxide and metal sulfide nanoparticles for water treatment and sensing applications, J. Ind. Eng. Chem. (2024). https://doi.org/10.1016/j.jiec.2024.09.039

[23]   V. Rocha, A. Lago, B. Silva, Ó. Barros, I.C. Neves, T. Tavares, Immobilization of biogenic metal nanoparticles on sustainable materials–green approach applied to wastewater treatment: a systematic review, Environ. Sci. Nano (2024). https://doi.org/10.1039/D3EN00623A

[24]   S.S. Salem, A.E. Mekky, Biogenic nanomaterials: Synthesis, characterization, and applications, in: Biog. Nanomater. Environ. Sustain. Princ. Pract. Oppor., Springer, 2024: pp. 13–43. https://doi.org/10.1007/978-3-031-45956-6_2

[25]   F. Samuel, Sustainability in Electrocoagulation for Water Treatment: Green innovations and eco-friendly applications., (2025).

[26]   S. Kaur, G. Setia, M. Sikenis, S. Kumar, Synthesis of Biogenic Nanomaterials, Their Characterization, and Applications, in: Biog. Nanomater. Environ. Sustain. Princ. Pract. Oppor., Springer, 2024: pp. 45–75. https://doi.org/10.1007/978-3-031-45956-6_3

[27]   D. Das, K. Verma, S. Tangjang, A Comprehensive Perspective of Conventional and Biogenic Nanoparticles, in: Biog. Nanomater. Heal. Environ., CRC Press, 2024: pp. 1–17. https://doi.org/10.1201/9781003430087-1

[28]   M. Kumar, A. Mathur, R.P. Singh, Biogenic Nanomaterials: Synthesis, Characterization and Its Potential in Dye Remediation, Green Technol. Ind. Waste Remediat. (2023) 221–245. https://doi.org/10.1007/978-3-031-46858-2_11

[29]   M.P. Shah, N. Bharadvaja, L. Kumar, Biogenic Nanomaterials for Environmental Sustainability: Principles, Practices, and Opportunities, Springer Nature, 2024. https://doi.org/10.1007/978-3-031-45956-6

[30]   R. Badru, Y. Singh, N. Singh, D. Dubal, Biogenic Nanomaterial for Health and Environment, CRC Press, 2023. https://doi.org/10.1201/9781003430087

[31]   U. Sundaresan, G. Kasi, Synthesis of ZnO nanoparticles using Sargassum wightii ethanol extract and their antibacterial and anticancer applications, Biomass Convers. Biorefinery 14 (2024) 26173–26191. https://doi.org/10.1007/s13399-023-04977-1

[32]   A. Aghababai Beni, M. Haghmohammadi, S. Delnabi Asl, S.M. Hakimzadeh, A. Nezarat, Algae application for treating wastewater contaminated with heavy metal ions, in: Algae as a Nat. Solut. Challenges Water-Food-Energy Nexus Towar. Carbon Neutrality, Springer, 2024: pp. 297–322. https://doi.org/10.1007/978-981-97-2371-3_12

[33]   L.Y. Martínez-Zamudio, R.B. González-González, R.G. Araújo, J.A.R. Hernández, E.A. Flores-Contreras, E.M. Melchor-Martínez, R. Parra-Saldívar, H.M.N. Iqbal, Emerging pollutants removal from leachates and water bodies by nanozyme-based approaches, Curr. Opin. Environ. Sci. Heal. 37 (2024) 100522. https://doi.org/10.1016/j.coesh.2023.100522

[34]   O. Emmanuel, T.C. Ezeji, Exploring the synergy of nanomaterials and microbial cell factories during biohydrogen and biobutanol production from different carbon sources, Sustain. Chem. Environ. 6 (2024) 100098. https://doi.org/10.1016/j.scenv.2024.100098

[35]   S. Rathod, S. Preetam, C. Pandey, S.P. Bera, Exploring synthesis and applications of green nanoparticles and the role of nanotechnology in wastewater treatment, Biotechnol. Reports (2024) e00830. https://doi.org/10.1016/j.btre.2024.e00830

[36]   S. Boonphan, S. Prachakiew, C. Nontakoat, Y. Keereeta, C. Boonruang, A. Klinbumrung, Crystallographic Defects Induced F-Center and Optical Enhancements in CeO2-TiO2 Nanocomposites, South African J. Chem. Eng. (2025). https://doi.org/10.1016/j.sajce.2025.01.010

[37]   N. El Messaoudi, Z. Ciğeroğlu, Z.M. Şenol, E.S. Kazan-Kaya, Y. Fernine, S. Gubernat, Z. Lopicic, Green synthesis of CuFe2O4 nanoparticles from bioresource extracts and their applications in different areas: a review, Biomass Convers. Biorefinery (2024) 1–22. https://doi.org/10.1007/s13399-023-

05264-9

[38]   K. Malik, A. Kazmi, T. Sultana, N.I. Raja, Y. Bibi, M. Abbas, I.A. Badruddin, M.M. Ali, M.N. Bashir, A Mechanistic Overview on Green Assisted Formulation of Nanocomposites and Their Multifunctional Role in Biomedical Applications, Heliyon (2025).

[39]   C. Demir, A. Aygun, M.K. Gunduz, B.Y. Altınok, T. Karahan, I. Meydan, E. Halvaci, R.N.E. Tiri, F. Sen, Production of plant-based ZnO NPs by green synthesis; anticancer activities and photodegradation of methylene red dye under sunlight, Biomass Convers. Biorefinery 2024 (2024) 1–16. https://doi.org/10.1007/S13399-024-06172-2

[40]   N.U. Uza, G. Dastagir, I. Ahmad, S. Ullah, I.U. Din, M. Suleman, Estimation of Secondary Metabolites, Nutrients, Minerals, and Anti-Inflammatory and Antidiarrheal Agents in Heliotropium rariflorum Stocks at Two Phenological Stages, Chem. Biodivers. (2025) e202402009. https://doi.org/10.1002/cbdv.202402009

[41]   S. Seena, A. Rai, S. Kumar, Nanoparticles and Plant-Microbe Interactions: An Environmental Perspective, Elsevier, 2023.

[42]   T. Gur, I. Meydan, H. Seckin, M. Bekmezci, F. Sen, Green synthesis, characterization and bioactivity of biogenic zinc oxide nanoparticles, Environ. Res. 204 (2022) 111897. https://doi.org/10.1016/J.ENVRES.2021.111897

[43]   M. Pirsaheb, T. Gholami, H. Seifi, E.A. Dawi, E.A. Said, A.-H.M. Hamoody, U.S. Altimari, M. Salavati-Niasari, Green synthesis of nanomaterials by using plant extracts as reducing and capping agents, Environ. Sci. Pollut. Res. 31 (2024) 24768–24787. https://doi.org/10.1007/s11356-024-32983-x

[44]   A. Islam, I. Rahat, C. Rejeeth, D. Sharma, A. Sharma, Recent advcances on plant-based bioengineered nanoparticles using secondary metabolites and their potential in lung cancer management, J. Futur. Foods 5 (2025) 1–20. https://doi.org/10.1016/j.jfutfo.2024.01.001

[45]   E.J. Mohammed, A.E.M. Abdelaziz, A.E. Mekky, N.N. Mahmoud, M. Sharaf, M.M. Al-Habibi, N.M. Khairy, A.A. Al-Askar, F.S. Youssef, M.A. Gaber, Biomedical promise of Aspergillus flavus-biosynthesized selenium nanoparticles: A green synthesis approach to antiviral, anticancer, anti-biofilm, and antibacterial applications, Pharmaceuticals 17 (2024) 915. https://doi.org/10.3390/ph17070915

[46]   S.S. Chan, K.S. Khoo, R. Abdullah, J.C. Juan, E.-P. Ng, R.J. Chin, T.C. Ling, Harnessing microalgae for metal nanoparticles biogenesis using heavy metal ions from wastewater as a metal precursor: A review, Sci. Total Environ. (2024) 176989. https://doi.org/10.1016/j.scitotenv.2024.176989

[47]   B. Şen, B. Demirkan, A. Savk, R. Kartop, M.S. Nas, M.H. Alma, S. Sürdem, F. Şen, High-performance graphite-supported ruthenium nanocatalyst for hydrogen evolution reaction, J. Mol. Liq. 268 (2018) 807–812. https://doi.org/10.1016/j.molliq.2018.07.117

[48]   S.H. Nguyen, N.T. Vu, H. Van Nguyen, B. Nguyen, T.T. Luong, Biologically synthesized Fe0-based nanoparticles and their application trends as catalysts in the treatment of chlorinated organic compounds: a review, Environ. Sci. Nano (2025). https://doi.org/10.1039/D4EN00843J

[49]   T.N. Vigil, L.C. Spangler, Understanding biomineralization mechanisms to produce size-controlled, tailored nanocrystals for optoelectronic and catalytic applications: a Review, ACS Appl. Nano Mater. 7 (2024) 18626–18654. https://doi.org/10.1021/acsanm.3c04277

[50]   T. Wang, Biologically Derived and Bio-inspired Nanomaterials, University of California, Irvine, 2022.

[51]   S.E. Wolf, Bioinorganic and bioinspired solid-state chemistry: From classical crystallization to nonclassical synthesis concepts, in: Synth. Inorg. Chem., Elsevier, 2021: pp. 433–490. https://doi.org/10.1016/B978-0-12-818429-5.00006-5

[52]   C. Wang, K. Jiao, J. Yan, M. Wan, Q. Wan, L. Breschi, J. Chen, F.R. Tay, L. Niu, Biological and synthetic template-directed syntheses of mineralized hybrid and inorganic materials, Prog. Mater. Sci. 116 (2021) 100712. https://doi.org/10.1016/j.pmatsci.2020.100712

[53]   S. Kumari, P. Kumar, EXPLORING ECO-FRIENDLY APPROACHES: A COMPREHENSIVE REVIEW ON GREEN SYNTHESIS OF NANOPARTICLES, (2023).

[54]   A. Karnwal, A.Y. Jassim, A.A. Mohammed, V. Sharma, A.R.M.S. Al-Tawaha, I. Sivanesan, Nanotechnology for Healthcare: Plant-Derived Nanoparticles in Disease Treatment and Regenerative Medicine, Pharmaceuticals 17 (2024) 1711. https://doi.org/10.3390/ph17121711

[55]   J. Jeevanandam, V. Vadanasundari, S. Pan, A. Barhoum, M.K. Danquah, Bionanotechnology and Bionanomaterials: Emerging Applications, Market, and Commercialization, Bionanotechnol. Emerg. Appl. Bionanomaterials (2022) 3–44. https://doi.org/10.1016/B978-0-12-823915-5.00009-5

[56]   A.K. Oyebamiji, S.A. Akintelu, S.O. Afolabi, O. Ebenezer, E.T. Akintayo, C.O. Akintayo, A Comprehensive Review on Mycosynthesis of Nanoparticles, Characteristics, Applications, and Limitations, Plasmonics (2025) 1–19. https://doi.org/10.1007/s11468-024-02755-x

[57]   K.V. Reddy, N.R.S. Sree, P.S. Kumar, P. Ranjit, Microbial enzymes in the biosynthesis of metal nanoparticles, in: Ecol. Interplays Microb. Enzymol., Springer, 2022: pp. 329–350. https://doi.org/10.1007/978-981-19-0155-3_15

[58]   J. Iqbal, S. Ijaz, N. Ijaz, B.A. Abbasi, T. Yaseen, Z. Ullah, R. Iqbal, G. Murtaza, Z. Ashraf, T. Mahmood, 5 Microbial Synthesis of Metal Nanoparticles for Nanomedicinal and Catalytic Applications, Expand. Nanobiotechnology Appl. Commer. Appl. Commer. (2025) 88. https://doi.org/10.1201/9781003378563-5

[59]   O. Strbak, P. Hnilicova, J. Gombos, A. Lokajova, P. Kopcansky, Magnetotactic bacteria: From evolution to biomineralization and biomedical applications, Minerals 12 (2022) 1403. https://doi.org/10.3390/min12111403

[60]   T. Ayodele, A. Tijani, M. Liadi, K. Alarape, C. Clementson, A. Hammed, Biomass-Based Microbial Protein Production: A Review of Processing and Properties, Front. Biosci. 16 (2024) 40. https://doi.org/10.31083/j.fbe1604040

[61]   T. Wu, J. Hou, J.C. White, D. Lin, Nanomaterials for soil contaminant remediation, in: Nano-Enabled Sustain. Precis. Agric., Elsevier, 2023: pp. 143–180. https://doi.org/10.1016/B978-0-323-91233-4.00017-X

[62]   L. Sreelatha, A.L. Ambili, S.C. Sreedevi, D. Achuthavarier, Metallothioneins: an unraveling insight into remediation strategies of plant defense mechanisms, Environ. Sci. Pollut. Res. (2024) 1–23. https://doi.org/10.1007/s11356-024-35790-6

[63]   N. El Messaoudi, Z. Ciğeroğlu, Z.M. Şenol, A. Bouich, E.S. Kazan-Kaya, L. Noureen, J.H.P. Américo-Pinheiro, Green synthesis of nanoparticles for remediation organic pollutants in wastewater by adsorption, in: Adv. Chem. Pollution, Environ. Manag. Prot., Elsevier, 2024: pp. 305–345. https://doi.org/10.1016/bs.apmp.2023.06.016

[64]   G. Fadillah, N.T.S. Alarifi, I.W.K. Suryawan, T.A. Saleh, Advances in designed reactors for water treatment process: A review highlighting the designs and performance, J. Water Process Eng. 63 (2024) 105417. https://doi.org/10.1016/j.jwpe.2024.105417

[65]   M.S. Akhtar, S. Ali, W. Zaman, Innovative adsorbents for pollutant removal: Exploring the latest research and applications, Molecules 29 (2024) 4317. https://doi.org/10.3390/molecules29184317

[66]   M.S. Shafeeyan, Application of photocatalytic and Fenton processes for the degradation of toxic pollutants from pulp and paper industry effluents, Water Resour. Ind. (2024) 100260. https://doi.org/10.1016/j.wri.2024.100260

[67]   A.A. Adesibikan, S.S. Emmanuel, C.O. Olawoyin, P. Ndungu, Cellulosic metallic nanocomposites for photocatalytic degradation of persistent dye pollutants in aquatic bodies: a pragmatic review, J. Organomet. Chem. (2024) 123087. https://doi.org/10.1016/j.jorganchem.2024.123087

[68]   Y. Miyah, N. El Messaoudi, M. Benjelloun, J. Georgin, D.S.P. Franco, Y. Acikbas, H.S. Kusuma, M. Sillanpää, MOF-derived magnetic nanocomposites as potential formulations for the efficient removal

of organic pollutants from water via adsorption and advanced oxidation processes: A review, Mater. Today Sustain. (2024) 100985. https://doi.org/10.1016/j.mtsust.2024.100985

[69]   S.R. Dhokpande, S.M. Deshmukh, A. Khandekar, A. Sankhe, A review outlook on methods for removal of heavy metal ions from wastewater, Sep. Purif. Technol. (2024) 127868. https://doi.org/10.1016/j.seppur.2024.127868

[70]   U. Sarma, M.E. Hoque, A. Thekkangil, N. Venkatarayappa, S. Rajagopal, Microalgae in removing heavy metals from wastewater–An advanced green technology for urban wastewater treatment, J. Hazard. Mater. Adv. 15 (2024) 100444. https://doi.org/10.1016/j.hazadv.2024.100444

[71]   Y.G. Ko, Hybrid method integrating adsorption and chemical precipitation of heavy metal ions on polymeric fiber surfaces for highly efficient water purification, Chemosphere (2024) 142909. https://doi.org/10.1016/j.chemosphere.2024.142909

[72]   A. Singh, S.S. Shah, C. Sharma, V. Gupta, A.K. Sundramoorthy, P. Kumar, S. Arya, An approach towards different techniques for detection of heavy metal ions and their removal from waste water, J. Environ. Chem. Eng. (2024) 113032. https://doi.org/10.1016/j.jece.2024.113032

[73]   N.O. Solomon, S. Kanchan, M. Kesheri, Nanoparticles as Detoxifiers for Industrial Wastewater, Water, Air, Soil Pollut. 235 (2024) 214. https://doi.org/10.1007/s11270-024-07016-5

[74]   D.S. Idris, A. Roy, Biogenic synthesis of Ag–CuO nanoparticles and its antibacterial, antioxidant, and catalytic activity, J. Inorg. Organomet. Polym. Mater. 34 (2024) 1055–1067. https://doi.org/10.1007/s10904-023-02873-9

[75]   D.B. Olawade, O.Z. Wada, B.I. Egbewole, O. Fapohunda, A.O. Ige, S.O. Usman, O. Ajisafe, Metal and metal oxide nanomaterials for heavy metal remediation: novel approaches for selective, regenerative, and scalable water treatment, Front. Nanotechnol. 6 (2024) 1466721. https://doi.org/10.3389/fnano.2024.1466721

[76]   S.-Y. Yu, Z.-H. Xie, X. Wu, Y.-Z. Zheng, Y. Shi, Z.-K. Xiong, P. Zhou, Y. Liu, C.-S. He, Z.-C. Pan, Review of advanced oxidation processes for treating hospital sewage to achieve decontamination and disinfection, Chinese Chem. Lett. 35 (2024) 108714. https://doi.org/10.1016/j.cclet.2023.108714

[77]   F. Seyedpour, J. Farahbakhsh, Z. Dabaghian, W. Suwaileh, M. Zargar, A. Rahimpour, M. Sadrzadeh, M. Ulbricht, Y. Mansourpanah, Advances and challenges in tailoring antibacterial polyamide thin film composite membranes for water treatment and desalination: A critical review, Desalination (2024) 117614. https://doi.org/10.1016/j.desal.2024.117614

[78]   L. Luo, W. Huang, J. Zhang, Y. Yu, T. Sun, Metal-Based Nanoparticles as Antimicrobial Agents: A Review, ACS Appl. Nano Mater. 7 (2024) 2529–2545. https://doi.org/10.1021/acsanm.3c05615

[79]   A. Saud, S. Gupta, A. Allal, H. Preud'Homme, B. Shomar, S.J. Zaidi, Progress in the sustainable development of biobased (nano) materials for application in water treatment technologies, ACS Omega 9 (2024) 29088–29113. https://doi.org/10.1021/acsomega.3c08883

[80]   M.K. Hussain, S. Khatoon, G. Nizami, U.K. Fatma, M. Ali, B. Singh, A. Quraishi, M.A. Assiri, S. Ahamad, M. Saquib, Unleashing the power of bio-adsorbents: efficient heavy metal removal for sustainable water purification, J. Water Process Eng. 64 (2024) 105705. https://doi.org/10.1016/j.jwpe.2024.105705

[81]   F. Gao, M. Zhang, S. Li, L. Liu, J. Tang, Effect of physical and chemical co-application of biochar and sulfidated nano scale zero valent iron on the NB degradation in soil: key roles of biochar, Sep. Purif. Technol. 353 (2025) 128546. https://doi.org/10.1016/j.seppur.2024.128546

[82]   M.P. Gomes, Nanophytoremediation: advancing phytoremediation efficiency through nanotechnology integration, Discov. Plants 2 (2025) 8. https://doi.org/10.1007/s44372-025-00090-x

[83]   I. Salahshoori, Q. Wang, M.A.L. Nobre, A.H. Mohammadi, E.A. Dawi, H.A. Khonakdar, Molecular simulation-based insights into dye pollutant adsorption: a perspective review, Adv. Colloid Interface Sci. (2024) 103281. https://doi.org/10.1016/j.cis.2024.103281

[84]   A. Badoni, S. Thakur, N. Vijayan, H.C. Swart, M. Bechelany, Z. Chen, S. Sun, Q. Cai, Y. Chen, J. Prakash, Recent progress in understanding the role of graphene oxide, TiO2 and graphene oxide-TiO2 nanocomposites as multidisciplinary photocatalysts in energy and environmental applications, Catal. Sci. Technol. (2025). https://doi.org/10.1039/D4CY01334D

[85]   M. Yang, X. Zhang, Y. Sun, Remediation of Cr (VI) Polluted Groundwater Using Zero-Valent Iron Composites: Preparation, Modification, Mechanisms, and Environmental Implications, Molecules 29 (2024) 5697. https://doi.org/10.3390/molecules29235697

[86]   M. Namakka, M.R. Rahman, K.A.B.M. Said, A. Muhammad, Insights into micro-and nano-zero valent iron materials: synthesis methods and multifaceted applications, RSC Adv. 14 (2024) 30411–30439. https://doi.org/10.1039/D4RA03507K

[87]   P.M. Sah, S.G. Gite, R. Sonawane, R.W. Raut, Biogenic nanomaterials as a catalyst for photocatalytic dye degradation, in: Biog. Nanomater. Environ. Sustain. Princ. Pract. Oppor., Springer, 2024: pp. 409–433. https://doi.org/10.1007/978-3-031-45956-6_16

[88]   J. Tripathy, A. Mishra, M. Pandey, R.R. Thakur, S. Chand, P.R. Rout, M.K. Shahid, Advances in Nanoparticles and Nanocomposites for Water and Wastewater Treatment: A Review, Water 16 (2024) 1481. https://doi.org/10.3390/w16111481

[89]   M. Summer, S. Ali, H.M. Tahir, R. Abaidullah, U. Fiaz, S. Mumtaz, H. Fiaz, A. Hassan, T.A. Mughal, M.A. Farooq, Mode of action of biogenic silver, zinc, copper, titanium and cobalt nanoparticles against antibiotics resistant pathogens, J. Inorg. Organomet. Polym. Mater. 34 (2024) 1417–1451. https://doi.org/10.1007/s10904-023-02935-y

[90]   K. Das Purkayastha, N. Gogoi, Prospects of biosynthesized nanoparticles in treating pharmaceutical wastewater in relation to human health, in: Nanotechnol. Hum. Heal., Elsevier, 2023: pp. 75–120. https://doi.org/10.1016/B978-0-323-90750-7.00013-2

[91]   E. Demir, A. Savk, B. Sen, F. Sen, A novel monodisperse metal nanoparticles anchored graphene oxide as Counter Electrode for Dye-Sensitized Solar Cells, Nano-Structures & Nano-Objects 12 (2017) 41–45. https://doi.org/10.1016/j.nanoso.2017.08.018

[92]   A. Şavk, K. Cellat, K. Arıkan, F. Tezcan, S.K. Gülbay, S. Kızıldağ, E.Ş. Işgın, F. Şen, Highly monodisperse Pd-Ni nanoparticles supported on rGO as a rapid, sensitive, reusable and selective enzyme-free glucose sensor, Sci. Rep. 9 (2019) 19228. https://doi.org/10.1038/s41598-019-55746-y

[93]   N.H. Khand, A.R. Solangi, S. Ameen, A. Fatima, J.A. Buledi, A. Mallah, S.Q. Memon, F. Sen, F. Karimi, Y. Orooji, A new electrochemical method for the detection of quercetin in onion, honey and green tea using Co3O4 modified GCE, J. Food Meas. Charact. 2021 154 15 (2021) 3720–3730. https://doi.org/10.1007/S11694-021-00956-0

[94]   W.H. Foo, W.Y. Chia, S. Ende, S.R. Chia, K.W. Chew, Nanomaterials in aquaculture disinfection, water quality monitoring, and wastewater remediation, J. Environ. Chem. Eng. (2024) 113947. https://doi.org/10.1016/j.jece.2024.113947

[95]   N. Liu, J. Zhao, J. Du, C. Hou, X. Zhou, J. Chen, Y. Zhang, Non-phytoremediation and phytoremediation technologies of integrated remediation for water and soil heavy metal pollution: A comprehensive review, Sci. Total Environ. (2024) 174237. https://doi.org/10.1016/j.scitotenv.2024.174237

[96]   B. Mughal, S.Z.J. Zaidi, X. Zhang, S.U. Hassan, Biogenic nanoparticles: Synthesis, characterisation and applications, Appl. Sci. 11 (2021) 2598. https://doi.org/10.3390/app11062598

[97]   B.H. Abdelmonem, L.T. Kamal, R.M. Elbaz, M. Khalifa, A. Abdelnaser, From contamination to detection: The growing threat of heavy metals, Heliyon (2025). https://doi.org/10.1016/j.heliyon.2025.e41713

[98]   V. Singh, G. Ahmed, S. Vedika, P. Kumar, S.K. Chaturvedi, S.N. Rai, E. Vamanu, A. Kumar, Toxic heavy metal ions contamination in water and their sustainable reduction by eco-friendly methods:

isotherms, thermodynamics and kinetics study, Sci. Rep. 14 (2024) 7595. https://doi.org/10.1038/s41598-024-58061-3

[99]   A. Thakur, A. Kumar, A. Singh, Adsorptive removal of heavy metals, dyes, and pharmaceuticals: Carbon-based nanomaterials in focus, Carbon N. Y. 217 (2024) 118621. https://doi.org/10.1016/j.carbon.2023.118621

[100]   Z. Chen, Y. Li, Y. Cai, S. Wang, B. Hu, B. Li, X. Ding, L. Zhuang, X. Wang, Application of covalent organic frameworks and metal–organic frameworks nanomaterials in organic/inorganic pollutants removal from solutions through sorption-catalysis strategies, Carbon Res. 2 (2023) 8. https://doi.org/10.1007/s44246-023-00041-9

[101]   S.O. Adewuyia, B.T. Ibigbamia, A.O. Mmuoegbulamb, F.S. Abimbadea, O.M. Abioduna, M.J. Klinkc, S.M. Nelanac, D. Malomoa, O.S. Ayandaa, Toxicity and health implications of pesticides and the need to remediate pesticide-contaminated wastewater through the advanced oxidation processes, Water Conserv. Manag. 8 (2024) 97–108. https://doi.org/10.26480/wcm.01.2024.97.108

[102]   W.-C. Chou, Y.-H. Cheng, J.E. Riviere, N.A. Monteiro-Riviere, W.G. Kreyling, Z. Lin, Development of a multi-route physiologically based pharmacokinetic (PBPK) model for nanomaterials: a comparison between a traditional versus a new route-specific approach using gold nanoparticles in rats, Part. Fibre Toxicol. 19 (2022) 1–19. https://doi.org/10.1186/s12989-022-00489-4

[103]   A. Karnwal, T. Malik, Nano-revolution in heavy metal removal: engineered nanomaterials for cleaner water, Front. Environ. Sci. 12 (2024) 1393694. https://doi.org/10.3389/fenvs.2024.1393694

[104]   A.E. Gahrouei, A. Rezapour, M. Pirooz, S. Pourebrahimi, From Classic to Cutting-Edge Solutions: A Comprehensive Review of Materials and Methods for Heavy Metal Removal from Water Bodies, Desalin. Water Treat. (2024) 100446. https://doi.org/10.1016/j.dwt.2024.100446

[105]   A. Saravanan, P.S. Kumar, S. Jeevanantham, S. Karishma, B. Tajsabreen, P.R. Yaashikaa, B. Reshma, Effective water/wastewater treatment methodologies for toxic pollutants removal: Processes and applications towards sustainable development, Chemosphere 280 (2021) 130595. https://doi.org/10.1016/j.chemosphere.2021.130595

[106]   R.S. Al-Habeeb, W.M. Al-Bishri, Synthesis of Bimetallic Copper oxide-silver Nanoparticles by Gum Arabic: Unveiling Antimicrobial, and Antibiofilm Potential against some Pathogenic Microbes Causing Wound Infection, J. Clust. Sci. 35 (2024) 2311–2327. https://doi.org/10.1007/s10876-024-02656-3

[107]   A. Nawaz, A. Farhan, F. Maqbool, H. Ahmad, W. Qayyum, E. Ghazy, A. Rahdar, A.M. Díez-Pascual, S. Fathi-karkan, Zinc Oxide Nanoparticles: Pathways to Micropollutant Adsorption, Dye Removal, and Antibacterial Actions-A Study of Mechanisms, Challenges, and Future Prospects, J. Mol. Struct. (2024) 138545. https://doi.org/10.1016/j.molstruc.2024.138545

[108]   H. Kamyab, S. Chelliapan, G. Hayder, M. Yusuf, M.M. Taheri, S. Rezania, M. Hasan, K.K. Yadav, M. Khorami, M. Farajnezhad, Exploring the potential of metal and metal oxide nanomaterials for sustainable water and wastewater treatment: a review of their antimicrobial properties, Chemosphere 335 (2023) 139103. https://doi.org/10.1016/j.chemosphere.2023.139103

[109]   S.P. Goutam, G. Saxena, Biogenic nanoparticles for removal of heavy metals and organic pollutants from water and wastewater: advances, challenges, and future prospects, Bioremediation Environ. Sustain. (2021) 623–636. https://doi.org/10.1016/B978-0-12-820524-2.00025-0

[110]   S. Akhai, T. Taneja, Liquid Gold: Safeguarding Our Health Through Pure Drinking Water and Rigorous Water Testing, in: Nat. Resour. Manag. Issues Human-Influenced Landscapes, IGI Global, 2024: pp. 129–148. https://doi.org/10.4018/978-1-6684-7051-0.ch007

[111]   S. Bakhtiari, M. Salari, M. Shahrashoub, A. Zeidabadinejad, G. Sharma, M. Sillanpää, A Comprehensive Review on Green and Eco-Friendly Nano-Adsorbents for the Removal of Heavy Metal Ions: Synthesis, Adsorption Mechanisms, and Applications, Curr. Pollut. Reports 10 (2024) 1–39. https://doi.org/10.1007/s40726-023-00290-7

[112]   D.I. Fertu, L. Bulgariu, M. Gavrilescu, SUSTAINABLE USE OF SOME NATURAL MATERIALS AND WASTE FOR THE DECONTAMINATION OF ENVIRONMENTAL COMPONENTS POLLUTED WITH HEAVY METALS., Environ. Eng. Manag. J. 23 (2024). https://doi.org/10.30638/eemj.2024.126

[113]   J. Yadav, M. Rani, U. Shanker, M. Sillanpaa, Forging the advances of iron-based nanomaterials by functionalizing charge carriers regions for eradication of heavy metal ion contamination, Inorg. Chem. Commun. (2024) 112440. https://doi.org/10.1016/j.inoche.2024.112440

[114]   S. Ganesan, P. Roshan, R. Mathur, Sustainable Approach for the Evacuation of Heavy Metal Pollutants by Bioremediation, in: Sustain. Water Treat. Ecosyst. Prot. Strateg., Apple Academic Press, 2024: pp. 207–235. https://doi.org/10.1201/9781003470014-6

[115]   N.A.S. Feisal, N.H. Kamaludin, M.A. Ahmad, T.N.B.T. Ibrahim, A comprehensive review of nanomaterials for efficient heavy metal ions removal in water treatment, J. Water Process Eng. 64 (2024) 105566. https://doi.org/10.1016/j.jwpe.2024.105566

[116]   M.J. Ibrahimova, V.M. Abbasov, P.I. Huseyn, F.A. Nasirov, A.M. Aslanbeyli, R.A. Rasulova, METHODS OF REMOVING HEAVY METALS FROM WASTEWATERS, Process. Petrochemistry Oil Refin. 26 (2025). https://doi.org/10.62972/1726-4685.2025.1.41

[117]   M.M. Youssif, H.G. El-Attar, V. Hessel, M. Wojnicki, Recent Developments in the Adsorption of Heavy Metal Ions from Aqueous Solutions Using Various Nanomaterials, Materials (Basel). 17 (2024) 5141. https://doi.org/10.3390/ma17215141

[118]   M. Qasim, M.I. Arif, A. Naseer, L. Ali, R. Aslam, S.A. Abbasi, Q. Ullah, Biogenic nanoparticles at the forefront: transforming industrial wastewater treatment with TiO2 and graphene, Sch J Agric Vet Sci 5 (2024) 56–76. https://doi.org/10.36347/sjavs.2024.v11i05.002

[119]   O. Altintas Yildirim, A.M. Bahadir, E. Pehlivan, The Role and Significance of Biochars for the Photodegradation of Selected Organic Dyes in Water and Wastewater, in: Pollut. Recent Trends Wastewater Treat., Springer, 2024: pp. 271–287. https://doi.org/10.1007/978-3-031-62054-6_16

[120]   F. Gol, A. Aygun, C. Ture, R.N.E. Tiri, Z.G. Sarıtaş, E. Kaçar, M. Arslan, F. Sen, Environmentally Friendly Synthesis of 2D Cu2O Nanoleaves: Morphological Evaluation, Their Photocatalytic Activity Against Azo Dye And Antibacterial Activity For Ceramic Structures, Bionanoscience (2024) 1–10. https://doi.org/10.1007/S12668-024-01440-X

[121]   S.S. Hashmi, M. Shah, W. Muhammad, A. Ahmad, M.A. Ullah, M. Nadeem, B.H. Abbasi, Potentials of phyto-fabricated nanoparticles as ecofriendly agents for photocatalytic degradation of toxic dyes and waste water treatment, risk assessment and probable mechanism, J. Indian Chem. Soc. 98 (2021) 100019. https://doi.org/10.1016/j.jics.2021.100019

[122]   E.E.A. Suarez, M.E.R. Jalil, M.A.F. Baldo, S.A. Cuozzo, Nanobiotechnology approaches for the remediation of persistent and emerging organic pollutants: strategies, interactions, and effectiveness, Environ. Sci. Nano (2025).

[123]   X. Li, L. Yang, J. Zhou, B. Dai, D. Gan, Y. Yang, Z. Wang, J. He, S. Xia, Biogenic palladium nanoparticles for wastewater treatment: formation, applications, limitations, and future directions, J. Water Process Eng. 64 (2024) 105641. https://doi.org/10.1016/j.jwpe.2024.105641

[124]   N.O. Etafo, M.O. Bamidele, A. Bamisaye, Y.A. Alli, Revolutionizing photocatalysis: Unveiling efficient alternatives to titanium (IV) oxide and zinc oxide for comprehensive environmental remediation, J. Water Process Eng. 62 (2024) 105369. https://doi.org/10.1016/j.jwpe.2024.105369

[125]   S. Rostami, S. Jafari, Z. Moeini, M. Jaskulak, L. Keshtgar, A. Badeenezhad, A. Azhdarpoor, M. Rostami, K. Zorena, M. Dehghani, Current methods and technologies for degradation of atrazine in contaminated soil and water: A review, Environ. Technol. Innov. 24 (2021) 102019. https://doi.org/10.1016/j.eti.2021.102019

[126]   A.A. Adesibikan, S.S. Emmanuel, S.A. Nafiu, M.J. Tachia, K.O. Iwuozor, E.C. Emenike, A.G.

Adeniyi, A review on sustainable photocatalytic degradation of agro-organochlorine and organophosphorus water pollutants using biogenic iron and iron oxide-based nanoarchitecture materials, Desalin. Water Treat. (2024) 100591. https://doi.org/10.1016/j.dwt.2024.100591

[127]   A. Al Miad, S.P. Saikatb, M.K. Alam, M.S. Hossain, N.M. Bahadur, S. Ahmed, Metal oxides-based photocatalyst for the efficient degradation of organic pollutants for a sustainable environment: a review, Nanoscale Adv. (2024). https://doi.org/10.1039/D4NA00517A

[128]   O.S. Ayanda, A.O. Mmuoegbulam, O. Okezie, N.I. Durumin Iya, S.E. Mohammed, P.H. James, A.B. Muhammad, A.A. Unimke, S.A. Alim, S.M. Yahaya, Recent progress in carbon-based nanomaterials: critical review, J. Nanoparticle Res. 26 (2024) 1–42. https://doi.org/10.1007/s11051-023-05905-0

[129]   A. Aygun, R.N.E. Tiri, R. Bayat, F. Sen, Hydrothermal synthesis of BCQD@g-C3N4 nanocomposites supporting environmental sustainability: Organic dye removal and bacterial inactivation, J. Hazard. Mater. Adv. 16 (2024) 100464. https://doi.org/10.1016/J.HAZADV.2024.100464

[130]   A.O. Adeniyi, M.O. Jimoh, Decontamination Potential of Ultraviolet Type C Radiation in Water Treatment Systems: Targeting Microbial Inactivation, Water 16 (2024) 2725. https://doi.org/10.3390/w16192725

[131]   P.B. Rathod, M.P. Singh, A.S. Taware, S.U. Deshmukh, C.K. Tagad, A. Kulkarni, A.B. Kanagare, Comprehensive insights into water remediation: chemical, biotechnological, and nanotechnological perspectives, Environ. Pollut. Bioavailab. 36 (2024) 2329660. https://doi.org/10.1080/26395940.2024.2329660

[132]   G. Bhavya, S.A. Belorkar, R. Mythili, N. Geetha, H.S. Shetty, S.S. Udikeri, S. Jogaiah, Remediation of emerging environmental pollutants: a review based on advances in the uses of eco-friendly biofabricated nanomaterials, Chemosphere 275 (2021) 129975. https://doi.org/10.1016/j.chemosphere.2021.129975

[133]   J. Zhang, P. Su, H. Chen, M. Qiao, B. Yang, X. Zhao, Impact of reactive oxygen species on cell activity and structural integrity of Gram-positive and Gram-negative bacteria in electrochemical disinfection system, Chem. Eng. J. 451 (2023) 138879. https://doi.org/10.1016/j.cej.2022.138879

[134]   G. Yesilay, O.A.L. Dos Santos, L.J. Hazeem, B.P. Backx, A.H. Kamel, M. Bououdina, Impact of pathogenic bacterial communities present in wastewater on aquatic organisms: application of nanomaterials for the removal of these pathogens, Aquat. Toxicol. 261 (2023) 106620. https://doi.org/10.1016/j.aquatox.2023.106620

[135]   S. Singh, R. Garg, A. Jana, C. Bathula, S. Naik, M. Mittal, Current developments in nanostructurally engineered metal oxide for removal of contaminants in water, Ceram. Int. 49 (2023) 7308–7321. https://doi.org/10.1016/j.ceramint.2022.10.183

[136]   L. Doan, N.N. Lam, K. Tran, K.G. Huynh, Fruit derived silver nanoparticles synthesis for beginners–a review, Nanocomposites 11 (2025) 20–51. https://doi.org/10.1080/20550324.2024.2442270

[137]   M. Nagpal, A. Mittal, Application of Biosurfactants in the Green Synthesis of Inorganic Nanoparticles, Bionanoscience 15 (2025) 1–17. https://doi.org/10.1007/s12668-024-01628-1

[138]   S.A. Mazari, E. Ali, R. Abro, F.S.A. Khan, I. Ahmed, M. Ahmed, S. Nizamuddin, T.H. Siddiqui, N. Hossain, N.M. Mubarak, Nanomaterials: Applications, waste-handling, environmental toxicities, and future challenges–A review, J. Environ. Chem. Eng. 9 (2021) 105028. https://doi.org/10.1016/j.jece.2021.105028

[139]   K. Naseem, A. Aziz, M.H. Tahir, A. Ameen, A. Ahmad, K. Ahmad, M. Arif, W. Hassan, J. Najeeb, E. Rao, Biogenic synthesized nanocatalysts and their potential for the treatment of toxic pollutants: environmental remediation, a review, Int. J. Environ. Sci. Technol. 21 (2024) 2163–2194. https://doi.org/10.1007/s13762-023-05166-3

[140]   S. Alimohammadvand, M.K. Zenjanab, M. Mashinchian, J. Shayegh, R. Jahanban-Esfahlan,

Recent advances in biomimetic cell membrane–camouflaged nanoparticles for cancer therapy, Biomed. Pharmacother. 177 (2024) 116951. https://doi.org/10.1016/j.biopha.2024.116951

[141]　K. Jiang, Q. Wang, X.-L. Chen, X. Wang, X. Gu, S. Feng, J. Wu, H. Shang, X. Ba, Y. Zhang, Nanodelivery Optimization of IDO1 Inhibitors in Tumor Immunotherapy: Challenges and Strategies, Int. J. Nanomedicine (2024) 8847–8882. https://doi.org/10.2147/IJN.S458086

[142]　E.A. Aboagye, M.L.S. Santos, P.V. dos Santos Lins, L. Meili, D.S.P. Franco, J. Georgin, R. Selvasembian, Green Synthesized Bio-nanomaterials for Pollutant Remediation, Nanotechnol. Environ. Manag. (2024) 83. https://doi.org/10.1201/9781003350941-9

[143]　M. Yıldırım, D.B. Demirkaya, C. Aydın, S. Yalçın, Cytotoxicity and Biocompatibility of Bionanomaterials, in: Bionanomaterials Ind. Appl., CRC Press, 2024: pp. 325–346. https://doi.org/10.1201/9781003432791-22

[144]　A. Ghosh, U. Saha, S. Roy, S. Roy, S.K. Verma, Green Synthesized Biocompatible Nanomaterial for Environmental Application, in: Sustain. Green Synth. Nano-Dimensional Mater. Energy Environ. Appl., CRC Press, 2024: pp. 241–270. https://doi.org/10.1201/9781003362241-13

[145]　N. Chatterjee, P. Dhar, Bio-nanocomposites: curse or miracle for the society?, in: Charact. Tech. Bionanocomposites, Elsevier, 2025: pp. 3–39. https://doi.org/10.1016/B978-0-443-22067-8.00001-0

[146]　M. Ehtesham, N. Ansari, G. Kumar, S. Kumar, P. Gaijon, S. Ghosh, M.R. Singh, A. Kant, Detoxification of Industrial Wastewater by Catalytic (Photo/Bio/Nano) Techniques, Sustain. Green Catal. Process. (2024) 377–396. https://doi.org/10.1002/9781394212767.ch16

[147]　Y. Bhaskar, I. Goel, V. Ranga, C.K. Singh, Challenges and Fate of Bio-nanomaterials in Industrial Applications, Bio-Nanomaterials Environ. Remediat. Ind. Appl. (2025) 273–294. https://doi.org/10.1002/9783527848546.ch11

[148]　V. Pandey, A. Sharma, D. Kumar, N. Samadhiya, S.S. Tomar, Industrial Application of Bio-nanomaterials in Agriculture, Bio-Nanomaterials Environ. Remediat. Ind. Appl. (2025) 225–254. https://doi.org/10.1002/9783527848546.ch9

[149]　S. Singh, K. Gupta, M. Sharma, H. Kaur, N.K. Sharma, Desalination of Wastewater Using Bio-nanomaterials, Bio-Nanomaterials Environ. Remediat. Ind. Appl. (2025) 105–132. https://doi.org/10.1002/9783527848546.ch4

[150]　A. Kishore, C. Singh, G. Kaur*, Bio-nanomaterials: An Introduction, Bio-Nanomaterials Environ. Remediat. Ind. Appl. (2025) 1–46. https://doi.org/10.1002/9783527848546.ch1

[151]　S.M. Mousavi, A. Badkoobeh, 10 Applications of Biomimetics, Bionanomaterials Ind. Appl. (2024) 181. https://doi.org/10.1201/9781003432791-13

Biogenic Nanomaterials: Synthesis, Characterization, Applications, and Future Remarks   Materials Research Forum LLC
Materials Research Foundations 180 (2025)   https://doi.org/21741/9781644903759

Chapter 17

Hydrogen/Energy Storage Applications Using Biogenic Nanomaterials

A.R. Kul[1]*, V. Benek[2], S. Celikozlu[3], A. Aygun[4], F. Sen[4]*

[1]Department of Chemistry, Faculty of Science, Van Yuzuncu Yil University, Van, Turkiye

[2]M.E.B. Vakıfbank Secondary School, Edremit, Van, Turkiye

[3]Department of Food Processing, Altıntaş Vocational School, Kütahya Dumlupınar University, Kutahya, Turkiye

[4]Sen Research Group, Department of Biochemistry, Kutahya Dumlupinar University, Kutahya, Turkiye.

alirizakul@yyu.edu.tr; fatih.sen@dpu.edu.tr

Abstract

The worldwide shift towards alternative energy sources has increased the potential of hydrogen as an energy carrier. However, expensive production methods and advanced storage technologies of hydrogen still pose major challenges. The use of biogenic nanomaterials offers a promising solution for hydrogen production and storage. These materials, designed at the nanoscale, have the potential to increase storage capacity and reduce costs. The integration of nanotechnology and biological systems offers new solutions for hydrogen storage, allowing an important step towards a clean energy future.

Keywords

Hydrogen Storage, Energy Efficiency, Greenhouse Gas Reduction, Sustainable Energy, Biogenic Nanomaterials

1.   Introduction

In the contemporary scientific landscape, the domain of renewable energy sources occupies a position of paramount importance, attracting fervent research endeavors across the globe [1]. Within this multi-faceted world, the supremacy of wind and sun power is the most critical. Inspired by a growing awareness of the destructive ecological consequences associated with the continued use of the planet's finite fossil fuel reserves, the search for alternative sources of energy has ignited an irrepressible force [2]. Amidst such fervent quest, hydrogen power has come to the forefront as a dominant topic in scientific debate. With the uncorrupted attributes, remarkable abundance, and natural bent towards ecological balance, hydrogen represents an intriguing option for future energy use [3–5]. In comparison to its carbon-based counterparts, hydrogen burning leaves behind no more than pristine water vapor and no permanent impact on the environment [6]. This inherent value, paired with the boundless reservoirs deep within the Earth's own framework, places hydrogen as a beacon of hope in the onset of dwindling supplies of fossil fuel. Though still challenging to overcome and harness is the whole nature of hydrogen as an energy force, the world of science cannot help but push forward regardless of the hurdles. By tireless investigation and unwavering perseverance, the vision of an energy future fueled by this pristine energy source can very well become an imminent reality [7,8].

However, hydrogen production is a major expense, thereby creating the pressing need for low-cost and high-efficiency storage methods. Conventional storage of hydrogen as a gas or liquid needs large volumes. Compression, the method employed in gas-phase storage, accounts for the use of nearly 15% of the stored hydrogen's energy, thereby adding significantly to associated costs. On the other hand, the liquefaction of hydrogen requires high-pressure vessels constructed from costly woven carbon nanofibers, resulting in the conceptualization of outsized equipment and vehicles in the hope of utilizing hydrogen as a significant fuel source [9].

In the contemporary scenario, strenuous efforts are being made relentlessly to alleviate the limitations associated with the production and storage of hydrogen. At the same time, one can discern a marked intensification of research activities for exploring new methods, which refuses to display any signs of abatement. These unwavering efforts, like a sustained siege laid on unyielding barriers, have the potential to unleash the hidden potential of this enigmatic fuel and lead it to a future that holds much promise.

2. What is Energy?

The energy ideology continues to be a dominant force within the rhetoric of this age, integral to the picture of our economic life. At the center of such debates is a point of attention: the primary sources powering our world and the powerful role that they play in shaping the destiny of our world. The argument here is one of all-encompassing attention towards the significance and sheer necessity of sources of clean, renewable energy. These sources have become a focus point, garnering significant awareness and attention in recent times, representing a landmark moment in our quest for green and sustainable sources to meet our energy needs [7]. The discourse around energy sources and their impact on the global economy stands as a testament to the evolving nature of our energy paradigms, underscoring the imperative need for transitioning towards renewable and cleaner sources to ensure a more sustainable future for generations to come.

At its core, energy embodies the capacity or capability to perform work. Any motion exhibited by animate or inanimate entities from one point to another, or any physical or chemical metamorphosis from one state to another, constitutes work and necessitates the utilization of energy. For modern humans, energy stands as a fundamental necessity for the sustenance of their daily lives. In earlier times, prior to the industrial revolution, energy requirements relied on elemental sources such as wood, wind, natural water resources, and human or animal labor. But with the advent of the coal-fired steam engine, the energy resource landscape underwent a dramatic transformation. The principal sources of energy now are coal, oil, natural gas, hydropower stations, and nuclear power stations [10].

The need for diversification and use of various sources of energy cannot be overstated. Yet even with this, energy consumption in the short term can be realistically problematic. Even while being produced, energy usage can be harmful to the environment. This issue is a pointer to the necessity of energy storage and underscores the need for increased efficiency in storage [11].

The development of effective energy storage technology is a breakthrough in bridging the gap of intermittency between renewable sources of energy and reducing the impact of energy consumption on our world. Storage efficiency not only guarantees a safe, stable, and balanced energy supply but also prevents energy waste considerably and reduces the energy consumption and production footprint. As we transition through the nuanced challenges of our energy needs, investing in technologies to hold onto energy becomes increasingly at the forefront of developing a more sustainable, stable energy future.

3. Storage of Energy

Increasing population along with development at the industrial scale has resulted in a runaway escalation of energy demands, whereas resources with us are constrained. Due to this mismatch, the difference between energy consumption and production has increased. In such a scenario, with the constant rise in energy needs, there is a more pressing need to maximize the use of our existing energy reservoirs. To meet the relentless increase in energy demand necessitates a crucial focus on not just maximizing the utilization of energy from renewable resources but also simplifying energy conversion methods.

The prime objective of this project is to fill the gap in energy demand and supply by optimally storing the energy derived from renewable resources and the remaining energy resources, which may in turn be available intermittently or geographically limited. This strategic move would assist in the creation of a stronger and more sustainable energy supply and thus address the increasing energy requirement of our world [11].

Greater efficiency in energy systems and concerted conservation can lead to dramatic changes. Consider, for instance, thermal energy storage and reuse in the winter, something that can result in savings of up to 22% in heating systems and a whopping 40% in air conditioning systems for hot summers. Such an approach not only minimizes reliance upon alternate energy sources but also maintains the purity of valuable fossil fuel reserves like coal, oil, and natural gas [11].

4. Energy Storage Methods

The capacity to access energy precisely when and where it is needed is most essential. The storage mechanism is the magic key that allows this flexibility, allowing us to use energy at our convenience. Energy storage technologies range across a broad spectrum, learning from natural ecologies where, for instance, biomass is an energy reservoir for other organisms and parasites.

The good storage techniques demand a special combination of the following properties: good storage capacity, efficient charging and discharging efficiencies, minimal loss through self-discharge or age-related loss of capacity, high longevity, cost-effectiveness, and high energy density—preferably stated in kilowatt-hours per kilogram or kilowatt-hours per liter. The importance is creating methods that are not only very effective at storing energy but also occupy minimal physical space and weight, thus being appropriate for different uses [12][11].

Energy storage in all its various forms is meant to meet these needs. The landscape has a myriad of storage technologies, both thermal and electrical. Electrical energy storage systems encompass a wide variety of solutions, including electrochemical systems such as batteries, kinetic energy storage methods through devices such as flywheels, and potential energy storage systems such as pumped hydro or compressed air, among others. Likewise, thermal energy storage systems take advantage of various technologies designed to fit particular applications and needs [12].

4.1   Usage Areas

The significance of energy storage is that it can hold energy generated during periods of maximum efficiency, where it can be saved for later use at varying times, hence avoiding energy wastage. This utility has been surprisingly used in a vast array of applications. Most significantly, it is significant in enhancing system flexibility, particularly in situations where primary resources are inadequate, basically reducing high transmission costs [13]. Moreover, it resolves the conundrum of portable energy in electric vehicles [14][15], ensures the provision of requisite energy for a safe aircraft landing in scenarios where onboard systems are disengaged during flight [16], and is instrumental in the optimization of lighting and engine operation in both marine and terrestrial vehicles [17]. Similarly, the storage of energy assumes critical importance in addressing the energy quandaries encountered by portable electronic devices such as mobile phones [17], as well as in furnishing the propulsive force for locomotives within mining operations [11].

The art of storing energy to be used in the future is no enigma peculiar to our own time; its history is traced through the ages. Processes like air energy storage or use and re-use of water energy [11] have been in use for centuries. But in our modern age, storage of different forms of energy for future use has not only formed the cornerstone of our civilization but has also reached an unprecedented level of significance. This growing significance has prompted a pressing need for innovative research to engage in researching new and efficient ways in the area of energy storage.

5.   Nanotechnology

Nanotechnology is a multidisciplinary science encompassing the examination of physical, chemical, and biological phenomena exhibited by materials within the small size range of 1 to 100 nanometers, along with their various applications. Nanomaterials are the foundation of nanotechnology with their distinctive optical, magnetic, mechanical, and electrical properties at this nanoscale. What is interesting in nanotechnology is the alteration in the behavior of materials at this scale from the macroscopic world. Power-to-weight ratio, electrical conductivity, and optical and magnetic properties are a few of the parameters that become radically different when moving from macro to nanoscale.

The father of modern nanotechnology is Richard Feynman, the renowned recipient of the 1965 Nobel Prize in Physics. In his seminal session titled "There's Plenty of Room at the Bottom," delivered at the American Physical Society meeting in 1959, Feynman suggested nanotechnology based on the manipulation at the atomic level. This groundbreaking idea was a paradigm shift in thinking, and Feynman's speculations have since been confirmed through empirical studies. It is because of these early efforts that Richard Feynman is generally considered the founder, and more popularly known as the "father," of nanotechnology today [18].

Nanotechnology has some key features characterizing its significant impact:

1. Unlimited Potential Technology: Nanotechnology is a technology of unlimited potential, revealing an entire world of possibilities, and unlocking doors towards the development of new

types of products previously impossible. It serves as a driving technology to create new products and processes in different industrial fields.

2. Disruptive Innovation: Changing old-school paradigms, nanotechnology is disrupting age-old conventions through new modes of production, hence yielding improved and superior grades of products. Slowly but surely, it disintegrates outdated technologies, transforming industrial environments to create a revolution.

3. Interdisciplinary Nature: Nanotechnology is an interdisciplinary phenomenon that transcends the traditional limits of science. It unifies various scientific disciplines so that it becomes simple for scientists from various disciplines to interact and communicate, thereby erasing boundaries between various scientific disciplines. The major subfields include nanostructured materials, nanoelectronics, nanophotonics, nanobiotechnology, and nanoanalytics.

4. Global Scope: Embraced globally, nanotechnology draws huge levels of government sponsorship within industrialized nations to finance gigantic research efforts. Further, companies and big firms across the world are investing heavily into furthering nanotechnology, bearing testimony to its global significance and appeal [19].

5.1 Biogenic Nanomaterials

Biometallic nanoparticles are now versatile agents with extensive uses cutting across several fields. They find applications on a wide spectrum of industries ranging from chemical engineering, tissue engineering, textile manufacture, nanomedicine, clinical diagnostics (such as nanorobots), electronics, to organ implantation [20]. Moreover, their significance extends to pivotal areas such as biosensors [21], bioimaging [22], biomarkers [23], and cell markers [24].

Amidst various synthetic methods available for manufacturing these nanoparticles, the innovative concept of employing biological organisms as potential nanofactories has gained traction. This approach introduces a novel means of generating a wide spectrum of nanoparticle variations, amplifying the scope and possibilities within the realm of biometallic nanoparticles.

In fact, the employment of many biological organisms in synthesizing nanoparticles demonstrates multiplicity of nature's capability in synthesizing nanoparticles. Some of them are:

1. *Klebsiella aerogenes:* The bacterium was employed in the intracellular synthesis of cadmium sulfide nanoparticles ranging in size from 20-200 nanometers [25].

2. *Verticillium dahliae*: This fungus was instrumental in the synthesis of nanosized gold and silver particles.

3. *Fusarium oxysporum*: A fungus, Fusarium oxysporum, produced silver nanocrystals of precise sizes and clear morphologies.

4. *Pseudomonas stutzeri:* It was able to produce silver nanocrystals of specific sizes and discrete morphologies [26][27].

These figures show the extensive range of biological systems used for their unique ability in the synthesis of nanoparticles of specific properties, size, and shape, paving the way for novel nanoparticle synthesis techniques.

In actuality, research has uncovered the remarkable role played by the Usnea longissima mushroom towards the development of antimicrobial nanoparticles predominantly composed of usnic acid. The nanoparticles have proven highly promising in combat against dermatophytic infections among humans, primarily through the use of the nanoemulsion method. The new frontier of biological entities such as this fungus being used in the synthesis of metallic nanoparticles is a burgeoning field with tremendous potential in a wide variety of applications. This field of study is still active and unfolding, marking the shifting landscape of nanoparticle synthesis by biological entities [28].

6.   Hydrogen

The course of global economic development since the start of the Industrial Revolution in 1750 has undergone a series of typical fluctuations, each propelled by technological advances. The stages of development have been characterized by five fundamental transitions.

1. 1750-1825 - Coal Period: The earliest period saw the rise of coal as the source of energy and forming the basis for the economic growth.

2. 1825-1860 - Era of Electricity: The second phase witnessed the advent of electricity, which was a stimulus for economic development.

3. 1860-1910 - Era of Electricity and Dominance of Oil: While electricity was prevailing, this era saw oil emerging as a fresh energy source to support economic growth.

4. 1910-1970 - Nuclear Energy Era: It was an era where nuclear energy held sway, remapping the economy in a vastly different manner.

5. 1970s-Present - The Era of Uncertainty: Currently, the world finds itself amidst a new fluctuation, the trajectory of which remains uncertain as it unfolds throughout the 21st century. This phase has been characterized by the emergent significance of hydrogen as a pivotal energy source, playing a transformative role in shaping the economy [29].

This ongoing phase underscores a critical juncture where hydrogen stands poised to potentially redefine the energy landscape, yet the ultimate outcome and ramifications of this phase remain to be fully understood as the century progresses.

The term "hydrogen" finds its roots in the Greek words "hydro," signifying water, and "genes," meaning generating. Positioned as the inaugural element on the periodic table, hydrogen reigns as the most prevalent and fundamental element in the universe. An estimated 75% or more of the universe's mass comprises hydrogen. However, despite its prevalence, hydrogen is seldom encountered in a free state on Earth due to its high reactivity.

Tightly interwoven with water, which covers roughly 60% of the Earth's surface, hydrogen emerges as a critical and fundamental component. Its intrinsic versatility spans across a multitude of domains, existing in diverse forms within living organisms, plants, fossil fuels, and an extensive array of chemical compounds [7]. This versatile element serves as an essential building block in numerous natural and industrial processes, showcasing its significance in powering various sectors, from energy production to chemical manufacturing. Moreover, hydrogen's role in the global pursuit of sustainable energy solutions is becoming increasingly prominent, particularly as a clean and renewable energy carrier that holds the potential to revolutionize our energy systems and reduce carbon emissions significantly. Its widespread availability and adaptable nature position hydrogen as a key player in the quest for a more sustainable and environmentally conscious future.

6.1 Production Hydrogen Methods

Hydrogen, often perceived as a synthetic fuel, is indeed not naturally occurring but is rather synthesized from various raw materials employing primary energy sources. Its production draws upon diverse sources, including water, fossil fuels, and biomass materials. Currently, the world utilizes approximately 500-600 billion cubic meters of hydrogen annually, primarily derived from fossil fuels, meeting the technological demands globally.

The fundamental principle guiding hydrogen fuel production revolves around obtaining hydrogen from water by harnessing renewable energies. Direct electrolysis of water stands as the foremost method for hydrogen production, with electricity required for electrolysis sourced from an array of energy streams: fossil fuels, hydroelectric power, nuclear power, geothermal energy, solar, wind, and sea wave energy. Among these, photovoltaic solar generators hold significant promise and garner substantial attention for future applications.

Additionally, hydrogen production methods encompass thermal cracking of water, steam reforming of natural gas and gaseous hydrocarbons, and coal gasification. The gasification process, notably, offers the advantage of easy sulfur elimination. On average, approximately 3,785 liters of hydrogen, equivalent to the energy of 6 kilograms of coal or 1 kilogram of hydrogen comparable to gasoline, is obtained through this method. Coal, recognized as the world's most abundant fossil fuel, holds a vast reserve estimated to last for about 200 years, with potential extensions to 400 years.

Moreover, solid waste and sewage materials also serve as viable raw materials for hydrogen production. Through the gasification process and subsequent reforming of syngas with air or oxygen, hydrogen can be derived. Notably, thermochemical cycles offer another avenue for hydrogen production from water, as do organometallic compounds or enzyme-water mixtures through photochemical treatment [29].

The diverse methodologies and raw materials utilized for hydrogen production underscore the versatility of this element and the breadth of possibilities in harnessing various sources to meet global energy demands while exploring cleaner, renewable avenues for sustainable energy production.

6.2 Use of Biogenic Nanomaterials for Storage of Hydrogen

Hydrogen stands as a compelling alternative to petroleum as an energy source, offering numerous advantages. However, unlike oil, hydrogen itself isn't a primary energy source; rather, it functions as an energy carrier, necessitating generation through various resources.

At present, a significant portion of the global hydrogen supply stems from methane, extracted via the steam methane reforming (SMR) process. Another prevalent method for hydrogen production is water electrolysis. Post-production, hydrogen requires transportation to end-users or distribution facilities, where it is stored for future utilization.

In terms of energy content, hydrogen outperforms gasoline significantly, boasting approximately three times the energy content on a mass basis (120 megajoules per kilogram compared to gasoline's 44 megajoules per kilogram). However, hydrogen's energy density in terms of volume is lower than gasoline, with liquid hydrogen measuring at 8 megajoules per liter compared to gasoline's 32 megajoules per liter.

Despite the challenges related to storage and transportation owing to its lower volumetric density, hydrogen showcases immense promise as a substitute for petroleum across various applications. Ongoing research endeavors are diligently focused on refining methods of hydrogen production, storage, and transportation, aiming to optimize its usage while simultaneously contributing to reducing reliance on petroleum-based energy sources [9].

Actually, hydrogen transport needs a good and sophisticated distribution system considering its unique characteristics. Hydrogen can be transported through various systems like pipelines, tube trailers, gas cylinders, or cryogenic tanks particularly if it is liquid.

However, one of the disadvantages of hydrogen is that it possesses relatively low energy density. Although hydrogen possesses high energy capacity, its energy density in unit volume is less than conventional fuels like gasoline. To enhance its energy density, compression and liquefaction operations can be employed. Compression compresses hydrogen gas to raise pressure, whereas liquefaction reduces its temperature to transform it into liquid. Both methods enable higher energy density but are expensive, thus slightly expensive to use [9].

Despite the impediments, development and research aspire to maximize vehicles of transport, as well as hydrogen infrastructure for utilizing it as a green as well as alternate energy carrier universally.

In fact, scientists acknowledge the enormous advantages of hydrogen as a source of power and have dedicated themselves to perfecting innovative methods for solving problems of storing and transporting it. In recent years, a great deal of interest has been attached to exploring whether biogenic matter can be employed to store hydrogen in an optimum way.

Researchers have conducted numerous studies exploring the feasibility of storing hydrogen within biogenic materials and evaluating the achievable yield levels. One such study, documented in a published paper [30], delves into the production of hydrogen from biogenic formic acid. This research not only investigates the production of hydrogen but also examines its potential utilization in rural areas. Insights garnered from this study shed light on the viability and practicality of storing hydrogen within biogenic formic acid.

Similarly, another notable study [31] emphasizes the significance of ethanol, a biologically derived compound, as a viable alternative for both obtaining and storing hydrogen. This research underscores the potential of ethanol derived from living organisms as an effective medium for hydrogen generation and storage.

These papers are landmark contributions to the field of scientific literature, offering vital information on the application of biogenic materials as a strategy for harvesting, storing, and using hydrogen, paving the way towards potentially innovative and sustainable hydrogen energy technologies.

7. Conclusion

Indeed, efficient, convenient, and dependable storage of hydrogen stands as a critical aspect across various applications. The current expense associated with hydrogen production underscores the necessity for cost-effective storage solutions. The future trajectory of available alternatives, whether based on physical or chemical methods, will likely hinge upon cost and capacity considerations, influenced by technological advancements within each method.

Hydrogen can be stored and transported either as a gas or in its liquid form, but both methods pose unique challenges. Gas-phase storage demands significantly large volumes due to hydrogen's low density. Moreover, compression of hydrogen gas incurs an energy cost, with up to 15% of the stored hydrogen's energy expended in this process. Additionally, the requisite high-pressure storage tanks, typically constructed from woven carbon nanofibers, contribute to elevated manufacturing expenses. Consequently, devices and vehicles utilizing hydrogen as a fuel may need to accommodate abnormally large storage units, presenting a challenge in terms of size and cost [9].

Absolutely, storing hydrogen in liquid form serves as an alternative method, offering certain advantages, although it comes with its own set of challenges and associated expenses.

Liquefying hydrogen requires subjecting it to extremely low temperatures, typically below -253 degrees Celsius (-423 degrees Fahrenheit), to transition it into a liquid state [32]. This process demands high pressure and consumes a considerable amount of energy, accounting for approximately 30% of the stored hydrogen's energy. Moreover, maintaining the integrity of liquid hydrogen tanks necessitates robust insulation to minimize heat transfer from the surroundings, as any rise in temperature can prompt the reversion of liquid hydrogen back into its gaseous state, leading to potential safety hazards and efficiency losses.

Despite these challenges and the associated costs, liquid hydrogen storage remains an important avenue in certain applications, particularly where higher energy density and reduced storage volume are imperative. Continued research and technological advancements aim to mitigate these challenges, striving to improve the efficiency and cost-effectiveness of storing hydrogen in its liquid form for various industrial and transportation applications.

Absolutely, solid-state hydrogen storage methods have garnered attention as a potential alternative to traditional gas and liquid storage due to their potential for safety and storage efficiency. However, these methods still face challenges related to their storage capacity, which tends to be relatively lower compared to other storage techniques like gas or liquid storage.

Despite these limitations, ongoing research and development efforts are diligently focused on advancing solid-state hydrogen storage technologies. Scientists and engineers are working towards enhancing the storage capacity, efficiency, and practicality of solid-state storage methods. Novel materials and innovative engineering approaches are being explored to overcome these limitations and maximize the potential of solid-state hydrogen storage.

The broader focus of research encompasses not only storage but also the improvement of hydrogen transportation methods and utilization as an alternative energy source. The utilization of hydrogen holds immense promise in significantly reducing greenhouse gas emissions and fostering sustainable energy production. Continued advancements in hydrogen storage and utilization technologies can potentially play a pivotal role in addressing environmental concerns and achieving a more sustainable energy future.

Indeed, various methodologies are being explored to overcome the challenges associated with hydrogen recovery and storage. Among these, studies focusing on biogenic nanomaterials have shown promising potential as an alternative for hydrogen storage.

Research into biogenic nanomaterials has unveiled their unique properties and their applicability in hydrogen storage. These materials derived from biological sources at the nanoscale exhibit distinctive characteristics that make them promising candidates for efficient and effective hydrogen storage.

The utilization of biogenic nanomaterials offers several advantages, including their high surface area, porosity, and tunable properties, which can contribute to improved hydrogen adsorption and storage capacities. Moreover, their environmentally friendly nature, abundance, and potential scalability make them appealing for sustainable hydrogen storage solutions.

Studies exploring the utilization of biogenic nanomaterials for hydrogen storage continue to advance, driven by the prospect of overcoming existing challenges and establishing innovative pathways for efficient, cost-effective, and eco-friendly hydrogen storage technologies. These endeavors hold promise in addressing limitations associated with conventional storage methods, contributing to the advancement of hydrogen as a viable and sustainable energy carrier.

Reference

[1]   A. Hojjati-Najafabadi, A. Aygun, R.N.E. Tiri, F. Gulbagca, M.I. Lounissaa, P. Feng, F. Karimi, F. Sen, Bacillus thuringiensis Based Ruthenium/Nickel Co-Doped Zinc as a Green Nanocatalyst: Enhanced Photocatalytic Activity, Mechanism, and Efficient H2Production from Sodium Borohydride Methanolysis, Ind. Eng. Chem. Res. (2022). https://doi.org/10.1021/ACS.IECR.2C03833

[2]   A. Aygun, F. Gulbagca, E.E. Altuner, M. Bekmezci, T. Gur, H. Karimi-Maleh, F. Karimi, Y. Vasseghian, F. Sen, Highly active PdPt bimetallic nanoparticles synthesized by one-step bioreduction method: Characterizations, anticancer, antibacterial activities and evaluation of their catalytic effect for hydrogen generation, Int. J. Hydrogen Energy (2022). https://doi.org/10.1016/J.IJHYDENE.2021.12.144

[3]   A. Cherif, R. Nebbali, J.W. Sheffield, N. Doner, F. Sen, Numerical investigation of hydrogen production via autothermal reforming of steam and methane over Ni/Al2O3 and Pt/Al2O3 patterned catalytic layers, Int. J. Hydrogen Energy (2021). https://doi.org/10.1016/j.ijhydene.2021.04.032

[4]   S. Günbatar, A. Aygun, Y. Karataş, M. Gülcan, F. Şen, Carbon-nanotube-based rhodium nanoparticles as highly-active catalyst for hydrolytic dehydrogenation of dimethylamineborane at room temperature, J. Colloid Interface Sci. 530 (2018) 321–327. https://doi.org/10.1016/J.JCIS.2018.06.100

[5]   H. Goksu, Y. Yildiz, B. Çelik, M. Yazici, B. Kilbas, F. Sen, Eco-friendly hydrogenation of aromatic aldehyde compounds by tandem dehydrogenation of dimethylamine-borane in the presence of a reduced graphene oxide furnished platinum nanocatalyst, Catal. Sci. Technol. 6 (2016) 2318–2324. https://doi.org/10.1039/C5CY01462J

[6]   H. Göksu, Y. Yıldız, B. Çelik, M. Yazıcı, B. Kılbaş, F. Şen, Highly Efficient and Monodisperse Graphene Oxide Furnished Ru/Pd Nanoparticles for the Dehalogenation of Aryl Halides via Ammonia Borane, ChemistrySelect 1 (2016) 953–958. https://doi.org/10.1002/slct.201600207

[7]   H. Goksu, Y. Yıldız, B. Çelik, M. Yazici, B. Kilbas, F. Şen, Eco-friendly hydrogenation of aromatic aldehyde compounds by tandem dehydrogenation of dimethylamine-borane in the presence of a reduced graphene oxide furnished platinum nanocatalyst, Catal. Sci. Technol. 6 (2016) 2318–2324. https://doi.org/10.1039/C5CY01462J

[8]   B. Sen, B. Demirkan, A. Şavk, S. Karahan Gülbay, F. Sen, Trimetallic PdRuNi nanocomposites decorated on graphene oxide: A superior catalyst for the hydrogen evolution reaction, Int. J. Hydrogen Energy 43 (2018) 17984–17992. https://doi.org/10.1016/J.IJHYDENE.2018.07.122

[9]   M.R. Usman, Hydrogen storage methods: Review and current status, Renew. Sustain. Energy Rev. 167 (2022) 112743. https://doi.org/10.1016/j.rser.2022.112743

[10]   Y. Wu, E.E. Altuner, R.N. El Houda Tiri, M. Bekmezci, F. Gulbagca, A. Aygun, C. Xia, Q. Van Le, F. Sen, H. Karimi-Maleh, Hydrogen generation from methanolysis of sodium borohydride using waste coffee oil modified zinc oxide nanoparticles and their photocatalytic activities, Int. J. Hydrogen

Energy 48 (2023) 6613–6623. https://doi.org/10.1016/J.IJHYDENE.2022.04.177

[11]   J. Mitali, S. Dhinakaran, A.A. Mohamad, Energy storage systems: a review, Energy Storage Sav. 1 (2022) 166–216. https://doi.org/10.1016/j.enss.2022.07.002

[12]   E. Sayed, A. Olabi, A. Alami, A. Radwan, A. Mdallal, A. Rezk, M. Abdelkareem, Renewable Energy and Energy Storage Systems, Energies 16 (2023) 1415. https://doi.org/10.3390/en16031415

[13]   M. Huang, W. He, A. Incecik, A. Cichon, G. Królczyk, Z. Li, Renewable energy storage and sustainable design of hybrid energy powered ships: A case study, J. Energy Storage 43 (2021) 103266. https://doi.org/10.1016/j.est.2021.103266

[14]   Ü. ÖZBALCI, E. KILIÇ, MODELING THE BATTERY SYSTEM OF AN ELECTRIC VEHICLE, Kahramanmaraş Sütçü İmam Üniversitesi Mühendislik Bilim. Derg. 22 (2019) 64–69. https://doi.org/10.17780/ksujes.600809

[15]   S. S. Rangarajan, S.P. Sunddararaj, A. Sudhakar, C.K. Shiva, U. Subramaniam, E.R. Collins, T. Senjyu, Lithium-Ion Batteries—The Crux of Electric Vehicles with Opportunities and Challenges, Clean Technol. 4 (2022) 908–930. https://doi.org/10.3390/cleantechnol4040056

[16]   M. Clarke, J.J. Alonso, Lithium–Ion Battery Modeling for Aerospace Applications, J. Aircr. 58 (2021) 1323–1335. https://doi.org/10.2514/1.C036209

[17]   Y. Wang, R. Xu, C. Zhou, X. Kang, Z. Chen, Digital twin and cloud-side-end collaboration for intelligent battery management system, J. Manuf. Syst. 62 (2022) 124–134. https://doi.org/10.1016/j.jmsy.2021.11.006

[18]   A. Hamisu, O. Khiter, S. Al-Zhrani, W.S.B. Haridh, Y. Al-Hadeethi, M.I. Sayyed, S.A. Tijani, The use of nanomaterial polymeric materials as ionizing radiation shields, Radiat. Phys. Chem. 216 (2024) 111448. https://doi.org/10.1016/j.radphyschem.2023.111448

[19]   G.J. Jordaan, W.J. vdM. Steyn, Practical Application of Nanotechnology Solutions in Pavement Engineering: Identifying, Resolving and Preventing the Cause and Mechanism of Observed Distress Encountered in Practice during Construction Using Marginal Materials Stabilised with New-Age (Nan, Appl. Sci. 12 (2022) 2573. https://doi.org/10.3390/app12052573

[20]   T. Santhoshkumar, A.A. Rahuman, G. Rajakumar, S. Marimuthu, A. Bagavan, C. Jayaseelan, A.A. Zahir, G. Elango, C. Kamaraj, Synthesis of silver nanoparticles using Nelumbo nucifera leaf extract and its larvicidal activity against malaria and filariasis vectors, Parasitol. Res. 108 (2011) 693–702. https://doi.org/10.1007/s00436-010-2115-4

[21]   H. Vasconcelos, L.C.C. Coelho, A. Matias, C. Saraiva, P.A.S. Jorge, J.M.M.M. de Almeida, Biosensors for Biogenic Amines: A Review, Biosensors 11 (2021) 82. https://doi.org/10.3390/bios11030082

[22]   L.-J. Tian, N.-Q. Zhou, X.-W. Liu, J.-H. Liu, X. Zhang, H. Huang, T.-T. Zhu, L.-L. Li, Q. Huang, W.-W. Li, Y.-Z. Liu, H.-Q. Yu, A sustainable biogenic route to synthesize quantum dots with tunable fluorescence properties for live cell imaging, Biochem. Eng. J. 124 (2017) 130–137. https://doi.org/10.1016/j.bej.2017.05.011.

[23]   R. Nagraik, A. Sharma, D. Kumar, S. Mukherjee, F. Sen, A.P. Kumar, Amalgamation of biosensors and nanotechnology in disease diagnosis: Mini-review, Sensors Int. 2 (2021) 100089. https://doi.org/10.1016/J.SINTL.2021.100089

[24]   A.T. Le, P.T. Huy, L.T. Tam, P.D. Tam, N. Van Hieu, T.Q. Huy, Novel silver nanoparticles: synthesis, properties and applications, Int. J. Nanotechnol. 8 (2011) 278. https://doi.org/10.1504/IJNT.2011.038205

[25]   J.D. Holmes, P.R. Smith, R. Evans-Gowing, D.J. Richardson, D.A. Russell, J.R. Sodeau, Energy-dispersive X-ray analysis of the extracellular cadmium sulfide crystallites of Klebsiella aerogenes, Arch. Microbiol. 163 (1995) 143–147. https://doi.org/10.1007/BF00381789

[26]   T. Klaus-Joerger, R. Joerger, E. Olsson, C.-G. Granqvist, Bacteria as workers in the living

factory: metal-accumulating bacteria and their potential for materials science, Trends Biotechnol. 19 (2001) 15–20. https://doi.org/10.1016/S0167-7799(00)01514-6

[27]   M. Sastry, V. Patil, S.R. Sainkar, Electrostatically Controlled Diffusion of Carboxylic Acid Derivatized Silver Colloidal Particles in Thermally Evaporated Fatty Amine Films, J. Phys. Chem. B 102 (1998) 1404–1410. https://doi.org/10.1021/jp9719873

[28]   A.R. Shahverdi, S. Minaeian, H.R. Shahverdi, H. Jamalifar, A.-A. Nohi, Rapid synthesis of silver nanoparticles using culture supernatants of Enterobacteria: A novel biological approach, Process Biochem. 42 (2007) 919–923. https://doi.org/10.1016/j.procbio.2007.02.005

[29]   E.I. Epelle, K.S. Desongu, W. Obande, A.A. Adeleke, P.P. Ikubanni, J.A. Okolie, B. Gunes, A comprehensive review of hydrogen production and storage: A focus on the role of nanomaterials, Int. J. Hydrogen Energy 47 (2022) 20398–20431. https://doi.org/10.1016/j.ijhydene.2022.04.227

[30]   P. Preuster, J. Albert, Biogenic Formic Acid as a Green Hydrogen Carrier, Energy Technol. 6 (2018) 501–509. https://doi.org/10.1002/ente.201700572

[31]   C. Mevawala, K. Brooks, M.E. Bowden, H.M. Breunig, B.L. Tran, O.Y. Gutiérrez, T. Autrey, K. Müller, The Ethanol–Ethyl Acetate System as a Biogenic Hydrogen Carrier, Energy Technol. 11 (2023). https://doi.org/10.1002/ente.202200892

[32]   C. Tarhan, M.A. Çil, A study on hydrogen, the clean energy of the future: Hydrogen storage methods, J. Energy Storage 40 (2021) 102676. https://doi.org/10.1016/j.est.2021.102676

Biogenic Nanomaterials: Synthesis, Characterization, Applications, and Future Remarks   Materials Research Forum LLC
Materials Research Foundations 180 (2025)   https://doi.org/21741/9781644903759

Chapter 18

Other Applications of Biogenic Nanomaterials: Applications of Fuel Cells Using Biogenic Nanomaterial

H. Elaibi[1,2*], F. Mutlag[1], R. Mahious[3], F. Sen[1*]

[1]Department of Biochemistry, Sen Research Group, Kutahya Dumlupinar University, 43000, Kutahya, Turkiye

[2]Department of Pharmacy. Al Safwa University College, Karbala, 56001, Iraq
[3]Department of Biology, Kutahya Dumlupinar University, 43000, Kutahya, Turkiye

Hussein.elaibi@ogr.dpu.edu.tr; fatih.sen@dpu.edu.tr

Abstract

Fuel cells are electrochemical reactions that convert chemical energy into electricity, rendering them potentially beneficial energy sources. Proton exchange membranes (PEMFCs), solid oxides (SOFCs), alkaline solutions (AFCs), direct methanol (DMFCs), biofuels (Biofuel Cells), and microorganisms (Microbial Fuel Cells) are all fuel cells that have their own set of advantages and weaknesses. The increased interest in biogenic nanomaterials and carbon-based electrode supports is boosting fuel cell efficiency, lowering prices, and promoting sustainability. This study focuses on the key properties, applications, and problems of these fuel cell technologies, as well as their significance in the shift to cleaner energy sources.

Keywords

Fuel cells, Proton Exchange Membrane Fuel Cells (PEMFCs), Solid Oxide Fuel Cells (SOFCs), Alkaline Fuel Cells (AFCs), Direct Methanol Fuel Cells (DMFCs), Biofuel Cells, Microbial Fuel Cells (MFCs)

1. Introduction

Pursuit of sustainable and efficient energy solutions has intensified in recent years, driven by increasing interest in renewable energy and the pressing need to mitigate pollution levels [1–5]. The adaptability, minimal emissions, and elevated efficiency of fuel cells have rendered them a very desirable green energy conversion technology [6]. Catalysts are crucial for the operation of fuel cells as they significantly improve the electrochemical reactions that produce energy [7].

Biogenic nanomaterials, nanostructures produced by biological systems such as plants, microbes, and enzymes, represent a recent advancement in this domain [8–12]. They could supplant conventional synthetic materials in a sustainable and economically viable manner [13,14].

Nanomaterials have unique benefits since they can be scaled, are made using environmentally acceptable processes, and have tunable features [15]. Their high conductivity, stability, and catalytic efficacy, these materials are the best choice for use in fuel cells, including microbial fuel cells (MFCs) and proton exchange membrane fuel cells (PEMFCs) [16,17] .That use of biological systems in synthesis makes it easier to produce materials with precise morphologies and compositions while also lowering energy consumption and minimizing the generation undesired byproducts [18]. Microorganisms like fungi and bacteria have attracted a lot of attention for their ability to produce metal nanoparticles, while plant-based methods have produced nanostructures with interesting surface properties [19,20]. Biogenic nanomaterials, made from microbes, plants, natural extracts in an eco-friendly manner, have attracted a lot of attention for a variety of uses, one of which being fuel cells [21]. Fuel cell technology stands to benefit greatly from their greater surface area, biocompatibility, and higher catalytic activity, among other unique properties [22,23].

Recent studies indicate that in fuel cells biogenic metal and metal oxide nanoparticles generated from plant extracts, microbial enzymes, or biomolecules can effectively catalyze hydrogen oxidation and oxygen reduction reactions. High electrical conductivity and stability carbon-based nanomaterials as graphene and biogenic carbon nanotubes help to enhance fuel cell performance even more [24,25].

Fuel cells that use biogenic nanoparticles not only minimize the demand for important raw materials, but they also make these energy sources more ecologically friendly and longer-lasting. Recent research has made it possible to create fuel cell devices that use biogenic nanomaterials. These devices could drastically change the next generation of green energy technology. These devices are more effective and have a smaller negative impact on the environment [26,27] .

1.1   Biogenic Nanomaterials in Fuel Cell Catalysts

Biogenic nanomaterials, made from microbes, plants, natural extracts in an eco-friendly manner, have attracted a lot of attention for a variety of uses, one of which being fuel cells [21]. Fuel cell technology stands to benefit greatly from their greater surface area, biocompatibility, and higher catalytic activity, among other unique properties [28]. As in Figure 1.

To improve the anode and cathode processes in fuel cells, which frequently involve oxygen and hydrogen, catalysts are used [29]. Traditional catalysts, those based on platinum, are both scarce and expensive. An eco-friendly and economical substitute would be biogenic nanoparticles [30].

Raising fuel cell efficiency depends on biogenic nanoparticles. Both of which depend on appropriate catalysts, two of the most major problems with fuel cells are the oxygen reduction reaction (ORR) and the hydrogen oxidation process (HOR) [31]. Platinum-based catalysts have long been employed since their great catalytic activity [32]. They do, however, have limited availability, costly prices, and degrade with time. One interesting substitute are biogenic nanomaterials.

Two bacteria that might lower metal ions to generate gold, silver, and platinum nanoparticles with improved catalytic activity are Pseudomonas aeruginosa and Escherichia coli [33,34]. Biogenic synthesis also produces metal oxides with great catalytic activity for fuel cell uses including Fe_3O_4, TiO_2, and CeO_2. Excellent substitutes for platinum-based catalysts include carbon nanotubes and graphene produced by bacterial and plant extracts since of their great conductivity and stability [35–37].

Biogenic Nanomaterials: Synthesis, Characterization, Applications, and Future Remarks   Materials Research Forum LLC
Materials Research Foundations 180 (2025)   https://doi.org/21741/9781644903759

Nanostructured supports enhance the electrical conductivity and endurance of catalyst dispersion. The anodes and cathodes of fuel cells are enhanced by biogenic materials, including reduced graphene oxide carbon nanofibres and biochar-derived carbon[38]. Biochar, which is carbon that is extracted from agricultural residue and subjected to microbial synthesis, provides a high surface area and porosity. This increases electrocatalytic activity. High-conductivity, defect-free materials enhanced by microbiological assistance from graphene and carbon nanotubes improve electron transport [39,40] .

Specifically, PEMFCs call for fuel cell membranes with strong proton conductivity, mechanical stability, and fuel crossover resistance. Ionic conductivity raised by biogenic nanoparticles aids to improve membranes [42]. While lowering methanol crossover, nanofibers of cellulose and biogenic silica can boost proton transfer channels. In direct methanol fuel cells, functionalized nanocellulose membranes increase selectivity, hence boosting their efficiency [43,44].

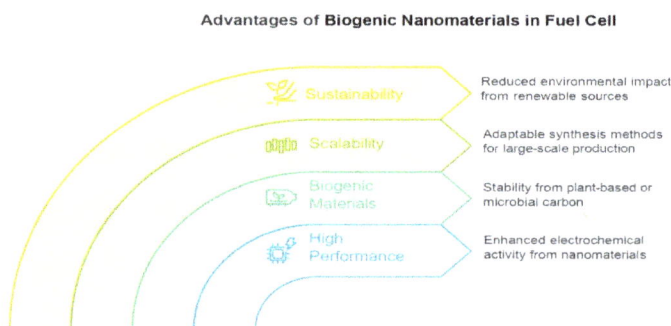

Figure 1. Advantages of biogenic nanomaterials in fuel cell.

1.2   Applications for Biogenic Nanomaterials in Fuel Cell Catalysts

Fuel cells are devices that use electrochemical processes to convert chemical energy into electricity in a clean and efficient manner [45]. The use of biogenic nanoparticles improves the cost-effectiveness, environmental sustainability, and efficiency of fuel cells [24]. These nanoparticles are made of biological components like plants, bacteria, fungi, and agricultural waste [46]. They are ideal for use in fuel cells because they have excellent conductivity, catalytic efficiency, and structural integrity [47,48].

Biogenic nanomaterials, derived from biological sources such as plants, fungi, and bacteria, present an innovative and sustainable method for fuel cell catalysis [49]. These materials are synthesized into nanoscale particles with remarkable catalytic capabilities [50]. These materials often include metals such platinum, silver, and gold in addition to nonmetallic elements include carbon-based structures that help fuel cell operations [51,52]. Increased interaction with reactants and effective electron transport made possible by their porous structure and large surface area help to improve fuel cell performance generally [53,54].

Its key benefit is that it may be manufactured without harmful chemicals or energy-intensive methods, unlike other catalysts [55,56]. This improves manufacturing efficiency and reduces environmental damage. Modifying the surfaces of biogenic nanomaterials made by biological processes makes them more stable and durable [57,58].

Fuel cell activities using hydrogen and oxygen are crucial for energy conversion[59,60]. Biogenic nanoparticles improve fuel cell power generation and longevity[61]. Scientists create more effective and long-lasting catalysts by merging tiny structures with natural nanomaterials [62–64]

Biogenic nanoparticles in fuel cell catalysts could create sustainable energy sources. Great process catalysts, low toxicity, and little environmental impact [65,66].

2. Key Fuel Cell Types Utilizing Biogenic Nanomaterials

Biogenic nanoparticles are efficient and sustainable, which helps several kinds of fuel cells [24,67]. Highly effective catalysts are required for proton exchange membrane fuel cells to reduce oxygen and oxidize hydrogen, therefore enabling energy conversion [68,69]. Large surface area biogenic nanomaterials provide increased interaction with reactants and electron movement. Their capacity to create composite buildings using different materials increases fuel cell lifetime by reducing catalyst degradation and therefore improving durability [70,71]. Moreover, their ecologically benign manufacture qualifies them as a more sustainable substitute for traditional metal-based catalysts [56].

Alkaline fuel cells operate in an alkaline atmosphere where oxygen reduction reaction kinetics are more favorable, so allowing the use of non-precious metal catalysts [72,73]. Good catalysts in this regard are biogenic nanomaterials including metal oxides, carbon-based nanostructures, and biomineralized nanoparticles. They increase response rate generally, conductivity, and effective charge transfer [74,75]. Their adjustable surface properties allow better interaction with reactants, hence increasing fuel cell performance and reducing dependency on costly platinum-based catalysts [76–78].

Solid oxide fuel cells, which run at high temperatures, depend on stable materials that can withstand severe conditions and maintain catalytic efficacy [79,80]. Biogenic nanomaterials much enhance electrode performance by raising ionic and electrical conductivity [81,82]. Gas diffusion, which their porous designs enable, is fundamental for both effective fuel oxidation and oxygen reducing. Moreover, composite materials and biogenic metal oxides can help solid oxide fuel cells to be stabilized, hence enhancing their resistance to degradation over time [83,84]

Using biogenic nanomaterials, microbial fuel cells in a unique approach enhance electron flow between the electrode surface and electro active bacteria. The enhancement of this electron transfer route is essential for the increase in power output, as microbial fuel cells generate electricity by decomposing organic materials. Carbon-based nanostructures and bio-synthesized metal nanoparticles are two biogenic nanostructures that enhance bacterial adhesion to the electrode and provide conductive channels for efficient charge transfer. These advances render microbial fuel cells more viable for applications including bioelectricity production and wastewater treatment, enhancing energy conversion efficiency [85,86]

Biogenic nanoparticles offer a renewable, high-performance, and cost-effective alternative to conventional catalysts, hence fostering a more sustainable energy ecosystem for various fuel cells. They are essential for the progression of fuel cell technology in future energy applications since they enhance stability, reaction kinetics, and environmental impact [87].

2.1   *Proton Exchange Membrane Fuel Cells (PEMFCs)*

Proton exchange membrane fuel cells (PEMFCs) are prevalent fuel cell varieties that generate electricity through an electrochemical reaction between hydrogen and oxygen [88,89]. They utilize a solid polymer membrane as their electrolyte. This lets protons move through it while making electrons flow through an external circuit, creating electric current. The reaction happens at low temperatures, which makes PEMFCs great for uses that need fast startup and quick response, like in cars, portable power devices, and backup energy sources [90,91]. PEMFCs are light and small, which makes them great for transportation, especially in hydrogen fuel cell cars. They are quite successful and do not generate any contaminants [92]. As shown in Table 1.

Despite its benefits, traditional PEMFCs are limited due to factors like as cost, material durability, and catalyst degradation. Biogenic nanomaterials, generated from microbes, plants, algae, and animals, are a viable alternative to traditional catalysts and electrode materials since they are inexpensive, last a long time, and improve the efficiency and longevity of PEMFCs. More Research into biogenic nanomaterials for fuel cell uses might pave the way for next-generation energy technologies that are less harmful to the environment, more cost-effective, and more efficient [93–95].

Table 1. Characteristics of Proton Exchange Membrane Fuel Cells (PEMFCs).

Feature	Proton Exchange Membrane Fuel Cells (PEMFCs)	References
Electrolyte Type	Solid polymer membrane (Proton Exchange Membrane)	[96]
Operating Temperature	Low (typically 60–80°C)	[97]
Fuel Source	Hydrogen and oxygen (high purity required)	[98]
Ion Transport	Protons (H$^+$) move from anode to cathode	[99]
Catalyst Type	Platinum-based catalysts	[100]
Applications	Hydrogen fuel cell vehicles, portable power, backup power, stationary power	[101]
Advantages	High efficiency, fast startup, compact design, zero emissions	[102]
Challenges	Expensive catalysts, hydrogen purity requirements, water management issues	[103]

Table 1. Shows there is a synopsis of the key features of proton exchange membrane fuel cells (PEMFCs), together includes the type of electrolyte they employ, operational parameters, fuel sources, applications, advantages, and disadvantages.

2.2   Alkaline Fuel Cells (AFCs)

Alkaline fuel cells (AFCs) are a class of electrochemical device designed to induce redox reactions converting chemical energy into electricity by means of an alkaline electrolyte. Using non-platinum catalysts allows AFCs—a less costly substitute for proton exchange membrane fuel cells (PEMFCs)—to be used[42,104].

Alkaline fuel cells (AFCs) usually use a potassium hydroxide (KOH) solution as an alkaline medium to help generate energy through an electrochemical process. These cells mix with hydrogen to produce water and electrons by allowing hydroxide ions (OH^-) to move from the cathode to the anode. As the electrons travel across an external circuit, an electric current is produced [42,105]. AFCs are especially useful in airplane applications since of their great efficiency and rapid reaction times. For instance, they have worked on space missions producing drinking water and energy.

Moreover, Electrolyte instability, electrode degradation, and carbonate toxicity (resulting from airborne CO_2 contamination) are all common issues in conventional AFCs [80].

The research of biogenic nanomaterials, which are nanostructured materials produced by biological processes, for the purpose of serving as catalysts, electrode materials, and alkaline membranes in AFCs is a solution to these limitations [72,106].

These nanoparticles exhibit significant potential for the development of next-generation AFCs due to their enhanced durability, environmental sustainability, and robust catalytic activity. they may run at rather low temperatures, which simplifies system design and lowers material requirements [107,108]. As shown in Table 2.

Table 2. Characteristics of alkaline fuel cells (AFCs)

Feature	Alkaline Fuel Cells (AFCs)	References
Electrolyte Type	Potassium hydroxide (KOH) solution or solid alkaline membrane	[109]
Operating Temperature	Low to moderate (typically 60–90°C)	[110]
Fuel Source	Hydrogen and oxygen (require high purity)	[111]
Ion Transport	Hydroxide ions (OH^-) move from cathode to anode	[112]
Catalyst Type	Non-precious metals (e.g., nickel)	[113]
Applications	Space missions, backup power, submarines, military uses	[114]
Advantages	High efficiency, fast reaction kinetics, low operating temperature	[115]
Challenges	CO_2 sensitivity, requires purified gases, potential carbonate formation, limited terrestrial applications	[116]

Table 2 gives a summary of Alkaline Fuel Cells (AFCs). It shows their electrolyte type, how they work, what fuels they use, where they are applied, their benefits, and their difficulties.

2.3   Solid Oxide fuel cells (SOFCs)

To convert chemical energy into electricity, solid oxide fuel cells (SOFCs) oxidize fuels such hydrogen, natural gas, biogas, or hydrocarbons[117]. Unlike some fuel cells using liquid or polymer electrolytes, SOFCs make use of a solid ceramic electrolyte most usually yttria-stabilized zirconia (YSZ). This chemical enhances energy conversion by increasing the conduction of oxygen ions ($O2^-$) at elevated temperatures (600–1000°C) [118], [119].

Fuel flexibility (compatible with hydrogen, syngas, methane, and other hydrocarbons) and high efficiency (50–65% for electricity generation and up to 85% in combined heat and power systems) are all essential components[120]. However, many people are not using SOFCs widely because there are significant obstacles, even though they have benefits [121]. This chemical helps convert energy better by allowing oxygen ions ($O2^-$) to move more easily at high temperatures (600–1000°C)[122].

The high operating temperature (800–1000°C) is a significant issue since it results in lengthy startup times, damage to materials, and thermal stress.The electrodes are deteriorating, which is causing the system to be less reliable and efficient. This deterioration is the result of sulfur, repetitive oxidation and reduction, and the accumulation of carbon. The performance of the fuel cell is also significantly impacted by the sluggish oxygen reduction reaction (ORR) of the cathode, which is another important factor. The use of rare-earth-doped ceramics and other costly materials raises production costs even more. In the end, mechanical stability problems arise because the materials that are used for the electrolyte and electrodes need to be able to withstand high temperatures without breaking or deteriorating [123,124].

Researchers are investigating biogenic nanomaterials as a low-cost, high-performance alternative to solid oxide fuel cell (SOFC) components in order to address these issues. Biogenic nanomaterials are nanostructured materials that are produced by biological processes. These materials have the potential to improve catalytic activity, ionic conductivity, electrode durability, and overall cost-effectiveness. Incorporating biogenic nanomaterials into SOFC electrodes, electrolytes, and interconnect coatings will help fuel cells to be more efficient, long-lasting, and financially feasible. More developments in biogenic material synthesis, nanotechnology, and electrochemical engineering are needed to completely use SOFCs in upcoming energy systems [24,125].

Common uses for these fuel cells in stationary power generation—that is, in combined heat and power (CHP) systems—where both useful heat and electricity are generated Additionally under investigation are uses for them in remote power systems and auxiliary power units for transportation [106]. Due to their effective utilization of hydrocarbon fuels and electricity generation capabilities, SOFCs are considered a viable solution for energy production in a dispersed and sustainable fashion. Critical domains requiring further inquiry and enhancement include material stability difficulties, prolonged system initiation times, and the associated costs of these systems [126]. As shown in Table 3.

Table 3. Characteristics of solid oxide fuel cells (SOFCS).

Feature	Solid Oxide Fuel Cells (SOFCs)	References
Electrolyte Type	Solid ceramic (e.g., zirconia-based)	[127]
Operating Temperature	High (typically above 600°C)	[128]
Fuel Source	Hydrogen, natural gas, biogas, syngas	[129]
Ion Transport	Oxygen ions (O^{2-}) move through the electrolyte	[130]
Catalyst Type	Non-precious metal oxides (no need for noble metals)	[131]
Applications	Stationary power plants, CHP systems, remote power, auxiliary power units	[132]
Advantages	High efficiency, fuel flexibility, no need for expensive catalysts, useful heat byproduct	[133]
Challenges	High operating temperature, material degradation, long startup time, high manufacturing cost	[128]

Table 3 summarizes the operating principle, materials, fuel sources, applications, benefits, and limitations of Solid Oxide Fuel Cells (SOFCs).

2.4 Microbial fuel cells (MFCs)

The microbial fuel cell (MFC) is a new bioelectrochemical device that produces energy by using the metabolic processes of microorganisms. MFCs use microorganisms as biocatalysts to drive electrochemical reactions, as opposed to typical fuel cells, which use chemical catalysts (such as platinum). MFCs have the potential to be an environmentally friendly technology for the production of renewable energy due to their inherent qualities [134,135].

The two primary components of an MFC are the anode and the cathode. Protons and electrons are generated when the anode microorganisms decompose organic materials, waste, and natural sugars. Protons go in the direction of the cathode and join forces with an electron acceptor, usually oxygen, to finish the process. At the same time as electrons flow through an external circuit, electricity is produced [136–138]. As shown in Table 4.

Microbial fuel cells (MFCs) find several possible uses. Wastewater treatment is one especially interesting use since it produces energy while helping to remove pollutants, therefore improving the water quality. Environmental monitoring benefits from their capacity to identify contaminants in soil and water. Moreover, they could energize small electrical equipment, especially in remote areas lacking conventional energy sources [139,140].

Using MFCs has various benefits. By way of naturally occurring biological processes, they present a sustainable and ecologically benign source of energy. The fact that they can react with many various kinds of organic molecules strengthens their versatility as an energy source. By converting rubbish into electricity, they also contribute to lessening of pollution [141,142]. As shown in Table 5.

Table 4. Key components of an MFC.

Component	Function	References
Anode	Site where bacteria grow and oxidize organic matter, releasing electrons.	[143]
Cathode	Accepts electrons and completes the circuit by reducing oxygen or another electron acceptor.	[144]
Proton Exchange Membrane (PEM)	Separates anode and cathode chambers, allowing protons to pass while preventing oxygen from reaching the anode.	[63]
External Circuit	Provides a pathway for electron flow, generating electricity.	[145]

Table 5 . Characteristics of microbial fuel cells (MFCs)

Feature	Microbial Fuel Cells (MFCs)	References
Electrolyte Type	Liquid medium (typically wastewater or organic solution)	[146]
Operating Temperature	Room temperature (ambient conditions)	[146]
Fuel Source	Organic matter (e.g., wastewater, food waste, natural sugars)	[147]
Ion Transport	Protons (H^+) move through the proton exchange membrane (PEM)	[148]
Catalyst Type	Microbial catalysts (bacteria or other microorganisms)	[149]
Applications	Wastewater treatment, environmental monitoring, power generation in remote areas	[150]
Advantages	Sustainable energy production, waste remediation, low operating costs	[151]
Challenges	Low power output, efficiency improvements needed, scaling issues	[152]

Table 5 gives a summary of the operating principle, materials, fuel sources, applications, benefits, and limitations of Microbial Fuel Cells (MFCs).

2.5   Direct Methanol Fuel Cells (DMFCs) and Biofuel Cells

Biofuel and direct methane fuel cells (DMFCs) generate energy from liquid fuels. They are healthier and greener than conventional power sources [153].

A Direct Methanol Fuel Cell (DMFC) anode chemically converts methanol into energy. During methanol breakdown, protons and electrons are released. Electrons travelling through an external circuit generate energy. Conversely, protons cross a membrane and mix with oxygen at the cathode to generate water. DMFCs are perfect for portable power and small electronics since they can be fuelled by liquid methanol [154,155].

Biofuel cells generate energy from glucose, ethanol, or waste using enzymes or microorganisms. Biofuel cells use metabolic processes to generate energy, while conventional fuel cells use metal catalysts [156,157] . They are used in medical implants that take glucose from the body to generate electricity and waste treatment systems that convert biological waste into power [158].

BFCs offer a biocompatible and sustainable energy source; nonetheless, poor power output and enzyme instability remain major obstacles [159]. Advances in biogenic nanomaterials, catalyst development, and bioengineering will be vital to break free from these restrictions and open the path for more affordable, ecologically friendly, and efficient fuel cell technology. As Table 6 reveals.

Table 6: Direct methanol fuel cells (DMFC) and biofuel cells compared.

Feature	Direct Methanol Fuel Cells (DMFCs)	Biofuel Cells	References
Energy Source	Methanol (liquid fuel)	Organic compounds (glucose, ethanol, wastewater)	[160,161]
Catalyst Type	Metal-based catalysts (e.g., platinum)	Biological catalysts (enzymes or microorganisms)	[162,163]
Working Principle	Methanol is broken down at the anode, releasing electrons and protons that generate electricity	Enzymes or microbes break down organic matter, producing electrons for electricity generation	[164,165]
Applications	Portable electronics, transportation, off-grid power	Medical implants, biosensors, wastewater treatment, bioenergy systems	[157,165]
Advantages	Compact, easy fuel storage, cleaner than fossil fuels	Renewable, operates in mild conditions, can utilize biological fluids	[156,166]
Challenges	Methanol toxicity, high catalyst cost, efficiency improvements needed	Low power output, enzyme/microbe degradation, limited large-scale application potential	[166,167]

Table 6 assesses the energy source, catalyst type, operating principle, applications, benefits, and drawbacks of Direct Methanol Fuel Cells (DMFCs) and Biofuel Cells.

3.   Conclusion

Fuel cells are a promising category of energy conversion technologies that possess high efficiency, numerous fuel adaptabilities, and minimal environmental impact. Among them are Direct Methanol Fuel Cells (DMFCs), Microbial Fuel Cells (MFCs), Solid Oxide Fuel Cells (SOFCs), Alkaline Fuel Cells (AFCs), Proton Exchange Membrane Fuel Cells (PEMFCs), and Biofuel Cells (BFCs). Each has advantages and disadvantages that make it suitable for a specific type of application. Both PEMFCs and AFCs generate power in an efficient and clean manner. PEMFCs' quick start and low running temperature make them perfect for portable and transportation applications, but AFCs' high efficiency and simple design make them ideal for space and aerospace applications. Both have disadvantages, such as being susceptible to unclean fuel and worn-out catalysts. SOFCs are advantageous for the generation of power in a single location and for systems that provide both heat and electricity due to their ability to operate at elevated temperatures. This enables them to operate at a high level of efficiency and utilize a variety of fuels. Nevertheless, their widespread application is restricted due to their high cost, extended startup time, and the potential for material degradation. New developments in biogenic nanoparticles could resolve these issues by improving the functionality of catalysts, increasing ionic conductivity, and extending the lifespan of electrodes. MFCs and BFCs offer green and sustainable energy by using biological processes. BFCs use enzymes or whole cells to speed up the breakdown of biofuels, while MFCs use special bacteria that create electricity as they break down organic materials. Both technologies can be used in biosensors, medical devices that can be implanted in the body, and treating garbage. Nevertheless, significant challenges persist, including inadequate power density, sluggish response rates, and scaling issues. Direct Methanol Fuel Cells (DMFCs) are ideal for military and portable technology applications due to their high energy density and ease of fuel storage. They operate on liquid methanol. Nonetheless, methanol crossover, deterioration of platinum catalysts, and CO_2 emissions constrain their economic feasibility. Recent advances in nanotechnology and catalysts, particularly natural nanomaterials, could help overcome these limitations. Biogenic nanomaterials, which come from natural sources, help make fuel cells work better, cost less, and be more environmentally friendly.

References

[1]   B. Mahjoub, C. Fersi, M. Bouteffeha, K. Kummerer, Green Chemistry and Sustainable Chemistry Related to Water Challenges: Solutions and Prospects in a Changing Climate, Curr. Opin. Green Sustain. Chem. (2025) 101000. https://doi.org/10.1016/j.cogsc.2025.101000

[2]   A. Hojjati-Najafabadi, A. Aygun, R.N.E. Tiri, F. Gulbagca, M.I. Lounissaa, P. Feng, F. Karimi, F. Sen, Bacillus thuringiensis Based Ruthenium/Nickel Co-Doped Zinc as a Green Nanocatalyst: Enhanced Photocatalytic Activity, Mechanism, and Efficient H2Production from Sodium Borohydride Methanolysis, Ind. Eng. Chem. Res. 62 (2023) 4655–4664. https://doi.org/10.1021/ACS.IECR.2C03833

[3]   Y. Wu, E.E. Altuner, R.N. El Houda Tiri, M. Bekmezci, F. Gulbagca, A. Aygun, C. Xia, Q. Van Le, F. Sen, H. Karimi-Maleh, Hydrogen generation from methanolysis of sodium borohydride using waste coffee oil modified zinc oxide nanoparticles and their photocatalytic activities, Int. J. Hydrogen Energy 48 (2023) 6613–6623. https://doi.org/10.1016/J.IJHYDENE.2022.04.177

[4]   A. Aygun, F. Gulbagca, E.E. Altuner, M. Bekmezci, T. Gur, H. Karimi-Maleh, F. Karimi, Y. Vasseghian, F. Sen, Highly active PdPt bimetallic nanoparticles synthesized by one-step bioreduction method: Characterizations, anticancer, antibacterial activities and evaluation of their catalytic effect for hydrogen generation, Int. J. Hydrogen Energy (2022). https://doi.org/10.1016/J.IJHYDENE.2021.12.144

[5]   S. Günbatar, A. Aygun, Y. Karataş, M. Gülcan, F. Şen, Carbon-nanotube-based rhodium

nanoparticles as highly-active catalyst for hydrolytic dehydrogenation of dimethylamineborane at room temperature, J. Colloid Interface Sci. 530 (2018) 321–327. https://doi.org/10.1016/J.JCIS.2018.06.100

[6]   Z. Abdin, Shaping the stationary energy storage landscape with reversible fuel cells, J. Energy Storage 86 (2024) 111354. https://doi.org/10.1016/j.est.2024.111354

[7]   H. Gao, H. Ishitobi, N. Nakagawa, Improved performance of a direct methanol fuel cell by the Highly-developed mesopores of the carbon nanofibers catalyst support, Carbon Resour. Convers. (2025) 100304. https://doi.org/10.1016/j.crcon.2025.100304

[8]   M. Padhiary, D. Roy, P. Dey, Mapping the Landscape of Biogenic Nanoparticles in Bioinformatics and Nanobiotechnology: AI-Driven Insights, in: Synth. Charact. Plant-Mediated Biocompatible Met. Nanoparticles, IGI Global, 2025: pp. 337–376. https://doi.org/10.4018/979-8-3693-6240-2.ch014

[9]   K. Nesrin, C. Yusuf, K. Ahmet, S.B. Ali, N.A. Muhammad, S. Suna, Ş. Fatih, Biogenic silver nanoparticles synthesized from Rhododendron ponticum and their antibacterial, antibiofilm and cytotoxic activities, J. Pharm. Biomed. Anal. 179 (2020) 112993. https://doi.org/10.1016/J.JPBA.2019.112993

[10]   A. Hojjati-Najafabadi, S. Salmanpour, F. Sen, P.N. Asrami, M. Mahdavian, M.A. Khalilzadeh, A Tramadol Drug Electrochemical Sensor Amplified by Biosynthesized Au Nanoparticle Using Mentha aquatic Extract and Ionic Liquid, Top. Catal. (2021). https://doi.org/10.1007/s11244-021-01498-x

[11]   F. Gulbagca, A. Aygün, M. Gülcan, S. Ozdemir, S. Gonca, F. Şen, Green Synthesis of Palladium Nanoparticles: Preparation, Characterization, and Investigation of Antioxidant, Antimicrobial, Anticancer, and DNA Cleavage Activities, Appl. Organomet. Chem. 35 (2021). https://doi.org/10.1002/aoc.6272

[12]   N. Korkmaz, Y. Ceylan, P. Taslimi, A. Karadağ, A.S. Bülbül, F. Şen, Biogenic nano silver: Synthesis, characterization, antibacterial, antibiofilms, and enzymatic activity, Adv. Powder Technol. 31 (2020) 2942–2950. https://doi.org/10.1016/j.apt.2020.05.020

[13]   R.A. Omar, D. Chauhan, N. Talreja, M. Ashfaq, Endophytic Fungi: A Biofactories for the Synthesis of Nanomaterials, in: Fungal Endophytes Vol. I Biodivers. Bioact. Mater., Springer, 2025: pp. 443–470. https://doi.org/10.1007/978-981-97-7312-1_16

[14]   S. Rashid, S. Islam, S. Qamer, M. Ali, M. Fatima, L. Javaid, A. Ali, A. Sarwar, A.S. Farooq, A review of green synthesized magnesium oxide nanoparticles coated textiles, J. Ind. Text. 55 (2025) 15280837251313518. https://doi.org/10.1177/15280837251313518

[15]   K.E. Bassey, From waste to wonder: Developing engineered nanomaterials for multifaceted applications, GSC Adv. Res. Rev. 20 (2024) 109–123. https://doi.org/10.30574/gscarr.2024.20.3.0326

[16]   A.K. Oyebamiji, S.A. Akintelu, S.O. Afolabi, O. Ebenezer, E.T. Akintayo, C.O. Akintayo, A Comprehensive Review on Mycosynthesis of Nanoparticles, Characteristics, Applications, and Limitations, Plasmonics (2025) 1–19. https://doi.org/10.1007/s11468-024-02755-x

[17]   H.E. Ali, B.A. Hemdan, M.E. El-Naggar, M.A. El-Liethy, D.A. Jadhav, H.H. El-Hendawy, M. Ali, G.E. El-Taweel, Harnessing the power of microbial fuel cells as pioneering green technology: advancing sustainable energy and wastewater treatment through innovative nanotechnology, Bioprocess Biosyst. Eng. (2025) 1–24. https://doi.org/10.1007/s00449-024-03115-z

[18]   S.R. Padhan, S.S. Rathore, S.L. Jat, S. Saini, A. Mohanty, K.K. Panigrahi, S. Ranjan, S. Sow, P. Mishra, K. Baral, Synthesis of Carbon Nanomaterials from Agro-Industrial Wastes and Their Extensive Applications, in: Waste-Derived Carbon Nanostructures Synth. Appl., Springer, 2025: pp. 71–106. https://doi.org/10.1007/978-3-031-75247-6_3

[19]   R. Pabbati, K. Chepuri, K.V. Reddy, N.R. Maddela, Plant-Based Nanoparticle Synthesis for Sustainable Agriculture, CRC Press, 2025. https://doi.org/10.1201/9781003477730

[20]   M.W. Alam, N. Dhanda, H.H. Almutairi, N.S. Al-Sowayan, S. Mushtaq, S.A. Ansari, Green Ferrites: Eco-Friendly Synthesis to Applications in Environmental Remediation, Antimicrobial Activity, and Catalysis—A Comprehensive Review, Appl. Organomet. https://doi.org/10.1002/aoc.7962Chem. 39

(2025) e7962.

[21] M.P. Shah, N. Bharadvaja, L. Kumar, Biogenic Nanomaterials for Environmental Sustainability: Principles, Practices, and Opportunities, Springer Nature, 2024. https://doi.org/10.1007/978-3-031-45956-6

[22] H.A. Alhadrami, H.M. Hassan, A.H. Alhadrami, M.E. Rateb, A.A. Hamed, Green synthesis and anticancer activity of titanium dioxide nanoparticles using the endophytic fungus Aspergillus sp., J. Radiat. Res. Appl. Sci. 18 (2025) 101229. https://doi.org/10.1016/j.jrras.2024.101229

[23] T. Tesfaye, Y. Shuka, S. Tadesse, T. Eyoel, M. Mengesha, Improving the power production efficiency of microbial fuel cell by using biosynthesized polyanaline coated Fe3O4 as pencil graphite anode modifier, Sci. Rep. 15 (2025) 587. https://doi.org/10.1038/s41598-024-84311-5

[24] A. Singh, A. Gangwar, S. Kumar, Prospects and limitations of existing biofuels and emerging trends in the utilization of nanoparticles for enhanced biofuel production and microbial fuel cell efficiency, Biofuels (2024) 1–26. https://doi.org/10.1080/17597269.2024.2431766

[25] S. Afzal, S. Ullah, M. Shahid, T. Najam, I.A. Shaaban, S.S.A. Shah, M.A. Nazir, Metal oxide nanoparticles: Synthesis and applications in energy, biomedical, and environment sector, J. Chinese Chem. Soc. (n.d.).

[26] A. Shahbaz, N. Hussain, M.Z. Saleem, M.U. Saeed, M. Bilal, H.M.N. Iqbal, Nanoparticles as stimulants for efficient generation of biofuels and renewables, Fuel 319 (2022) 123724. https://doi.org/10.1016/j.fuel.2022.123724

[27] M.S. Nas, M.H. Calimli, Recent development of nanoparticle by green-conventional methods and applications for corrosion and fuel cells, Curr. Nanosci. 17 (2021) 525–539. https://doi.org/10.2174/1573413716999200925163316

[28] J. Li, Z. Chen, Revitalizing microbial fuel cells: A comprehensive review on the transformative role of iron-based materials in electrode design and catalyst development, Chem. Eng. J. (2024) 151323. https://doi.org/10.1016/j.cej.2024.151323

[29] M. Chitt, S. Thangavel, V. Verma, A. Kumar, Green hydrogen productions: Methods, designs and smart applications, in: Highly Effic. Therm. Renew. Energy Syst., CRC Press, 2024: pp. 261–276. https://doi.org/10.1201/9781003472629-16

[30] L. Devi, P. Kushwaha, T.M. Ansari, A. Kumar, A. Rao, Recent trends in biologically synthesized metal nanoparticles and their biomedical applications: a review, Biol. Trace Elem. Res. 202 (2024) 3383–3399. https://doi.org/10.1007/s12011-023-03920-9

[31] S. Zaman, L. Huang, A.I. Douka, H. Yang, B. You, B.Y. Xia, Oxygen reduction electrocatalysts toward practical fuel cells: progress and perspectives, Angew. Chemie 133 (2021) 17976–17996. https://doi.org/10.1002/ange.202016977

[32] M.O. Ojemaye, A.I. Okoh, Global research direction on Pt and Pt based electro-catalysts for fuel cells application between 1990 and 2019: A bibliometric analysis, Int. J. Energy Res. 45 (2021) 15783–15796. https://doi.org/10.1002/er.6907

[33] D.Z. Khater, R.S. Amin, M.O. Zhran, Z.K. Abd El-Aziz, M. Mahmoud, H.M. Hassan, K.M. El-Khatib, The enhancement of microbial fuel cell performance by anodic bacterial community adaptation and cathodic mixed nickel–copper oxides on a graphene electrocatalyst, J. Genet. Eng. Biotechnol. 20 (2022) 12. https://doi.org/10.1186/s43141-021-00292-2

[34] R. Darabi, F.E.D. Alown, A. Aygun, Q. Gu, F. Gulbagca, E.E. Altuner, H. Seckin, I. Meydan, G. Kaymak, F. Sen, H. Karimi-Maleh, Biogenic platinum-based bimetallic nanoparticles: Synthesis, characterization, antimicrobial activity and hydrogen evolution, Int. J. Hydrogen Energy (2023). https://doi.org/10.1016/j.ijhydene.2022.12.072

[35] U. Shanker, Vipin, M. Rani, Metal oxides–based nanomaterials: green synthesis methodologies and sustainable environmental applications, in: Handb. Green Sustain. Nanotechnol. Fundam. Dev. Appl.,

Springer, 2023: pp. 1–27. https://doi.org/10.1007/978-3-030-69023-6_80-2

[36] S.G. Peera, T. Maiyalagan, C. Liu, S. Ashmath, T.G. Lee, Z. Jiang, S. Mao, A review on carbon and non-precious metal based cathode catalysts in microbial fuel cells, Int. J. Hydrogen Energy 46 (2021) 3056–3089. https://doi.org/10.1016/j.ijhydene.2020.07.252

[37] N.H. Khand, A.R. Solangi, S. Ameen, A. Fatima, J.A. Buledi, A. Mallah, S.Q. Memon, F. Sen, F. Karimi, Y. Orooji, A new electrochemical method for the detection of quercetin in onion, honey and green tea using Co3O4 modified GCE, J. Food Meas. Charact. 2021 154 15 (2021) 3720–3730. https://doi.org/10.1007/S11694-021-00956-0

[38] S. Parida, D.P. Dutta, Nanostructured materials from biobased precursors for renewable energy storage applications, in: Biorenewable Nanocomposite Mater. Vol. 1 Electrocatal. Energy Storage, ACS Publications, 2022: pp. 307–366. https://doi.org/10.1021/bk-2022-1410.ch013

[39] P. Aiswaria, S.N. Mohamed, D.L. Singaravelu, K. Brindhadevi, A. Pugazhendhi, A review on graphene/graphene oxide supported electrodes for microbial fuel cell applications: Challenges and prospects, Chemosphere 296 (2022) 133983. https://doi.org/10.1016/j.chemosphere.2022.133983

[40] R. Ulus, Y. Yıldız, S. Eriş, B. Aday, F. Şen, M. Kaya, Functionalized Multi-Walled Carbon Nanotubes (f-MWCNT) as Highly Efficient and Reusable Heterogeneous Catalysts for the Synthesis of Acridinedione Derivatives, ChemistrySelect 1 (2016) 3861–3865. https://doi.org/10.1002/SLCT.201600719

[41] K. Elangovan, P. Saravanan, C.H. Campos, F. Sanhueza-Gómez, M.M.R. Khan, S.Y. Chin, S. Krishnan, R. Viswanathan Mangalaraja, Outline of microbial fuel cells technology and their significant developments, challenges, and prospects of oxygen reduction electrocatalysts, Front. Chem. Eng. 5 (2023) 1228510. https://doi.org/10.3389/fceng.2023.1228510

[42] N.A.A. Qasem, G.A.Q. Abdulrahman, A recent comprehensive review of fuel cells: history, types, and applications, Int. J. Energy Res. 2024 (2024) 7271748. https://doi.org/10.1155/2024/7271748

[43] X. Guo, J. Zhang, L. Yuan, B. Xi, F. Gao, X. Zheng, R. Pan, L. Guo, X. An, T. Fan, Biologically assisted construction of advanced electrode materials for electrochemical energy storage and conversion, Adv. Energy Mater. 13 (2023) 2204376. https://doi.org/10.1002/aenm.202204376

[44] R.K. Sharma, S. Yadav, S. Dutta, H.B. Kale, I.R. Warkad, R. Zbořil, R.S. Varma, M.B. Gawande, Silver nanomaterials: synthesis and (electro/photo) catalytic applications, Chem. Soc. Rev. 50 (2021) 11293–11380. https://doi.org/10.1039/D0CS00912A

[45] S. Zhang, S. Hess, H. Marschall, U. Reimer, S. Beale, W. Lehnert, openFuelCell2: A new computational tool for fuel cells, electrolyzers, and other electrochemical devices and processes, Comput. Phys. Commun. 298 (2024) 109092. https://doi.org/10.1016/j.cpc.2024.109092

[46] R. Mostafazade, L. Arabi, Z. Tazik, M. Akaberi, B.S.F. Bazzaz, Green synthesis of gold, copper, zinc, iron, and other metal nanoparticles by fungal endophytes; characterization, and their biological activity: a review, Biocatal. Agric. Biotechnol. (2024) 103307. https://doi.org/10.1016/j.bcab.2024.103307

[47] S. Yu, S. Xu, R. Khan, H. Zhao, C. Li, Research developments in the application of electrospun nanofibers in direct methanol fuel cells, Catal. Sci. Technol. 14 (2024) 820–834. https://doi.org/10.1039/D3CY01509B

[48] F. Sun, Y. Chen, Q. Wen, Y. Yang, Poly (3, 4-ethylenedioxythiophene): poly (styrenesulfonate) bioanodes in Co-doped modified microbial fuel cell promote sulfamethoxine degradation with high enrichment of electroactive bacteria and extracellular electron transfer, Renew. Energy 232 (2024) 121091. https://doi.org/10.1016/j.renene.2024.121091

[49] B.S.M. Kumar, K.J.R. Kumar, S.J. Patil, S.B.N. Krishna, Advances in biogenic synthesis of metal sulfide nanomaterials, in: Met. Sulfide Nanomater. Environ. Appl., Elsevier, 2025: pp. 107–134. https://doi.org/10.1016/B978-0-443-13464-7.00003-7

[50]   S. Iravani, A. Zarepour, A. Khosravi, A. Zarrabi, Environmental and biomedical applications of 2D transition metal borides (MBenes): recent advancements, Nanoscale Adv. (2025). https://doi.org/10.1039/D4NA00867G

[51]   R. Tenchov, K.J. Hughes, M. Ganesan, K.A. Iyer, K. Ralhan, L.M. Lotti Diaz, R.E. Bird, J.M. Ivanov, Q.A. Zhou, Transforming Medicine: Cutting-Edge Applications of Nanoscale Materials in Drug Delivery, ACS Nano (2025). https://doi.org/10.1021/acsnano.4c09566

[52]   N.B. Singh, B. Kumar, U.L. Usman, M.A.B.H. Susan, Nano revolution: exploring the frontiers of nanomaterials in science, technology, and society, Nano-Structures & Nano-Objects 39 (2024) 101299. https://doi.org/10.1016/j.nanoso.2024.101299

[53]   C. Zhao, Y. Song, H. Chen, H. Chen, Y. Li, A. Lei, Q. Wu, L. Zhu, Improving the performance of microbial fuel cell stacks via capacitive-hydrogel bioanodes, Int. J. Hydrogen Energy 97 (2025) 708–717. https://doi.org/10.1016/j.ijhydene.2024.11.424

[54]   A.A.A. Abdelazeez, M. Rabia, F. Hasan, V. Mahanta, E.R. Adly, Polymer Nanocomposites: Catalysts for Sustainable Hydrogen Production from Challenging Water Sources, Adv. Energy Sustain. Res. 5 (2024) 2400077. https://doi.org/10.1002/aesr.202400077

[55]   D.D. Shah, M.R. Chorawala, M.K.A. Mansuri, P.S. Parekh, S. Singh, B.G. Prajapati, Biogenic metallic nanoparticles: from green synthesis to clinical translation, Naunyn. Schmiedebergs. Arch. Pharmacol. 397 (2024) 8603–8631. https://doi.org/10.1007/s00210-024-03236-y

[56]   O. Awogbemi, D.V. Von Kallon, Recent advances in the application of nanomaterials for improved biodiesel, biogas, biohydrogen, and bioethanol production, Fuel 358 (2024) 130261. https://doi.org/10.1016/j.fuel.2023.130261

[57]   D. Das, K. Verma, S. Tangjang, A Comprehensive Perspective of Conventional and Biogenic Nanoparticles, in: Biog. Nanomater. Heal. Environ., CRC Press, 2024: pp. 1–17. https://doi.org/10.1201/9781003430087-1

[58]   T.S. Da Costa, M.R. Assalin, M.C. Lacerda, N. Durán, L. Tasic, Biogenic Nanomaterials for Water Detoxification and Disinfection, in: Circ. Econ. Appl. Water Secur., CRC Press, n.d.: pp. 61–81. https://doi.org/10.1201/9781003441007-7

[59]   P. Gupta, B. Toksha, M. Rahaman, A Critical Review on Hydrogen Based Fuel Cell Technology and Applications, Chem. Rec. 24 (2024) e202300295. https://doi.org/10.1002/tcr.202300295

[60]   Z. Jia, X. Lin, C. Li, A review of fuel cell cathode catalysts based on hollow porous materials for improving oxygen reduction performance, Catal. Sci. Technol. 14 (2024) 5505–5524. https://doi.org/10.1039/D4CY00830H

[61]   S.A. Bhat, V. Kumar, D.S. Dhanjal, Y. Gandhi, S.K. Mishra, S. Singh, T.J. Webster, P.C. Ramamurthy, Biogenic nanoparticles: pioneering a new era in breast cancer therapeutics—a comprehensive review, Discov. Nano 19 (2024) 121. https://doi.org/10.1186/s11671-024-04072-y

[62]   M.H. Saleem, U. Ejaz, M. Vithanage, N. Bolan, K.H.M. Siddique, Synthesis, characterization, and advanced sustainable applications of copper oxide nanoparticles: a review, Clean Technol. Environ. Policy (2024) 1–26. https://doi.org/10.1007/s10098-024-02774-6

[63]   H. Duman, E. Akdaşçi, F. Eker, M. Bechelany, S. Karav, Gold Nanoparticles: Multifunctional Properties, Synthesis, and Future Prospects, Nanomaterials 14 (2024) 1805. https://doi.org/10.3390/nano14221805

[64]   M. Ehtesham, N. Ansari, G. Kumar, S. Kumar, P. Gaijon, S. Ghosh, M.R. Singh, A. Kant, Detoxification of Industrial Wastewater by Catalytic (Photo/Bio/Nano) Techniques, Sustain. Green Catal. Process. (2024) 377–396. https://doi.org/10.1002/9781394212767.ch16

[65]   N.A.I.M. Ishak, S.K. Kamarudin, M. Mansor, N. Yahya, R. Bahru, S. Rahman, Effect of tunable composition-shape of bio-inspired Pt NPs electrocatalyst in direct methanol fuel cell: Process optimization and kinetic studies, J. Clean. Prod. 440 (2024) 140637.

https://doi.org/10.1016/j.jclepro.2024.140637

[66] K.B. Malunga-Makatu, S.A. Rahimi, E.M.N. Chirwa, S.M. Tichapondwa, Performance Enhancement of an Air-cathode Microbial Fuel Cell with Subsequent Carbon Source Removal, Chem. Eng. Trans. 110 (2024) 349–354.

[67] H. Wang, H. Zhu, Y. Zhang, Y. Li, Boosting electricity generation in biophotovoltaics through nanomaterials targeting specific cellular locations, Renew. Sustain. Energy Rev. 202 (2024) 114718. https://doi.org/10.1016/j.rser.2024.114718

[68] I. Martinaiou, M.K. Daletou, Enhancing Electrode Efficiency in Proton Exchange Membrane Fuel Cells with PGM-Free Catalysts: A Mini Review, Energies 17 (2024) 3443. https://doi.org/10.3390/en17143443

[69] O. Agboola, A. Adeyanju, L.M. Mavhungu, E.C. Igbokwe, O. Oladokun, A.O. Ayeni, D. Omole, R. Sadiku, O.S.I. Fayomi, The Role of Nanotechnology in Proton Exchange Membrane Fuel Cell and Microbial Fuel Cell: The Insight of Nanohybrid, J. Membr. Sci. Res. 10 (2024).

[70] A. Kaur, S. Kumar, H. Kaur, G.S. Lotey, P.P. Singh, G. Singh, S. Kumar, J. Dalal, G. Bouzid, M. Misra, Enhanced photocatalytic degradation and antimicrobial activities of biogenic Co 3 O 4 nanoparticles mediated by fenugreek: sustainable strategies, Mater. Adv. 5 (2024) 8111–8131. https://doi.org/10.1039/D4MA00795F

[71] F. Şen, G. Gökağaç, Improving Catalytic Efficiency in the Methanol Oxidation Reaction by Inserting Ru in Face-Centered Cubic Pt Nanoparticles Prepared by a New Surfactant, tert-Octanethiol, Energy and Fuels 22 (2008) 1858–1864. https://doi.org/10.1021/EF700575T

[72] T.I. Awan, S. Afsheen, A. Mushtaq, Role of Noble Metals in Fuel Cells and Batteries, in: Influ. Noble Met. Nanoparticles Sustain. Energy Technol., Springer, 2025: pp. 197–221. https://doi.org/10.1007/978-3-031-80983-5_9

[73] Z. Yuhang, W. ZHANG, Y. Luo, H.A.O. Peixuan, S.H.I. Yixiang, Progress of elevated-temperature alkaline electrolysis hydrogen production and alkaline fuel cells power generation, J. Fuel Chem. Technol. 53 (2025) 231–247. https://doi.org/10.1016/S1872-5813(24)60503-7

[74] K.S. Anantharaju, Nanomaterials for fuel cell and corrosion inhibition: A comprehensive review, Curr. Nanosci. 17 (2021) 591–611. https://doi.org/10.2174/1573413716666210101121907

[75] B. Ramasubramanian, R.P. Rao, V. Chellappan, S. Ramakrishna, Towards sustainable fuel cells and batteries with an AI perspective, Sustainability 14 (2022) 16001. https://doi.org/10.3390/su142316001

[76] X. Hu, B. Yang, S. Ke, Y. Liu, M. Fang, Z. Huang, X. Min, Review and perspectives of carbon-supported platinum-based catalysts for proton exchange membrane fuel cells, Energy & Fuels 37 (2023) 11532–11566. https://doi.org/10.1021/acs.energyfuels.3c01265

[77] S.S. Gwebu, Materials for Alkaline Direct Alcohol Fuel Cells (ADAFC) and Perovskite Solar Cells (PSC): Synthesis, Characterisation and Application, University of Johannesburg (South Africa), 2021.

[78] S. Ertan, F. Şen, S. Şen, G. Gökağaç, Platinum nanocatalysts prepared with different surfactants for C1-C3 alcohol oxidations and their surface morphologies by AFM, J. Nanoparticle Res. 14 (2012) 1–12. https://doi.org/10.1007/S11051-012-0922-5

[79] O. Chun, F. Jamshaid, M.Z. Khan, O. Gohar, I. Hussain, Y. Zhang, K. Zheng, M. Saleem, M. Motola, M.B. Hanif, Advances in low-temperature solid oxide fuel cells: An explanatory review, J. Power Sources 610 (2024) 234719. https://doi.org/10.1016/j.jpowsour.2024.234719

[80] J. Li, J. Cheng, Y. Zhang, Z. Chen, M. Nasr, M. Farghali, D.W. Rooney, P. Yap, A.I. Osman, Advancements in Solid Oxide Fuel Cell Technology: Bridging Performance Gaps for Enhanced Environmental Sustainability, Adv. Energy Sustain. Res. 5 (2024) 2400132. https://doi.org/10.1002/aesr.202400132

[81] S.L. Tripathi, K. Arora, C. Lwendi, A Review of Effective Biomass, Chemical, Recycling and Storage Processes for Electrical Energy Generations, Carbon-based Nanomater. Green Appl. (2024) 331–

353. https://doi.org/10.1002/9781394243426.ch13

[82] M.A. Hefnawy, R. Abdel-Gaber, S.M. Gomha, M.E.A. Zaki, S.S. Medany, Synthesis of Nickel-Manganese Spinel Oxide Supported on Carbon-Felt Surface to Enhance Electrochemical Capacitor Performance, Electrocatalysis (2025) 1–13. https://doi.org/10.1007/s12678-025-00932-y

[83] S. Kumari, K. Sharma, S. Korpal, J. Dalal, A. Kumar, S. Kumar, S. Duhan, A comprehensive study on photocatalysis: materials and applications, CrystEngComm (2024). https://doi.org/10.1039/D4CE00630E

[84] A. Navaee, A. Salimi, Review on CO2 Management: From CO2 Sources, Capture, and Conversion to Future Perspectives of Gas-Phase Electrochemical Conversion and Utilization, Energy & Fuels 38 (2024) 2708–2742. https://doi.org/10.1021/acs.energyfuels.3c04269

[85] M.A. Raheem, M.A. Rahim, I. Gul, X. Zhong, C. Xiao, H. Zhang, J. Wei, Q. He, M. Hassan, C.Y. Zhang, D. Yu, V. Pandey, K. Du, R. Wang, S. Han, Y. Han, P. Qin, Advances in nanoparticles-based approaches in cancer theranostics, OpenNano (2023) 100152. https://doi.org/https://doi.org/10.1016/j.onano.2023.100152

[86] S.S.S. Garimella, S.V. Rachakonda, S.S. Pratapa, G.D. Mannem, G. Mahidhara, From cells to power cells: harnessing bacterial electron transport for microbial fuel cells (MFCs), Ann. Microbiol. 74 (2024) 19. https://doi.org/10.1186/s13213-024-01761-y

[87] S. Kaur, A. Rani, A. Sharma, N. Luhakhra, V. Karol, Green Nanomaterials in Energy Storage: Advancements and Challenges, in: Mater. Boost. Energy Storage. Vol. 3 Adv. Sustain. Energy Technol., ACS Publications, 2024: pp. 281–307. https://doi.org/10.1021/bk-2024-1488.ch012

[88] A. Samris, H. Mounir, Three-dimensional computational fluid dynamics (3D-CFD) simulation of hydrogen transport to investigate the effect of output voltage and inlet anode velocity on proton exchange membrane fuel cell performances, Results Chem. 13 (2025) 101929. https://doi.org/10.1016/j.rechem.2024.101929

[89] Y. Chen, J. Lu, Z. Liu, Y. Liu, T. Huang, X. Ren, X. Wang, Z. Wan, Comprehensive evaluation on a heat self-balanced low-temperature ammonia reforming-based high-power hybrid power generation system combined with proton exchange membrane fuel cell and internal combustion engine, J. Clean. Prod. (2025) 144755. https://doi.org/10.1016/j.jclepro.2025.144755

[90] B. Govind, S. Rattan, P. Singhal, B. Ameduri, A. Tyagi, A. Modak, Polyethylene Functionalized with Imidazolium and Pyridinium Moieties through Radiation-Induced Grafting for Alkaline Solid Polymer Electrolyte Membranes, Chem. Eng. Technol. (n.d.) e202400279.

[91] M.T. Mohamed, O.O. Khalifa, A.H.A. Hashim, African Journal of Advanced Pure and Applied Sciences (AJAPAS), (n.d.).

[92] S. Jana, A. Parthiban, W. Rusli, Polymer materials innovations for green hydrogen economy, Chem. Commun. (2025). https://doi.org/10.1039/D4CC05750C

[93] D.S. Chormey, B.T. Zaman, T.B. Kustanto, S.E. Bodur, S. Bodur, Z. Tekin, O. Nejati, S. Bakırdere, Biogenic synthesis of novel nanomaterials and their applications, Nanoscale 15 (2023) 19423–19447. https://doi.org/10.1039/D3NR03843B

[94] R. Lepikash, D. Lavrova, D. Stom, V. Meshalkin, O. Ponamoreva, S. Alferov, State of the Art and Environmental Aspects of Plant Microbial Fuel Cells' Application, Energies 17 (2024) 752. https://doi.org/10.3390/en17030752

[95] H.K. Elaibi, F.F. Mutlag, E. Halvaci, A. Aygun, F. Sen, Review: Comparison of traditional and modern diagnostic methods in breast cancer, Measurement 242 (2025) 116258. https://doi.org/10.1016/J.MEASUREMENT.2024.116258

[96] S. Samanta, B. Maity, A. Ghosh, S.B. Kuila, B. Mandal, S. Bhattacharya, Performance Evaluation of Graphene-Supported Electro-Catalysts in Proton Exchange Membrane (PEM) Fuel Cells, (2025). https://doi.org/10.21203/rs.3.rs-5775883/v1

[97] Y. Choi, M. Kim, J. Park, Y. Goo, Proton Exchange Membrane Fuel Cell Stack Durability Prediction Using Arrhenius-Based Accelerated Degradation Model, Appl. Sci. 15 (2025) 1300. https://doi.org/10.3390/app15031300

[98] Z.-L. Chen, B.-X. Zhang, C.-L. Zhang, J.-H. Xu, X.-Y. Zheng, K.-Q. Zhu, Y.-L. Wang, Z. Bo, Y.-R. Yang, X.-D. Wang, Deep learning-based fault diagnosis of high-power PEMFCs with ammonia-based hydrogen sources, J. Power Sources 629 (2025) 236018. https://doi.org/10.1016/j.jpowsour.2024.236018

[99] G. Chao, H. Gao, T. Guo, L. Wu, X. Zhou, D. Cai, K. Geng, N. Li, The effect of ion exchange capacity values of sulfonated poly (oxindole biphenylene) ionomer in cathode for proton exchange membrane fuel cells, J. Power Sources 631 (2025) 236246. https://doi.org/10.1016/j.jpowsour.2025.236246

[100] B. Cao, Research Progress on Oxygen Reduction Catalyst Materials for Proton Exchange Membrane Fuel Cells, in: E3S Web Conf., EDP Sciences, 2025: p. 2015. https://doi.org/10.1051/e3sconf/202560602015

[101] M. Sung, H. Yi, J. Han, J.B. Lee, S.-H. Yoon, J.-I. Park, Enhanced Performance and Durability of Proton Exchange Membrane Fuel Cell Catalyst Supports via Nanodrilling-Induced Selective Mesoporous Carbon Structures, ACS Appl. Energy Mater. (2025). https://doi.org/10.1021/acsaem.4c02713

[102] I. Sebbani, M.K. Ettouhami, M. Boulakhbar, Cleaner Energy Systems, (n.d.).

[103] E. Harkou, H. Wang, G. Manos, A. Constantinou, J. Tang, Advances in catalysts and reactors' design for methanol steam reforming and PEMFC applications, Chem. Sci. (2025). https://doi.org/10.1039/D4SC06526C

[104] M. Wang, J. Ruan, J. Zhang, Y. Jiang, F. Gao, X. Zhang, E. Rahman, J. Guo, Modeling, thermodynamic performance analysis, and parameter optimization of a hybrid power generation system coupling thermogalvanic cells with alkaline fuel cells, Energy 292 (2024) 130557. https://doi.org/10.1016/j.energy.2024.130557

[105] H. Lei, X. Yang, Z. Chen, D. Rawach, L. Du, Z. Liang, D. Li, G. Zhang, A.C. Tavares, S. Sun, Multiscale understanding of anion exchange membrane fuel cells: Mechanisms, electrocatalysts, polymers, and cell management, Adv. Mater. (2025) 2410106. https://doi.org/10.1002/adma.202410106

[106] A.S. Deepi, S. Dharani Priya, A. Samson Nesaraj, A.I. Selvakumar, Component fabrication techniques for solid oxide fuel cell (SOFC)–A comprehensive review and future prospects, Int. J. Green Energy 19 (2022) 1600–1612. https://doi.org/10.1080/15435075.2021.2018320

[107] Y. Lai, B. Wang, B. Zu, C. Tongsh, Z. Qin, Q. Du, S. Jiang, K. Jiao, EIS equivalent circuit modeling of direct ammonia fuel cell (DAFC) and mass transfer characteristics for anode diffusion layers with different hydrophobicity, Chem. Eng. J. (2025) 159200. https://doi.org/10.1016/j.cej.2024.159200

[108] J.W. Haverkort, Electrolysers, Fuel Cells and Batteries, (2024).

[109] Ş. Genç, N. Ayas, Upgrading hydrogen production rate and energy efficiency of alkaline water electrolysis under effect of magnetic field, Int. J. Hydrogen Energy 98 (2025) 820–832. https://doi.org/10.1016/j.ijhydene.2024.12.076

[110] L. Han, S. Gong, H. Zhang, M. Yang, O. Javed, X. Yan, G. He, F. Zhang, High performance anion exchange membrane containing large, rigid branching structural unit for fuel cell and electrodialysis applications, J. Memb. Sci. 717 (2025) 123552. https://doi.org/10.1016/j.memsci.2024.123552

[111] M.A. Mustaghfirin, M. Santoso, M. Hakam, R.A. Heriyansyah, A. Sa'diyah, E. Novianarenti, The Effect of Increasing Catalyst Concentration of Fabricated Hydrogen Generator on Proton Exchange Membrane Fuel Cell Performance, in: Int. Conf. Marit. Technol. Its Appl., 2025: pp. 1–7.

[112] Z. Yao, M. Hu, L. Xiao, G. Wang, L. Zhuang, Eliminating the Sudden Voltage Drop of Alkaline Polymer Membrane Fuel Cells at the Initial Discharge Stage, Energy & Fuels (2025). https://doi.org/10.1021/acs.energyfuels.4c05199

[113] C. Huang, F. Wang, X. Chen, J. Li, M. Shao, Z. Wei, Innovative strategies for designing and

constructing efficient fuel cell electrocatalysts, Chem. Commun. (2025).
https://doi.org/10.1039/D4CC05928J

[114] S.A. Mvokwe, O.O. Oyedeji, M.A. Agoro, E.L. Meyer, N. Rono, A Critical Review of the
Hydrometallurgy and Pyrometallurgical Recovery Processes of Platinum Group Metals from End-of-Life
Fuel Cells, Membranes (Basel). 15 (2025) 13. https://doi.org/10.3390/membranes15010013

[115] J. Zhang, R. Shao, T. Yin, D. Chu, X. Zhang, N. Li, L. Liu, Impact of side chain length on the
properties and alkaline fuel cell performance of OEG-grafted poly (terphenyl piperidinium) anion
exchange membranes, J. Memb. Sci. 713 (2025) 123375. https://doi.org/10.1016/j.memsci.2024.123375

[116] S. Merouani, O. Hamdaoui, Hydrogen Production, Storage and Utilization: Thermochemical,
Electrochemical, Sonochemical, Biolohttps://doi.org/10.1515/9783111623863gical and Photocatalytic
Processes, Walter de Gruyter GmbH & Co KG, 2025.

[117] S. Zhai, J. Cai, I.T. Bello, X. Chen, N. Yu, R. Zhao, X. Cai, Y. Jiang, M. Ni, H. Xie, Boosting
Direct-Ethane Solid Oxide Fuel Cell Efficiency with Anchored Palladium Nanoparticles on Perovskite-
Based Anode, Adv. Appl. Energy (2025) 100206. https://doi.org/10.1016/j.adapen.2025.100206

[118] M.A. Gordeeva, A.P. Tarutin, N.A. Danilov, D.A. Medvedev, Technological achievements in the
fabrication of tubular-designed protonic ceramic electrochemical cells, Mater. Futur. 3 (2024) 42102.
https://doi.org/10.1088/2752-5724/ad7872

[119] K. Abouemara, M. Shahbaz, G. Mckay, T. Al-Ansari, The review of power generation from
integrated biomass gasification and solid oxide fuel cells: current status and future directions, Fuel 360
(2024) 130511. https://doi.org/10.1016/j.fuel.2023.130511

[120] V. Marcantonio, L. Scopel, Thermodynamic Models of Solid Oxide Fuel Cells (SOFCs): A
Review, Sustainability 16 (2024) 10773. https://doi.org/10.3390/su162310773

[121] T. Gechev, Progress in fuel cell usage as an auxiliary power unit in heavy-duty vehicles, in: AIP
Conf. Proc., AIP Publishing, 2024. https://doi.org/10.1063/5.0198806

[122] A. Samreen, M.S. Ali, M. Huzaifa, N. Ali, B. Hassan, F. Ullah, S. Ali, N.A. Arifin, Advancements
in perovskite-based cathode materials for solid oxide fuel cells: a comprehensive review, Chem. Rec. 24
(2024) e202300247. https://doi.org/10.1002/tcr.202300247

[123] N. Shah, X. Xu, J. Love, H. Wang, Z. Zhu, L. Ge, Mitigating thermal expansion effects in solid
oxide fuel cell cathodes: A critical review, J. Power Sources 599 (2024) 234211.
https://doi.org/10.1016/j.jpowsour.2024.234211

[124] M. Li, J. Wang, Z. Chen, X. Qian, C. Sun, D. Gan, K. Xiong, M. Rao, C. Chen, X. Li, A
Comprehensive Review of Thermal Management in Solid Oxide Fuel Cells: Focus on Burners, Heat
Exchangers, and Strategies, Energies 17 (2024) 1005. https://doi.org/10.3390/en17051005

[125] B. Jaleh, M. Nasrollahzadeh, A. Nasri, M. Eslamipanah, J. Advani, P. Fornasiero, M.B. Gawande,
Application of Biowaste and Nature-Inspired (Nano) Materials in Fuel Cells, J. Mater. Chem. A (2023).
https://doi.org/10.1039/D2TA09732J

[126] A. Hussain, N. Jabeen, A. Tabassum, J. Ali, 3D-printed conducting polymers for solid oxide fuel
cells, in: 3D Print. Conduct. Polym., CRC Press, 2024: pp. 179–195.
https://doi.org/10.1201/9781003415985-12

[127] M. Dziubaniuk, R. Piech, B. Paczosa-Bator, Electrochemical Impedance Spectroscopy Study of
Ceria-and Zirconia-Based Solid Electrolytes for Application Purposes in Fuel Cells and Gas Sensors,
Materials (Basel). 17 (2024) 5224. https://doi.org/10.3390/ma17215224

[128] M. Yousaf, Y. Lu, M. Akbar, L. Lei, S. Jing, Y. Tao, Advances in solid oxide fuel cell
technologies: lowering the operating temperatures through material innovations, Mater. Today Energy
(2024) 101633. https://doi.org/10.1016/j.mtener.2024.101633

[129] W. Lee, M. Lang, R. Costa, I.-S. Lee, Y.-S. Lee, J. Hong, Enhancing uniformity and performance
in Solid Oxide Fuel Cells with double symmetry interconnect design, Appl. Energy 381 (2025) 125178.

https://doi.org/10.1016/j.apenergy.2024.125178

[130] S. Shin, X. Huang, M.Y. Oh, Y.J. Ye, J. Lee, J.T.S. Irvine, S. Lee, T.H. Shin, Strontium-free Ruddlesden–Popper cuprates (La1. 7Ca0. 3Cu0. 75M0. 25O4+ δ, M= Fe, Co, Ni) as cathode materials for high-performance solid oxide fuel cells, J. Eur. Ceram. Soc. 45 (2025) 117138. https://doi.org/10.1016/j.jeurceramsoc.2024.117138

[131] J. Wang, W. Chen, Y. Wang, J. Wei, W. Zhang, C. Sun, S. Peng, Recent Advances in Proton-Conducting Solid Oxide Electrolysis Cells, Recent Adv. Proton-Conducting Solid Oxide Electrolysis Cells (n.d.).

[132] M. Portarapillo, A. Bellucci Sessa, D. Russo, A. di Benedetto, Ammonia as a Hydrogen Carrier: Energetic Assessment of Processes Integrated with Fuel Cells for Power Generation, Energy & Fuels (2025). https://doi.org/10.1016/j.rser.2025.116134

[133] A. Lahrichi, Y. El Issmaeli, S.S. Kalanur, B.G. Pollet, Advancements, strategies, and prospects of solid oxide electrolysis cells (SOECs): Towards enhanced performance and large-scale sustainable hydrogen production, J. Energy Chem. (2024). https://doi.org/10.1016/j.jechem.2024.03.020

[134] R.S. Pandya, T. Kaur, R. Bhattacharya, D. Bose, D. Saraf, Harnessing microorganisms for bioenergy with Microbial Fuel Cells: Powering the future, Water-Energy Nexus 7 (2024) 1–12. https://doi.org/10.1016/j.wen.2023.11.004

[135] A.Y. Radeef, A.A. Najim, Microbial fuel cell: The renewable and sustainable magical system for wastewater treatment and bioenergy recovery, Energy 360 (2024) 100001. https://doi.org/10.1016/j.energ.2024.100001

[136] M. Sharma, S. Sharma, A.A.M. Alkhanjaf, N.K. Arora, B. Saxena, A. Umar, A.A. Ibrahim, M.S. Akhtar, A. Mahajan, S. Negi, Microbial fuel cells for azo dye degradation: A perspective review, J. Ind. Eng. Chem. (2024). https://doi.org/10.1016/j.jiec.2024.07.031

[137] M. Hassan, S. Kanwal, R.S. Singh, M.A. SA, M. Anwar, C. Zhao, Current challenges and future perspectives associated with configuration of microbial fuel cell for simultaneous energy generation and wastewater treatment, Int. J. Hydrogen Energy 50 (2024) 323–350. https://doi.org/10.1016/j.ijhydene.2023.08.134

[138] R.A. Almasri, N.A.M. Barakat, O.M. Irfan, Enhancing sustainability and power generation from sewage-driven air-cathode microbial fuel cells through innovative anti-biofouling spacer and biomass-derived anode, J. Chem. Technol. Biotechnol. 100 (2025) 344–359. https://doi.org/10.1002/jctb.7776

[139] W. Cui, S. Espley, W. Liang, S. Yin, X. Dong, Microbial Fuel Cells for Power Generation by Treating Mine Tailings: Recent Advances and Emerging Trends, Sustainability 17 (2025) 466. https://doi.org/10.3390/su17020466

[140] C.-T. Wang, P. Pal, X.-C. Wang, EM waves-based microbial fuel cells integrated to improve performance, Appl. Energy 377 (2025) 124412. https://doi.org/10.1016/j.apenergy.2024.124412

[141] S.Z. Abbas, S. Beddu, N.L.M. Kamal, M. Rafatullah, D. Mohamad, A Review on Recent Advancements in Wearable Microbial Fuel Cells, J. Environ. Chem. Eng. (2024) 112977. https://doi.org/10.1016/j.jece.2024.112977

[142] M. Kwofie, B. Amanful, S. Gamor, F. Kaku, Comprehensive Analysis of Clean Energy Generation Mechanisms in Microbial Fuel Cells, Int. J. Energy Res. 2024 (2024) 5866657. https://doi.org/10.1155/2024/5866657

[143] Y. Wang, W. Wang, X. Qi, D. Li, Y. Liu, X. Song, X. Cao, Magnetite-equipped algal-rich sediments for microbial fuel cells: Remediation of sediment organic matter pollution and mechanisms of remote electron transfer, Sci. Total Environ. 912 (2024) 169545. https://doi.org/10.1016/j.scitotenv.2023.169545

[144] M. Eryılmaz, J. Otuzoğlu, U. Tezel, O. Demircan, The influence of ZIF-L in a microbial fuel cell (MFC) cathode for oxygen reduction reaction (ORR), Biotechnol. Lett. 47 (2025) 5.

https://doi.org/10.1007/s10529-024-03548-2

[145] D. Liu, C. Xu, W. Fang, C. Li, Revealed mechanism of 3D-open-microarray boosting exoelectrogens Geobacter enrichment and extracellular electron transfer for high power generation in microbial fuel cells, Bioresour. Technol. (2025) 132049. https://doi.org/10.1016/j.biortech.2025.132049

[146] N. Altın, B. Uyar, Increasing power generation and energy efficiency with modified anodes in algae-supported microbial fuel cells, Biomass Convers. Biorefinery (2025) 1–13. https://doi.org/10.1007/s13399-025-06536-2

[147] K.P. Shabangu, M. Chetty, B.F. Bakare, Metagenomic Insights into Pollutants in Biorefinery and Dairy Wastewater: rDNA Dominance and Electricity Generation in Double Chamber Microbial Fuel Cells, Bioengineering 12 (2025) 88. https://doi.org/10.3390/bioengineering12010088

[148] Z.İ. Özyörü, F.U. Nigiz, Electricity production from dairy wastewater using phosphotungstic acid-Poly (vinylidene fluoride) Membrane supported microbial fuel cell, Int. J. Hydrogen Energy (2025). https://doi.org/10.1016/j.ijhydene.2025.01.019

[149] M. Latif, L.B. Chan, M. Muharam, A comparative analysis of catalyst addition in microbial fuel cells using palm oil mill effluent (POME) for electricity generation, in: AIP Conf. Proc., AIP Publishing, 2025. https://doi.org/10.1063/5.0243232

[150] X. Qi, R. Liu, T. Cai, Z. Huang, X. Wang, X. Wang, Harnessing electroactive microbial community for energy recovery from refining wastewater in microbial fuel cells, Int. J. Hydrogen Energy 102 (2025) 874–886. https://doi.org/10.1016/j.ijhydene.2025.01.087

[151] N. Kundu, S. Yadav, A. Bhattacharya, G.K. Aseri, N. Jain, Constructed wetland-microbial fuel cell (CW-MFC) mediated bio-electrodegradation of azo dyes from textile wastewater, Lett. Appl. Microbiol. (2025) ovaf010

[152] S.S. Satpathy, P.C. Ojha, R. Ojha, J. Dash, D. Pradhan, Recent Modifications of Anode Materials and Performance Evaluation of Microbial Fuel Cells: A Brief Review, J. Energy Eng. 151 (2025) 3125001. https://doi.org/10.1061/JLEED9.EYENG-5732

[153] S.H. Osman, O.H. Chyuan, S.K. Kamarudin, N. Shaari, I.H. Hanapi, Z. Zakaria, N.H. Ahmad Zaidi, S.A. Adnan, R. Bahru, Nanocatalysts in direct liquid fuel cells: Advancements for superior performance and energy sustainability, Int. J. Green Energy 21 (2024) 3654–3674. https://doi.org/10.1080/15435075.2024.2396070

[154] M.G. Buonomenna, J. Bae, Block Copolymer-Based Symmetric Membranes for Direct Methanol Fuel Cells, Symmetry (Basel). 16 (2024) 1079. https://doi.org/10.3390/sym16081079

[155] Y. Zhu, Z. Ma, Y. Li, Y. Zhang, Electrochemical simulation of direct methanol solid oxide fuel cells, J. Solid State Electrochem. (2024) 1–14. https://doi.org/10.1007/s10008-024-06135-7

[156] D. Suri, L.M. Aeshala, T. Palai, Microbial electrosynthesis of valuable chemicals from the reduction of CO_2: a review, Environ. Sci. Pollut. Res. (2024) 1–24. https://doi.org/10.1007/s11356-024-33678-z

[157] V.J. Reddy, N.P. Hariram, R. Maity, M.F. Ghazali, S. Kumarasamy, Sustainable Vehicles for Decarbonizing the Transport Sector: A Comparison of Biofuel, Electric, Fuel Cell and Solar-Powered Vehicles, World Electr. Veh. J. 15 (2024) 93. https://doi.org/10.3390/wevj15030093

[158] Y. Dessie, E. Tilahun, T.H. Wondimu, Functionalized carbon electrocatalysts in energy conversion and storage applications: A review, Heliyon (2024). https://doi.org/10.1016/j.heliyon.2024.e39395

[159] S. ul Haque, M. Yasir, S. Cosnier, Recent advancements in the field of flexible/wearable enzyme fuel cells, Biosens. Bioelectron. (2022) 114545. https://doi.org/10.1016/j.bios.2022.114545

[160] P. Boruah, A. Prasad, 14 Bio-based materials in advance separation, Sustain. Bio-Based Compos. Biomed. Eng. Appl. 20 (2024) 297. https://doi.org/10.1515/9783111321530-014

[161] G. Iakovidou, A. Itziou, A. Tsiotsias, E. Lakioti, P. Samaras, C. Tsanaktsidis, V. Karayannis, Application of microalgae to wastewater bioremediation, with CO_2 biomitigation, health product and

biofuel development, and environmental biomonitoring, Appl. Sci. 14 (2024) 6727. https://doi.org/10.3390/app14156727

[162] Z. Khorsandi, M. Nasrollahzadeh, B. Kruppke, A. Abbasi, H.A. Khonakdar, Research progress in the preparation and application of Lignin-and Polysaccharide-Carbon nanotubes for renewable energy conversion reactions, Chem. Eng. J. (2024) 150725. https://doi.org/10.1016/j.cej.2024.150725

[163] S.B. Ummalyma, T. Bhaskar, Recent advances in the role of biocatalyst in biofuel cells and its application: An overview, Biotechnol. Genet. Eng. Rev. (2023) 1–39.

[164] M.H. Hasan, M.H. Jihad, M.A.B.H. Susan, Biopolymers for Fuel Cells, in: Bio-Based Polym. Farm to Ind. Vol. 3 Emerg. Trends Appl., ACS Publications, 2024: pp. 121–142. https://doi.org/10.1021/bk-2024-1487.ch007

[165] I. Chakraborty, R.T. Olsson, R.L. Andersson, A. Pandey, Glucose-based biofuel cells and their applications in medical implants: a review, Heliyon (2024). https://doi.org/10.1016/j.heliyon.2024.e33615

[166] A. Mishra, R. Bhatt, J. Bajpai, A.K. Bajpai, Nanomaterials based biofuel cells: A review, Int. J. Hydrogen Energy 46 (2021) 19085–19105. https://doi.org/10.1016/j.ijhydene.2021.03.024

[167] P. Joghee, J.N. Malik, S. Pylypenko, R. O'Hayre, A review on direct methanol fuel cells–In the perspective of energy and sustainability, MRS Energy Sustain. 2 (2015) E3. https://doi.org/10.1557/mre.2015.4

Chapter 19

Sensor / Diagnostic Applications Using Biogenic Nanomaterials

S. Cecen Sular[1*], Y. Sular[2*], S. Kaptanoğlu[1], F. N. Maran[3], A. Aygun[3], F. Sen[3*]

[1]Vocational School of Health Care, Van Yuzuncu Yil University, Van, Turkiye

[3]Sen Research Group, Biochemistry Department, Faculty of Arts and Science, Dumlupinar University, Evliya Celebi Campus, 43100, Kutahya, Turkiye

sabriyececensular@yyu.edu.tr, yunussular@gmail.com, fatih.sen@dpu.edu.tr)

Abstract

Nanotechnological studies are making surprising progress in the field of health as in many other fields. This section includes information on the use of biosensors such as biochips, respirocytes, nanorobots, which represent innovative approaches in the diagnosis and treatment of diseases, in many application areas including drug delivery systems, microbiology, diagnosis of cancer cells, diagnosis and treatment of vascular occlusions, ophthalmology, endodontics, anesthesia, obesity, orthopedics and dermatology.

Keywords

Biosensor, Nanoparticles, Diagnostic Applications, Nanotechnology, Nano Application

1.   Introduction

Nanotechnology works on techniques that enable the production, research, and integration of miniature materials, which are small enough to be seen only with a microscope, into many branches of science. The advancement of nanotechnology in medical applications has led to the emergence of a new scientific field. This technology, called nano medicine, not only provides medical diagnosis, but also manages the monitoring of treatment and the prevention of infectious diseases. For this purpose, it is possible to study health at the molecular level using nano devices and nano structures. Revolutionizing diagnosis and treatment in the field of medicine, this branch of science is used in many fields, from medical and biomedical applications to environmental pollution control, cosmetics, energy, materials science, optics, textiles, and electronic engineering [1–10]. These nanoscale materials range in size from 1 to 100 nanometers (nm), obtained by using nanotechnology, are made functional by applying different surface modification methods and their biocompatibility is increased. When biocompatible nanomaterials, nanodevices and nanosystems produced today are examined, it is seen that diagnostic devices, chips, and sensors include popular and innovative approaches for medical applications [11]. Nanotechnology, which offers promising improvements for many pathological conditions in the health field; It is a powerful alternative that is used especially in terms of diagnosis, treatment and prevention methods [12]. This study has brought to the fore unprecedented methods and opportunities in recognizing the pathological condition in early diagnosis methods within the nanomedicine fields of nanotechnology.

2.   Nano Diagnosis

Defined as a field that combines medical and diagnostic imaging methods, theranostics offers therapeutic drugs and diagnostic imaging agents to a system simultaneously. In the theranostic approach, first diagnostic methods are applied to the target area to determine whether the therapeutic agent will be effective. Then, the nano-treatment method is applied to the patients whose diagnostic test is positive. If the diagnostic test is negative, patients are referred to other treatment options [13].

Since many medical methods used today are not effective enough, there are difficulties in the treatment and diagnosis of diseases and injuries. There are many materials such as smart drug carriers, biosensors, nano-biorobots, implants, artificial organs and tissues developed for this purpose [14,15].

By loading diagnostic agents into nanoparticles, a concept we call "nanotheranostic" has emerged in medicine. Nanotheranostics provide a convenient screen to monitor the chemokinetics and chemotodynamics of the drug injected inside the cell or tissue. With the help of nanoparticles, the drug can achieve appropriate diagnostic and therapeutic effect by argeting diseased cells at the same time after systemic blood flow, imaging diseased areas, avoiding the immune system [16].

3.   Nanoparticles in a Diagnosis

In addition to many applications of nanotechnology, nano-scale materials and biosensors at the nano level are being used for diagnosis, monitoring, therapy, treat and disease prevention from its applications in the health field known as nanomedicine. While many illnesses such as cancer and diabetes, Alzheimer's, Parkinson's, which are among the diseases of the modern age, pose a great danger to human health, the correct diagnosis must be made first in order for the treatment to be effective. The role of nanoscale sensors produced with the use of nanotechnology is very essential to help diagnose and treat in a timely manner of the diagnosis. By injecting drugs loaded with nano-sensors into the bodies of the sick and injured, these nanoparticles allow the detection of pathological cells that are foreign to the structure of the organism. The drugs carried by the nanosensors target the sick cells, ensuring that only the sick cells are destroyed without damaging the healthy cells. For this purpose, nanosensors loaded with chemotherapy drugs are currently used for cancer treatment [17,18].

With the new imaging and monitoring process offered by nanotechnology, the surface of semiconductor nanocrystals called quantum dots is coated with antibodies of 5-25 nm size, allowing precise imaging of the target cell, tissue or organ. These quantum dots are currently mainly used for imaging cancer cells (Figure 1).

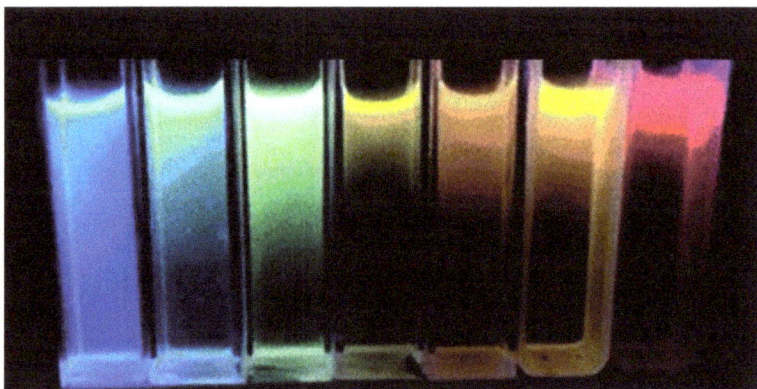

Figure 1. Quantum dots [19]

4.   Nano-scale Controlled Swing Systems

The biggest advantage gained by placing the drugs used for diagnosis and treatment in nanoscale capsules or spheres is the following. These drugs are able to exit the blood through the capillary vessel and act directly on the tissue. The second major advantage is that it has a local effect instead of a systemic effect. Another advantage is that they can create a higher tissue concentration and are eliminated by the phagocytic system as a result of the process [20].

A few areas where nanoparticles are combined and used with drugs are shown in the table below.

Table 1. Nanoparticles used in medicine

Dendrimer	Targeting cancer cells, imaging, drug delivery
Ceramic nanoparticle	passively targeting cancer cells
liposomes	Targeting cancer cells, gene therapy, drug delivery
quantum dots	Texture imaging
carbon nanotubes	As an electronic biosensor

Systemic effects caused by the use of corticosteroids and cancer drugs are expected to be a thing of the past. Serious pan-effects resulting from the administration of non-steroidal anti-inflammatory drugs, which are frequently used for muscular system diseases, at the tissue level will no longer be a problem. For this purpose, studies continue on whether ibuprofen is successfully placed into nanoparticles and whether it is superior to standard forms [21].

Since the serious systemic side effects of anti-TNF, which is a biological agent, will disappear, their use will become widespread rapidly. By developing more effective anticholinergic drugs with nanotechnology, neurological hyperactive bladder will be taken under control, and patients will be able to get rid of side effects such as dryness and constipation. Thanks to the new release systems, the problem of severe spasticity originating from the central or spinal cord will be eliminated. For example, spasticity in the arms without affecting the knees or spasticity in the knees without affecting the arms can be reduced. The bioavailability of the drug may be close to one hundred percent. Free radical-induced tissue damage caused by brain or spinal cord injuries due to paralysis will be eliminated thanks to nanoparticles called fullerene, so that neurological and motor losses will be minimized, and the healing process will be shortened. Big pharmaceutical companies are working on billions of molecules. Those with suitable solubility properties are included in pre-clinical studies. Insoluble molecules, on the other hand, can reach target tissues thanks to high-resolution nanoparticles, and many drugs can be developed in this way [22],[23].

5.   Nanomicrobiology

Thanks to extremely sensitive small, portable nanosensor kits in the field of nanomicrobiology, it will be possible to analyze body fluids such as blood, urine and sputum in seconds using nanosensors at the bedside before they are sent for culture [24]. Thanks to nanotechnology, it will be able to inform the antibiotic resistance that develops in some disease conditions in the early period. Materials called nanobiotics, which are more lethal than all antibiotics, will be developed. For example, urinary catheters coated with nanoscale silver crystals, dressing materials, bedding will significantly reduce hospital infections due to long stays in the clinic [25],[26].

6.   Nanorobots

Small miniature structures are needed to find or repair damaged structures within the cell, to carry the necessary materials into or remove them from the cells in order to check chemical reactions. The materials size we have makes it impossible to intervene inside the cell, however, thanks to the nanorobots produced, it has become easier to enter the cell, enter the bloodstream and intervene in the damaged area like a doctor. Nanorobots are structures of 1-100 nanometers in size and consist of several different atoms. The carbon atom is the most commonly used atom in nanorobot production. When the carbon atoms are arranged as in diamond, they become very strong structures. In addition to the carbon atom, hydrogen, sulphur, silicon, nitrogen, and oxygen atoms are also used. Considering that the capillary diameter is three microns, the size of nanorobots should exceed this scale. In order to prevent the loads carried by nanorobots from being affected by body fluids, their outer surfaces are modified by making them durable with special proteins in a way that they are waterproof and airtight. Nanorobots developed in the form of spheres can reach all organs and tissues by joining the blood circulation intravenously. In this way, 106 -1012 nanorobots are sent to the body in 1-2 cubic centimetres of liquid. Nanorobots do not stay in the body for long. The nanorobots that complete the task are excreted with urine or faces. It is also possible to intervene and control nanorobots from the outside. Thanks to sound waves in the frequency range of 1-10 MHz, acoustic messages can be sent and received. Messages received from tumor tissue can be detected with special ultrasonography devices. In this way, necessary chemical substances can be sent to every cell of the body. Nanorobots, which reach their goal and complete their task, provide the treatment of the disease by throwing out the chemical substance inside (Figure 2). Nanorobots are planned to be used initially to deliver drugs to treat cancer [27].

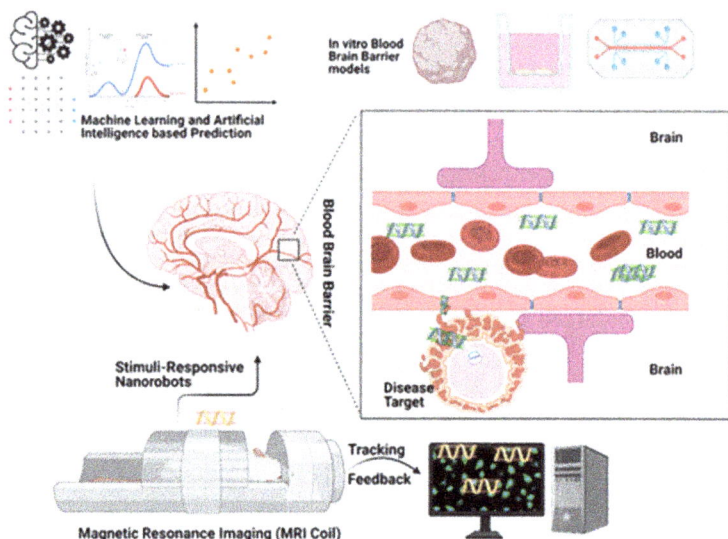

Figure 2. Nanorobots [28].

7. Nano Diagnosis of Cancer

Early diagnosis is the most important factor in the success of cancer treatment. Traditional diagnostic methods are insufficient to detect cancer at an early stage. Can be reduced to nano size; When the surface of the semiconductor features of 5-25 nm is coated with antibodies, the expected sensitive imaging of the target tissue, cell, organ can be obtained, and diagnosis can be facilitated. Thanks to this imaging technique, cancer can be diagnosed at an early stage and it is possible to start treatment effectively [29].

The biggest advantage of diagnosing cancer at an early stage is that the number of cells in the cancerous area is decisive. While it is expected that one one billion cancer cells must be formed in order to diagnose breast cancer with the traditional diagnostic imaging technique, even fewer than 100 cells are sufficient for the diagnosis of breast cancer with the imaging technique using quantum particles, thanks to nanotechnology [30].

Another application of nanobot-science in health is using quantum dots to diagnose and treat tumors in the human body. This technique shows great promising approach to cancer treatment. Again, with iron oxide nanoparticles, the locations of tumor tissues can be detected, and cancer can be diagnosed. Whichever tumor is sought in the body, antigens on the tumor surface are targeted by special antibodies labeled with iron oxide nanoparticles. The MRI (Magnetic Resonance Imaging) machine detects tumors on account of the magnetic emitted by the iron oxide particles in the antibodies pile up in the tumor tissue through the MRI device. Thus, even the smallest tissue can be detected [31].

The cancer is one of the most common causes of death in the world today. Early detection of cancer will significantly affect mortality and morbidity [32]. Diagnosing cancer in its early stages with conventional diagnostic methods is very difficult these days. For this reason, nanotech nuclear medicine methods offer important possibilities in diagnosing and treating cancer. Cancer is abnormal cell divisions that occur with uncontrolled cell proliferation in different parts of the body and can quickly spread to other organs in the body [33]. Unfortunately, radiological methods used to diagnose cancer can be identified only when tumor size reaches 0.5-1 cm^3 in diameter. In this case, no matter how early the diagnosis is made, in some cases it may be too late. The vast majority of cancers are diagnosed once they become terminal [34]. Thanks to nanoparticles, tumor cell tissues can be diagnosed at an early stage. Imaging analyzes can be performed by introducing nanoparticles into a single tumor cell. For example; while clinical breast cancer can be diagnosed by finding at least 1,000,000 tumor cells with traditional methods in mammography, it is probable to diagnose the breast cancer even with fewer than 100 tumor cells with the help of nanotechnology [30]. The cells in our body are 10,000–20,000 nm in diameter, while the hemoglobin circulating in the vascular system is 5 nm in diameter. Nanoparticles smaller than 50 nm in size are can easily enter. If the nanoparticles are smaller than 20 nm, they can be transported through the bloodstream. Thus, drugs increase intracellular concentrations in tumor cells and reduce side effects to healthy cells, increasing the role of nanoparticles in cancer diagnosis and treatment [35].

6.1   Cancer Diagnosis with Nanomagnet

Iron oxide nanoparticles with high magnetic properties are used for cancer diagnosis. Thanks to the superior magnetic power in these particles, it is very easy to locate the tumor tissues. Killer cells or special antibodies developed for the target tumor in the body are labeled with iron oxide nanoparticles. If the labeled nanoparticles are introduced into the body and the searched tumor is found in the body, the labeled antibodies or killer cells adhere to the antigens on the tumor surface. Antibodies accumulated in tumor tissue or iron oxide nanoparticles in killer cells are detected by magnetic resonance (MR) device by generating signals. Thus, it was easier to detect even the smallest tumor cell [36].

7.   Respirosides

Considered one of the most significant advances in nanomedicine, respirocytes are nanorobots that resemble red blood cells created in the laboratory and can carry oxygen inside them. Since they are one micron in diameter, they can easily move through the blood circulation system. Respirocytes are spherical and consist of 18 billion atoms. Since the arrangement of carbon atoms is similar to that of diamond, nine billion oxygen (O_2) and carbon dioxide (CO_2) can be fit into the respirocytes under 1000 atmospheres of pressure, so it can be compared to a pressurized gas tank. After respirocytes enter the body, they release (O_2) and (CO_2) in a controlled manner. There are special mechanisms that detect the amount of gas on the respirocyte surface and provide gas exchange. When respirocytes come to the lungs, they sense the high O_2 and low CO_2 levels into the external environment, absorb excess oxygen and expel the carbon dioxide inside. If the carbon dioxide is high in the lungs and the oxygen is low, it is planned to absorb the carbon dioxide in a controlled manner and to release the required amount of oxygen. Thus, it will fulfil the same task that natural red blood cells (erythrocytes) do. Since respirocytes are resistant to high pressure thanks to their diamond-coated surfaces, they can carry high amounts of gas in a small volume. In other words, a five cubic centimeter fluid containing 50 percent respirocytes can carry the same amount of oxygen and carbon dioxide as 5400 cubic centimeters of blood. Thus, it is thought to be useful for the treatment of many diseases in clinical use. In other words, if this application is successful in the clinic, it will increase the lung capacity. Thanks to respirators, it will be possible to stay breathless in water for four hours or to run for 15 minutes without breathing. Since respirators also send acoustic signals thanks to gas exchange, they control gas exchange at any time and at any place [37].

8.   Nano Diagnosis in Drug and Gene Carrier Systems

Drug and gene carrier nanosystems aid diagnostic imaging and therapy by delivering drugs, genes, or imaging agents. Drug and gene carrier nanosystems provide safe, controlled and effective access to various anatomical and biological structures, target organs or tissues like the respiratory system bronchioles, the blood-brain barrier and tight junctions in the skin [38]. The size of vectors carrying DNA is critical in nanocarrier systems. What makes nanoparticles suitable carriers is that they easily attach to the target cell and tissue and are sized to prevent DNA fragmentation by avoiding the endozone-lysozone process. Nanoparticles used for this purpose produce an immune response when they encounter viral vectors [39]. (Figure 3).

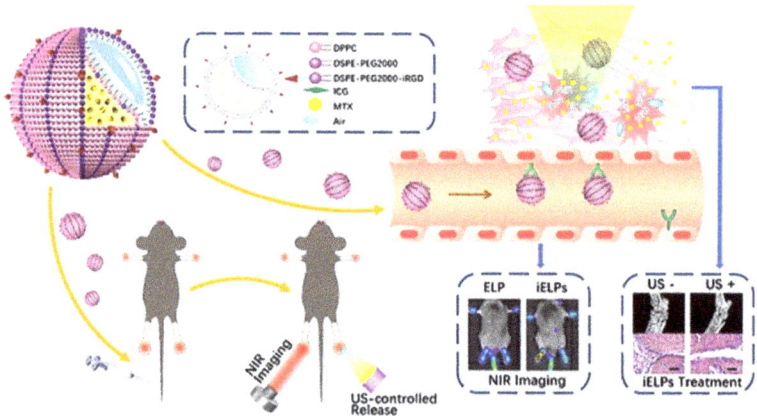

Figure 3. Drug release of nanotechnology [40].

9.   With Transportation of Growth Factors Nanocarriers

Growth factors play an essential role in the process of bone regeneration. VEGF (Vascular Endothelial Growth Factor), TGF (Transforming Growth Factor) and BMP (Bone Morphogenetic Protein) are important growth factors. These growth factors, acting alone or together, affect the bone structure, which we call osteogenic differentiation. The use of these factors is mandatory in tissue engineering applications. Growth factors are placed on nanocarriers so that a controlled and long-term release of growth factors takes place [41].

10.   Nano Diagnosis in Orthopaedics

Hydroxyapatite, a calcium compound, is the molecule that forms bone. It is theoretically possible for a nano-sized change in the crystalline structure of this molecule to make bone as durable as steel, and its production is planned. These 100 nm crystal structures, called nanoss, are a trade secret. It will appear in the near future. It will be possible to reproduce nanoss crystals of desired size, shape and purity in bone cells. The cell swallows these crystals and forms durable bones. After a while, nanosses will be able to be used instead of bone with its feature that is not different from natural bone. Thanks to nanoss transplantation, conditions such as fractures and cracks in old or young bones will be eliminated [42].

Today, thanks to nanotechnology, a hundred times more durable than steel, six times lighter and highly flexible structure (nanotube) has been developed (Figure 4).

It has been shown that hundreds of times more durable nanoorthoses and nanoprostheses will replace the orthoses and prostheses that we frequently use in the medical field [43].

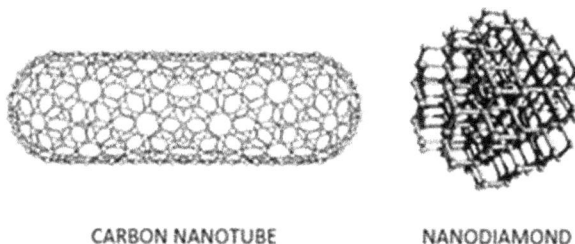

CARBON NANOTUBE   NANODIAMOND

Figure 4. Carbon nanotube sample [44].

11.   Nano Diagnosis in Dermatology

The main focus of nanotech is the treatment and diagnosis of metastatic melanoma, the deadliest of skin cancers [45]. Identification of specific proteins in the diagnosis of melanoma, development of new therapeutic agents, and delivery of these agents to target tumor cells will assist in treatment regulation. Low dose minimal side effect in such treatments is to increase the drug concentration in the tumor tissue [46].

In squamous cell carcinomas, the growth of the tumor is prevented by using gold nanoparticles with photothermal therapy method [47]. Nanoparticles accumulate in tumor cells and undesirable side effects can be prevented in healthy tissues around the tumor [48].

12.   Other Applications of Nano Diagnosis

Nanoparticles can be magnetic, fluorescent and electrical conductivity structures. Thanks to these features, various diseases can be diagnosed. Useful results can be obtained in a short time by using a small amount of tissue samples together with structures such as pathogen-specific antibodies, toxins, nucleic acid added to nanoparticle surfaces [49],[50],[51].

Size of the skin lesions can be determined, and nevus mapping and follow-up can be performed with clothing made of fiber optic fabrics. By following the heat inflammation, diseases such as atopic dermatitis and psoriasis can be diagnosed, body surface area calculation and diseases can be followed [51],[52],[53]. Gold nanoparticles are used for diagnostic purposes by combining them with different techniques due to their light emission [54]. Gold nanoparticles are the best structures for sensors and can be used in DNA analysis [55]. They act as detectors in DNA nano-PCR sequencing (sequencing) methods with analytical technologies such as fluorescence emissions and optical absorbance. Gold nanoparticles also function as a contrast enhancer in the imaging of cancer [45],[54],[24].

Quantum Dots (QD) are used to locate the tumor without the use of radioactive materials. By using QDs topically, sentinel lymph node evaluation can be performed without impairing both skin and tumor development [45],[51],[54],[56].

Contrast media used in the magnetic resonance imaging in hospitals form basis of diagnosis for imaging. This function is performed by nanoparticles such as gold, silver, iron oxide, gadolinium [56]. In tumor cells and ligands, magnetic particles loaded with diagnostic agents to detect the tumor accumulate densely in the tissue where the tumor is located, and magnetic particles create an image and give a topographic location, making early diagnosis easier [54],[57]. Positron-emitting silica nanoparticles have been successfully used in medicine for tumor of cell targeting and mapping of nodal sites. Human clinical trials have been approved in the areas of nodal metastasis, lymphatic drainage, differential tumor burden and imaging. Biodegradation of silica to silica and excretion by kidneys are the biggest advantages of this method [45]. Nano punch, on the other hand, is a very new application for taking biopsy material from areas such as the nail matrix [51].

13. Biochips

Biochip structures are formed by a combination of glass, a porous gel, and a polymer. Biochips have the ability to be programmed to perform a desired biological function. Today, biochips, which are used as biosensors, enable people who have lost their sight and hearing to regain their lost abilities. They have a wide area of use in hospitals, clinics, universities and laboratories. The most promising feature of biochip technology is that it allows simultaneous multi-diagnosis [58] (Figure 5).

Microstructures include systems that are invisible to the naked eye but can be characterized by a microscope. Mechanical materials can be brought to micro size and the size, shape, coating of drugs, type and thickness of the particles can be determined, and the desired macro-scale tasks can be loaded into the micro-size systems. Microstructures, microsensors, microactuators and microelectronic components can be integrated into a silicon chip. Although microsensors are versatile structures, they measure mechanical, thermal, magnetic, chemical or electromagnetic phenomena and information, and determine changes in the environment. Thanks to biosensors, biological and chemical threats are detected early [60]

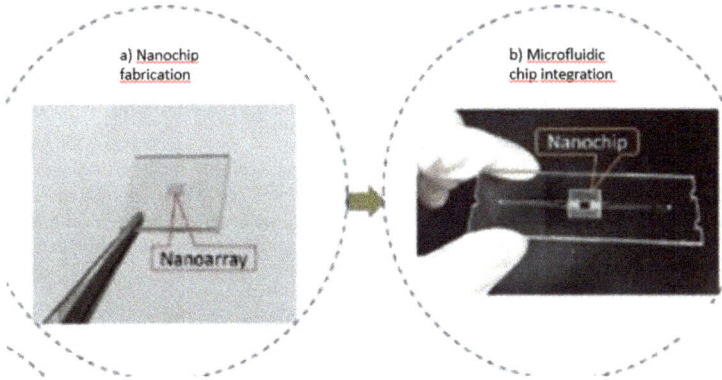

Figure 5. Nanochips [59].

14. Nano-Ophthalmology

In the field of nanomedicine, many studies have been carried out in corneal diseases, imaging, prevention and reduction of corneal opacities [61], [62]. A large number of studies have been carried out on nanoparticles and drug delivery technology in the diagnosis and local application of diseases of the eye. Nanoparticles in the 200 to 2000 nm range can remain in the eye for a minimum of two months. They can also remain in the anterior corneal tissue for a extend period of time by adhering to the mucous membrane. Ophthalmic drugs can be encapsulated in nanoparticles for absorption and conjugation [63]. (Figure 6).

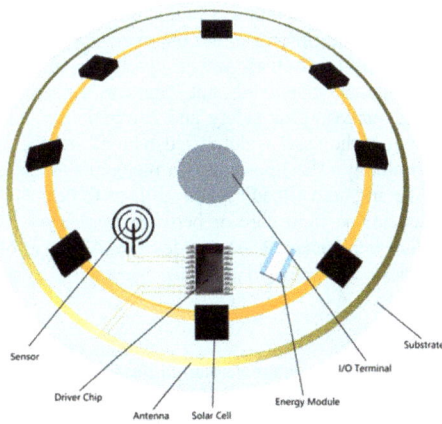

Figure 6. Smart lenses [64].

15. Nano Teeth

It has many uses ranging from implants to oral hygiene and oral care products containing nanoparticles, from the diagnostic and the medical treatment of oral pathologies to prevention of oral illnesses, from relieving toothache to protecting dental health [65],[66].

It is aimed to increase mechanical strength and to minimize wear resistance with nanoparticles added to dental materials [67]. In nanotechnological applications in dentistry, nanodiagnostic devices, nano robots can be implanted in patients' bodies to identify tumor cells and toxic substances in the diagnosis of oral cancer and other oral diseases at the molecular and cellular levels [68]. It is envisaged to produce dentifrobots that can control local and general plaque deposition in the oral region [69].

16. Atherosclerosis

Nanocarriers used in the diagnosis of atherosclerosis are used for imaging to monitor inflammation and progression of heart disease and narrowed vascular structures. Most of the nanomedicine molecules are used in endothelial cell and macrophage targeting, interventions for iodine content, angiogenesis and thrombosis phenomena [70].

17. Nanoanesthesia

In the field of anaesthesiology and reanimation, nanotechnology applications at device and drug level are used in general anesthesia, regional anesthesia, local anesthesia. Today, more than 100 FDA (Food and Drug Administration) certified nanomolecules are being used in medical applications. And three of them are in the experimental stage. Drug molecules containing nanoparticles improve drug bioavailability by delaying drug metabolism and clearance in specific cells and tissues. Through the nanoparticles in the halothane molecule, it can increase the benefits in the field of medicine as an anesthetic agent in the cardiovascular system and lungs and also prevent liver toxicity. It has been observed that the use of propofol-loaded nanomicelles in medicine for diagnostic purposes improves the pharmacokinetic properties of propofol, thus improving the quality of anesthesia with lower cytotoxic and hemolytic effects, higher permeability properties. Nanomicelles did not prevent the accumulation of excessive concentrations of propofol, increasing its safety and compatibility. The antidote used in toxic doses of bupivacaine, which is the most widely used type of anesthesia in regional anesthesia, was created with nanotechnology. The liposome is a nanometer-sized phospholipid-based lipid vesicle. Nanodrugs; Passive and active loading allows them to be freely released into the target tissue by being transported in the fluid core or between lipid layers in several ways, such as double emulsion. Lipozolam bupivacaine can provide up to 72 hours of analgesia with a single dose injection for postoperative analgesia [71].

18. Nano Obesity

Polyethylene glycol (PEG) nanoparticles are versatile linear synthetic polymer used because they provide high solubility in water and are compatible with many therapeutic substances. Coating nanoparticles with PEG is called "PEGylation".". With that process, nanoparticle surfaces are protected, and it is widely used in the diagnosis and treatment of obesity by increasing therapeutic efficacy of bio active components and delivery of the gene to the target tissue.

Prohibitin protein nano sized particles encapsulated in a proapoptotic petite have been shown to significantly reduce unwanted ectopic fat accumulating in muscles and liver. These results indicate the potential advantages of vascular-targeted nanoparticles in controlling ectopic fat deposition associated with obesity and metabolic syndrome [72].

Chitosan nanoparticles are used as an anti-obesity agent by inhibiting adipocyte formation [73].

19. Magnetic Nanoparticles in the Detection of Antidepressants and Certain Drugs

Nowadays, the number of people using antidepressant drugs is quite high. The effects of these drugs in water and terrestrial environment are not fully known [74]. Mistakes such as inconsiderate flushing of expired antidepressant drugs and their derivatives down the toilet at home, accidents during the production of such drugs, and the spread of hospital wastewater into the environment cause the spread of antidepressant drugs into the environment. Some of the antidepressant drugs detected in wastewater facilities are as follows: Diazepam, Oxazepam, Fluoxetine, Carbamazepine, Lorazepam and similar drugs. As a result, polluted environmental waters affect many areas of living life. Recently, many pharmaceutical methods have been used to detect these pollutants that affect human health. Nano filtration processes are used together with classical methods to eliminate this situation, which poses a great risk to living life [75]. For this purpose, the adsorption capacity of artificially produced magnetic nanoparticles is quite high. Thus, foreign agents in water, especially organic and inorganic contaminants of antidepressant, anti-inflammatory, analgesic and antibacterial drugs, adhere to their surfaces and purify the water from the harmful effects of these drugs [76,77].

Antidepressant drugs, which are highly detected in hospital wastewater, threaten health, but 90% of the negative effects are eliminated by nanofiltration. Using the nanofiltration method, wastewater goes through many stages such as adsorption, sieving, electrostatic repulsion, primary and secondary treatment. For example, in cases where secondary treatment is not sufficient, pharmaceutical compounds detected in pharmaceutical contaminated waters have been successfully removed from the environment using nanofiltration osmosis process. With the help of nanofiltration membranes, multivalent ions and very small organic pollutants can be easily separated [78].

Adsorption, one of the nanotechnological methods to eliminate pharmaceutical effects in water, is used. Adsorption process is defined as the adhesion of molecules, ions and atoms on a solid surface in a fluid surface. It is a very effective method used to eliminate pharmaceutical effects and reduce the toxic levels of these drugs. Two magnetic ion exchange resins remove the negative effects of ibuprofen and diclofenac group drugs used as anti-inflammatory and analgesic drugs, as well as Sulfadiazine group drugs used as antibiotics, in water. These are MIER1 and MIER2, which are magnetic ion exchange resins (MIER). It has been observed that sulfadiazine and diclofenac group drugs create high adsorption on MIER 2 resin. A high amount of adsorption has been recorded in the ibuprofen group of drugs due to the large pore group of MIER1 and its specific surface area [79],[80].

20. Conclusion

Nowadays, it is possible to frequently come across nanotechnology products that have a very old history. However, the nanoparticles on the basis of these products were investigated in the middle of the 19th century. The saying of Richart Feynman, who is considered the father of nanotechnology, "There is a lot of ground at the bottom" is very meaningful. Nanosized particles have many uses. In this article, we focused on the production, design, efficiency of nanomaterials in every field, as well as their functionality in diagnosing diseases, especially in the field of health. Nanotechnology's sensor designs, nanorobots, respirocytes, biochips, especially in biodiagnosis, nanoscale-controlled release system, microbiology, cancer diagnosis, gene transport systems, transport of growth factors, ophthalmology, dentology, anesthesiology, atherosclerosis, obesity, orthopedics, dermatology and many more. It has been shown that accurate studies have been carried out in the early diagnosis of diseases. Based on the unique characteristics of nanomaterials, the sensors produced by modifying their surfaces have made the studies advantageous because they only lock on the target tissue. Electrochemical sensors that respond to molecules and ions containing electroactive properties create electrical signals depending on the concentration of the nanoparticle component and have become a part of our lives as devices that provide qualitative and quantitative information. Biosensor applications of nanoparticles working with the principle of accurate, simple, fast and sensitive detection of diseases have achieved successful results. Every country in the world strategically shows great interest and support to this technology. As an indispensable technology of the future, it should be assured that it will be useful in improving public health and extending life.

References

[1]   H. Göksu, H. Burhan, S.D. Mustafov, F. Şen, Oxidation of Benzyl Alcohol Compounds in the Presence of Carbon Hybrid Supported Platinum Nanoparticles (Pt@CHs) in Oxygen Atmosphere, Sci. Reports 2020 101 10 (2020) 1–8. https://doi.org/10.1038/s41598-020-62400-5

[2]   A. Hojjati-Najafabadi, S. Salmanpour, F. Sen, P.N. Asrami, M. Mahdavian, M.A. Khalilzadeh, A Tramadol Drug Electrochemical Sensor Amplified by Biosynthesized Au Nanoparticle Using Mentha aquatic Extract and Ionic Liquid, Top. Catal. 65 (2022) 587–594. https://doi.org/10.1007/s11244-021-01498-x

[3]   A. Şavk, K. Cellat, K. Arıkan, F. Tezcan, S.K. Gülbay, S. Kızıldağ, E.Ş. Işgın, F. Şen, Highly monodisperse Pd-Ni nanoparticles supported on rGO as a rapid, sensitive, reusable and selective enzyme-free glucose sensor, Sci. Rep. 9 (2019) 19228. https://doi.org/10.1038/s41598-019-55746-y

[4]   N.H. Khand, A.R. Solangi, S. Ameen, A. Fatima, J.A. Buledi, A. Mallah, S.Q. Memon, F. Sen, F. Karimi, Y. Orooji, A new electrochemical method for the detection of quercetin in onion, honey and green tea using Co3O4 modified GCE, J. Food Meas. Charact. 2021 154 15 (2021) 3720–3730. https://doi.org/10.1007/S11694-021-00956-0

[5]   A. Cherif, R. Nebbali, J.W. Sheffield, N. Doner, F. Sen, Numerical investigation of hydrogen production via autothermal reforming of steam and methane over Ni/Al2O3 and Pt/Al2O3 patterned catalytic layers, Int. J. Hydrogen Energy (2021). https://doi.org/10.1016/j.ijhydene.2021.04.032

[6]   R. Ulus, Y. Yıldız, S. Eriş, B. Aday, F. Şen, M. Kaya, Functionalized Multi-Walled Carbon Nanotubes (f-MWCNT) as Highly Efficient and Reusable Heterogeneous Catalysts for the Synthesis of Acridinedione Derivatives, ChemistrySelect 1 (2016) 3861–3865. https://doi.org/10.1002/SLCT.201600719

[7]   F. Şen, G. Gökağaç, Improving catalytic efficiency in the methanol oxidation reaction by inserting Ru in face-centered cubic Pt nanoparticles prepared by a new surfactant, tert-octanethiol, Energy and Fuels 22 (2008) 1858–1864. https://doi.org/10.1021/ef700575t

[8]   J.T. Abrahamson, B. Sempere, M.P. Walsh, J.M. Forman, F. Şen, S. Şen, S.G. Mahajan, G.L.C. Paulus, Q.H. Wang, W. Choi, M.S. Strano, Excess Thermopower and the Theory of Thermopower Waves, ACS Nano 7 (2013) 6533–6544. https://doi.org/10.1021/nn402411k

[9]   F. Gulbagca, A. Aygün, M. Gülcan, S. Ozdemir, S. Gonca, F. Şen, Green Synthesis of Palladium Nanoparticles: Preparation, Characterization, and Investigation of Antioxidant, Antimicrobial, Anticancer, and DNA Cleavage Activities, Appl. Organomet. Chem. 35 (2021). https://doi.org/10.1002/aoc.6272

[10]   S. Ertan, F. Şen, S. Şen, G. Gökağaç, Platinum nanocatalysts prepared with different surfactants for C1-C3 alcohol oxidations and their surface morphologies by AFM, J. Nanoparticle Res. 14 (2012) 1–12. https://doi.org/10.1007/S11051-012-0922-5

[11]   H. Hong, Y. Zhang, J. Sun, W. Cai, Molecular imaging and therapy of cancer with radiolabeled nanoparticles, Nano Today 4 (2009) 399–413. https://doi.org/10.1016/J.NANTOD.2009.07.001

[12]   İ. Meydan, H. Seçkin, Green synthesis, characterization, antimicrobial and antioxidant activities of zinc oxide nanoparticles using Helichrysum arenarium extract, Int. J. Agric. Environ. Food Sci. 5 (2021) 33–41. https://doi.org/10.31015/JAEFS.2021.1.5

[13]   H. Durak, Personalized Therapy and Theranostic Approaches in Oncology/ Onkolojide Kisisellestirilmis Tedavi ve Teranostik Yaklasimlar, Nucl. Med. Semin. 1 (2015) 80–85.

[14]   J.K. Patra, G. Das, L.F. Fraceto, E.V.R. Campos, M.D.P. Rodriguez-Torres, L.S. Acosta-Torres, L.A. Diaz-Torres, R. Grillo, M.K. Swamy, S. Sharma, S. Habtemariam, H.S. Shin, Nano based drug delivery systems: recent developments and future prospects, J. Nanobiotechnology 2018 161 16 (2018) 1–33. https://doi.org/10.1186/S12951-018-0392-8

[15]   Z.U. Arif, M.Y. Khalid, A. Zolfagharian, M. Bodaghi, 4D bioprinting of smart polymers for biomedical applications: recent progress, challenges, and future perspectives, React. Funct. Polym. 179 (2022) 105374. https://doi.org/10.1016/J.REACTFUNCTPOLYM.2022.105374

[16]   R.S. Kalash, V.K. Lakshmanan, C.S. Cho, I.K. Park, Theranostics, Biomater. Nanoarchitectonics (2016) 197–215. https://doi.org/10.1016/B978-0-323-37127-8.00012-1

[17]   R. Nagraik, A. Sharma, D. Kumar, S. Mukherjee, F. Sen, A.P. Kumar, Amalgamation of biosensors and nanotechnology in disease diagnosis: Mini-review, Sensors Int. 2 (2021) 100089. https://doi.org/10.1016/J.SINTL.2021.100089

[18]   K. Elumalai, S. Srinivasan, A. Shanmugam, Review of the efficacy of nanoparticle-based drug delivery systems for cancer treatment, Biomed. Technol. 5 (2024) 109–122. https://doi.org/10.1016/J.BMT.2023.09.001

[19]   S. Li, T. Zha, X. Gong, Q. Hu, M. Yu, J. Wu, R. Li, J. Wang, Y. Chen, Cu–Cd–Zn–S/ZnS core/shell quantum dot/polyvinyl alcohol flexible films for white light-emitting diodes, RSC Adv. 10 (2020) 24425–24433. https://doi.org/10.1039/D0RA03540H

[20]   V. Hasirci, N. Hasirci, Fundamentals of Biomaterials, Fundam. Biomater. (2018) 1–338. https://doi.org/10.1007/978-1-4939-8856-3

[21]   B. Jiang, L. Hu, C. Gao, J. Shen, Ibuprofen-loaded nanoparticles prepared by a co-precipitation method and their release properties, Int. J. Pharm. 304 (2005) 220–230. https://doi.org/10.1016/J.IJPHARM.2005.08.008

[22]   K.L. Quick, L.L. Dugan, Superoxide stress identifies neurons at risk in a model of ataxia-telangiectasia, Ann. Neurol. 49 (2001) 627–635. https://doi.org/10.1002/ANA.1005

[23]   L.L. Dugan, K.L. Quick, Reactive Oxygen Species and Aging: Evolving Questions, Sci. Aging Knowl. Environ. 2005 (2005). https://doi.org/10.1126/SAGEKE.2005.26.PE20

[24]   K.K. Jain, Nanotechnology in clinical laboratory diagnostics, Clin. Chim. Acta 358 (2005) 37–54. https://doi.org/10.1016/J.CCCN.2005.03.014

[25]   R. Ramakrishnan, W. Buckingham, M. Domanus, L. Gieser, K. Klein, G. Kunkel, A. Prokhorova, P. V. Riccelli, Sensitive Assay for Identification of Methicillin-Resistant Staphylococcus aureus, Based on Direct Detection of Genomic DNA by Use of Gold Nanoparticle Probes, Clin. Chem. 50 (2004) 1949–

1952. https://doi.org/10.1373/CLINCHEM.2004.036723

[26]   H. Seçkin, I. Meydan, Synthesis and characterization of *Sophora alopecuroides* L. green synthesized of Ag nanoparticles for the antioxidant, antimicrobial and DNA damage prevention activity, Brazilian J. Pharm. Sci. 58 (2022) e20992. https://doi.org/10.1590/S2175-97902022E20992

[27]   X. Kong, P. Gao, J. Wang, Y. Fang, K.C. Hwang, Advances of medical nanorobots for future cancer treatments, J. Hematol. Oncol. 16 (2023) 74. https://doi.org/10.1186/S13045-023-01463-Z

[28]   A.V. Singh, V. Chandrasekar, P. Janapareddy, D.E. Mathews, P. Laux, A. Luch, Y. Yang, B. Garcia-Canibano, S. Balakrishnan, J. Abinahed, A. Al Ansari, S.P. Dakua, Emerging Application of Nanorobotics and Artificial Intelligence to Cross the BBB: Advances in Design, Controlled Maneuvering, and Targeting of the Barriers, ACS Chem. Neurosci. 12 (2021) 1835–1853. https://doi.org/10.1021/ACSCHEMNEURO.1C00087

[29]   N. Rashidi, M. Davidson, V. Apostolopoulos, K. Nurgali, Nanoparticles in cancer diagnosis and treatment: Progress, challenges, and opportunities, J. Drug Deliv. Sci. Technol. 95 (2024) 105599. https://doi.org/10.1016/J.JDDST.2024.105599

[30]   K.K. Singh, Nanotechnology in Cancer Detection and Treatment, Http://Dx.Doi.Org/10.1177/153303460500400601 4 (2005) 583. https://doi.org/10.1177/153303460500400601

[31]   A.P. Nikalje, Nanotechnology and its Applications in Medicine, Med. Chem. (Los. Angeles). 5 (2015). https://doi.org/10.4172/2161-0444.1000247

[32]   F. Sauter-Starace, D. Ratel, C. Cretallaz, M. Foerster, A. Lambert, C. Gaude, T. Costecalde, S. Bonnet, G. Charvet, T. Aksenova, C. Mestais, A.L. Benabid, N. Torres-Martinez, Long-Term Sheep Implantation of WIMAGINE®, a Wireless 64-Channel Electrocorticogram Recorder, Front. Neurosci. 13 (2019) 451890. https://doi.org/10.3389/FNINS.2019.0084

[33]   D. Pal, A.K. Nayak, Editorial Article NANOTECHNOLOGY FOR TARGETED DELIVERY IN CANCER THERAPEUTICS, Int. J. Pharm. Sci. Rev. Res. 1 (n.d.).

[34]   L. V. Wang, Ultrasound-mediated biophotonic imaging: A review of acousto-optical tomography and photo-acoustic tomography, Dis. Markers 19 (2004) 123–138.

[35]   Ö. Oylar, İ. Tekin, Uludağ Üniversitesi Mühendislik-Mimarlık Fakültesi Dergisi, Cilt 16 (2011).

[36]   C.E. Neumaier, G. Baio, S. Ferrini, G. Corte, A. Daga, MR and Iron Magnetic Nanoparticles. Imaging Opportunities in Preclinical and Translational Research, Https://Doi.Org/10.1177/030089160809400215 94 (2018) 226–233. https://doi.org/10.1177/030089160809400215

[37]   Tübitak, Bilim ve Teknik Tübitak Aylık Popüler Bilim Dergisi- Nisan- Sayı 497, Tübitak (2009) 30–37.

[38]   T.J. Wickham, Ligand-directed targeting of genes to the site of disease, Nat. Med. 2003 91 9 (2003) 135–139. https://doi.org/10.1038/nm0103-135

[39]   Y. Guo, X. Cao, X. Zheng, S.J. Abbas, J. Li, W. Tan, Construction of nanocarriers based on nucleic acids and their applications in nanobiology delivery systems, Natl. Sci. Rev. 9 (2022) nwac006. https://doi.org/10.1093/NSR/NWAC006

[40]   H. Wu, Y. He, H. Wu, M. Zhou, Z. Xu, R. Xiong, F. Yan, H. Liu, Near-infrared fluorescence imaging-guided focused ultrasound-mediated therapy against Rheumatoid Arthritis by MTX-ICG-loaded iRGD-modified echogenic liposomes, Theranostics 10 (2020) 10092–10105. https://doi.org/10.7150/THNO.44865

[41]   S. Zhang, H. Uludağ, Nanoparticulate systems for growth factor delivery, Pharm. Res. 26 (2009) 1561–1580. https://doi.org/10.1007/S11095-009-9897-Z

[42]   Y. Korucu, Nanotüp Çeşitleri ve Uygulamaları, Yüksek Lisans Tezi. Fizik Anabilim Dalı, Eskişehir Osmangazi Üniversitesi, (2010).

[43]   W.R. Smith, P.W. Hudson, B.A. Ponce, S.R. Rajaram Manoharan, Nanotechnology in orthopedics: a clinically oriented review, BMC Musculoskelet. Disord. 19 (2018) 67. https://doi.org/10.1186/S12891-018-1990-1

[44]   M. Caffo, A. Curcio, K. Rajiv, G. Caruso, M. Venza, A. Germanò, Potential Role of Carbon Nanomaterials in the Treatment of Malignant Brain Gliomas, Cancers (Basel). 15 (2023). https://doi.org/10.3390/CANCERS15092575

[45]   L.A. Delouise, Applications of nanotechnology in dermatology, J. Invest. Dermatol. 132 (2012) 964–975. https://doi.org/10.1038/jid.2011.425

[46]   M.D. Wang, D.M. Shin, J.W. Simons, S. Nie, Nanotechnology for targeted cancer therapy, Expert Rev. Anticancer Ther. 7 (2007) 833–837. https://doi.org/10.1586/14737140.7.6.833

[47]   E.B. Dickerson, E.C. Dreaden, X. Huang, I.H. El-Sayed, H. Chu, S. Pushpanketh, J.F. McDonald, M.A. El-Sayed, Gold nanorod assisted near-infrared plasmonic photothermal therapy (PPTT) of squamous cell carcinoma in mice, Cancer Lett. 269 (2008) 57–66. https://doi.org/10.1016/J.CANLET.2008.04.026

[48]   D.K. Chatterjee, L.S. Fong, Y. Zhang, Nanoparticles in photodynamic therapy: An emerging paradigm, Adv. Drug Deliv. Rev. 60 (2008) 1627–1637. https://doi.org/10.1016/J.ADDR.2008.08.003

[49]   Tasleem Arif Dar As Sihha Medical centre Nanomedicine & Biotherapeutic Discovery Arif et al, (2015). https://doi.org/10.4172/2155-983X.1000134

[50]   D. Schairer, M. Chouake, M. Nasir, A. Friedman, Nanotechnology Applications in Dermatology, Clin. Nanomedicine Handb. (2013) 85–194. https://doi.org/10.1201/B15642-3

[51]   M. Adil, T. Arif, Nanotechnology-Dermatological Perspective, Int. J. Nanomedicine Nanosurgery Open Access Nanotechnology-Dermatological Perspect. Artic. 2 (2016). https://doi.org/10.16966/2470

[52]   K.B. Paz, A. Friedman, Nanotechnology and the diagnosis of dermatological infectious disease., J. Drugs Dermatol. 11 (2012) 846–851.

[53]   J.G. Eden, S.J. Park, N.P. Ostrom, K.F. Chen, Recent advances in microcavity plasma devices and arrays: a versatile photonic platform, J. Phys. D. Appl. Phys. 38 (2005) 1644. https://doi.org/10.1088/0022-3727/38/11/002

[54]   J.K. Kim, A. Nasir, K.C. Nelson, Nanotechnology and the Diagnosis of Cutaneous Malignancies, Nanotechnol. Dermatology (2013) 127–132. https://doi.org/10.1007/978-1-4614-5034-4_12

[55]   L. Zuo, W. Wei, M. Morris, J. Wei, M. Gorbounov, C. Wei, New Technology and Clinical Applications of Nanomedicine, Med. Clin. North Am. 91 (2007) 845–862. https://doi.org/10.1016/J.MCNA.2007.05.004

[56]   R. Saraceno, A. Chiricozzi, M. Gabellini, S. Chimenti, Emerging applications of nanomedicine in dermatology, Ski. Res. Technol. 19 (2013) e13–e19. https://doi.org/10.1111/J.1600-0846.2011.00601.X

[57]   A. Solanki, J.D. Kim, K.B. Lee, Nanotechnology for Regenerative Medicine: Nanomaterials for Stem Cell Imaging, Nanomedicine 3 (2008) 567–578. https://doi.org/10.2217/17435889.3.4.567

[58]   Z. Tüylek, Nanotıp Alanında Kullanılan Sistemler, Arch. Med. Rev. J. 28 (2019) 119–129. https://doi.org/10.17827/aktd.412772

[59]   L.L. Sun, Y.S. Leo, X. Zhou, W. Ng, T.I. Wong, J. Deng, Localized surface plasmon resonance based point-of-care system for sepsis diagnosis, Mater. Sci. Energy Technol. 3 (2020) 274–281. https://doi.org/10.1016/J.MSET.2019.10.007

[60]   N. Maluf, An Introduction to Microelectromechanical Systems Engineering, Meas. Sci. Technol. 13 (2002) 229. https://doi.org/10.1088/0957-0233/13/2/701

[61]   S.S. Chaurasia, R.R. Lim, R. Lakshminarayanan, R.R. Mohan, Nanomedicine Approaches for Corneal Diseases, J. Funct. Biomater. 2015, Vol. 6, Pages 277-298 6 (2015) 277–298. https://doi.org/10.3390/JFB6020277

[62]   M. Gómez-Garzón, M.A. Martínez-Ceballos, A. Gómez-López, A. Rojas-Villarraga, Nanotechnology applications in ophthalmology: An update, Rev. Mex. Oftalmol. 94 (2020) 199–210. https://doi.org/10.24875/RMOE.M20000127

[63]   J. Zhang, J. Jiao, M. Niu, X. Gao, G. Zhang, H. Yu, X. Yang, L. Liu, Ten years of knowledge of nano-carrier based drug delivery systems in ophthalmology: Current evidence, challenges, and future prospective, Int. J. Nanomedicine 16 (2021) 6497–6530. https://doi.org/10.2147/IJN.S329831

[64]   Y. Xia, M. Khamis, F.A. Fernandez, H. Heidari, H. Butt, Z. Ahmed, T. Wilkinson, R. Ghannam, State-of-the-Art in Smart Contact Lenses for Human Machine Interaction, IEEE Trans. Human-Machine Syst. 53 (2021) 187–200. https://doi.org/10.1109/THMS.2022.3224683

[65]   I. Kürkçüoğlu, A. Köroğlu, S. Emre Özkır, M. Ateş, S. Demirel Üniversitesi, D. Hekimliği Fakültesi, P.A. Diş Tedavisi, Nanoteknoloji kavramı ve diş hekimliğindeki uygulamaları The nanotechnology concept and its applications to dentistry, (n.d.).

[66]   S.S. Mantri, S. Mantri, The nano era in dentistry, J. Nat. Sci. Biol. Med. 4 (2013) 39. https://doi.org/10.4103/0976-9668.107258

[67]   G. Naguib, A.A. Maghrabi, A.I. Mira, H.A. Mously, M. Hajjaj, M.T. Hamed, Influence of inorganic nanoparticles on dental materials' mechanical properties. A narrative review, BMC Oral Heal. 2023 231 23 (2023) 1–18. https://doi.org/10.1186/S12903-023-03652-1

[68]   N. Beyth, I. Yudovin-Farber, M. Perez-Davidi, A.J. Domb, E.I. Weiss, Polyethyleneimine nanoparticles incorporated into resin composite cause cell death and trigger biofilm stress in vivo, Proc. Natl. Acad. Sci. U. S. A. 107 (2010) 22038–22043. https://doi.org/10.1073/PNAS.1010341107

[69]   R.A. Freitas, Nanodentistry, J. Am. Dent. Assoc. 131 (2000) 1559–1565. https://doi.org/10.14219/JADA.ARCHIVE.2000.0084

[70]   A.M. Flores, J. Ye, K.U. Jarr, N. Hosseini-Nassab, B.R. Smith, N.J. Leeper, Nanoparticle Therapy for Vascular Diseases, Arterioscler. Thromb. Vasc. Biol. 39 (2019) 635–646. https://doi.org/10.1161/ATVBAHA.118.311569

[71]   SAĞLIK & BİLİM 2022: NANOTIP - İlter DEMİRHAN - Google Kitaplar, (n.d.).

[72]   M.N. Hossen, K. Kajimoto, H. Akita, M. Hyodo, H. Harashima, Vascular-targeted nanotherapy for obesity: Unexpected passive targeting mechanism to obese fat for the enhancement of active drug delivery, J. Control. Release 163 (2012) 101–110. https://doi.org/10.1016/J.JCONREL.2012.09.002

[73]   Chitosan Oligosaccharides Inhibit Adipogenesis in 3T3-L1 Adipocytes, (n.d.).

[74]   J. Argaluza, S. Domingo-Echaburu, G. Orive, J. Medrano, R. Hernandez, U. Lertxundi, Environmental pollution with psychiatric drugs, World J. Psychiatry 11 (2021) 791. https://doi.org/10.5498/WJP.V11.I10.791

[75]   Y. Luo, W. Guo, H.H. Ngo, L.D. Nghiem, F.I. Hai, J. Zhang, S. Liang, X.C. Wang, A review on the occurrence of micropollutants in the aquatic environment and their fate and removal during wastewater treatment, Sci. Total Environ. 473–474 (2014) 619–641. https://doi.org/10.1016/J.SCITOTENV.2013.12.065

[76]   D. Aksu, Türkiye'de Kullanimi En Fazla Olan Seçilmiş Antibiyotik İlaçlarinin Atiksudan Biyo-Sentez Demir Nanopartikülleri İle Giderimi, Cevre_Muhendisligi_Anabilim_Dali_-_TC_E (accessed February 17, 2025).

[77]   Nishu, S. Kumar, Smart and innovative nanotechnology applications for water purification, Hybrid Adv. 3 (2023) 100044. https://doi.org/10.1016/J.HYBADV.2023.100044

[78]   P. Verlicchi, A. Galletti, M. Petrovic, D. BarcelÓ, Hospital effluents as a source of emerging pollutants: An overview of micropollutants and sustainable treatment options, J. Hydrol. 389 (2010) 416–428. https://doi.org/10.1016/J.JHYDROL.2010.06.005

[79]   H. L., Tarımsal atık karışımlarından hazırlanan aktif karbonla sulu ortamlardan bazı farmasötik bileşiklerin giderimi / Removal of some pharmaceutical compounds from aqueous solution by using activated carbon prepared from agricultural waste mixtures, Sustain. 11 (2019) 1–14.

[80]   S. Cheng, L. Zhang, A. Ma, H. Xia, J. Peng, C. Li, J. Shu, Comparison of activated carbon and iron/cerium modified activated carbon to remove methylene blue from wastewater, J. Environ. Sci. 65 (2018) 92–102. https://doi.org/10.1016/J.JES.2016.12.027

Biogenic Nanomaterials: Synthesis, Characterization, Applications, and Future Remarks Materials Research Forum LLC
Materials Research Foundations 180 (2025) https://doi.org/21741/9781644903759

Chapter 20

Healing Applications Using Biogenic Nanomaterials

S. Cecen Sular[1*], Y. Sular[1], I. Turk[2], F. Sen[2*]

[1]Vocational School of Health Services, Van Yuzuncu Yil University, Van, Turkiye

[3]Sen Research Group, Biochemistry Department, Faculty of Arts and Science, Dumlupinar University, Evliya Celebi Campus, 43100, Kutahya, Turkiye

sabriyececensular@yyu.edu.tr; fatihsen1980@gmail.com

Abstract

Nanotechnology is making innovations in the world of science every day. Biomedical drugs modified with nanoparticles have been used in many fields of medical science and significant progress has been made in the treatment of diseases. In this chapter, the reader will have a broad perspective and detailed information about which diseases nanotherapies are curing.

Keywords

Healing Applications, Nanotreatment, Innovative Therapy, Biogenic Nanomaterials

1. Introduction

Scientific and technological developments make life easier in every area where people live. Nanotechnology has found a place in many application areas such as health, environment, energy, sensors, fuel cells, food, and agriculture [1–10]. Nanotechnology, which is one of the branches of science used extensively in the field of health, is used effectively in the treatment of patients. Nanotechnology is the control of matter the atomic and molecular scale. In other words, nanotechnology can be expressed as controlling the substance in a dimension ranging from 1 to 100 nanometers [11].

Using nanotechnology in the health sciences includes production of molecules tools for the treatment and protection of many pathological conditions such as disease prevention, early diagnosis and treatment, wound repair and monitoring, controlling cell functions, nanosurgical interventions, normalizing hormonal imbalances. Nanotechnology principles such as bioactivity and artificial intelligence are applied in the field of health [12].

Nanotechnology is a promising and fast- developing area that has interested a lot of attention in recent years, particularly in the fields of health [13]. Studies show that nanotechnology in health has excellent results in treatment of numerous diseases such as neurodegenerative diseases in musculoskeletal system, cancer, gene therapy, bacterial and viral infections, obesity and diabetes. In this study, the developments of nanotechnology in the field of health will be discussed in detail.

Biogenic Nanomaterials: Synthesis, Characterization, Applications, and Future Remarks   Materials Research Forum LLC
Materials Research Foundations 180 (2025)   https://doi.org/21741/9781644903759

1.   Nanotechnology and Health

Nanotechnology, which has proven successful applications in every field, plays an important role in people's lives, also including in field of health, from targeting and killing only cancerous cells without harming the body, to creams whose effect continues for days, from sensor structures that can detect only anthrax microbes, to socks that do not smell because it kills bacteria, and from microbe-free refrigerators to food. Light, which can keep fresh for up to 14 days, has entered our lives by adding color to nano light systems [14,15].

Today, nanotechnology, which is still rapidly developing, has led to significant advances in drug development, gene and skin transfer, and many different types of treatment, prevention of wounds caused by diseases and traumas, protection and prevention of human health by providing appropriate device production [16]. Robots that can be manufactured at nanoscale are called nanobots, artificial nerve cells are called nanobots, and the process of bringing nerve cells together with microelectronic circuits is called micro-noroelectronic network (Figure 1). It has been shown that natural neurons can communicate normally with these nanobot neurons. There are about 15 billion neurons in the human brain, and each of them makes about 5-10 thousand connections with other neurons. This perfect network is created with artificial nanobot neurons. It seems possible to intervene in the area where the damaged nerve structure is in the very near future [17].

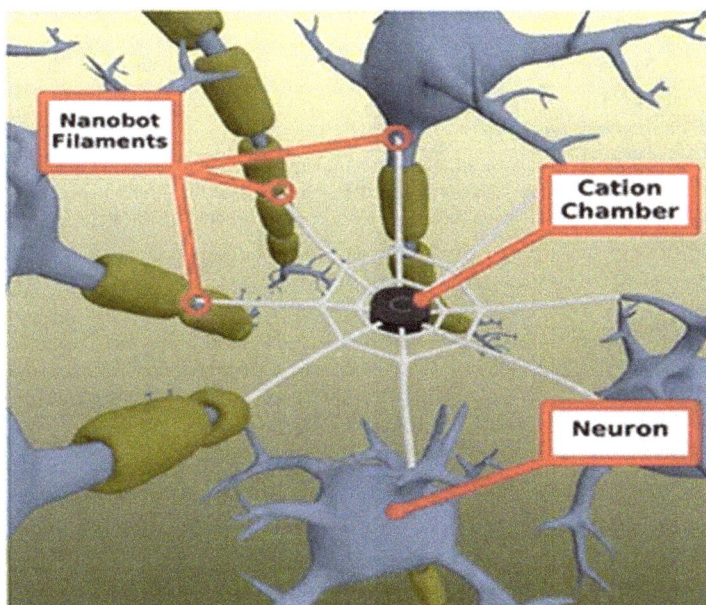

Figure. 1 Micro-neuroelectronic network consisting of nanobut neurons [18].

1.1   Tissue Engineering and Nano

Tissue engineering is one of the most popular fields in healthcare today. In tissue engineering, an organ or tissue is created in a laboratory environment to be used in case of tissue and organ loss or damage. Thus, thanks to the tissue and organ created in the near future, there will be no need for organ transplantation from the donor. It takes a few seconds to a few weeks for cells to form tissues in culture medium in the laboratory. This period is related to the type and size of the tissue. It is possible to produce tissue in varying sizes from 0.0001 cm to 10 cm. Clinical studies are ongoing on cartilage, bone, pancreas, temporary liver-support systems, cornea, nerves, heart valves, vessels and soft tissue. The aim in tissue engineering is to repair or reconstruct living tissues by using biomaterial, cell and biosignal molecules together or separately. Accordingly, it is possible to talk about four approaches for tissue engineering. In the first approach, only biomaterial is used in the formation of tissue. In the second approach, treatment is carried out using cells called "cell transplantation". These cells can be cells isolated from living tissue or genetically engineered cells. In the third approach, biomaterials containing adhesion and growth factors and biosignal molecules are used. The fourth approach is the most studied approach. Tissue formation is aimed by using all three of the cell biomaterials and biosignal molecules together. These four components scaffold, functional cell, biosignal molecule and dynamic force are essential components of nano tissue engineering. Nanoparticles smaller than 100 nm used in tissue engineering move in the bloodstream without precipitation and can easily pass through microvascular structures. Because of these properties, it is preferred for the release and transport of growth factors in tissue studies. Because of their size, these nanoparticles are used for marking cells because they can move easily inside the cell. Thus, interaction with target cells and tissues can be achieved [19].

1.2   Bone Implants and Nanotechnology

Titanium is the most widely used and best-known bone repair material in dentistry and orthopaedics. It is highly break-resistant, machinable and has the strength to withstand high weight, but does not support cell growth and adhesion because it does not have sufficient bioactivity [20].

1.3   Plevra and Nanotechnology

Carbon nanotubes (CNT), which has made a name for itself in the research carried out in the last few years, is a new type of technological crystalline carbon. CNT, with its unique physicochemical features, is used in a wide selection from electronics to healthcare. To date, the effects of CNTs on the lungs and many organs have been studied, but few animal and in vitro studies have been conducted for their effects on the pleura [21]. The biggest thing about CNTs is that they have the potential to cause a situation similar to asbestos (asbestos stone cotton is a carcinogenic mineral in the rock fiber structure) [22]. Because CNTs show asbestos-like structural properties. Asbestosis is a dust disease caused by breathing in asbestos. This disease is also called pneumoconiosis. In a new clinical study conducted at the People's Republic of China, pleural fluids and pleural tissue of seven women who were exposed to nanomaterial working at the factory for 5-17 months were examined under an electron microscope and nanoparticles of 30 nm size were found [23]. These studies suggest that nanomaterials may adhere to the pleura by skin route or inhalation and cause disease. There are no studies yet on whether the exposure of human mesothelial cells to CNTs has an effect on cell and tissue proliferation. An aggressive tumor, malignant mesothelioma (MM) is derived from surfaces such as epithelium, pleural mesothelium and formed by exposure to asbestos mineral fibers [21]. Today, it is not known exactly which mechanisms are associated with asbestos exposure and lead to MM [24].

1.4　Nanodermotology

Applying nanotechnology to the dermatology has revolutionized both treatment and diagnostic methods of many skin diseases [25].

With the method we call topical treatment, that is, the method of applying drugs to the skin, many substances are slightly soluble or insoluble in water. These substances can be denatured and spoiled before reaching the target. At the same time, undesirable side effects may occur with nonspecific distribution in tissues and organs. When the desired effect on the target tissues and organs does not occur, dermatology needs new treatment agents. Therefore, it is exciting to control the skin barrier with nanotechnology and allow continuous drug release [26].

Recently, sunscreens, moisturizers, cosmetics and lasers have been used for treatment and diagnosis in the field of nanomolecular dermatology [25]. Lipophilic nanoparticles designed to deliver transcutaneous drugs are undergoing clinical trials [27].

Nanoparticles can be used as a drug carrier vector, as a contrast agent in field of medical image processing, at the gene therapy, in treatment and diagnosis of skin carcinoma. There are many studies on its use as a drug in the treatment of melanoma, which is one of the promising developments in nanodermatology, especially a type of skin cancer. With nanotechnology, it allows to improve the circulation time of these drugs and to minimize drug toxicity [28].

1.4.1　Nanotechnology in Cosmetics

Many cosmetic elements are embedded in nanoparticles to help protect active compounds from oxidation for effective penetration. For example; Metal oxide and metal nanoparticles are often applied in antifungal, antibacterial, photoprotective and cosmetic products to accelerate cell regeneration [29].

Gammaamino-butyric acid (GABA), botulinum toxin and nano hyaluronic acid nanoparticles, which are especially used as muscle relaxants, have been introduced to reduce or prevent wrinkles. The amount of cosmetic products containing nanoparticles is increasing every day [25], [30], [31].

1.4.2　Humidifiers and Nano

The biggest problem in treatment of many dermatoses, especially atopic dermatitis, is inability of the skin to maintain sufficient moisture. Conventional moisturizers unfortunately deliver ceramide and its derivatives to the active area. Nanoemulsions, SLNs and liposomes are widely used because they increase the therapeutic effect. With the use of these nanoparticles, many positive features are offered, such as small particle size, proper transport, ensuring the invisibility of the preparation used, and most importantly, the absence of oily feeling [25], [32].

1.4.3　Nano in Hair Diseases

When many products such as minoxidil, cyclosporine, finasteride, flutamide used in the treatment of hair diseases are encapsulated with nanoparticles, it has been shown that there is a good penetration and reduction of their side effects [33].

1.4.4   Sunscreens and Nano

Zinc oxide (ZnO) and titanium dioxide (TiO$_2$) are used into sunscreen products due to their ability to reflect, break and absorb nanoparticles. Diamond nanoparticles are shown to be very effective in UVB protection in cell culture and mouse models [34].

In addition, studies in human keratinocyte cultures have shown that it has the same effect on protection from UVB [35], [36].

1.4.5   Nano in Acne Treatment

In many studies, it has been shown that these agents used in acne treatment such as benzoyl peroxide, salicylic acid, clindamycin, and finasteride use nanoparticles to increase treatment efficacy and decrease their side effects [37], [38], [39], [40], [41].

It has been shown that fullerenes create an anti-inflammatory effect by collecting carbon and oxygen radicals, reducing P. acnes lipase activity and reducing sebun production [40].

1.5   Nano as Antiseptic and Antibacterial

Nanoparticles such as silver, TiO$_2$, ZnO, platinum, palladium, gold are recommended not to be used as antiseptic and antibacterial, as they cause reactive oxygen radicals to increase in the cell, disrupting protein synthesis, and damaging the cell wall and DNA [29,42], [43].

1.6   Antiparasitic Nano

In the treatment of leishmania, studies have been conducted showing that it reduces promastigote proliferation and inhibits amastigote growth. Silver nanoparticles have been shown to reduce local use, lesion healing and spleen parasites with UVB [44], [45], [46].

It has been shown that palladium and platinum can be used as a catalytic effect, especially for vitiligo and anti-aging [47]. It has been reported to be effective and strategically important in treatment of antifungal, inflammatory skin diseases, and topical vaccination [48], [49], [50], [51], [52], [53], [54], [55], [56], [57].

1.7   Nano in Wound Healing

In order to accelerate healing of wounds as nanotechnological agents, antimicrobial and anti-inflammatory drugs and creams have been produced by using silver nanoparticles for the development of antibacterial drugs using chitosan nanofibrils. Copper-containing nanoparticles have antimicrobial and angiogenetic effects [58], [59], [60].

Drugs are developed by adding growth factors, bacteriostatic antibiotics, gene therapies, analgesics, stem cells and many other agents to the structure of biomaterials in wound treatments, reconstruction of tissue defects, and burn treatment. Thus, successful results have been obtained in nerve healing, bone healing, shrinkage of cancerous tissues and lymphedema treatments. It has been shown that biodegradable chitin nanofibrils have positive effects on fibroblast proliferation and collagen synthesis, as well as accelerate wound healing by regulating macrophage and cytokine secretions [61], [62], [63]. It has been shown that silver nanoparticles are effective in wound healing by creating fibroblast proliferation, collagen increase, anti-inflammatory effect, DNA and RNA denaturation and antimicrobial effect. Titanium nanoparticles have been shown to have an antimicrobial effect as well as have a positive effect on wound healing [64], [65], [66].

1.8   Nanotechnology at Cancer Treatment

The goal of cancer therapy with nanoparticles is to deliver selective drugs to malignant cells, to alleviate drug-related side effects, and to increase drug concentration in tumor cells with low but effective doses [25], [67]. In cancer treatments, many drugs fail clinically due to their insolubility, but this problem is overcome by transporting many types of drugs in high concentrations thanks to nanoparticles [68]. Systemic drug use, especially corticosteroids and cancer drugs, is expected to become a thing of the past. For example, it has been shown that the application of dexamethasone in nanoparticles suppresses cell proliferation more and for longer [69].

The prevalence of success in cancer treatments is that the given drugs are directed to cancer cells as soon as possible and locked to the target area without damaging healthy tissues. When this situation is successful, a substantial advancement in life span and quality of patient can be achieved. When nanoparticle-loaded cancer drugs reach the target tissue, they can increase the drug concentration and minimize its toxicity on the surrounding healthy tissue [70].

One of the biggest limitations in cancer treatment is that drugs not only affect cancer cells but also damage other healthy cells. In today's studies, treatments that target only cancer cells have yielded positive results. Special carrier systems developed using nanotechnology make it possible not to harm healthy cells while targeting and killing cancerous cells. In cancer treatments, the desired drug is placed in these capsule-like particles with a scale of 10 to 100 nanometers, allowing it to circulate easily in the blood and reach all parts of the body. Nanocapsules smaller than 10 nanometers are administered intravenously to the patient and can be excreted immediately as they pass through the kidneys. It is tough for nanoparticles bigger than 100 nm to enter the tumor structure. The drug nanocapsules that come into contact with the cancerous tissue are taken into the cell by "endocytosis", that is, by absorption. Thanks to special proteins, the cell's rejection of this structure is prevented. Thus, when the drug in the nanocapsule leaves the capsule, it begins to destroy the cancerous cells. Liposomal doxorubicin is a nanodrug used for cancer treatments in clinics for this purpose [71].

1.9   Nano in Diabetes Treatment

Diabetes patients need to constantly calculate insulin doses (sometimes it can be looked at 8-10 times a day) and apply them at certain intervals. In today's application technique, for the control of blood glucose, patients take a blood sample by pricking their fingers using a small insulin needle and perform glucose monitoring themselves. It is aimed that glucose monitoring in diabetic patients is painless. Enzyme electrodes and microdialysis probes in the form of injectors that can be implanted into the subcutaneous tissue are commercially available for periodic glucose control. The disadvantage of these products is that they have a short life and some problems such as lack of sensors. Repositioning the sensor probes, which is a semi-invasive procedure, can disturb patients. Fluorescent-based nanosensors that respond to glucose, called "Smart Tattoo", have been developed. These sensors, which are thinner than paper, can be implanted under the skin and manipulated from the outside. Researches on reading the values to computers or smart phones are continuing [72].

Nanotechnology has developed remote monitoring systems to monitor diabetes patients. Using smart nanodevices, researchers were able to remotely monitor patients' glucose levels. The goal of this study is to monitor the patient's blood glucose levels by introducing the nanoparticles, which are inhaled into the body, into the patient's bloodstream. These miniature nanochips recognize the patient's glucose molecule and measure the glucose level. Thus, fluctuations in blood glucose level are prevented by mimicking the pancreas. It is planned to eliminate the side effects of this method [73].

1.10   Nano in Endodontics

The most important of the endodontic treatment steps is the disinfection of the root canal. Many antibiotic pastes are used for effective disinfection in regenerative treatments. However, these pastes have been reported to cause tooth discoloration, cervical root fractures and insufficient pulp-dentin formation. In order to eliminate or reduce to overcome these very serious disadvantages of pastes, studies are conducted on using antibiotic-containing nanoscaffolds or nanoencapsulated gels [74]. The effect of structures containing antibiotics on regeneration is being evaluated. They showed that the use of an antibiotic-containing electrospin scaffold in regenerative endodontics can be a biologically very safe antimicrobial drug delivery system [75].

1.11   Nanoophtalmology

Classical drug forms such as drops, gels and ointments are commonly used in treatment of eye diseases. 90% of these medications come in the form of eye drops or ophthalmic solutions. 95% of the drugs used are lost through eye drainage [76].

Many ocular barriers such as corneal barrier, tear film barrier, conjunctival-scleral barrier, blood-retina barrier, poor permeability caused by rapid tear filtration, ineffective drug distribution, insufficient bioavailability, and intravitreal (delivering the drug directly to the vitreous fluid in the eye) route of drug administration, eye drugs clinically may limit its effects. Therefore, the treatments given for the eye should be applied frequently [77], [78], [79].

Many nanoscale materials can be designed with different properties in ophthalmic drug applications. Nanocarriers such as liposomes and dendrimers, which are easy to apply in the form of eye drops, are not affected by physiological barrier mechanisms of the eye, increasing the targeted effect and the residence time on the ocular surface after instillation into eye. Nanodrugs increase interactions between cornea and conjunctival epithelium, increasing drug delivery efficiency and bioavailability [76].

Nanoparticle-loaded contact lenses for delivery of acetazolamide for the management of glaucoma [80], degradable subconjunctival implants to apply cyclosporine at the treating dry eye syndrome [81], diclofenacin hydrogel-based nanocolloidal systems used in ocular release [82] many nano-treatment systems such as nano-structured lipid carriers [83] are used for the controlling release of drugs in eye infections.

Phosphate-containing dendrimers have been developed for treatment of eye hypertension and glycone [84]. By putting ophthalmic drugs inside the lens, nanoparticle-loaded contact lenses have been developed. As the developed contact lenses come into contact with the eye fluid, the drug release begins. The solution that makes up the lens is the mixture of molecules that make up the nanochannels. These nano channels make the lens nanoporous so that tears and gases can easily pass into and out of the lens. This makes the lens more compatible with the human eye [85].

1.12   Obesity and Nano

In the process of thermogenesis, it is brown adipose tissue, one of the most important body fats that burns energy. In studies conducted to ensure that the drugs used in the treatment of obesity reach the target tissue, it has been determined that it is difficult for the active substances to reach the white adipose tissue. For this reason, due to the difficulties in reaching the white adipose tissue, the white adipose tissue is transformed into brown tissue by using certain drugs. Presence of a high number of blood vessels in brown adipose tissue is a great advantage. For this reason, drugs containing nanoparticles that were strengthened and made effective during transformation were developed [86].

In an attempt to increase the effectiveness of anti-obesity treatment, therapeutic activities of nanocarrier drug systems are being investigated in order to provide drug release into adipose tissue. In a study conducted in 2021, white adipose tissue was transformed into brown adipose tissue using dibenzazepine-loaded nanoparticles, which trigger the transformation in adipocytes and reduce adiposity. Using polymer nanoparticles of brown adipose biogenesis adding versatility to obesity treatment opens new avenues for therapeutic treatment [86].

Poria cocos gold nanoparticles have been shown to be a powerful anti-obesity drug and protection against metabolic disorders due to obesity. Liposomes, microspheres, Chitosan microspheres, floating microspheres, self-nanoemulsifying drug delivery systems, therapeutic systems coated with other nanoparticles used in the treatment of obesity [87].

1.13   Nanotechnology in the Treatment of Atherosclerosis

With the use of nanotechnology in management of atherosclerosis, the door to a new treatment method has been opened [88]. It consists of strategies to complement existing treatments. Thanks to these strategies, atherosclerosis plaques can be visualized and a contribution to the treatment is provided by targeting the cells with nanoparticles in the atherosclerosis process. Two drugs, Sirolimus (Rapamycin) and Paclitaxel, have important effects in nanomedicine applications in the treatment of atherosclerosis. It has been determined that nanoparticles of these two drugs have antiprolative effects [89]. Sirolimus nanoparticles appear to be good candidates for the treatment of atherosclerosis in plaques, especially since they show a higher efficacy in hypoxia. In a studying by [90] they developed nanoparticles filled with Annexin A1 molecules and injected them into the bloodstream of mice together with collagen IV. At the end of the study, more than 70% of the nanoparticles reached the target lesions, and it was determined that chronic atherosclerosis and plaque necrosis decreased, the protective collagen layer increased, and the oxidative stress in plaque properties decreased [90]. Another treatment modality for atherosclerosis is the active targeting of macrophages. For example, a treatment method is planned by magnetic resonance imaging of atherosclerosis plaques in ApoE knockout mice by means of iron oxide nanoparticles [91]. Collagen IV, which helps form the basis of the blood vessels in the body, is the extracellular matrix that is released when atherosclerosis progresses. Atherosclerosis treatment strategy is based on targeting this extracellular matrix. Theragnostic nanoparticles can reach atherosclerotic plaques caused by the collagen IV binding peptides [92].

1.14   Respirosides

One of the subjects on which many studies have been done recently is artificial blood (Figure 2). Artificial blood cells called reciprocides can carry 236 times more oxygen to the tissues [17]. When examined theoretically, it has been shown that if only one liter of human blood is replaced with reciprocides, it can be sustained for 4 hours without breathing. Considering the role played by ischemia due to hypoxia in tissue damage that develops after heart attack and stroke, reciprosides provide a great deal of benefit in treatment [93,94].

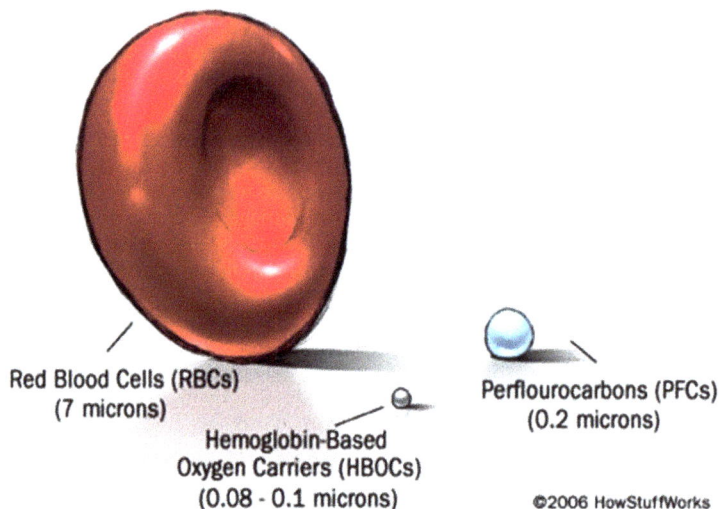

Red Blood Cells (RBCs)
(7 microns)

Hemoglobin-Based
Oxygen Carriers (HBOCs)
(0.08 - 0.1 microns)

Perflourocarbons (PFCs)
(0.2 microns)

©2006 HowStuffWorks

Figure. 2 Artificial blood cell (reciproside) [95].

1.15   Nanotechnology in Aesthetic Surgery

The aim of surgical treatments is to make the diagnosis and treatment fast and safe, to cause minimal trauma to the patient, to leave no scars after the treatment and to minimize the complications for the postoperative period [96].

Nerve injuries are one of the most common conditions in plastic surgery. Understanding the location, injury and severity of the nerve in nerve defects is very important for correct intervention. Current imaging methods do not have sufficient quality effect in the diagnosis of nerve injury and in the follow-up of regeneration. And it has been shown that iron oxide nanoparticles added to the collagen-roofed conduits used increase the contrast visible quality in Magnetic Resonance Imaging, and provide detailed imaging of nerve regeneration [97].
Perfluorocarbon nanoemulsions, which are used to evaluate nerve regeneration, are also

effectively used in the theragnostic field, where different drugs can be bound [98].

1.16   Vaccine and Nano

Nanoparticle (NP)-based vaccine applications have a great role in reducing infectious diseases in cellular and humoral immunity events. Although it is a disadvantage that there is no effective vaccine for every disease today, thanks to the nanoparticles in new generation vaccines, premature deterioration of vaccines can be prevented, and the recovery period of diseases is shortened. Nanoparticles with adjuvant properties are used to deliver many antigens to certain organs and tissues. These biodegradable particles have a low toxic effect and represent an alternative to traditional vaccination methods [99].

NP-based vaccine studies offer effective protective immunity against potent pathogens [100].

In addition to providing controlled antigen release, NPs extend the life of many vaccine types. Thus, they increase humoral and cellular immunity and create an immune response [101].

Physical and morphological properties of NPs such as their composition, charge amount, size, mode of action; It may affect the toxic effects of NPs and the general immune response [102]. For example, the surface charge of NPs creates electrostatic interaction with surface molecules on target cells [103].

The shape of the NPs determines the rate of antigen release. When spherical gold NPs were compared with rod-shaped NPs, it was found that gold NPs had a more effective penetration power and induced strong immune responses [104].

Small-sized NPs (20-200 nm) can be easily taken into the cell with the help of dendritic cells. In recent studies, many NP-containing therapeutic enhancements have been developed to activate T cell activation against bacterial, viral and fungal infections [105].

Many nanoparticles have found use in vaccines. Inorganic NPs include gold, silica, aluminium, carbon, calcium phosphate and magnetic NPs. They enhance the immune response. For example, bacterial and viral antigens given to mice with the use of gold-based NPs have been found to create effective immune responses in the host body against influenza, foot, mouth and tuberculosis diseases and immunodeficiency virus in mice [106],[107].

Silica-loaded NPs facilitate delivery to target cells by recognizing certain groups thanks to silanol-containing molecules on their surfaces [108].

Alums, one of the most commonly used excipients in vaccines, activate immune responses to antigens such as tetanus, influenza and diphtheria [101].

The biocompatibility of polymeric NPs in vaccine applications has an important place in terms of surface modification in all shapes and sizes, safety and non-toxicity. Using Chitosan, a polymeric NP, delivery of tuberculosis lipids to the target cell generates strong humoral and cellular responses [109].

Synthetic Polymer NPs: Synthetic Polymer NPs have been tested in vaccine applications, especially in many types of antigens such as Hepatitis-B antigens, bacillus anthracis, tetenosis, and have been supported by the US Food and Drug Administration (FDA) by forming mechanically strong polymer chains [110], [111], [112].

Detriments are another nanoparticle used in vaccine applications. Antigens modified with dentrimenes have been proven to create strong antibodies against Ebola virus, H1N1 virus and Toxoplasma gondi parasite and to induce cellular T response. In the vaccines developed for SARS-CoV-2 (Covit-19), one of the epidemics that poses a major threat worldwide, lipid nanoparticles of 80-100 nm in size with nanoparticle properties were used. It has been determined that mRNAs modified with positively charged lipid are more stable and resistant to degradation. The antibody responses they generate are recognized as promising vaccine approaches on different Corona viruses [113].

The application of NPs in the treatment of diseases is supported as an effective method for both immune response and infective therapeutic activation [99].

2.   Conclusion

Nanotechnology has gained great momentum in all branches of science in recent years, especially in the field of medicine. The treatment options offered by this technology have improved and improved in the parameters of almost all diseases. Studies on treatment and prevention of possible pathological conditions together with early diagnosis will make a great contribution to the treatment and protection of health. In this study, promising studies of nanotechnology in many fields such as tissue, bone implants, pleural diseases, dermatology, wound healing, cancer treatment, diabetes, obesity, endodontics, ophthalmology, plastic surgery and atherosclerosis were demonstrated. Man's quest to find perfection will always continue. Revolutionary nanotechnology is progressing day by day. It is certain that it will enter our lives not only in the field of health, but also in every field. Thanks to this technology, the ease of use and practicality of all kinds of items in our house, from the car we will ride in the future, to the clothes we will wear, from a simple painkiller to cooking utensils. Studies on nanotechnology, which is accepted in the field of health, continue without losing speed. A new technique that can be developed in nanotechnology will bring a new perspective in many different fields. For the long-term, side effects of nanoparticles in the human body should be investigated. It should be a necessity for scientists to find and use the correct, fast and effective method without harming human health and ecological balance. The universe is for humans, human life is the most precious thing in the universe. As long as nanotechnology serves this value, it will bring convenience in our lives.

References

[1]   B. Şen, B. Demirkan, A. Savk, R. Kartop, M.S. Nas, M.H. Alma, S. Sürdem, F. Şen, High-performance graphite-supported ruthenium nanocatalyst for hydrogen evolution reaction, J. Mol. Liq. 268 (2018) 807–812. https://doi.org/10.1016/j.molliq.2018.07.117

[2]   R. Nagraik, A. Sharma, D. Kumar, S. Mukherjee, F. Sen, A.P. Kumar, Amalgamation of biosensors and nanotechnology in disease diagnosis: Mini-review, Sensors Int. 2 (2021) 100089. https://doi.org/10.1016/J.SINTL.2021.100089

[3]   B. Şen, N. Lolak, Ö. Paralı, M. Koca, A. Şavk, S. Akocak, F. Şen, Bimetallic PdRu/graphene oxide based Catalysts for one-pot three-component synthesis of 2-amino-4H-chromene derivatives, Nano-Structures and Nano-Objects 12 (2017) 33–40. https://doi.org/10.1016/j.nanoso.2017.08.013

[4]   A. Hojjati-Najafabadi, A. Aygun, R.N.E. Tiri, F. Gulbagca, M.I. Lounissaa, P. Feng, F. Karimi, F. Sen, Bacillus thuringiensis Based Ruthenium/Nickel Co-Doped Zinc as a Green Nanocatalyst: Enhanced Photocatalytic Activity, Mechanism, and Efficient H2Production from Sodium Borohydride Methanolysis, Ind. Eng. Chem. Res. (2022). https://doi.org/10.1021/ACS.IECR.2C03833

[5]   E. Demir, B. Sen, F. Sen, Highly efficient Pt nanoparticles and f-MWCNT nanocomposites based counter electrodes for dye-sensitized solar cells, Nano-Structures & Nano-Objects 11 (2017) 39–45. https://doi.org/10.1016/J.NANOSO.2017.06.003

[6]   Y. Wu, E.E. Altuner, R.N. El Houda Tiri, M. Bekmezci, F. Gulbagca, A. Aygun, C. Xia, Q. Van Le, F. Sen, H. Karimi-Maleh, Hydrogen generation from methanolysis of sodium borohydride using waste coffee oil modified zinc oxide nanoparticles and their photocatalytic activities, Int. J. Hydrogen Energy 48 (2023) 6613–6623. https://doi.org/10.1016/J.IJHYDENE.2022.04.177

[7]   Y. Yıldız, İ. Esirden, E. Erken, E. Demir, M. Kaya, F. Şen, Microwave (Mw)-assisted Synthesis

of 5-Substituted 1H-Tetrazoles via [3+2] Cycloaddition Catalyzed by Mw-Pd/Co Nanoparticles Decorated on Multi-Walled Carbon Nanotubes, ChemistrySelect 1 (2016) 1695–1701. https://doi.org/10.1002/SLCT.201600265

[8]   K. Nesrin, C. Yusuf, K. Ahmet, S.B. Ali, N.A. Muhammad, S. Suna, Ş. Fatih, Biogenic silver nanoparticles synthesized from Rhododendron ponticum and their antibacterial, antibiofilm and cytotoxic activities, J. Pharm. Biomed. Anal. 179 (2020) 112993. https://doi.org/10.1016/J.JPBA.2019.112993

[9]   R. Ayranci, G. Baskaya, M. Guzel, S. Bozkurt, M. Ak, A. Savk, F. Sen, Enhanced optical and electrical properties of PEDOT via nanostructured carbon materials: A comparative investigation, Nano-Structures & Nano-Objects 11 (2017) 13–19. https://doi.org/10.1016/J.NANOSO.2017.05.008

[10]   B. Sen, B. Demirkan, A. Şavk, S. Karahan Gülbay, F. Sen, Trimetallic PdRuNi nanocomposites decorated on graphene oxide: A superior catalyst for the hydrogen evolution reaction, Int. J. Hydrogen Energy 43 (2018) 17984–17992. https://doi.org/10.1016/J.IJHYDENE.2018.07.122

[11]   A. Aygun, G. Sahin, R.N.E. Tiri, Y. Tekeli, F. Sen, Colorimetric sensor based on biogenic nanomaterials for high sensitive detection of hydrogen peroxide and multi-metals, Chemosphere 339 (2023) 139702. https://doi.org/10.1016/J.CHEMOSPHERE.2023.139702

[12]   W. Zhang, Y. Wang, B.T.K. Lee, C. Liu, G. Wei, W. Lu, A novel nanoscale-dispersed eye ointment for the treatment of dry eye disease, Nanotechnology 25 (2014) 125101. https://doi.org/10.1088/0957-4484/25/12/125101

[13]   H. Seckin, R.N.E. Tiri, I. Meydan, A. Aygun, M.K. Gunduz, F. Sen, An environmental approach for the photodegradation of toxic pollutants from wastewater using Pt–Pd nanoparticles: Antioxidant, antibacterial and lipid peroxidation inhibition applications, Environ. Res. 208 (2022) 112708. https://doi.org/10.1016/J.ENVRES.2022.112708

[14]   M. Nasrollahzadeh, S.M. Sajadi, M. Sajjadi, Z. Issaabadi, Applications of Nanotechnology in Daily Life, Interface Sci. Technol. 28 (2019) 113–143. https://doi.org/10.1016/B978-0-12-813586-0.00004-3

[15]   D. Schaming, H. Remita, Nanotechnology: from the ancient time to nowadays, Found. Chem. 17 (2015) 187–205. https://doi.org/10.1007/S10698-015-9235-Y

[16]   S.K. Sahoo, S. Parveen, J.J. Panda, The present and future of nanotechnology in human health care, Nanomedicine Nanotechnology, Biol. Med. 3 (2007) 20–31. https://doi.org/10.1016/J.NANO.2006.11.008

[17]   X. Navarro, M. Vivó, A. Valero-Cabré, Neural plasticity after peripheral nerve injury and regeneration, Prog. Neurobiol. 82 (2007) 163–201. https://doi.org/10.1016/J.PNEUROBIO.2007.06.005

[18]   A. Saniotis, M. Henneberg, A.R. Sawalma, Integration of nanobots into neural circuits as a future therapy for treating neurodegenerative disorders, Front. Neurosci. 12 (2018) 307073. https://doi.org/10.3389/FNINS.2018.00153

[19]   S. Barua, S. Mitragotri, Challenges associated with penetration of nanoparticles across cell and tissue barriers: A review of current status and future prospects, Nano Today 9 (2014) 223–243. https://doi.org/10.1016/J.NANTOD.2014.04.008.

[20]   Z. Sheikh, S. Najeeb, Z. Khurshid, V. Verma, H. Rashid, M. Glogauer, Biodegradable Materials for Bone Repair and Tissue Engineering Applications, Materials (Basel). 8 (2015) 5744. https://doi.org/10.3390/MA8095273

[21]   M. Rahamathulla, R.R. Bhosale, R.A.M. Osmani, K.C. Mahima, A.P. Johnson, U. Hani, M. Ghazwani, M.Y. Begum, S. Alshehri, M.M. Ghoneim, F. Shakeel, H. V. Gangadharappa, Carbon Nanotubes: Current Perspectives on Diverse Applications in Targeted Drug Delivery and Therapies, Materials (Basel). 14 (2021) 6707. https://doi.org/10.3390/MA14216707

[22]   V. Kumar, A.K. Abbas, N. Fausto, J.C. Aster, Robbins and Cotran Pathological Basis of Disease. 8th edn. Saunders ElsevierKumar, V. et al. (2010) Robbins and Cotran Pathological Basis of Disease. 8th

edn. Saunders Elsevier, Elsevier Health Sciences., Elsevier Heal. Sci. 2 (2010) 897–957.

[23]   Y. Song, X. Li, X. Du, Exposure to nanoparticles is related to pleural effusion, pulmonary fibrosis and granuloma, Eur. Respir. J. 34 (2009) 559–567. https://doi.org/10.1183/09031936.00178308

[24]   Ö. Dikensoy, G. Üniversitesi, T. Fakültesi, G. Hastalıkları, A. Dalı, T. Gaziantep, NANOPARTİKÜLLER VE PLEVRA, (n.d.).

[25]   L.A. Delouise, Applications of nanotechnology in dermatology, J. Invest. Dermatol. 132 (2012) 964–975. https://doi.org/10.1038/jid.2011.425

[26]   D. Schairer, M. Chouake, M. Nasir, A. Friedman, Nanotechnology Applications in Dermatology, Clin. Nanomedicine Handb. (2013) 85–194. https://doi.org/10.1201/B15642-3.

[27]   S.M. Moghimi, A.C. Hunter, J.C. Murray, Nanomedicine: current status and future prospects, FASEB J. 19 (2005) 311–330. https://doi.org/10.1096/FJ.04-2747REV

[28]   M.A. Tran, R.J. Watts, G.P. Robertson, Use of liposomes as drug delivery vehicles for treatment of melanoma, Pigment Cell Melanoma Res. 22 (2009) 388–399. https://doi.org/10.1111/J.1755-148X.2009.00581.X

[29]   J.R. Antonio, C.R. Antônio, I.L.S. Cardeal, J.M.A. Ballavenuto, J.R. Oliveira, Nanotechnology in dermatology, An. Bras. Dermatol. 89 (2014) 126–136. https://doi.org/10.1590/abd1806-4841.20142228

[30]   S.A. Nasrollahi, H. Hassanzade, A. Moradi, M. Sabouri, A. Samadi, M.N. kashani, A. Firooz, Safety Assessment of Tretinoin Loaded Nano Emulsion and Nanostructured Lipid Carriers: A Non-invasive Trial on Human Volunteers, Curr. Drug Deliv. 14 (2017). https://doi.org/10.2174/1567201813666160512145954

[31]   A. Mihranyan, N. Ferraz, M. Strømme, Current status and future prospects of nanotechnology in cosmetics, Prog. Mater. Sci. 57 (2012) 875–910. https://doi.org/10.1016/J.PMATSCI.2011.10.001

[32]   H.B. Sibel, M. Akyol, Nanoteknolojinin Dermatoloji Alaninda Kullanımı/Nanotechnology Use in Dermatology, GüncelDeermatolojiDergisi (2018) 44–55.

[33]   H. Hamishehkar, S. Ghanbarzadeh, S. Sepehran, Y. Javadzadeh, Z.M. Adib, M. Kouhsoltani, Histological assessment of follicular delivery of flutamide by solid lipid nanoparticles: potential tool for the treatment of androgenic alopecia, Drug Dev. Ind. Pharm. 42 (2016) 846–853. https://doi.org/10.3109/03639045.2015.1062896

[34]   M.S. Wu, D.S. Sun, Y.C. Lin, C.L. Cheng, S.C. Hung, P.K. Chen, J.H. Yang, H.H. Chang, Nanodiamonds protect skin from ultraviolet B-induced damage in mice, J. Nanobiotechnology 13 (2015) 1–12. https://doi.org/10.1186/S12951-015-0094-4

[35]   F. Chirico, C. Fumelli, A. Marconi, A. Tinari, E. Straface, W. Malorni, R. Pellicciari, C. Pincelli, Carboxyfullerenes localize within mitochondria and prevent the UVB-induced intrinsic apoptotic pathway, Exp. Dermatol. 16 (2007) 429–436. https://doi.org/10.1111/J.1600-0625.2007.00545.X

[36]   C. Fumelli, A. Marconi, S. Salvioli, E. Straface, W. Malorni, A.M. Offidani, R. Pellicciari, G. Schettini, A. Giannetti, D. Monti, C. Franceschi, C. Pincelli, Carboxyfullerenes Protect Human Keratinocytes from Ultraviolet-B-Induced Apoptosis, J. Invest. Dermatol. 115 (2000) 835–841. https://doi.org/10.1046/J.1523-1747.2000.00140.X

[37]   G.A. Castro, L.A.M. Ferreira, Novel vesicular and particulate drug delivery systems for topical treatment of acne, Expert Opin. Drug Deliv. 5 (2008) 665–679. https://doi.org/10.1517/17425247.5.6.665

[38]   M. Qin, A. Landriscina, J.M. Rosen, G. Wei, S. Kao, W. Olcott, G.W. Agak, K.B. Paz, J. Bonventre, A. Clendaniel, S. Harper, B.L. Adler, A.E. Krausz, J.M. Friedman, J.D. Nosanchuk, J. Kim, A.J. Friedman, Nitric Oxide–Releasing Nanoparticles Prevent Propionibacterium acnes–Induced Inflammation by Both Clearing the Organism and Inhibiting Microbial Stimulation of the Innate Immune Response, J. Invest. Dermatol. 135 (2015) 2723–2731. https://doi.org/10.1038/JID.2015.277

[39]   H.R. Kelidari, M. Saeedi, Z. Hajheydari, J. Akbari, K. Morteza-Semnani, J. Akhtari, H. Valizadeh, K. Asare-Addo, A. Nokhodchi, Spironolactone loaded nanostructured lipid carrier gel for

effective treatment of mild and moderate acne vulgaris: A randomized, double-blind, prospective trial, Colloids Surfaces B Biointerfaces 146 (2016) 47–53. https://doi.org/10.1016/J.COLSURFB.2016.05.042

[40] S. Inui, K. Kokubo, Inhibition of sebum production and Propionibacterium acnes lipase activity by fullerenol, a novel polyhydroxylated fullerene: Potential as a therapeutic reagent for acne, Artic. J. Cosmet. Sci. (2012).

[41] S. Inui, H. Aoshima, A. Nishiyama, S. Itami, Improvement of acne vulgaris by topical fullerene application: unique impact on skin care, Nanomedicine Nanotechnology, Biol. Med. 7 (2011) 238–241. https://doi.org/10.1016/J.NANO.2010.09.005

[42] A. Aygun, F. Gulbagca, E.E. Altuner, M. Bekmezci, T. Gur, H. Karimi-Maleh, F. Karimi, Y. Vasseghian, F. Sen, Highly Active PdPt Bimetallic Nanoparticles Synthesized By One Step Bioreduction Method: Characterizations, Anticancer, Antibacterial Activities and Evaluation of Their Catalytic Effect for Hydrogen Generation, Int. J. Hydrogen Energy 48 (2023) 6666–6679. https://doi.org/10.1016/j.ijhydene.2021.12.144

[43] D. Papakostas, F. Rancan, W. Sterry, U. Blume-Peytavi, A. Vogt, Nanoparticles in dermatology, Arch. Dermatol. Res. 303 (2011) 533–550. https://doi.org/10.1007/S00403-011-1163-7

[44] A. Sazgarnia, A.R. Taheri, S. Soudmand, A.J. Parizi, O. Rajabi, M.S. Darbandi, Antiparasitic effects of gold nanoparticles with microwave radiation on promastigotes and amastigotes of Leishmania major, Int. J. Hyperth. 29 (2013) 79–86. https://doi.org/10.3109/02656736.2012.758875

[45] F. Ramezani, A. Jebali, B. Kazemi, A green approach for synthesis of gold and silver nanoparticles by leishmania sp., Appl. Biochem. Biotechnol. 168 (2012) 1549–1555. https://doi.org/10.1007/S12010-012-9877-3

[46] K. Mayelifar, A.R. Taheri, O. Rajabi, A. Sazgarnia, Ultraviolet B efficacy in improving antileishmanial effects of silver nanoparticles, Iran. J. Basic Med. Sci. 18 (2015) 677.

[47] G. Tsuji, A. Hashimoto-Hachiya, M. Takemura, T. Kanemaru, M. Ichihashi, M. Furue, Palladium and Platinum Nanoparticles Activate AHR and NRF2 in Human Keratinocytes-Implications in Vitiligo Therapy, J. Invest. Dermatol. 137 (2017) 1582–1586. https://doi.org/10.1016/J.JID.2017.02.981

[48] G.M. Soliman, Nanoparticles as safe and effective delivery systems of antifungal agents: Achievements and challenges, Int. J. Pharm. 523 (2017) 15–32. https://doi.org/10.1016/J.IJPHARM.2017.03.019

[49] A.A. Tawfik, I. Noaman, H. El-Elsayyad, N. El-Mashad, M. Soliman, A study of the treatment of cutaneous fungal infection in animal model using photoactivated composite of methylene blue and gold nanoparticle, Photodiagnosis Photodyn. Ther. 15 (2016) 59–69. https://doi.org/10.1016/J.PDPDT.2016.05.010

[50] M.E. Amin, M.M. Azab, A.M. Hanora, S. Abdalla, Antifungal activity of silver nanoparticles on Fluconazole resistant Dermatophytes identified by (GACA)4 and isolated from primary school children suffering from Tinea Capitis in Ismailia – Egypt, Cell. Mol. Biol. 63 (2017) 63–67. https://doi.org/10.14715/CMB/2017.63.11.12

[51] B. Mordorski, C.B. Costa-Orlandi, L.M. Baltazar, L.J. Carreño, A. Landriscina, J. Rosen, M. Navati, M.J.S. Mendes-Giannini, J.M. Friedman, J.D. Nosanchuk, A.J. Friedman, Topical nitric oxide releasing nanoparticles are effective in a murine model of dermal Trichophyton rubrum dermatophytosis, Nanomedicine Nanotechnology, Biol. Med. 13 (2017) 2267–2270. https://doi.org/10.1016/J.NANO.2017.06.018

[52] C.B. Costa-Orlandi, B. Mordorski, L.M. Baltazar, M.J.S. Mendes-Giannini, J.M. Friedman, J.D. Nosanchuk, A.J. Friedman, Nitric Oxide Releasing Nanoparticles as a Strategy to Improve Current Onychomycosis Treatments., J. Drugs Dermatol. 17 (2018) 717–720

[53] M.F. Pinto, C.C. Moura, C. Nunes, M.A. Segundo, S.A. Costa Lima, S. Reis, A new topical formulation for psoriasis: Development of methotrexate-loaded nanostructured lipid carriers, Int. J.

Pharm. 477 (2014) 519–526. https://doi.org/10.1016/J.IJPHARM.2014.10.067

[54]   H. Bessar, I. Venditti, L. Benassi, C. Vaschieri, P. Azzoni, G. Pellacani, C. Magnoni, E. Botti, V. Casagrande, M. Federici, A. Costanzo, L. Fontana, G. Testa, F.F. Mostafa, S.A. Ibrahim, M.V. Russo, I. Fratoddi, Functionalized gold nanoparticles for topical delivery of methotrexate for the possible treatment of psoriasis, Colloids Surfaces B Biointerfaces 141 (2016) 141–147. https://doi.org/10.1016/J.COLSURFB.2016.01.021

[55]   H. Pischon, M. Radbruch, A. Ostrowski, P. Volz, C. Gerecke, M. Unbehauen, S. Hönzke, S. Hedtrich, J.W. Fluhr, R. Haag, B. Kleuser, U. Alexiev, A.D. Gruber, L. Mundhenk, Stratum corneum targeting by dendritic core-multishell-nanocarriers in a mouse model of psoriasis, Nanomedicine Nanotechnology, Biol. Med. 13 (2017) 317–327. https://doi.org/10.1016/J.NANO.2016.09.004

[56]   K. Yu, Y. Wang, T. Wan, Y. Zhai, S. Cao, W. Ruan, C. Wu, Y. Xu, Tacrolimus nanoparticles based on chitosan combined with nicotinamide: Enhancing percutaneous delivery and treatment efficacy for atopic dermatitis and reducing dose, Int. J. Nanomedicine 13 (2018) 129–142. https://doi.org/10.2147/IJN.S150319

[57]   S. Jung, A.P. M.D., N. Otberg, G. Thiede, W.S. M.D., J.M. Lademann, Strategy of topical vaccination with nanoparticles, Https://Doi.Org/10.1117/1.3080714 14 (2009) 021001. https://doi.org/10.1117/1.3080714.

[58]   J. Parks, M. Kath, K. Gabrick, J.P. Ver Halen, Nanotechnology applications in plastic and reconstructive surgery: A review, Plast. Surg. Nurs. 32 (2012) 156–164. https://doi.org/10.1097/PSN.0B013E3182701824

[59]   K. Amin, R. Moscalu, A. Imere, R. Murphy, S. Barr, Y. Tan, R. Wong, P. Sorooshian, F. Zhang, J. Stone, J. Fildes, A. Reid, J. Wong, The Future Application of Nanomedicine and Biomimicry in Plastic and Reconstructive Surgery, Nanomedicine 14 (2019) 2679–2696. https://doi.org/10.2217/nnm-2019-0119

[60]   H. Pangli, S. Vatanpour, S. Hortamani, R. Jalili, A. Ghahary, Incorporation of Silver Nanoparticles in Hydrogel Matrices for Controlling Wound Infection, J. Burn Care Res. 42 (2021) 785–793. https://doi.org/10.1093/JBCR/IRAA205

[61]   Z. Xie, C.B. Paras, H. Weng, P. Punnakitikashem, L.C. Su, K. Vu, L. Tang, J. Yang, K.T. Nguyen, Dual growth factor releasing multi-functional nanofibers for wound healing, Acta Biomater. 9 (2013) 9351–9359. https://doi.org/10.1016/J.ACTBIO.2013.07.030

[62]   R. Izumi, S. Komada, K. Ochi, L. Karasawa, T. Osaki, Y. Murahata, T. Tsuka, T. Imagawa, N. Itoh, Y. Okamoto, H. Izawa, M. Morimoto, H. Saimoto, K. Azuma, S. Ifuku, Favorable effects of superficially deacetylated chitin nanofibrils on the wound healing process, Carbohydr. Polym. 123 (2015) 461–467. https://doi.org/10.1016/J.CARBPOL.2015.02.005

[63]   P. Mezzana, Clinical efficacy of a new chitin nanofibrils-based gel in wound healing., Acta Chir. Plast. 50 (2008) 81–84.

[64]   The role of RNA interference in dermatology: Current perspectives and future directions, (n.d.).

[65]   P.S. Rabbani, A. Zhou, Z.M. Borab, J.A. Frezzo, N. Srivastava, H.T. More, W.J. Rifkin, J.A. David, S.J. Berens, R. Chen, S. Hameedi, M.H. Junejo, C. Kim, R.A. Sartor, C.F. Liu, P.B. Saadeh, J.K. Montclare, D.J. Ceradini, Novel lipoproteoplex delivers Keap1 siRNA based gene therapy to accelerate diabetic wound healing, Biomaterials 132 (2017) 1–15. https://doi.org/10.1016/J.BIOMATERIALS.2017.04.001

[66]   S.D. Smith, A. Dodds, S. Dixit, A. Cooper, Role of nanocrystalline silver dressings in the management of toxic epidermal necrolysis (TEN) and TEN/Stevens–Johnson syndrome overlap, Australas. J. Dermatol. 56 (2015) 298–302. https://doi.org/10.1111/AJD.12254

[67]   M.D. Wang, D.M. Shin, J.W. Simons, S. Nie, Nanotechnology for targeted cancer therapy, Expert Rev. Anticancer Ther. 7 (2007) 833–837. https://doi.org/10.1586/14737140.7.6.833

[68]   D.K. Chatterjee, L.S. Fong, Y. Zhang, Nanoparticles in photodynamic therapy: An emerging paradigm, Adv. Drug Deliv. Rev. 60 (2008) 1627–1637. https://doi.org/10.1016/J.ADDR.2008.08.003

[69]   J. Panyam, V. Labhasetwar, Sustained cytoplasmic delivery of drugs with intracellular receptors using biodegradable nanoparticles., Mol. Pharm. 1 (2004) 77–84. https://doi.org/10.1021/MP034002C

[70]   R. Misra, S. Acharya, S.K. Sahoo, Cancer nanotechnology: Application of nanotechnology in cancer therapy, Drug Discov. Today (2010). https://doi.org/10.1016/j.drudis.2010.08.006

[71]   F. Şenel, D. Sami Ulus Çocuk Hastanesi, Nanotıp Michigan Center for Biologic Nanotechnology, (n.d.).

[72]   Nanotechnology and the future of diabetes management, (n.d.).

[73]   N. Staggers, T. McCasky, N. Brazelton, R. Kennedy, Nanotechnology: The coming revolution and its implications for consumers, clinicians, and informatics, Nurs. Outlook 56 (2008) 268–274. https://doi.org/10.1016/J.OUTLOOK.2008.06.004

[74]   S.N. Kaushik, J. Scoffield, A. Andukuri, G.C. Alexander, T. Walker, S. Kim, S.C. Choi, B.C. Brott, P.D. Eleazer, J.Y. Lee, H. Wu, N.K. Childers, H.W. Jun, J.H. Park, K. Cheon, Evaluation of ciprofloxacin and metronidazole encapsulated biomimetic nanomatrix gel on Enterococcus faecalis and Treponema denticola, Biomater. Res. 19 (2015) 1–10. https://doi.org/10.1186/S40824-015-0032-4/FIGURES/6

[75]   M.C. Bottino, K. Kamocki, G.H. Yassen, J.A. Platt, M.M. Vail, Y. Ehrlich, K.J. Spolnik, R.L. Gregory, Bioactive Nanofibrous Scaffolds for Regenerative Endodontics, Http://Dx.Doi.Org/10.1177/0022034513505770 92 (2013) 963–969. https://doi.org/10.1177/0022034513505770

[76]   J. Vandervoort, A. Ludwig, Ocular drug delivery: nanomedicine applications, Https://Doi.Org/10.2217/17435889.2.1.11 2 (2007) 11–21. https://doi.org/10.2217/17435889.2.1.11

[77]   L.M. Ensign, R. Cone, J. Hanes, Nanoparticle-based drug delivery to the vagina: A review, J. Control. Release 190 (2014) 500–514. https://doi.org/10.1016/J.JCONREL.2014.04.033

[78]   L. DeSantis, Preclinical Overview of Brinzolamide, Surv. Ophthalmol. 44 (2000) S119–S129. https://doi.org/10.1016/S0039-6257(99)00108-3.

[79]   G.R. Adelli, P. Bhagav, P. Taskar, T. Hingorani, S. Pettaway, W. Gul, M.A. ElSohly, M.A. Repka, S. Majumdar, Development of a Δ9-Tetrahydrocannabinol Amino Acid-Dicarboxylate Prodrug With Improved Ocular Bioavailability, Invest. Ophthalmol. Vis. Sci. 58 (2017) 2167. https://doi.org/10.1167/IOVS.16-20757

[80]   M. Prakash, R. Shankaran Dhesingh, Nanoparticle Modified Drug Loaded Biodegradable Polymeric Contact Lenses for Sustainable Ocular Drug Delivery, Curr. Drug Deliv. 14 (2017). https://doi.org/10.2174/1567201813666161018153547

[81]   S.B. Pehlivan, B. Yavuz, S. Çalamak, K. Ulubayram, A. Kaffashi, I. Vural, H.B. Çakmak, M.E. Durgun, E.B. Denkbaş, N. Ünlü, Preparation and In Vitro/In Vivo Evaluation of Cyclosporin A-Loaded Nanodecorated Ocular Implants for Subconjunctival Application, J. Pharm. Sci. 104 (2015) 1709–1720. https://doi.org/10.1002/JPS.24385

[82]   X. Li, Z. Zhang, H. Chen, Development and evaluation of fast forming nano-composite hydrogel for ocular delivery of diclofenac, Int. J. Pharm. 448 (2013) 96–100. https://doi.org/10.1016/J.IJPHARM.2013.03.024

[83]   N. Üstündağ-Okur, E.H. Gökçe, D.I. Bozbiyik, S. Eğrilmez, Ö. Özer, G. Ertan, Preparation and in vitro–in vivo evaluation of ofloxacin loaded ophthalmic nano structured lipid carriers modified with chitosan oligosaccharide lactate for the treatment of bacterial keratitis, Eur. J. Pharm. Sci. 63 (2014) 204–215. https://doi.org/10.1016/J.EJPS.2014.07.013

[84]   G. Spataro, F. Malecaze, C.O. Turrin, V. Soler, C. Duhayon, P.P. Elena, J.P. Majoral, A.M. Caminade, Designing dendrimers for ocular drug delivery, Eur. J. Med. Chem. 45 (2010) 326–334.

https://doi.org/10.1016/J.EJMECH.2009.10.017

[85]   K.K. Jain, The handbook of nanomedicine, third edition, Handb. Nanomedicine, Third Ed. (2017) 1–659. https://doi.org/10.1007/978-1-4939-6966-1

[86]   C. Jiang, L. Kuang, M.P. Merkel, F. Yue, M.A. Cano-Vega, N. Narayanan, S. Kuang, M. Deng, Biodegradable polymeric microsphere-based drug delivery for inductive browning of fat, Front. Endocrinol. (Lausanne). 6 (2015) 169564. https://doi.org/10.3389/FENDO.2015.00169

[87]   W. Li, H. Wan, S. Yan, Z. Yan, Y. Chen, P. Guo, T. Ramesh, Y. Cui, L. Ning, Gold nanoparticles synthesized with Poria cocos modulates the anti-obesity parameters in high-fat diet and streptozotocin induced obese diabetes rat model, Arab. J. Chem. 13 (2020) 5966–5977. https://doi.org/10.1016/J.ARABJC.2020.04.031

[88]   L.J.F. Peters, A. Jans, M. Bartneck, E.P.C. van der Vorst, Immunomodulatory Nanomedicine for the Treatment of Atherosclerosis, J. Clin. Med. 2021, Vol. 10, Page 3185 10 (2021) 3185. https://doi.org/10.3390/JCM10143185

[89]   S. Hossaini Nasr, X. Huang, Nanotechnology for Targeted Therapy of Atherosclerosis, Front. Pharmacol. 12 (2021) 755569. https://doi.org/10.3389/FPHAR.2021.755569

[90]   G. Fredman, N. Kamaly, S. Spolitu, J. Milton, D. Ghorpade, R. Chiasson, G. Kuriakose, M. Perretti, O. Farokhzad, I. Tabas, Targeted nanoparticles containing the proresolving peptide Ac2-26 protect against advanced atherosclerosis in hypercholesterolemic mice, Sci. Transl. Med. 7 (2015). https://doi.org/10.1126/SCITRANSLMED.AAA1065

[91]   C. Tarin, M. Carril, J.L. Martin-Ventura, I. Markuerkiaga, D. Padro, P. Llamas-Granda, J.A. Moreno, I. García, N. Genicio, S. Plaza-Garcia, L.M. Blanco-Colio, S. Penades, J. Egido, Targeted gold-coated iron oxide nanoparticles for CD163 detection in atherosclerosis by MRI, Sci. Reports 2015 51 5 (2015) 1–9. https://doi.org/10.1038/srep17135

[92]   M. Yu, J. Amengual, A. Menon, N. Kamaly, F. Zhou, X. Xu, P.E. Saw, S.J. Lee, K. Si, C.A. Ortega, W. Il Choi, I.H. Lee, Y. Bdour, J. Shi, M. Mahmoudi, S. Jon, E.A. Fisher, O.C. Farokhzad, Targeted Nanotherapeutics Encapsulating Liver X Receptor Agonist GW3965 Enhance Antiatherogenic Effects without Adverse Effects on Hepatic Lipid Metabolism in Ldlr−/− Mice, Adv. Healthc. Mater. 6 (2017) 1700313. https://doi.org/10.1002/ADHM.201700313

[93]   S. Sarkar, Artificial blood, Indian J. Crit. Care Med. 12 (2008) 140. https://doi.org/10.4103/0972-5229.43685

[94]   F. Khan, K. Singh, M.T. Friedman, Artificial Blood: The History and Current Perspectives of Blood Substitutes, Discoveries 8 (2020) e104. https://doi.org/10.15190/D.2020.1.

[95]   Singh Neelam*, Semwal B. C, Maurya Krishna, Khatoon Ruqsana, Paswan Shravan, Various Techniques For The Modification of Starch and The Applications of Its Derivatives, Int. Res. J. Pharm. 3 (2012) 25–31.

[96]   Z. Salehahmadi, F. Hajiliasgari, Nanotechnology Tolls the Bell for Plastic Surgeons, World J. Plast. Surg. 2 (2013) 71.

[97]   I.D. Luzhansky, L.C. Sudlow, D.M. Brogan, M.D. Wood, M.Y. Berezin, Imaging in the Repair of Peripheral Nerve Injury, Nanomedicine 14 (2019) 2659–2677. https://doi.org/10.2217/NNM-2019-0115

[98]   R.E. Stratford, N. Dave, R.F. Bergstrom, J.M. Janjic, V.S. Gorantla, Peripheral Nerve Nanoimaging: Monitoring Treatment and Regeneration, AAPS J. 2017 195 19 (2017) 1304–1316. https://doi.org/10.1208/S12248-017-0129-X

[99]   Dönmez, H. Dolgun Yüksel, Ş. Kırkan, Journal of Anatolian Environmental and Animal Sciences (Anadolu Çevre ve Hayvancılık Bilimleri Dergisi) Nanopartiküler Aşılar, Nanoparticular Vaccines. J. Anatol. Env. Anim. Sci. 6 (2021) 578–584. https://doi.org/10.35229/jaes.970713

[100]   A.F. Altenburg, J.H.C.M. Kreijtz, R.D. de Vries, F. Song, R. Fux, G.F. Rimmelzwaan, G. Sutter, A. Volz, Modified Vaccinia Virus Ankara (MVA) as Production Platform for Vaccines against Influenza

and Other Viral Respiratory Diseases, Viruses 6 (2014) 2735. https://doi.org/10.3390/V6072735

[101]   M. kheirollahpour, M. Mehrabi, N.M. Dounighi, M. Mohammadi, A. Masoudi, Nanoparticles and Vaccine Development, Pharm. Nanotechnol. 8 (2019) 6–21. https://doi.org/10.2174/2211738507666191024162042

[102]   S.T. Reddy, A.J. Van Der Vlies, E. Simeoni, V. Angeli, G.J. Randolph, C.P. O'Neil, L.K. Lee, M.A. Swartz, J.A. Hubbell, Exploiting lymphatic transport and complement activation in nanoparticle vaccines, Nat. Biotechnol. 2007 2510 25 (2007) 1159–1164. https://doi.org/10.1038/nbt1332

[103]   C. Foged, B. Brodin, S. Frokjaer, A. Sundblad, Particle size and surface charge affect particle uptake by human dendritic cells in an in vitro model, Int. J. Pharm. 298 (2005) 315–322. https://doi.org/10.1016/J.IJPHARM.2005.03.035

[104]   K. Niikura, T. Matsunaga, T. Suzuki, S. Kobayashi, H. Yamaguchi, Y. Orba, A. Kawaguchi, H. Hasegawa, K. Kajino, T. Ninomiya, K. Ijiro, H. Sawa, Gold nanoparticles as a vaccine platform: Influence of size and shape on immunological responses in vitro and in vivo, ACS Nano 7 (2013) 3926–3938. https://doi.org/10.1021/NN3057005/SUPPL_FILE/NN3057005_SI_001.PDF

[105]   E. Kolaczkowska, P. Kubes, Neutrophil recruitment and function in health and inflammation, Nat. Rev. Immunol. 2013 133 13 (2013) 159–175. https://doi.org/10.1038/nri3399

[106]   L. Xu, Y. Liu, Z. Chen, W. Li, Y. Liu, L. Wang, Y. Liu, X. Wu, Y. Ji, Y. Zhao, L. Ma, Y. Shao, C. Chen, Surface-engineered gold nanorods: Promising DNA vaccine adjuvant for HIV-1 treatment, Nano Lett. 12 (2012) 2003–2012. https://doi.org/10.1021/NL300027P

[107]   W. Tao, H.S. Gill, M2e-immobilized gold nanoparticles as influenza A vaccine: Role of soluble M2e and longevity of protection, Vaccine 33 (2015) 2307–2315. https://doi.org/10.1016/J.VACCINE.2015.03.063

[108]   M. Yu, S. Jambhrunkar, P. Thorn, J. Chen, W. Gu, C. Yu, Hyaluronic acid modified mesoporous silica nanoparticles for targeted drug delivery to CD44-overexpressing cancer cells, Nanoscale 5 (2012) 178–183. https://doi.org/10.1039/C2NR32145A

[109]   I. Das, A. Padhi, S. Mukherjee, D.P. Dash, S. Kar, A. Sonawane, Biocompatible chitosan nanoparticles as an efficient delivery vehicle for Mycobacterium tuberculosis lipids to induce potent cytokines and antibody response through activation of γδ T cells in mice, Nanotechnology 28 (2017) 165101. https://doi.org/10.1088/1361-6528/AA60FD

[110]   S.L. Demento, W. Cui, J.M. Criscione, E. Stern, J. Tulipan, S.M. Kaech, T.M. Fahmy, Role of sustained antigen release from nanoparticle vaccines in shaping the T cell memory phenotype, Biomaterials 33 (2012) 4957–4964. https://doi.org/10.1016/J.BIOMATERIALS.2012.03.041

[111]   M. Manish, A. Rahi, M. Kaur, R. Bhatnagar, S. Singh, A Single-Dose PLGA Encapsulated Protective Antigen Domain 4 Nanoformulation Protects Mice against Bacillus anthracis Spore Challenge, PLoS One 8 (2013) e61885. https://doi.org/10.1371/JOURNAL.PONE.0061885

[112]   C. Thomas, A. Rawat, L. Hope-Weeks, F. Ahsan, Aerosolized PLA and PLGA nanoparticles enhance humoral, mucosal and cytokine responses to hepatitis B vaccine, Mol. Pharm. 8 (2011) 405–415. https://doi.org/10.1021/MP100255C

[113]   A.A. Cohen, P.N.P. Gnanapragasam, Y.E. Lee, P.R. Hoffman, S. Ou, L.M. Kakutani, J.R. Keeffe, H.J. Wu, M. Howarth, A.P. West, C.O. Barnes, M.C. Nussenzweig, P.J. Bjorkman, Mosaic nanoparticles elicit cross-reactive immune responses to zoonotic coronaviruses in mice, Science (80-.). 371 (2021) 735–741. https://doi.org/10.1126/SCIENCE.ABF6840

Biogenic Nanomaterials: Synthesis, Characterization, Applications, and Future Remarks   Materials Research Forum LLC
Materials Research Foundations 180 (2025)   https://doi.org/21741/9781644903759

Chapter 21

Application of Coated/Hybrid Biogenic Nanomaterials as Advanced Theranostically

M. Koldemir Gündüz[1]*, E. Aydın[2], F. Sen[3]

[1]Department of Basic Sciences of Engineering, Faculty of Engineering and Natural Sciences, Kutahya Health Sciences University, Kutahya, Turkiye

[2]Faculty of Medicine, Agrı Ibrahim Çeçen University, Igdır, Turkiye

[3]Sen Research Group, Department of Biochemistry, Dumlupinar University, Kutahya, Turkiye.

meliha.koldemirgunduz@ksbu.edu.tr; fatih.sen@dpu.edu.tr)

Abstract

The utilization of novel approaches to create hybrid nanoparticles which combine many capabilities into a single nanocomposite system is becoming more and more popular at the moment. Hybrid nanomaterials present a special chance for use in a variety of domains, ranging from diagnostic to treatment. Under the umbrella of "green chemistry," approaches for synthesizing nanoparticles are shifting toward improving the process by using plant resources. These nanoparticles displays superior anticancer, antibacterial, anti-inflammatory, antioxidant, and sensing qualities. The focus of this section is to describe in detail the applications of biogenic nanoparticles in diagnosis and treatment.

Keywords

Theragnostic, Biogenic NPs, Cancer, Hybrid NPs, Treatment

1.   Introduction

The term Theragnostic, which emerged by combining diagnosis and treatment methods in the disease management process, consists of the words treatment and diagnosis [1]. Theragnostic is a new field of medicine in which radioactive drugs and/or techniques are used together for both diagnosis and treatment [2]. Theragnostic agents are mostly used in oncology but are also used in the diagnosis and treatment of other diseases. Combining the diagnosis and treatment method combines the results obtained from diagnosis with target-specific treatment [3]. This method, which provides personalized treatment, is economical and less time-consuming [4]. Theragnostic covers a wide area such as developing effective new targeted treatment methods and developing a good molecular understanding of how to optimize drug selection [5].

In a theragnostic system, the diagnosis includes a radioactive contrast agent and a particle system with inherent physical properties such as optical properties, magnetic properties, or acquired physical properties. Examples of the diagnostic component particles are gold nanoparticles, gold nanorods, and nanoshells [6]. The medication molecule linked to the carrier system or diagnostic component is part of the theragnostic system's therapeutic component. Examples of the therapeutic component include peptide-based medicines for the somatostatin receptor that target neuroendocrine cancers and Ga-68 labeled bisphosphonates for osteoblastic bone metastases [7].

The fields of physics, chemistry, and biology all make substantial use of nanoscale particles. These days, chemical and biological processes are used to create adaptable nanoparticles [8–10]. Nanoparticle production is carried out by top-down and bottom-up approaches [11–17]. Both of these approaches differ in their synthesis principles, but eventually NPs with the desired properties are produced. In the top-down approach, photolithographic techniques, milling, sputtering and milling techniques are used [18]. The bottom-up approach, which includes atom-to-atom, molecule-to-molecule, cluster-to-cluster combinations, or aggregations, is an additional technique for NP synthesis. Methods include sol-gel processing, laser pyrolysis, chemical vapor deposition (CVD), plasma or flame spray synthesis, chemical or electrochemical nanostructural deposition, self-assembly of monomer/polymer molecules, and bio-assisted synthesis [19].

Metallic nanoparticles have promising industrial applications with scientific value [20]. Nevertheless, in light of the advantages and disadvantages of metallic nanoparticle synthesis techniques, scientists are increasingly using plant resources to synthesize nanoparticles rather than chemicals and reagents. Superior antibacterial, anticancer, wound-healing, antioxidant, and sensing qualities are displayed by these biogenic nanoparticles. The process of creating nanoparticles using biological resources as a template is known as "green chemistry" [21].

In order to create novel NPs utilizing a green chemical technique, biogenic nanoparticles have evolved. A novel approach to creating NPs with natural stabilizing and reducing agents is offered by biogenic NPs. It is a cost-effective and eco-friendly substitute that doesn't require harmful chemicals or energy. Simple unicellular to large multicellular biological entities, including bacteria, actinomycetes, fungi, algae, and plant materials, are used in the bottom-up process of biological production of nanoparticles [22].

In general, microbes such bacteria, fungus, yeasts, and actinomycetes that are involved in the microbial creation of nanoparticles are able to withstand metals and flourish in the most extreme environments [23]. Microbes use these inherent qualities to withstand, amass, and transform metal into the appropriate metal ions. The first bacterial gold nanoparticles were produced using *Bacillus subtilis* [24]. Similar to this, microorganisms were used to create the varied face of metallic nanoparticles made of iron, gold, silver, zinc, copper, selenium and platinum.

In the creation of nanoparticles, plants are also utilized as sources. Alkaloids, phenols, terpenoids, flavonoids, polyphenols, saponins, polysaccharides, tannins and vitamins are only a few of the many active ingredients found in plants. Benefits of using plants as a source for synthesis include secondary metabolites, laxative qualities, low cost, easy scaling for scale manufacturing, and plant availability. Nanoparticles derived from plants are stable, reproducible, and eco-friendly [21,25].

This review's main emphasis is to describe in detail the applications of biogenic nanoparticles in diagnosis and treatment.

2.   Coated Biogenic Nanomaterials

In the digital age, despite all scientific developments, we are witnessing the emergence of new diseases and problems in diagnosis and treatment. Theragnostic' application of biogenic nanoparticles and polymers has enhanced disease diagnosis and treatment outcomes [26]. A promising method for functionalization in the diagnosis and treatment of numerous illnesses is the use of biocoatings. A broad variety of natural and synthetic materials are referred to as "biocoatings" and can be used to immobilize and protect biological components, enhance their functionality, and stop non-specific interactions with the sample matrix [27]. Coated biogenic nanoparticles can be utilized in the medical profession to alter biosensor surfaces so they bind specifically to pathogens or disease indicators. When used to identify particular proteins in blood, saliva, or urine, coated biogenic nanoparticles containing peptides or antibodies can help with early disease diagnosis. Furthermore, coated biogenic nanoparticles can be employed to improve implantable biosensors' stability and biocompatibility for long-term health parameter monitoring [28].

According to the literature, there are three different ways to create metallic nanoparticles: physical, chemical, and biological (or biogenic) [29]. The most conventional method, chemical synthesis, involves regulating the atmosphere and temperature, adding organic solvents and/or toxic substances that could increase biological and environmental risks, and adding capping agents to stop particle aggregation [30]. On the other hand, biological materials including bacteria, fungi, yeast, and plant extracts can be used to produce metallic nanoparticles by biogenic synthesis, which offers both financial and environmental benefits [31].

One common type of nanomaterial is silver nanoparticles (AgNP). Because of these NPs' antibacterial qualities against various harmful bacteria and fungus, as well as their antibiofilm activity, they are widely used in a variety of industries, including food science, medicine, and cosmetics. Due to its antiproliferative action on various tumor types, including lung, leukaemia, colorectal, and hepatic cancer, AgNP has recently been suggested as a possible option in the treatment of cancer [32].

Because of its low cost, quick reaction, and efficiency, plant extract-mediated synthesis of AgNPs may be a promising approach among the biogenic ways to create metallic nanoparticles [33]. A common plant extract for metal ions is green tea (*Camellia sinensis*). The primary ingredient in tea extract, catechin, serves as a capping agent that stabilizes nanoparticles and inhibits their oxidation and aggregation in addition to being a potent reducing agent that converts Ag^+ to Ag° [34]. On the surface of biogenically produced AgNPs, polyphenols function as capping agents [35]. It has been shown that using inorganic substances like silicon dioxide (SiO_2) to modify the surface of metallic nanoparticles can take advantage of their optical, stabilizing, and catalytic qualities, which offer good biocompatibility and comparatively easy core-shell structure formation [36]. By increasing the material's surface area, silica coating of nanoparticles may hold promise for biomedical applications [37].

3.   Hybrid Biogenic Nanomaterials: Varieties and Functional Properties

Combinations of nano-sized elements with different compositions and categories such as nanoparticles, nanoplates, nanofibers and nanotubes constitute hybrid nanoparticles [38]. Hybrid nanoparticles, which are formed by combining two or more functional nanomaterials with the same scale at the nanoscale, constitute an important part of nanotechnology [39–42]. Hybrid nanoparticles are formed by two or more inorganic compounds, two or more organic compounds, or at least one of organic and inorganic compounds [43]. Hybrid nanoparticles exhibit synergistic properties of two particles [44]. When many components are combined in the same nanostructure, the qualities of the individual components are collected, and the hybrid structure also gains additional features as a result of their synergistic actions [45]. Hybrid nanoparticles are important in fields such as biology, medicine, sensor technology, and are also suitable for the emergence of new and powerful analytical procedures that can overcome problems such as low bioavailability, poor immunogenicity, water solubility and high toxicity [38].

The characteristics of one component often dictate the application range of hybrid nanoparticles, while other components enhance functional capabilities in the same region or boost overall system efficiency, according to research on the subject [46]. When used as contrast agents for magnetic resonance imaging or for targeted medication delivery, nanostructures containing magnetic nanoparticles can be controlled by an external magnetic field, whereas hybrid nanoparticles can be used to predict the specific location of the distribution with the addition of luminescent nanostructures to magnetic nanoparticles. Therefore, the shell material improves efficiency while the core material establishes the qualities [38].

The increasing interest in biogenic nanofibrils is directly proportional to the improvement of the characterization of the nanostructure of other biological materials [47]. High surface density functional groups, including amines on chitin, C6-hydroxyls on cellulose, amino acids on collagen, or cysteine groups on viruses, are present in biogenic nanofibers. Control of the affinity of medications or enzymes is made possible by these functions [48]. With the production of metal nanoparticles utilizing plants, plant bodies, plant extracts, flowers, and leaves, hybrid nanoparticles may be produced by mixing various components (organic-inorganic or organic-organic). This creates a nanostructure with unique qualities by combining the attributes of each material [49]. In addition to plants, researchers have created hybrid nanoparticles (NPs) utilizing bacteria, yeast, fungus, algae, and other creatures [13,50–52]. Organic/inorganic hybrid nanoparticles have attracted a lot of attention in the development of various drugs and medical devices, such as drug delivery nanocarriers, tissue engineering frameworks, cardiovascular stents, and dental implants [53]. Nanoparticles can be directed to particular parenchymal regions, such as tumor locations, in the body by combining them with targeting ligands. It is possible to create hybrid biogenic nanoparticles with therapeutic and imaging components for real-time medication uptake and/or treatment effectiveness monitoring [54].

Upon analyzing the studies, it is noteworthy that hybrid nanoparticles consisting of at least two distinct materials are utilized to improve the properties, overcome the limitations of individual components, obtain multiple functions for single nanoarchitectures, and/or obtain new properties not possible for single-component nanoparticles [55]. Numerous hybrid nanostructures, such as yolkhell, core-shell, heterodimer, dot in nanotube, Janus, dot on nanorod, nanobranches, etc., have been created and produced [56]. With the advent of metallic nanoparticles with polymer-functionalized surfaces and the hybridization of silica with micelles, liposomes, and polymers at the start of the twenty-first century, the creation of hybrid nanosystems increased dramatically [57].

4.   Areas of Use of Biogenic Coated/Hybrid Nanoparticles

After reaching the nanoscale, there is a miraculous change in the physicochemical properties of organic/inorganic particles, and they show extraordinary biomedical activity. Organic/inorganic nanoparticles' special properties increase their antifungal, antibacterial, and anticancer therapy efficacy [58]. Among the numerous types of nanoparticles today, nanoparticles with core-shell architecture combine multiple functions into a single hybrid nanoparticle with surprising properties [50]. Because silica is thermally and chemically stable, as well as biocompatible, hydrophilic, and optically transparent, the combination of silver and silica nanoparticles results in unique effects. Silica's surface modification and adaptability provide it a significant edge in therapeutic applications, particularly medical ones.

Cancer is an unpredictable disease with an unpredictable onset, progression and aftermath. There are many drugs available for the treatment of cancer. Most of these drugs are inadequate in systematically distributing the expected pharmacological result at the desired concentrations. They cause irreversible damage to healthy cells and tissues until they reach the target cells and tissues. However, molecules such as silver nanoparticles can overcome the disadvantages of traditional therapeutic applications. When silver nanoparticles are coated with a specific biocompatible nanomaterial, they can be designed for a longer half-life, drug loading, non-degradable release and selective distribution in the body. These properties can be achieved by changing their size, surface chemistry, composition, and morphology. Recently, the rapid increase in resistance against pathogenic fungi and bacteria has caused a major problem in antimicrobial chemotherapy [59]. Silver nanoparticles can help pathogenic microbes overcome this resistance mechanism. Silver nanoparticles have provided great results for the detection and treatment of microbial infections through successful antibacterial vaccination, selection of target pathogens, reactive and combined delivery of antibiotics, and rapid detection of pathogens. Recently, many immunoassay methods have been studied to diagnose various tumor biomarkers to better manage cancer. Evaluation of serum tumor markers has become a way to determine cancer. Electrochemical immunoassay has gradually been accepted and is used to find tumor markers due to its intrinsic benefits such as high selectivity, sensitivity, convenient label-free manipulation, low cost, and minute size [60].

4.1   Coated Biosensors and their Applications in Healthcare

Increasing bacterial resistance to antibiotics and viral infections is causing approximately 2 million deaths per year [61]. As a result, new diagnostic methods need to be developed to better detect pathogens. Furthermore, the creation of sustainable biosensors can be useful for treatment and diagnosis by continuously-monitoring body signals and biomarkers.

Nanomaterials' low toxicity and biocompatibility make their green chemistry crucial for developing environmentally friendly sensors for biomedical applications. Zamarchi and Vieira [62] discovered that AgNPs synthesized using *Araucaria angustifolia* extract formed an electrochemical biosensor to determine paracetamol. Then, the synthesized silver nanoparticles modified the surface of the glassy carbon electrode to form a sensor that could determine paracetamol in solution. The sensor's functionality was tested using the modified electrode's current response at various paracetamol concentrations. The sensor, which is sensitive and selective to paracetamol, is also stable and reproducible over time [62]. In order to create a lactose biosensor, Bollella et al. [63] used quercetin to create AgNPs and AuNPs utilizing green chemistry. Green synthesis-derived gold nanoparticles [64] can be employed as biosensors for medical purposes, including figuring out how much glucose is in commercial glucose injections.

Because there are so many problems that need to be solved, research on biodegradable and environmentally friendly sensors for medical applications is still in its infancy. Even if coated biosensors are obtained through green synthesis methods, their biocompatibility, safety and performance need to be examined at the molecular level when detected using biopolymer nanocomposites or nanoparticles.

4.2   Use of Hybrid NPs in Detection

Nanomaterials can improve the effectiveness of diagnostic methods and address issues including tumor resistance, systemic distribution and solubility of different medications, and more [65]. Biosensor experts are conducting research on the synthesis of new materials to increase the activity of biosensors [66]. Biomarkers should be efficient at the level of differential diagnosis and should not give false positive results. Metal nanoparticles provide a suitable environment that improves the electrical signal by increasing the surface-to-volume ratio of a biomolecule [67]. AgNPs are a very good material for detection due to their high conductivity, catalytic activity, and plasmonic properties. Thanks to these properties, they enable biosensors to work better [38]. The sensitivity of the sensor must be quite sensitive for the detection of a biological analyte at a minimum concentration. AgNPs increase the electron transfer rate due to the electroactive field of the electrodes and high sensitivity biosensors can be synthesized [68]. Compared to single nanoparticles, hybrid nanomaterials can increase the transfer rate of electrons by creating a constant electric field.

Researchers developed a sensor to label the target analyte captured via the electrode using hybrid AgNPs for the detection of alpha-fetoprotein [69]. Chen et al. synthesized hybrid AgNPs to detect telomerase activity. AgNPs were hybridized with functional telomerase domain by integrating into telomerase binding sites of DNA. As a result, conjugate cohesion increased, and salt-induced aggregation decreased. The liquid becomes yellow when the telomerase enzyme is active. The hybridized solution cluster turns gray when telomerase is not active [70]. Dewangan et al. developed colorimetric probes using hybrid AgNPs for the detection of cholesterol [71]. Cheng et al. doubled the signal amplification using hybrid AgNPs [72]. By investigating the physicochemical characteristics of AgNPs in the creation of several sensor classes, mostly for biomedical applications, this research has advanced the field of point-of-care disease detection [72].

4.3   Antimicrobial Properties of Hybrid NPs

In the past, civilizations have benefited from the antimicrobial activity of different metals and metal salts for wound care. The use of antimicrobial metals has decreased in the biomedical field with the discovery of antibiotics. Recently, antibiotic resistance has become a major problem [73]. AgNPs can enter bacteria and interact with their DNA, causing DNA destruction [74]. Hybrid AgNPs are a good option to fight antibiotic-resistant bacteria by showing higher activity against multidrug-resistant strains.

Silver ions can become extremely harmful to many human pathogens by interacting with negatively charged molecules such as DNA, RNA and proteins. The shape, concentration, size, presence of Ag^+ and type of bacteria of the nanoparticle contribute to the inactivation of bacteria. Large surface areas per unit volume increase the antibacterial activity of AgNPs [73].

Various biocidal effects against gram-positive and gram-negative bacteria and eukaryotes have been demonstrated by metallic nanoparticles derived from plant sources [75]. Furthermore, resistant organisms including methicillin-resistant Escherichia coli, erythromycin-resistant Streptococcus pyogenes, and ampicillin-resistant Escherichia coli are effectively combatted by metallic nanoparticles. Pseudomonas aeruginosa, vancomycin-resistant Staphylococcus aureus and Staphylococcus aureus [76]. To combat microbial activity, various metallic nanoparticles employ a variety of strategies. The bactericidal action of biogenic metallic/metal oxide nanoparticles (NPs) is achieved by the release of metal ions, which interact with the cell membrane to cause damage and subsequent lysis [77].

AgNPs impact functioning enzymatic groups and proteins necessary for bacterial cell metabolism and release their ions under specific physiological circumstances. Enzymes become inactive, membrane structure is disrupted and as a result bacteria become inactive [78]. Ag is used as an antimicrobial and antiseptic against both Gram- and Gram$^+$ bacteria due to its low cytotoxicity. Bryaskova et al. reported that AgNPs/PVP exhibited superior antibacterial properties against various bacteria such as Escherichia coli, Bacillus subtilis, Staphylococcus aureus and Pseudomonas aeruginosa [79]. Metal-organic hybrid nanoparticles exhibit excellent antimicrobial properties while exhibiting minimal toxicity especially against eukaryotes. Thiagamani et al. found that cellulose/banana peel powder/AgNP hybrid nanoparticles were effective on P. aeruginosa, B. licheniformis, S. aureus and E. coli [80].

Metallic nanoparticles engage in cell wall interaction and start the membrane diffusion of metal ions in antifungal activity. Next, a crucial part of the fungal cell wall, N-acetylglucosamine synthase, is blocked [81]. Then, with the induction of ROS, NPs interact with macromolecules such as DNA, RNA or protein and cause cell death [82].

4.4   Anticancer Effects of Hybrid NPs

Despite numerous studies and advances related to cancer, treatment failure rates are quite high, mainly due to dose-limiting toxicity, drug resistance, and serious side effects. As a result, new research is needed for cancer treatment without ignoring the patient's quality of life [83]. Metallic nanoparticles' biological properties are being studied in cancer cells to trigger autophagy and encourage cell death. Furthermore, biological metallic nanoparticles are efficient cytotoxic agents for a variety of cancer types [21].

AgNPs and their related hybrid nanomaterials stand out due to their wide range of applications [84]. Many researchers discovered that AgNPs exhibited strong antitumoral properties in in vitro and in vivo tumor models, and AgNPs began to attract great interest in the nanomedical field. These results indicate that AgNPs may be promising in developing diagnostic tools and for various oncotherapy approaches [38]. Hybrid nanoparticles with multiple functionalities have surprising properties against cancer cells. Hybrid nanoparticles exhibit cytotoxic effects by disrupting mitochondrial membrane potential, increasing ROS coverage, and decreasing intracellular glutathione levels. Hybrid AgNPs can cause inflammation in epithelial lung cells. According to studies, it has been reported that the cytotoxicity of hybrid AgNP depends on the size and coating material [85]. Because the coating material used to create hybrid nanoparticles can significantly alter their shape, surface area, and physical characteristics, it can have an impact on the cytotoxicity process [86]. The surface coating of hybrid AgNPs affects the cytotoxic activity on cancer cells and antimicrobial activities on bacterial cells. The antimicrobial effect of the same coating may vary depending on the starting metals [38]. The cytotoxic effect of silver NPs (PgAgNPs) synthesized with Panax ginseng leaves on some cancer cell lines (HepG2, A549 and MCF7) has been reported to be via oxidative stress [87]. In addition, PgAgNPs increase the apoptosis rate by changing the morphology of the cell nucleus. It has been reported that PdNPs synthesized from Evolvulus alsinoides promote ROS production, disrupt autophagy and mitochondrial membrane potential in cancer cells [88]. Biological NPs are a developing field in the treatment of cancer due to their large surface area and small size, providing drug delivery and therapeutic activity [21].

4.5   Anti-Inflammatory Effects of Hybrid NPs

Localized conditions caused by stress, injury and infection are called inflammation. This begins with the attraction of macrophages, killer cells, cytokines such as IL-1, IL-1β and TNF-α to the desired area [89]. Steroids are used in traditional treatment. However, they have many negative side effects [90]. Many studies have reported that NPs synthesized from plants have anti-inflammatory effects [21]. AgNPs synthesized from *Selaginella myosurus* have been reported to have anti-inflammatory effects under in vivo/in vitro conditions [91]. In another study, the effect of gold NPs synthesized from *Prunus serrulata* on LPS-induced RAW264.7 macrophages was investigated. The findings showed that AuNPs inhibited NF-κB activation, which in turn decreased pro-inflammatory cytokines and inflammatory mediators [92]. According to Nagajyothi et al. [93], zinc oxide nanoparticles produced from Polygala tenuifolia root extract exhibited anti-inflammatory properties by suppressing the production of TNF-α, iNOS, COX-2, IL-1b, and IL-6 proteins. The results suggest that biogenic metallic NPs can efficiently reduce inflammation by inhibiting pro-inflammatory cytokines.

5.   Theragnostic Importance of Biogenic Nanoparticles

Biogenic nanoparticles include organic and inorganic nanomaterials found in nature or produced by biotechnological methods. For example, lipid-based nanoparticles, protein-based nanomaterials, and structures carrying genetic material such as DNA/RNA can be included in this group. These particles are used in applications such as targeted drug delivery, sensitive imaging, and molecular-level diagnosis [94].

Human health is greatly impacted by the relatively new but quickly evolving science of nanomedicine. Biomedical and pharmaceutical sciences can be integrated with nanotechnology-based technologies and procedures [95]. Nanotechnology can thus play a major role in the future in the monitoring, control, diagnosis, and treatment of biological systems, which can contribute to the development of customized medicine [96]. Nanomedicine encompasses diagnosis and therapy, commonly known as theragnostic, sensing, nanopharmaceuticals, and nanoimaging [97]. Research on nanotechnology and nanomedical transdisciplinary sectors is being carried out by academic institutions, the pharmaceutical industry, clinical organizations, and several national and international funding and regulatory authorities [98].

Theragnostic approaches offer more personalized and effective solutions compared to traditional medical applications. Biogenic nanoparticles have been shown to be of great importance in the diagnosis and treatment of complex diseases, especially cancer. These nanoparticles can be used in the imaging of diseased tissues thanks to their ability to bind to certain biomarkers with high specificity. At the same time, their capacity to release therapeutic agents in a controlled manner increases the success of the treatment by minimizing side effects. For example, gold nanoparticles provide both energy transfer for photothermal therapy and imaging opportunities by being functionalized with biomarkers targeting cancer cells [99].

One of the most important advantages of biogenic nanoparticles is that they are biocompatible and biodegradable. These features allow for long-term use in the body and reduced toxic effects. In addition, the fact that their surfaces can be easily functionalized allows them to be made compatible with different biomolecules. For example, polymer-based nanoparticles are being developed as drug delivery systems, while magnetic nanoparticles are used for magnetic resonance imaging (MRI) and targeted drug delivery. Thus, a nanoparticle can serve both as a contrast agent for the detection of a disease and as a drug carrier for treatment [98].

As a result, biogenic nanoparticles have become one of the cornerstones of theragnostic approaches in modern medicine. The development of these materials makes diagnostic and therapeutic processes more efficient and effective in many areas, from cancer to neurological diseases. In the future, biogenic nanoparticles integrated with gene editing technologies and artificial intelligence are expected to play a greater role in personalized medicine. Therefore, new research in the fields of biotechnology and nanoscience will continue to pave the way for theragnostic applications [100].

6. Conclusion

Today, biogenic hybrid nanoparticles are used in medical research, diagnosis and treatment to help millions of patients worldwide. Preclinical and clinical studies are increasing the targeted design of biogenic hybrid nanoparticles through treatment methods. Hybrid biogenic nanoparticles' high bioavailability, low toxicity, and cost have made them successful theragnostic agents. With the potential to offer encouraging advantages in clinical settings, hybrid biogenic nanoparticles are becoming a research hotspot in the field of nanomedicine. based on these characteristics and the high degree of controllability they provide because of green, biogenic pathways for their synthesis and modification. Hybrid biogenic nanoparticles are quite suitable for use in biomedical and therapeutic fields due to their lack of toxic or hazardous components. Hybrid biogenic nanoparticles' remarkable physicochemical and biological characteristics open the door to a range of biomedical contributions, such as imaging, therapeutic, pharmacological, and diagnostic uses. As a result, studies have been conducted showing that hybrid biogenic nanoparticles combat diseases such as cancer, infections, antimicrobial treatments, neurological diseases and drug-induced cyto- and geno-toxicity. Diagnostics and therapeutics consisting of hybrid biogenic nanoparticles are still very new and are preparing for clinical trials. At this point, many of these compounds may provide new mechanistic insights. In-depth analysis of structure-activity relationships may lead to the intelligent design of new therapeutics. Therefore, further preclinical safety and selectivity studies should be conducted to safely introduce new environmentally friendly hybrid biogenic nanoparticles into clinical practice.

References

[1]   F. Nurili, G.U. Vural, Ö. Aras, Teranostik Platformlarda Moleküler Görüntüleme Yöntemleri, in: Nucl. Med. Semin. Tıp Semin., 2015: pp. 120–127. https://doi.org/10.4274/nts.019

[2]   R.S. Kalash, V.K. Lakshmanan, C.S. Cho, I.K. Park, Theranostics, Biomater. Nanoarchitectonics (2016) 197–215. https://doi.org/10.1016/B978-0-323-37127-8.00012-1

[3]   H. OKŞAŞ, M. EKİNCİ, D. İLEM ÖZDEMİR, Theranostic Applications in Nuclear Medicine:

Traditional Review, J. Lit. Pharm. Sci. 11 (2022) 49–60. https://doi.org/10.5336/pharmsci.2021-86197

[4]   S.M. Janib, A.S. Moses, J.A. MacKay, Imaging and drug delivery using theranostic nanoparticles, Adv. Drug Deliv. Rev. 62 (2010) 1052–1063. https://doi.org/https://doi.org/10.1016/j.addr.2010.08.004

[5]   D.Y. Lee, K.C.P. Li, Molecular Theranostics: A Primer for the Imaging Professional, Am. J. Roentgenol. 197 (2011) 318–324. https://doi.org/10.2214/AJR.11.6797

[6]   X. Ma, Y. Zhao, X.-J. Liang, Theranostic Nanoparticles Engineered for Clinic and Pharmaceutics, Acc. Chem. Res. 44 (2011) 1114–1122. https://doi.org/10.1021/ar2000056

[7]   F.R. Vogenberg, C. Isaacson Barash, M. Pursel, Personalized medicine: part 1: evolution and development into theranostics., P T 35 (2010) 560–76

[8]   B. Şen, B. Demirkan, A. Savk, R. Kartop, M.S. Nas, M.H. Alma, S. Sürdem, F. Şen, High-performance graphite-supported ruthenium nanocatalyst for hydrogen evolution reaction, J. Mol. Liq. 268 (2018) 807–812. https://doi.org/10.1016/j.molliq.2018.07.117

[9]   A. Hojjati-Najafabadi, S. Salmanpour, F. Sen, P.N. Asrami, M. Mahdavian, M.A. Khalilzadeh, A Tramadol Drug Electrochemical Sensor Amplified by Biosynthesized Au Nanoparticle Using Mentha aquatic Extract and Ionic Liquid, Top. Catal. (2021). https://doi.org/10.1007/s11244-021-01498-x

[10]   K. Nesrin, C. Yusuf, K. Ahmet, S.B. Ali, N.A. Muhammad, S. Suna, Ş. Fatih, Biogenic silver nanoparticles synthesized from Rhododendron ponticum and their antibacterial, antibiofilm and cytotoxic activities, J. Pharm. Biomed. Anal. 179 (2020) 112993. https://doi.org/10.1016/J.JPBA.2019.112993

[11]   S. Gwo, H.-Y. Chen, M.-H. Lin, L. Sun, X. Li, Nanomanipulation and controlled self-assembly of metal nanoparticles and nanocrystals for plasmonics, Chem. Soc. Rev. 45 (2016) 5672–5716. https://doi.org/10.1039/C6CS00450D.

[12]   B. Şen, N. Lolak, Ö. Paralı, M. Koca, A. Şavk, S. Akocak, F. Şen, Bimetallic PdRu/graphene oxide based Catalysts for one-pot three-component synthesis of 2-amino-4H-chromene derivatives, Nano-Structures and Nano-Objects 12 (2017) 33–40. https://doi.org/10.1016/j.nanoso.2017.08.013

[13]   A. Hojjati-Najafabadi, A. Aygun, R.N.E. Tiri, F. Gulbagca, M.I. Lounissaa, P. Feng, F. Karimi, F. Sen, Bacillus thuringiensis Based Ruthenium/Nickel Co-Doped Zinc as a Green Nanocatalyst: Enhanced Photocatalytic Activity, Mechanism, and Efficient H2Production from Sodium Borohydride Methanolysis, Ind. Eng. Chem. Res. 62 (2023) 4655–4664. https://doi.org/10.1021/ACS.IECR.2C03833

[14]   E. Demir, B. Sen, F. Sen, Highly efficient Pt nanoparticles and f-MWCNT nanocomposites based counter electrodes for dye-sensitized solar cells, Nano-Structures & Nano-Objects 11 (2017) 39–45. https://doi.org/10.1016/J.NANOSO.2017.06.003

[15]   Y. Wu, E.E. Altuner, R.N. El Houda Tiri, M. Bekmezci, F. Gulbagca, A. Aygun, C. Xia, Q. Van Le, F. Sen, H. Karimi-Maleh, Hydrogen generation from methanolysis of sodium borohydride using waste coffee oil modified zinc oxide nanoparticles and their photocatalytic activities, Int. J. Hydrogen Energy 48 (2023) 6613–6623. https://doi.org/10.1016/J.IJHYDENE.2022.04.177

[16]   E. Demir, A. Savk, B. Sen, F. Sen, A novel monodisperse metal nanoparticles anchored graphene oxide as Counter Electrode for Dye-Sensitized Solar Cells, Nano-Structures & Nano-Objects 12 (2017) 41–45. https://doi.org/10.1016/j.nanoso.2017.08.018

[17]   N. Korkmaz, Y. Ceylan, P. Taslimi, A. Karadağ, A.S. Bülbül, F. Şen, Biogenic nano silver: Synthesis, characterization, antibacterial, antibiofilms, and enzymatic activity, Adv. Powder Technol. 31 (2020) 2942–2950. https://doi.org/10.1016/J.APT.2020.05.020

[18]   M.A. Meyers, A. Mishra, D.J. Benson, Mechanical properties of nanocrystalline materials, Prog. Mater. Sci. 51 (2006) 427–556. https://doi.org/10.1016/j.pmatsci.2005.08.003

[19]   C. Dhand, N. Dwivedi, X.J. Loh, A.N. Jie Ying, N.K. Verma, R.W. Beuerman, R. Lakshminarayanan, S. Ramakrishna, Methods and Strategies for the Synthesis of Diverse Nanoparticles and Their Applications: A Comprehensive Overview, RSC Adv. 5 (2015) 105003–105037. https://doi.org/10.1039/C5RA19388E

[20]   R. Nagraik, A. Sharma, D. Kumar, S. Mukherjee, F. Sen, A.P. Kumar, Amalgamation of biosensors and nanotechnology in disease diagnosis: Mini-review, Sensors Int. 2 (2021) 100089. https://doi.org/10.1016/J.SINTL.2021.100089

[21]   S. Patil, R. Chandrasekaran, Biogenic nanoparticles: a comprehensive perspective in synthesis, characterization, application and its challenges, J. Genet. Eng. Biotechnol. 2020 181 18 (2020) 1–23. https://doi.org/10.1186/S43141-020-00081-3

[22]   K. Kalishwaralal, S. BarathManiKanth, S.R.K. Pandian, V. Deepak, S. Gurunathan, Silver nanoparticles impede the biofilm formation by Pseudomonas aeruginosa and Staphylococcus epidermidis, Colloids Surfaces B Biointerfaces 79 (2010) 340–344. https://doi.org/10.1016/j.colsurfb.2010.04.014

[23]   X. Li, H. Xu, Z.-S. Chen, G. Chen, Biosynthesis of Nanoparticles by Microorganisms and Their Applications, J. Nanomater. 2011 (2011) 1–16. https://doi.org/10.1155/2011/270974

[24]   G. Southam, T.J. Beveridge, The in vitro formation of placer gold by bacteria, Geochim. Cosmochim. Acta 58 (1994) 4527–4530. https://doi.org/10.1016/0016-7037(94)90355-7

[25]   Y. Liang, H. Demir, Y. Wu, A. Aygun, R.N. Elhouda Tiri, T. Gur, Y. Yuan, C. Xia, C. Demir, F. Sen, Y. Vasseghian, Facile synthesis of biogenic palladium nanoparticles using biomass strategy and application as photocatalyst degradation for textile dye pollutants and their in-vitro antimicrobial activity, Chemosphere 306 (2022) 135518. https://doi.org/10.1016/J.CHEMOSPHERE.2022.135518

[26]   M. Popescu, C. Ungureanu, Biosensors in Food and Healthcare Industries: Bio-Coatings Based on Biogenic Nanoparticles and Biopolymers, Coatings 13 (2023) 486. https://doi.org/10.3390/coatings13030486

[27]   S. Pradhan, A.K. Brooks, V.K. Yadavalli, Nature-derived materials for the fabrication of functional biodevices, Mater. Today Bio 7 (2020) 100065. https://doi.org/10.1016/j.mtbio.2020.100065

[28]   Y. Onuki, U. Bhardwaj, F. Papadimitrakopoulos, D.J. Burgess, A Review of the Biocompatibility of Implantable Devices: Current Challenges to Overcome Foreign Body Response, J. Diabetes Sci. Technol. 2 (2008) 1003–1015. https://doi.org/10.1177/193229680800200610

[29]   K.S. Siddiqi, A. Husen, R.A.K. Rao, A review on biosynthesis of silver nanoparticles and their biocidal properties, J. Nanobiotechnology 16 (2018) 14. https://doi.org/10.1186/s12951-018-0334-5

[30]   Q. Sun, X. Cai, J. Li, M. Zheng, Z. Chen, C.-P. Yu, Green synthesis of silver nanoparticles using tea leaf extract and evaluation of their stability and antibacterial activity, Colloids Surfaces A Physicochem. Eng. Asp. 444 (2014) 226–231. https://doi.org/10.1016/j.colsurfa.2013.12.065

[31]   M.C. Rodrigues, W.R. Rolim, M.M. Viana, T.R. Souza, F. Gonçalves, C.J. Tanaka, B. Bueno-Silva, A.B. Seabra, Biogenic synthesis and antimicrobial activity of silica-coated silver nanoparticles for esthetic dental applications, J. Dent. 96 (2020) 103327. https://doi.org/10.1016/j.jdent.2020.103327

[32]   M. Mondéjar-López, A.J. López-Jimenez, O. Ahrazem, L. Gómez-Gómez, E. Niza, Chitosan coated - biogenic silver nanoparticles from wheat residues as green antifungal and nanoprimig in wheat seeds, Int. J. Biol. Macromol. 225 (2023) 964–973. https://doi.org/10.1016/j.ijbiomac.2022.11.159

[33]   A. Ebrahiminezhad, A. Zare-Hoseinabadi, A.K. Sarmah, S. Taghizadeh, Y. Ghasemi, A. Berenjian, Plant-Mediated Synthesis and Applications of Iron Nanoparticles, Mol. Biotechnol. 60 (2018) 154–168. https://doi.org/10.1007/s12033-017-0053-4

[34]   A.B. Seabra, N. Manosalva, B. de Araujo Lima, M.T. Pelegrino, M. Brocchi, O. Rubilar, N. Duran, Antibacterial activity of nitric oxide releasing silver nanoparticles, J. Phys. Conf. Ser. 838 (2017) 012031. https://doi.org/10.1088/1742-6596/838/1/012031

[35]   W.R. Rolim, M.T. Pelegrino, B. de Araújo Lima, L.S. Ferraz, F.N. Costa, J.S. Bernardes, T. Rodrigues, M. Brocchi, A.B. Seabra, Green tea extract mediated biogenic synthesis of silver nanoparticles: Characterization, cytotoxicity evaluation and antibacterial activity, Appl. Surf. Sci. 463 (2019) 66–74. https://doi.org/10.1016/j.apsusc.2018.08.203

[36]   A. Ghasemi, N. Rabiee, S. Ahmadi, S. Hashemzadeh, F. Lolasi, M. Bozorgomid, A. Kalbasi, B.

Nasseri, A. Shiralizadeh Dezfuli, A.R. Aref, M. Karimi, M.R. Hamblin, Optical assays based on colloidal inorganic nanoparticles, Analyst 143 (2018) 3249–3283. https://doi.org/10.1039/C8AN00731D

[37]   D. Aguilar-García, A. Ochoa-Terán, F. Paraguay-Delgado, M.E. Díaz-García, G. Pina-Luis, Water-compatible core–shell Ag@SiO2 molecularly imprinted particles for the controlled release of tetracycline, J. Mater. Sci. 51 (2016) 5651–5663. https://doi.org/10.1007/s10853-016-9867-x

[38]   P. Singh, S. Singh, B. Maddiboyina, S. Kandalam, T. Walski, R.A. Bohara, Hybrid silver nanoparticles: Modes of synthesis and various biomedical applications, Electron 2 (2024). https://doi.org/10.1002/elt2.22

[39]   B.A. Thomas-Moore, C.A. del Valle, R.A. Field, M.J. Marín, Recent advances in nanoparticle-based targeting tactics for antibacterial photodynamic therapy, Photochem. Photobiol. Sci. 21 (2022) 1111–1131. https://doi.org/10.1007/s43630-022-00194-3

[40]   H. Seckin, R.N.E. Tiri, I. Meydan, A. Aygun, M.K. Gunduz, F. Sen, An environmental approach for the photodegradation of toxic pollutants from wastewater using Pt–Pd nanoparticles: Antioxidant, antibacterial and lipid peroxidation inhibition applications, Environ. Res. 208 (2022) 112708. https://doi.org/10.1016/J.ENVRES.2022.112708

[41]   F. Gulbagca, A. Aygun, E.E. Altuner, M. Bekmezci, T. Gur, F. Sen, H. Karimi-Maleh, N. Zare, F. Karimi, Y. Vasseghian, Facile bio-fabrication of Pd-Ag bimetallic nanoparticles and its performance in catalytic and pharmaceutical applications: Hydrogen production and in-vitro antibacterial, anticancer activities, and model development, Chem. Eng. Res. Des. 180 (2022) 254–264. https://doi.org/10.1016/J.CHERD.2022.02.024

[42]   J. Lin, F. Gulbagca, A. Aygun, R.N. Elhouda Tiri, C. Xia, Q. Van Le, T. Gur, F. Sen, Y. Vasseghian, Phyto-mediated synthesis of nanoparticles and their applications on hydrogen generation on NaBH4, biological activities and photodegradation on azo dyes: Development of machine learning model, Food Chem. Toxicol. 163 (2022) 112972. https://doi.org/10.1016/J.FCT.2022.112972

[43]   S.J. Lee, T. Begildayeva, S. Yeon, S.S. Naik, H. Ryu, T.H. Kim, M.Y. Choi, Eco-friendly synthesis of lignin mediated silver nanoparticles as a selective sensor and their catalytic removal of aromatic toxic nitro compounds, Environ. Pollut. 269 (2021) 116174. https://doi.org/10.1016/j.envpol.2020.116174

[44]   G.K. Podagatlapalli, S. Hamad, S.V. Rao, Trace-Level Detection of Secondary Explosives Using Hybrid Silver–Gold Nanoparticles and Nanostructures Achieved with Femtosecond Laser Ablation, J. Phys. Chem. C 119 (2015) 16972–16983. https://doi.org/10.1021/acs.jpcc.5b03958

[45]   A. Fatima, I. Younas, M.W. Ali, An Overview on Recent Advances in Biosensor Technology and its Future Application, Arch. Pharm. Pract. 13 (2022) 5–10. https://doi.org/10.51847/LToGI43jil

[46]   C. Sun, L. Ma, Q. Qian, S. Parmar, W. Zhao, B. Zhao, J. Shen, A chitosan-Au-hyperbranched polyester nanoparticles-based antifouling immunosensor for sensitive detection of carcinoembryonic antigen, Analyst 139 (2014) 4216–4222. https://doi.org/10.1039/C4AN00479E

[47]   N.E.-A. El-Naggar, M.H. Hussein, A.A. El-Sawah, Bio-fabrication of silver nanoparticles by phycocyanin, characterization, in vitro anticancer activity against breast cancer cell line and in vivo cytotxicity, Sci. Rep. 7 (2017) 10844. https://doi.org/10.1038/s41598-017-11121-3

[48]   B. Calderón-Jiménez, M.E. Johnson, A.R. Montoro Bustos, K.E. Murphy, M.R. Winchester, J.R. Vega Baudrit, Silver Nanoparticles: Technological Advances, Societal Impacts, and Metrological Challenges, Front. Chem. 5 (2017). https://doi.org/10.3389/fchem.2017.00006

[49]   S. Bochicchio, A. Dalmoro, P. Bertoncin, G. Lamberti, R.I. Moustafine, A.A. Barba, Design and production of hybrid nanoparticles with polymeric-lipid shell–core structures: conventional and next-generation approaches, RSC Adv. 8 (2018) 34614–34624. https://doi.org/10.1039/C8RA07069E

[50]   S. Gobalakrishnan, N. Chidhambaram, M. Chavali, Role of greener syntheses at the nanoscale, in: Handb. Greener Synth. Nanomater. Compd., Elsevier, 2021: pp. 107–134. https://doi.org/10.1016/B978-

0-12-821938-6.00004-9

[51]   R.N.E. Tiri, F. Gulbagca, A. Aygun, A. Cherif, F. Sen, Biosynthesis of Ag–Pt bimetallic nanoparticles using propolis extract: Antibacterial effects and catalytic activity on NaBH4 hydrolysis, Environ. Res. 206 (2022) 112622. https://doi.org/10.1016/J.ENVRES.2021.112622

[52]   A. Aygun, F. Gulbagca, E.E. Altuner, M. Bekmezci, T. Gur, H. Karimi-Maleh, F. Karimi, Y. Vasseghian, F. Sen, Highly active PdPt bimetallic nanoparticles synthesized by one-step bioreduction method: Characterizations, anticancer, antibacterial activities and evaluation of their catalytic effect for hydrogen generation, Int. J. Hydrogen Energy (2022). https://doi.org/10.1016/J.IJHYDENE.2021.12.144

[53]   W. Park, H. Shin, B. Choi, W.-K. Rhim, K. Na, D. Keun Han, Advanced hybrid nanomaterials for biomedical applications, Prog. Mater. Sci. 114 (2020) 100686. https://doi.org/10.1016/j.pmatsci.2020.100686

[54]   K. Cho, X. Wang, S. Nie, Z. Chen, D.M. Shin, Therapeutic nanoparticles for drug delivery in cancer, Clin. Cancer Res. (2008). https://doi.org/10.1158/1078-0432.CCR-07-1441

[55]   D. Ma, Hybrid Nanoparticles, in: Noble Met. Oxide Hybrid Nanoparticles, Elsevier, 2019: pp. 3–6. https://doi.org/10.1016/B978-0-12-814134-2.00001-2

[56]   F. Yang, A. Skripka, A. Benayas, X. Dong, S.H. Hong, F. Ren, J.K. Oh, X. Liu, F. Vetrone, D. Ma, An Integrated Multifunctional Nanoplatform for Deep-Tissue Dual-Mode Imaging, Adv. Funct. Mater. 28 (2018). https://doi.org/10.1002/adfm.201706235

[57]   A. Rahman, M.A. Chowdhury, N. Hossain, Green synthesis of hybrid nanoparticles for biomedical applications: A review, Appl. Surf. Sci. Adv. 11 (2022) 100296. https://doi.org/10.1016/j.apsadv.2022.100296

[58]   D. Kovács, N. Igaz, M.K. Gopisetty, M. Kiricsi, Cancer Therapy by Silver Nanoparticles: Fiction or Reality?, Int. J. Mol. Sci. 23 (2022) 839. https://doi.org/10.3390/ijms23020839

[59]   P. Prasher, M. Sharma, H. Mudila, G. Gupta, A.K. Sharma, D. Kumar, H.A. Bakshi, P. Negi, D.N. Kapoor, D.K. Chellappan, M.M. Tambuwala, K. Dua, Emerging trends in clinical implications of bio-conjugated silver nanoparticles in drug delivery, Colloid Interface Sci. Commun. 35 (2020) 100244. https://doi.org/10.1016/j.colcom.2020.100244

[60]   M. Kasithevar, P. Periakaruppan, S. Muthupandian, M. Mohan, Antibacterial efficacy of silver nanoparticles against multi-drug resistant clinical isolates from post-surgical wound infections, Microb. Pathog. 107 (2017) 327–334. https://doi.org/10.1016/j.micpath.2017.04.013

[61]   B.G. Andryukov, I.N. Lyapun, E.V. Matosova, L.M. Somova, Biosensor Technologies in Medicine: from Detection of Biochemical Markers to Research into Molecular Targets (Review), Sovrem. Tehnol. v Med. 12 (2020) 70. https://doi.org/10.17691/stm2020.12.6.09

[62]   F. Zamarchi, I.C. Vieira, Determination of paracetamol using a sensor based on green synthesis of silver nanoparticles in plant extract, J. Pharm. Biomed. Anal. 196 (2021) 113912. https://doi.org/10.1016/j.jpba.2021.113912

[63]   P. Bollella, C. Schulz, G. Favero, F. Mazzei, R. Ludwig, L. Gorton, R. Antiochia, Green synthesis and characterization of gold and silver nanoparticles and their application for development of a third generation lactose biosensor, Electroanalysis 29 (2017) 77-86. https://doi.org/10.1002/elan.201600476

[64]   B. Zheng, L. Qian, H. Yuan, D. Xiao, X. Yang, M.C. Paau, M.M.F. Choi, Preparation of gold nanoparticles on eggshell membrane and their biosensing application, Talanta 82 (2010) 177–183. https://doi.org/10.1016/j.talanta.2010.04.014

[65]   A.A. Gabizon, R.T.M. de Rosales, N.M. La-Beck, Translational considerations in nanomedicine: The oncology perspective, Adv. Drug Deliv. Rev. 158 (2020) 140–157. https://doi.org/10.1016/j.addr.2020.05.012

[66]   W. Najahi-Missaoui, R.D. Arnold, B.S. Cummings, Safe Nanoparticles: Are We There Yet?, Int. J. Mol. Sci. 22 (2020) 385. https://doi.org/10.3390/ijms22010385

[67]   L. Lu, A. Kobayashi, Y. Kikkawa, K. Tawa, Y. Ozaki, Oriented Attachment-Based Assembly of Dendritic Silver Nanostructures at Room Temperature, J. Phys. Chem. B 110 (2006) 23234–23241. https://doi.org/10.1021/jp063978c

[68]   H. Zhou, X. Gan, J. Wang, X. Zhu, G. Li, Hemoglobin-Based Hydrogen Peroxide Biosensor Tuned by the Photovoltaic Effect of Nano Titanium Dioxide, Anal. Chem. 77 (2005) 6102–6104. https://doi.org/10.1021/ac050924a

[69]   J. Lin, H. Ju, Electrochemical and chemiluminescent immunosensors for tumor markers, Biosens. Bioelectron. 20 (2005) 1461–1470. https://doi.org/10.1016/j.bios.2004.05.008

[70]   W. Chen, L. Wang, R. He, X. Xu, W. Jiang, Convertible DNA ends-based silver nanoprobes for colorimetric detection human telomerase activity, Talanta 178 (2018) 458–463. https://doi.org/10.1016/j.talanta.2017.09.057

[71]   L. Dewangan, J. Korram, I. Karbhal, R. Nagwanshi, V.K. Jena, M.L. Satnami, A colorimetric nanoprobe based on enzyme-immobilized silver nanoparticles for the efficient detection of cholesterol, RSC Adv. 9 (2019) 42085–42095. https://doi.org/10.1039/C9RA08328F

[72]   Z. Cheng, M. Li, R. Dey, Y. Chen, Nanomaterials for cancer therapy: current progress and perspectives, J. Hematol. Oncol. 14 (2021) 85. https://doi.org/10.1186/s13045-021-01096-0

[73]   A.G. Arranja, V. Pathak, T. Lammers, Y. Shi, Tumor-targeted nanomedicines for cancer theranostics, Pharmacol. Res. 115 (2017) 87–95. https://doi.org/10.1016/j.phrs.2016.11.014

[74]   J.L. Elechiguerra, J.L. Burt, J.R. Morones, A. Camacho-Bragado, X. Gao, H.H. Lara, M.J. Yacaman, Interaction of silver nanoparticles with HIV-1, J. Nanobiotechnology 3 (2005) 6. https://doi.org/10.1186/1477-3155-3-6

[75]   M. Eltarahony, S. Zaki, M. ElKady, D. Abd-El-Haleem, Biosynthesis, Characterization of Some Combined Nanoparticles, and Its Biocide Potency against a Broad Spectrum of Pathogens, J. Nanomater. 2018 (2018) 1–16. https://doi.org/10.1155/2018/5263814

[76]   P. V. Baptista, M.P. McCusker, A. Carvalho, D.A. Ferreira, N.M. Mohan, M. Martins, A.R. Fernandes, Nano-Strategies to Fight Multidrug Resistant Bacteria—"A Battle of the Titans," Front. Microbiol. 9 (2018). https://doi.org/10.3389/fmicb.2018.01441

[77]   C. Rajkuberan, K. Sudha, G. Sathishkumar, S. Sivaramakrishnan, Antibacterial and cytotoxic potential of silver nanoparticles synthesized using latex of Calotropis gigantea L., Spectrochim. Acta Part A Mol. Biomol. Spectrosc. 136 (2015) 924–930. https://doi.org/10.1016/j.saa.2014.09.115

[78]   S. Egger, R.P. Lehmann, M.J. Height, M.J. Loessner, M. Schuppler, Antimicrobial Properties of a Novel Silver-Silica Nanocomposite Material, Appl. Environ. Microbiol. 75 (2009) 2973–2976. https://doi.org/10.1128/AEM.01658-08

[79]   R. Bryaskova, D. Pencheva, S. Nikolov, T. Kantardjiev, Synthesis and comparative study on the antimicrobial activity of hybrid materials based on silver nanoparticles (AgNps) stabilized by polyvinylpyrrolidone (PVP), J. Chem. Biol. 4 (2011) 185–191. https://doi.org/10.1007/s12154-011-0063-9

[80]   S.M.K. Thiagamani, N. Rajini, S. Siengchin, A. Varada Rajulu, N. Hariram, N. Ayrilmis, Influence of silver nanoparticles on the mechanical, thermal and antimicrobial properties of cellulose-based hybrid nanocomposites, Compos. Part B Eng. 165 (2019) 516–525. https://doi.org/10.1016/j.compositesb.2019.02.006

[81]   M. Malekzadeh, K.L. Yeung, M. Halali, Q. Chang, Preparation and antibacterial behaviour of nanostructured Ag@SiO 2 –penicillin with silver nanoplates, New J. Chem. 43 (2019) 16612–16620. https://doi.org/10.1039/C9NJ03727F

[82]   Y. Wang, P. Chen, M. Liu, Synthesis of hollow silver nanostructures by a simple strategy, Nanotechnology 19 (2008) 045607. https://doi.org/10.1088/0957-4484/19/04/045607

[83]   R.R. Miranda, I. Sampaio, V. Zucolotto, Exploring silver nanoparticles for cancer therapy and

diagnosis, Colloids Surfaces B Biointerfaces 210 (2022) 112254. https://doi.org/10.1016/j.colsurfb.2021.112254

[84]   S. Dawadi, S. Katuwal, A. Gupta, U. Lamichhane, R. Thapa, S. Jaisi, G. Lamichhane, D.P. Bhattarai, N. Parajuli, Current Research on Silver Nanoparticles: Synthesis, Characterization, and Applications, J. Nanomater. 2021 (2021) 1–23. https://doi.org/10.1155/2021/6687290

[85]   S. Kokura, O. Handa, T. Takagi, T. Ishikawa, Y. Naito, T. Yoshikawa, Silver nanoparticles as a safe preservative for use in cosmetics, Nanomedicine Nanotechnology, Biol. Med. 6 (2010) 570–574. https://doi.org/10.1016/j.nano.2009.12.002

[86]   M. Azizi, H. Ghourchian, F. Yazdian, S. Bagherifam, S. Bekhradnia, B. Nyström, Anti-cancerous effect of albumin coated silver nanoparticles on MDA-MB 231 human breast cancer cell line, Sci. Rep. 7 (2017) 5178. https://doi.org/10.1038/s41598-017-05461-3

[87]   V. Castro-Aceituno, S. Ahn, S.Y. Simu, P. Singh, R. Mathiyalagan, H.A. Lee, D.C. Yang, Anticancer activity of silver nanoparticles from Panax ginseng fresh leaves in human cancer cells, Biomed. Pharmacother. 84 (2016) 158–165. https://doi.org/10.1016/j.biopha.2016.09.016

[88]   S. Gurunathan, E. Kim, J. Han, J. Park, J.-H. Kim, Green Chemistry Approach for Synthesis of Effective Anticancer Palladium Nanoparticles, Molecules 20 (2015) 22476–22498. https://doi.org/10.3390/molecules201219860

[89]   H. Agarwal, A. Nakara, V.K. Shanmugam, Anti-inflammatory mechanism of various metal and metal oxide nanoparticles synthesized using plant extracts: A review, Biomed. Pharmacother. 109 (2019) 2561–2572. https://doi.org/10.1016/j.biopha.2018.11.116

[90]   C.K.S. Ong, P. Lirk, C.H. Tan, R.A. Seymour, An Evidence-Based Update on Nonsteroidal Anti-Inflammatory Drugs, Clin. Med. Res. 5 (2007) 19–34. https://doi.org/10.3121/cmr.2007.698

[91]   P.B.E. Kedi, F.E. Meva, L. Kotsedi, E.L. Nguemfo, C.B. Zangueu, A.A. Ntoumba, H.E.A. Mohamed, A.B. Dongmo, M. Maaza, Eco-friendly synthesis, characterization, in vitro and in vivo anti-inflammatory activity of silver nanoparticle-mediated Selaginella myosurus aqueous extract, Int. J. Nanomedicine 13 (2018) 8537–8548. https://doi.org/10.2147/IJN.S174530

[92]   P. Singh, S. Ahn, J.-P. Kang, S. Veronika, Y. Huo, H. Singh, M. Chokkaligam, M. El-Agamy Farh, V.C. Aceituno, Y.J. Kim, D.-C. Yang, In vitro anti-inflammatory activity of spherical silver nanoparticles and monodisperse hexagonal gold nanoparticles by fruit extract of Prunus serrulata : a green synthetic approach, Artif. Cells, Nanomedicine, Biotechnol. (2017) 1–11. https://doi.org/10.1080/21691401.2017.1408117

[93]   P.C. Nagajyothi, S.J. Cha, I.J. Yang, T.V.M. Sreekanth, K.J. Kim, H.M. Shin, Antioxidant and anti-inflammatory activities of zinc oxide nanoparticles synthesized using Polygala tenuifolia root extract, J. Photochem. Photobiol. B Biol. 146 (2015) 10–17. https://doi.org/10.1016/j.jphotobiol.2015.02.008

[94]   M.C. Zambonino, E.M. Quizhpe, L. Mouheb, A. Rahman, S.N. Agathos, S.A. Dahoumane, Biogenic Selenium Nanoparticles in Biomedical Sciences: Properties, Current Trends, Novel Opportunities and Emerging Challenges in Theranostic Nanomedicine, Nanomaterials 13 (2023) 424. https://doi.org/10.3390/nano13030424

[95]   V. Morigi, A. Tocchio, C. Bellavite Pellegrini, J.H. Sakamoto, M. Arnone, E. Tasciotti, Nanotechnology in Medicine: From Inception to Market Domination, J. Drug Deliv. 2012 (2012) 1–7. https://doi.org/10.1155/2012/389485

[96]   S. Mura, P. Couvreur, Nanotheranostics for personalized medicine, Adv. Drug Deliv. Rev. 64 (2012) 1394–1416. https://doi.org/10.1016/j.addr.2012.06.006

[97]   J.M. Caster, A.N. Patel, T. Zhang, A. Wang, Investigational nanomedicines in 2016: a review of nanotherapeutics currently undergoing clinical trials, WIREs Nanomedicine and Nanobiotechnology 9 (2017). https://doi.org/10.1002/wnan.1416

[98]   P. Satalkar, B.S. Elger, D.M. Shaw, Defining Nano, Nanotechnology and Nanomedicine: Why

Should It Matter?, Sci. Eng. Ethics 22 (2016) 1255–1276. https://doi.org/10.1007/s11948-015-9705-6

[99]   R. Bayford, T. Rademacher, I. Roitt, S.X. Wang, Emerging applications of nanotechnology for diagnosis and therapy of disease: a review, Physiol. Meas. 38 (2017) R183–R203. https://doi.org/10.1088/1361-6579/aa7182

[100]   D. Mundekkad, W.C. Cho, Nanoparticles in Clinical Translation for Cancer Therapy, Int. J. Mol. Sci. 23 (2022) 1685. https://doi.org/10.3390/ijms23031685

Chapter 22

COVID-19 and Biogenic Nanomaterials

E. Aydin*[1], M. Koldemir Gunduz[2], F. Sen[3*]

[1]Faculty of Medicine, Agrı Ibrahim Çeçen University, Igdır, Turkiye

[2]Department of Basic Sciences of Engineering, Faculty of Engineering and Natural Sciences, Kutahya Health Sciences University, Kutahya, Turkiye

[3]Sen Research Group, Department of Biochemistry, Dumlupinar University, Kutahya, Turkiye.

elifkn@hotmail.com; fatih.sen@dpu.edu.tr

Abstract

The COVID-19 pandemic affected millions of people worldwide, causing health and social changes, with the elderly and individuals with chronic diseases being especially at risk. During this process, biogenic nanomaterials emerged with significant potential in areas like vaccine development, diagnostic tests, and surface disinfection. These nanomaterials, obtained from natural sources, were used in biosensors for rapid and sensitive COVID-19 detection and as carrier systems to enhance vaccine effectiveness. Additionally, self-disinfecting nanomaterials helped prevent the virus spread on surfaces. In the future, sustainable production methods and clinical trials will be crucial.

Keywords

COVID-19, Nanoparticles, Nanotechnology, Biogenic Nanomaterials, Vaccine Development

1. Introduction

One of the major causes of illness and mortality globally, as well as a major contributor to large financial losses, are viral infections. Vaccination and medications that target important steps in the viral life cycle are the mainstays of conventional treatment methods. Nevertheless, a lot of viruses change as a result of selection pressures and frequently acquire treatment resistance, which calls for further funding to create new medications [1].

A novel virus that is genetically linked to the coronavirus that caused the 2003 epidemic of severe acute respiratory syndrome (SARS); the Severe Acute Respiratory Syndrome Coronavirus 2 (SARS-CoV-2) was the cause of the coronavirus infectious disease (COVID-19) pandemic that began at the end of 2019 [2]. Following reporting of the initial instances, the World Health Organization (WHO) launched its incident management support team on January 1, 2020, and on January 5, 2020, it published an official statement. The first COVID-19 case outside of China was reported on January 22, 2020. On February 4, 2020, the WHO made its first public statement on the appearance of the COVID-19 sickness. They presented some initial findings, stating that there were 20,471 instances in China and 425 fatalities (99% of all cases recorded), and they alerted the world to the possibility of a worldwide pandemic. Only 176 instances had been documented outside of China at that point [3].

Finding answers quickly is a major problem for academics because of the COVID-19 epidemic. SARS-CoV-2, the virus that causes COVID-19, has been detected in a variety of habitats, including specimens, blood, feces, and solid waste from infected individuals, according to studies [4]. Since disposable personal protective equipment is made of plastic material, the increase in plastic waste production has also led to an increase in the amount of chemical waste [5].

The COVID-19 pandemic has led to major changes in health, social dynamics, economic systems, and psychological well-being worldwide. The rapid spread of the SARS-CoV-2 virus has presented healthcare systems with new challenges and forced the rethinking of public health and medical response plans worldwide. Innovative solutions have become increasingly important due to the heavy burden on countries' healthcare services. Due to the COVID-19 pandemic, biogenic nanomaterials are crucial to prepare for future pandemics [6].

Nanotechnology has demonstrated that it can bridge the gap between biological and physical sciences by using nanostructures and nanophases in various scientific fields [7–16]. This is especially true for nanomedicine and nano-based drug delivery systems, where these particles are highly desired [17]. Nanoparticles are typically small nano-spheres because they are made from materials created at the atomic or molecular level [18]. As a result, they have greater mobility within the human body than larger materials. Nanoscale particles have special mechanical, chemical, electrical, magnetic, biological, and structural properties. The creation and use of various materials and technologies with at least one dimension smaller than 100 nanometers is called nanotechnology. Nanomedicine is the term used to describe the use of nanotechnology in the medical profession, including the use of nanomaterials for disease diagnosis, treatment, control and prevention [19]. Nanoparticles have been widely used and researched over the years due to their special properties such as small size, improved solubility, surface adaptability and multifunctionality [20,21]. This has led to the creation of safer and better drugs, tissue-targeted therapies, customized nanomedicines and early disease detection and prevention. Therefore, in the near future, nano-based methods seem to be the best option to create the best treatments for various diseases [22].

Nanotechnology likely holds great promise for COVID-19 diagnosis, treatment, and prevention. Development of infection-safe personal protective equipment (PPE) to improve the safety of healthcare workers, creation of antiviral disinfectants and surface coatings that can neutralize the virus and stop its spread, development of highly specific and sensitive nano-based sensors to quickly identify infection or immunological response, development of new drugs targeted to tissues such as the lungs with improved activity, reduced toxicity, and sustained release, and creation of a nano-based vaccine to enhance humoral and cellular immune responses are just a few of the ways nanotechnology can help combat COVID-19.

A well-established and well-known application of nanomaterials in vaccine design stems from pandemics. Modern vaccine design benefits from nanotechnology because nanoparticles are excellent for adjuvants, antigen delivery, and viral structural mimicry. In fact, an mRNA vaccine delivered via lipid nanoparticles is the first vaccine candidate to reach clinical testing. An effective vaccination platform should facilitate rapid discovery, scalable manufacturing, and global dissemination to eliminate pandemics both now and in the future [23].

For bioanalytical techniques that can be used to diagnose COVID-19, nanoparticles offer a number of benefits. High sensitivity is achieved due to the large surface area of nanoparticles. The modest SARS-COV2 payload can be pre-concentrated and enriched using nanoparticles. The surface can be modified to guarantee greater specificity [24].

The addition of biogenic nanoparticles to surface disinfection has also been helpful. The SARS-CoV-2 virus can spread to live on a variety of surfaces. While they work well, traditional disinfection techniques lack durability and are sometimes very labor-intensive. When nanoparticles with antimicrobial properties are added to surfaces, they can inactivate microorganisms for long periods of time. This innovative surface disinfection method could improve hospital safety and result in greater public health efforts.

This book began by examining how the COVID-19 pandemic has changed the world in terms of health and social interaction. It also examines the multiple uses and promise of biogenic nanomaterials in the field of health. By carefully examining current research, applications, and state-of-the-art technology, this study aims to reveal the important role that biogenic nanomaterials play in combating pandemics and in public health protection plans.

Therefore, there is an immediate need for new medical and healthcare solutions due to the COVID-19. Biogenic nanomaterials are studied for their use in novel, sustainable and efficient methods of diagnosis, treatment or prevention. Covered in ten chapter, this book will provide an overview of approaches to using biogenic nanomaterials in current public health problems and how they may affect the world healthcare situation globally with a desire towards remote sensing for currently far-away major diseases. These materials can serve as important tools for continued research and studies to better prepare us all for the next pandemic, a reminder of why science in service of humanity is so critical.

2.   COVID-19 Pandemic: Developments and Impacts

One of the biggest health disasters to hit humanity in recent memory is COVID-19 (WHO, 2021). COVID-19, a disease that has spread around the world through a range of transmission channels, including direct contact and airborne particles, is caused by the type 2 coronavirus that causes severe acute respiratory syndrome (SARS-CoV-2). The epidemic affected more so too much strain on the health care system along with daily life [25].

Infection rates and death statistics are used to measure the impact of COVID-19. Millions of people worldwide have contracted the virus, with several deaths reported. It has also profoundly changed how people work, communicate with each other, and consume [26]. The populations most affected by the virus were those classified as high-risk, such as the elderly and those with chronic diseases [27].

Researchers have created a number of COVID-19 preventive and treatment plans during the pandemic. In this context, biogenic nanomaterials have shown significant promise. Effective vaccines and antiviral treatment techniques have been developed using nanotechnology [28]. Currently, bio-based nanomaterials point to more sustainable development and better environmental health. Better and environmentally friendly nms have less harmful effects thanks to the recently developed bio-based nanotechnology. Microbes, for example plants, bacteria, algae and plant substances, have been used. Biological method uses materials like polysaccharides, microbial proteins, peptides, enzymes, and agricultural wastes [29].

COVID-19 reflects a significant deterioration in health, as well as a profound impact on the economy, education and psychology. While educational institutions resort to online platforms, facilities have adopted a remote working model. Seen amidst major changes in the economic landscape and many businesses have had to quickly adapt to the remote working paradigm. This shift was initially driven by necessity, as social distancing and restrictions on physical contact were required due to lockdowns and health regulations. Technologies such as video conferencing, project management software, and cloud-based solutions have become widely used as a result of organizations rapidly adopting digital tools to enable remote collaboration. Beyond the way work is performed, this shift has pushed businesses to re-evaluate their engagement strategies at large and also alters operational plans as well as workforce make-up for some companies, that has meant they can work remotely and gain flexibility while saving money on overhead with the added surprise benefit of a potential increase in productivity. However, these changes have also spawned worries around the new work-life balance normal; employees experiencing greater levels of loneliness and difficulty in adjusting to a remote corporate culture [30]. We have yet to understand the concrete, long-term effects of these changes and whether or not societies will readapt after this pandemic.

The emotional effect of the pandemic cannot be overstated, either. Forced social banishment, safety and health worries, as well as economic instability have taken a toll on mental health. Long-lasting uncertainty and daily life interference can be major stressors, feelings of anxiety, sadness or disconnecting socially. The pandemic has also brought to light mental health care and alerted us by provoking discussions around stronger resilience-building mechanisms in personal as well as professional contexts.

Consequently, the impacts of COVID-19 have hugely affected not only the health sector but also the economy, education, and mental health. Changes in online learning and working from home have not only changed our way of communication and working but also exposed social inequalities and vulnerabilities. We need to take a holistic approach to the complex issues, ensuring that lessons learned from the pandemic contribute to developing a resilient and more equitable future in all aspects of life.

3.   Biogenic Nanomaterials: Description and Properties

Biogenic nanomaterials are materials from nature with a nanoscale structure. This material can be obtained through biological means and is usually featured for being suitable for the ecosystem and sustainable. Recent scientific and technological advances in the field of nanotechnology have enhanced the potential of biogenic nanomaterials in various applications [31].

Most of these compounds are usually derived from plants, microorganisms, and other biological sources. For instance, certain microorganisms and plant extracts can combine to form nanoparticles with antibacterial, antiviral, and antifungal properties [32]. Due to such properties, biogenic nanomaterials become ideal options for applications in vaccine development, drug delivery systems, and biosensors in the medical industry.

High surface area, low toxicity, and biocompatibility are characteristics of biogenic nanomaterials. The properties enhance the efficiency of the materials in health and medical applications. For instance, the immune system recognizes nanoparticles having properties that mimic those of biological systems more efficiently for targeting [33]. Another advantage of these materials is their biodegradability, which lessens the potential negative impact on the environment and enhances the sustainable applications of the materials.

Biogenic nanomaterials can be applied in almost all sectors relevant to their purpose, especially to food safety, agriculture, and environmental engineering. The materials can be used, for example, in food safety to reduce the effect of microbes or to improve food quality, and in agriculture to increase the effectiveness of fertilizers and pesticides [34].

In the final analysis, biogenic nanomaterials are unique in the diversity of applications in the fields of environment and health, besides being materials produced from nature. The materials are expected to be very important in addressing some of the most serious issues, such as pandemics facing humankind in the near future as research and development in these materials are extended.

4. The Role of Biogenic Nanomaterials in Combating COVID-19

Most COVID-19 infections are thought to be from the inhalation of virus droplets. They tend to float or stick to objects in their immediate environment, such as doorknobs, desktop computers, and even protective clothing worn by healthcare personnel. According to the National Institutes of Health (NIH) COVID-19 treatment guidelines, healthy people can get COVID-19 by coming into contact with contaminated surfaces. Seven days after exposure, researchers found live virus on the surface of surgical masks in a study of COVID-19 survival [35].

Many hospitals and medical facilities use traditional disinfection techniques, including hydrogen peroxide spraying or UV disinfection, to remove surface contaminants. However, these disinfection techniques are usually one-time. The virus can irrationally attach to the surface of treated materials as soon as the UV lamp is turned off or the hydrogen peroxide decomposes. As nanoscience and technology develop, self-disinfection of nanomaterial surfaces may provide a viable solution to this problem. Copper, silver, and other heavy metal ions are well-liked in the medical field because of their antiviral properties. This might occur as a result of heavy metal coatings' ability to progressively release heavy metal ions when coming into contact with bacterial viruses. When these ions encounter the protein shell of the bacterial fungus, they render the proteins inactive, enabling the contaminants to self-disinfect. A coronavirus study found that coronavirus 229E (HUCOV-229E) lived much less time on copper-containing surfaces than on copper-free materials [36]. Other heavy metals showed potent antiviral properties in addition to copper and silver [37]. Several studies have shown that adding nanoparticles to a polluted surface not only disinfects it but also offers sustained viral protection [38]. Balagna et al. demonstrated the efficacy of a filtered facepiece-3 mask with silver nanocluster in silica composite coating against COVID-19. This study demonstrates that the silver nanocluster/silica composite coating of the mask has strong antiviral properties [39]. Additionally, graphene-silver nanocomposite-coated masks have been shown to have antiviral properties [40].

Biogenic nanomaterials also play a key role in the procedures involved in vaccine development. Medical equipment and instruments, nanometer adjuvant vaccines, nanometer-coated filters, and drug delivery systems built using nanometer carriers are included on the list. Accordingly, many studies had proven before the fight against COVID-19 that nanomaterials can be used in combating a range of viruses, including coronaviruses. Some practical solutions to preventing the spread of the COVID-19 virus may come from reports on nanoparticles. These include the development of a rapid and low-cost test that will be able to detect COVID-19 in all the populations worldwide, using nanomaterials to stop viral replication and RNA assembly through hindering COVID-19 interactions with the cellular receptor ACE-2, and developing a new vaccine that uses nanoparticles to help patients infected with the virus regain their innate immunity. In some areas, nanoparticles have also been found effective [41,42]. These substances can be used as carriers in order to enhance the vaccine efficacy. For instance, antigens could be delivered in a better way to their target cells by nanoparticles made from different biopolymers. Hence, the vaccine immunological response is enhanced, and protection is offered in the right way [43].

In conclusion, biogenic nanomaterials are one of the most crucial areas of research against COVID-19, considering their antiviral properties and potential applications in vaccine preparation. Further exploration of these materials might lead to more successful future pandemic prevention plans.

5.   Biogenic Nanomaterial-Based Diagnostic Techniques for COVID-19

Rapid and precise diagnostic methods are now more important than ever, and the COVID-19 pandemic has brought attention to the promise of biogenic nanomaterials in this field. The ecologically favorable qualities of biogenic nanomaterials, which are nanoscale materials derived from biological sources, draw attention. These resources are crucial to the creation of cutting-edge virus detection techniques.

Nanotechnology is among the advanced approaches utilized to mitigate the risks associated with COVID-19. The antiviral capabilities of nanomaterials play a crucial role in the diagnosis of viral infections. This chapter provides a detailed examination of the role of nanomaterials in diagnosing viral infections, highlighting their antiviral properties, particularly in the context of SARS-CoV-2 [44].

Gold-based nanoparticles (AuNPs) are widely known for their outstanding physical and chemical characteristics, including their resistance to chemical degradation, ability to dissolve in water, and flexibility in structural design. Their high surface free electron density contributes to their unique electrical, optical, and catalytic behaviors. These distinctive features make AuNPs a favored choice for various biological applications [45]. Gold-based nanoparticles transducers, optical signal amplifiers, and current amplifiers were used as signals in terms of resonant light scattering [46]. Localized surface plasmon resonance is the fundamental physical phenomenon that drives a category of nanoparticle-based biosensors. It describes the synchronized, non-dispersive oscillations of free electrons within metal nanoparticles [47]. The localized surface plasmon resonance phenomenon in two-dimensional gold nano-islands (AuNI) was harnessed to develop a dual-function plasmonic device capable of diagnosing COVID-19 with exceptional sensitivity, speed, and accuracy. In this approach, AuNIs were modified with complementary DNA receptors specific to SARS-CoV-2, facilitating the detection of targeted viral sequences during the nucleic acid hybridization process. By activating plasmonic resonance, the hybridization process becomes more efficient, resulting in enhanced detection accuracy for different gene sequences. Thus, the designed dual-function plasmonic system enables immediate and marker-free identification of viral sequences, achieving a detection threshold of 0.22 pM [48]. A colorimetric biosensor technique utilizing AuNPs functionalized with sulfur-linked antisense oligonucleotides (AuNPs-ASOs) has been developed for identifying SARS-CoV-2.

The colorimetric biosensor allows the visible detection of infectious diseases in viruses without the use of any sophisticated techniques. The presence of the SARS-CoV-2 nucleocapsid phosphorylation protein (N-gene) is measured after isolating the subject's RNA sample. The nanostructures selectively cluster in the presence of a SARS-CoV-2 RNA sequence, causing a shift toward longer wavelengths in the UV absorbance spectrum as a result of the surface plasmon resonance phenomenon. In less than ten minutes, the colorimetric biosensor test thereby validates the COVID-19 diagnosis [49].

To detect RdRp-specific genetic sequences, a rapid and cost-effective colorimetric test utilizing AuNPs was developed. When RNA samples from COVID-19-infected individuals were combined with the RdRp oligo probe, the gold colloid exhibited a color shift from pink to blue, whereas the absence of hybridization preserved the pink hue. The colorimetric evaluation demonstrated a sensitivity of 85.29% and a specificity of 94.12%, enabling an affordable and efficient COVID-19 detection within approximately 30 minutes [50,51]. Pramanik et al. developed a novel spike protein-targeting antibody conjugated with gold nanoparticles for rapid SARS-CoV-2 detection and inhibition. A rapid visual color shift occurring in under five minutes highlights the effectiveness and efficiency of anti-spike antibody-conjugated AuNPs in identifying SARS-CoV-2 within human specimens. Antibody-conjugated AuNPs demonstrated complete viral suppression efficacy and hold significant potential for both the prevention and rapid detection of COVID-19 [52]. A nanosensor system was designed using streptavidin-assisted binding of epitopes to AuNPs, enabling the detection of SARS-CoV-2 IgG antibodies in human specimens [53].

Alternative diagnostic approaches utilize carbon nanomaterials, particularly for SARS-CoV-2 detection. Researchers developed a semiconductor-grade single-walled carbon nanotube field-effect transistor, integrating it with SARS-CoV-2 antibodies to identify viral antigens in under five minutes. In one approach, a SARS-CoV-2 spike protein-targeting antibody (SAb) detected the S antigen (SAg), while a nucleocapsid protein-specific antibody identified the N antigen (NAg) [54]. A field-effect transistor (FET) utilizing modified graphene oxide and coated with SARS-CoV-2-specific monoclonal antibodies (mAbs) achieved an ultra-low detection threshold of 0.002 fM, representing a major advancement. As a result, even trace amounts of SARS-CoV-2 can be identified [55].

In summary, the diagnostic techniques using biogenic nanomaterials against COVID-19 have been of great importance in the management of the pandemic, as they provide a rapid, accurate, and sensitive diagnosis. It is expected that future virus detection and healthcare development will benefit from further studies and innovations in this field.

6.   Biogenic Nanomaterials for COVID-19: Future Perspectives and Research Opportunities

The COVID-19 pandemic has exposed weaknesses and subjected the world's healthcare systems to an unprecedented and challenging test, therefore raising the great need for the creation of strong public health infrastructure. As a result of the rapid spread of the virus, resources and staff have been stretched, making it extremely difficult for healthcare practitioners to manage large numbers of patients. In reaction to this catastrophe, there was a pressing need to cure people who were afflicted as well as to stop the disease's spread by implementing efficient preventative measures. The development of efficient treatment techniques and the prevention of transmission are two needs that have greatly boosted interest in the study and use of nanomaterials.

6.1 Biogenic Nanomaterials' use in Combating COVID-19

In the battle against COVID-19, biogenic nanomaterials—which are nanoscale engineering produced from natural sources—have shown to be an effective weapon. They are indispensable in a number of crucial fields, such as surface disinfectants, drug delivery systems, diagnostic testing, and vaccine development, due to their special qualities and broad range of uses. Here is a more thorough examination of their application.

Development of Vaccines: The most thrilling application of biogenic nanomaterials is probably in creating vaccines. These agents can act as adjuvants or carrier systems to greatly improve the efficacy of vaccine formulations. Biogenic nanomaterials help stabilize antigens by encapsulating them in nanoparticles, therefore ensuring that these components are active and efficient upon delivery into the immune system. By enhancing the delivery of antigens to antigen-presenting cells, nanoparticles can increase the immune response. More neutralizing antibodies are formed due to the more effective stimulation of T cells and B cells. The specificity of the immune response toward SARS-CoV-2—the virus behind COVID-19—can be enhanced by using tailoring in the size and surface properties of the nanoparticles to ensure targeted delivery. Several studies have proven that biogenic nanoparticle-based vaccines may bring improved safety profiles and longer-lasting protection, thus making them a critical tool in the continued development of powerful COVID-19 vaccines.

Diagnostic Examinations: Moreover, biogenic nanomaterials play a very crucial role in developing diagnostic tests for COVID-19. Researchers have harnessed their properties to come up with rapid and accurate diagnostic methods that lead to the effective identification of SARS-CoV-2. For instance, assays capable of detecting viral particles or even specific viral RNA sequences with high specificity and sensitivity have been developed using biosensors based on gold or silver nanoparticles. The advantages of nanodots, a type of fluorescent nanoparticles, give efficient results in viral detection through visually readable colorimetric or luminous changes. Incorporating biogenic nanomaterials in the diagnostic tools offers faster findings than traditional techniques and is of great essence in controlling the spread of outbreaks for the timely implementation of public health measures. All this enhances our ability to track and handle the spread of COVID-19 effectively.

Drug Delivery Systems: Biogenic nanomaterials hold great promise as parts of specialized drug delivery systems for the treatment of COVID-19. These nanomaterials can encapsulate antiviral drugs, ensuring that they reach their target tissues or cells. It is possible to design nanoparticles that will interact preferentially with certain cell types by modulating their surface properties, enhancing the accuracy of treatment. The therapeutic benefits of antiviral treatments can be maximized by directly delivering them to lung cells using biogenic nanomaterials; this will allow the lowering of doses and reduce systemic exposure. This small, focused strategy is valuable in treating COVID-19, especially for patients with co-occurring conditions that may be more prone to the negative side effects of traditional medications

Disinfectants for surfaces: By acting as antimicrobial surface disinfectants, biogenic nanomaterials have demonstrated efficacy in stopping the spread of COVID-19. Silver and zinc oxide nanoparticles have strong antibacterial qualities that allow them to destroy a variety of diseases, including the SARS-CoV-2 virus. The danger of viral transmission can be decreased in public areas, medical institutions, and personal protective equipment (PPE) by adding these nanoparticles to surface coatings or cleaning solutions. Because biogenic nanoparticles have long-lasting effects, surfaces continue to offer protection for lengthy periods of time after application. This is increasing their efficacy, and reduce their adverse effects. Researchers can improve the especially important in high-touch locations where there is a higher chance of disease transmission, such hospital rooms, transportation networks, and public spaces.

6.2   Future Perspectives

The significance of public health initiatives is becoming more and more clear in the post-COVID-19 world as communities start to rebuild and recover from the severe effects of the epidemic. The crises' lessons have brought to light the pressing need for creative solutions that can successfully handle new health risks. Because of this need, research into biogenic nanomaterials which have special qualities and uses that might completely transform healthcare procedures must continue. In this particular context:

Methods of Sustainable Production: Creating sustainable processes for producing nanomaterials derived from biological sources is crucial.

Clinical Trials: To assess the effectiveness of biogenic nanomaterials in the therapy of COVID-19, more clinical studies are required.

Law and Regulations: It is necessary to develop laws and a regulatory framework pertaining to the effectiveness and safety of novel goods in this industry.

6.3   Research Opportunities

Biogenic nanomaterials can be used in the following research areas:

Novel Vaccine Compositions: Examination of the possible use of nanoparticles derived from various biogenic sources in vaccine formulations.

Applications for Nicrosate: Examining and evaluating the antiviral properties of biogenic nanomaterials against SARS-CoV-2.

Development of Biosensors: Biosensor design and optimization for quicker and more accurate COVID-19 diagnostic testing.

7.   Conclusion

Global health systems have been profoundly affected by the COVID-19 epidemic, which has also fundamentally changed social and economic institutions. The critical need for new methods and tools in health research has been well illustrated by this process. In this regard, interest in the potential of biogenic nanomaterials is growing. These naturally occurring materials with nanoscale structure stand out for their potential medical applications and ecologically favorable qualities.

Biogenic nanomaterials have played an important part in the fight against COVID-19 through their wide use across different fields such as vaccine, viral detection and surface sterilization. Because of its antibacterial, antiviral and antifungal properties, it is an amazing alternative for healthcare uses. Gold and silver nanoparticles, for instance, make for good adjuvants in vaccine formulations as well as biosensors that facilitate faster diagnosis with high sensitivity. This upgrades the breaking down ability of our immune system to pathogenic microorganisms and also makes vaccines more potent.

The diagnostic tests for rapid confirmation of the COVID-19 effectively rely upon utilizing biogenic nanomaterials and contribute to fulfilling a lead time between required clinical treatments during pandemic outbreaks. The answers these nanomaterials offer are of paramount importance as traditional methods for testing blood infection often require delays, more laboratory resources and only making it possible to pick up infections with limited sensitivity. Since the interactive biosensors and colorimetric tests provide a rapid COVID-19 detection that allows immediate action.

Now, biogenic nanoparticles also present excellent prospects for green manufacturing approaches. Because these are natural resources, they have a better footprint in the environment. These nanoparticles are currently being investigated for more ecologically friendly manufacturing methods. For example, making biologically derived nanomaterials less environmentally harmful to return them more into favor. For industrial and agricultural uses this has major advantages, apart from just health.

More research on clinical efficacy and safety is necessary so as to improve the notion of these biogenic nanomaterials for health. It is important to note that clinical studies are needed to examine the impact of these materials on human health and possible adverse outcomes. Also importantly, identifying the potential of biogenic nanomaterials in developing therapies against distinct diseases may be crucial to steer further investigation into this regard.

In conclusion, the COVID-19 pandemic has placed an impactful focus again on biogenic nanomaterials in health. These are great opportunities for generating brand new solutions and exploring completely uncharted work paths. These studies, combined with new evidence of clinical efficacy and safety as well as sustainable production practices are the frontier to translating biogenic nanoparticles into effective tools against COVID-19 and other pandemics. This will benefit long-term health development and the creation of healthier, more durable societies.

References

[1]   Watkins, K. (2018). Emerging Infectious Diseases: a Review. *Current Emergency and Hospital Medicine Reports*. https://doi.org/10.1007/s40138-018-0162-9

[2]   Sivasankarapillai, V.S. *et al.* (2020). On Facing the SARS-CoV-2 (COVID-19) with Combination of Nanomaterials and Medicine: Possible Strategies and First Challenges. *Nanomaterials*. https://doi.org/10.3390/nano10050852

[3]   Velavan, T.P. and Meyer, C.G. (2020). The COVID-19 epidemic. *Tropical Medicine & International Health*. https://doi.org/10.1111/tmi.13383

[4]   Wang, D. *et al.* (2020). Clinical Characteristics of 138 Hospitalized Patients With 2019 Novel Coronavirus–Infected Pneumonia in Wuhan, China. *JAMA*. https://doi.org/10.1001/jama.2020.1585

[5]   Klemeš, J.J. *et al.* (2020). Minimising the present and future plastic waste, energy and environmental footprints related to COVID-19. *Renewable and Sustainable Energy Reviews*. https://doi.org/10.1016/j.rser.2020.109883

[6]   Zamora-Ledezma, C. *et al.* (2020). Biomedical Science to Tackle the COVID-19 Pandemic: Current Status and Future Perspectives. *Molecules*. https://doi.org/10.3390/molecules25204620

[7]   Liu, Z. *et al.* (2009). Carbon nanotubes in biology and medicine: In vitro and in vivo detection,

imaging and drug delivery. *Nano Research*. https://doi.org/10.1007/s12274-009-9009-8

[8] Hojjati-Najafabadi, A. *et al.* (2021). A Tramadol Drug Electrochemical Sensor Amplified by Biosynthesized Au Nanoparticle Using Mentha aquatic Extract and Ionic Liquid. *Topics in Catalysis*. https://doi.org/10.1007/s11244-021-01498-x

[9] Şavk, A. *et al.* (2019). Highly monodisperse Pd-Ni nanoparticles supported on rGO as a rapid, sensitive, reusable and selective enzyme-free glucose sensor. *Scientific Reports*. https://doi.org/10.1038/s41598-019-55746-y

[10] Zhao, P. *et al.* (2019). A novel ultrasensitive electrochemical quercetin sensor based on MoS2 - carbon nanotube @ graphene oxide nanoribbons / HS-cyclodextrin / graphene quantum dots composite film. *Sensors and Actuators, B: Chemical*. https://doi.org/10.1016/j.snb.2019.126997

[11] Cherif, A. *et al.* (2021). Numerical investigation of hydrogen production via autothermal reforming of steam and methane over Ni/Al2O3 and Pt/Al2O3 patterned catalytic layers. *International Journal of Hydrogen Energy*. https://doi.org/10.1016/j.ijhydene.2021.04.032

[12] Ulus, R. *et al.* (2016). Functionalized Multi-Walled Carbon Nanotubes (f-MWCNT) as Highly Efficient and Reusable Heterogeneous Catalysts for the Synthesis of Acridinedione Derivatives. *ChemistrySelect*. https://doi.org/10.1002/SLCT.201600719

[13] Gulbagca, F. *et al.* (2021). Green Synthesis of Palladium Nanoparticles: Preparation, Characterization, and Investigation of Antioxidant, Antimicrobial, Anticancer, and DNA Cleavage Activities. *Applied Organometallic Chemistry*. https://doi.org/10.1002/aoc.6272

[14] Ertan, S. *et al.* (2012). Platinum nanocatalysts prepared with different surfactants for C1-C3 alcohol oxidations and their surface morphologies by AFM. *Journal of Nanoparticle Research*. https://doi.org/10.1007/S11051-012-0922-5

[15] Günbatar, S. *et al.* (2018). Carbon-nanotube-based rhodium nanoparticles as highly-active catalyst for hydrolytic dehydrogenation of dimethylamineborane at room temperature. *Journal of Colloid and Interface Science*. https://doi.org/10.1016/J.JCIS.2018.06.100

[16] Askari, M.B. *et al.* (2021). Enhanced electrochemical performance of MnNi2O4/rGO nanocomposite as pseudocapacitor electrode material and methanol electro-oxidation catalyst. *Nanotechnology*. https://doi.org/10.1088/1361-6528/abfded

[17] Zafar Razzacki, S. (2004). Integrated microsystems for controlled drug delivery. *Advanced Drug Delivery Reviews*. https://doi.org/10.1016/j.addr.2003.08.012

[18] Rudramurthy, G. *et al.* (2016). Nanoparticles: Alternatives Against Drug-Resistant Pathogenic Microbes. *Molecules*. https://doi.org/10.3390/molecules21070836

[19] Choi, Y.H. and Han, H.-K. (2018). Nanomedicines: current status and future perspectives in aspect of drug delivery and pharmacokinetics. *Journal of Pharmaceutical Investigation*. https://doi.org/10.1007/s40005-017-0370-4

[20] Aygun, A. *et al.* (2022). Highly active PdPt bimetallic nanoparticles synthesized by one-step bioreduction method: Characterizations, anticancer, antibacterial activities and evaluation of their catalytic effect for hydrogen generation. *International Journal of Hydrogen Energy*. https://doi.org/10.1016/J.IJHYDENE.2021.12.144

[21] Nesrin, K. *et al.* (2020). Biogenic silver nanoparticles synthesized from Rhododendron ponticum and their antibacterial, antibiofilm and cytotoxic activities. *Journal of Pharmaceutical and Biomedical Analysis*. https://doi.org/10.1016/J.JPBA.2019.112993

[22] Soares, S. *et al.* (2018). Nanomedicine: Principles, Properties, and Regulatory Issues. *Frontiers in Chemistry*. https://doi.org/10.3389/fchem.2018.00360

[23] Shin, M.D. *et al.* (2020). COVID-19 vaccine development and a potential nanomaterial path forward. *Nature Nanotechnology*. https://doi.org/10.1038/s41565-020-0737-y

[24] Abdelhamid, H.N. and Badr, G. (2021). Nanobiotechnology as a platform for the diagnosis of

COVID-19: a review. *Nanotechnology for Environmental Engineering.* https://doi.org/10.1007/s41204-021-00109-0

[25] West, R. *et al.* (2020). Applying principles of behaviour change to reduce SARS-CoV-2 transmission. *Nature Human Behaviour.* https://doi.org/10.1038/s41562-020-0887-9

[26] Ioannidis, J.P.A. (2020). Global perspective of COVID-19 epidemiology for a full-cycle pandemic. *European Journal of Clinical Investigation.* https://doi.org/10.1111/eci.13423

[27] Naveed, M. *et al.* (2021). Review of potential risk groups for coronavirus disease 2019 (COVID-19). *New Microbes and New Infections.* https://doi.org/10.1016/j.nmni.2021.100849

[28] Xu, C. *et al.* (2022). Nanotechnology for the management of COVID-19 during the pandemic and in the post-pandemic era. *National Science Review.* https://doi.org/10.1093/nsr/nwac124

[29] Saraswat, P. *et al.* (2023). Applications of bio-based nanomaterials in environment and agriculture: A review on recent progresses. *Hybrid Advances.* https://doi.org/10.1016/j.hybadv.2023.100097

[30] Duden, G.S. *et al.* (2022). Global impact of the COVID-19 pandemic on mental health services: A systematic review. *Journal of Psychiatric Research.* https://doi.org/10.1016/j.jpsychires.2022.08.013

[31] Chormey, D.S. *et al.* (2023). Biogenic synthesis of novel nanomaterials and their applications. *Nanoscale.*

[32] Kulkarni, D. *et al.* (2023). Biofabrication of nanoparticles: sources, synthesis, and biomedical applications. *Frontiers in Bioengineering and Biotechnology.* https://doi.org/10.3389/fbioe.2023.1159193

[33] Joseph, T.M. *et al.* (2023). Nanoparticles: Taking a Unique Position in Medicine. *Nanomaterials 2023, Vol. 13, Page 574.* https://doi.org/10.3390/NANO13030574

[34] Mariyam, S. *et al.* (2024). Nanotechnology, a frontier in agricultural science, a novel approach in abiotic stress management and convergence with new age medicine-A review. *Science of The Total Environment.* https://doi.org/10.1016/j.scitotenv.2023.169097

[35] Hsiao, T.-C. *et al.* (2020). COVID-19: An Aerosol's Point of View from Expiration to Transmission to Viral-mechanism. *Aerosol and Air Quality Research.* https://doi.org/10.4209/aaqr.2020.04.0154

[36] Warnes, S.L. *et al.* (2015). Human Coronavirus 229E Remains Infectious on Common Touch Surface Materials. *mBio.* https://doi.org/10.1128/mBio.01697-15

[37] Pelgrift, R.Y. and Friedman, A.J. (2013). Nanotechnology as a therapeutic tool to combat microbial resistance. *Advanced Drug Delivery Reviews.* https://doi.org/10.1016/j.addr.2013.07.011

[38] Cheng, Y. *et al.* (2011). Deep Penetration of a PDT Drug into Tumors by Noncovalent Drug-Gold Nanoparticle Conjugates. *Journal of the American Chemical Society.* https://doi.org/10.1021/ja108846h

[39] Balagna, C. *et al.* (2020). Virucidal effect against coronavirus SARS-CoV-2 of a silver nanocluster/silica composite sputtered coating. *Open Ceramics.* https://doi.org/10.1016/j.oceram.2020.100006

[40] Chen, Y.-N. *et al.* (2016). Antiviral Activity of Graphene–Silver Nanocomposites against Non-Enveloped and Enveloped Viruses. *International Journal of Environmental Research and Public Health.* https://doi.org/10.3390/ijerph13040430

[41] Itani, R. *et al.* (2020). Optimizing use of theranostic nanoparticles as a life-saving strategy for treating COVID-19 patients. *Theranostics.* https://doi.org/10.7150/thno.46691

[42] Alphandéry, E. (2020). The Potential of Various Nanotechnologies for Coronavirus Diagnosis/Treatment Highlighted through a Literature Analysis. *Bioconjugate Chemistry.* https://doi.org/10.1021/acs.bioconjchem.0c00287

[43] Pereira, L.F.T. *et al.* (2024). Advanced biopolymeric materials and nanosystems for RNA/DNA vaccines: a review. *Nanomedicine.* https://doi.org/10.1080/17435889.2024.2382077

[44] Mujawar, M.A. *et al.* (2020). Nano-enabled biosensing systems for intelligent healthcare: towards COVID-19 management. *Materials Today Chemistry.* https://doi.org/10.1016/j.mtchem.2020.100306

[45] Draz, M.S. and Shafiee, H. (2018). Applications of gold nanoparticles in virus detection. *Theranostics*. https://doi.org/10.7150/thno.23856

[46] Srivastava, M. *et al.* (2021). Prospects of nanomaterials-enabled biosensors for COVID-19 detection. *Science of The Total Environment*. https://doi.org/10.1016/j.scitotenv.2020.142363

[47] Chen, Y. and Ming, H. (2012). Review of surface plasmon resonance and localized surface plasmon resonance sensor. *Photonic Sensors*. https://doi.org/10.1007/s13320-011-0051-2

[48] Qiu, G. *et al.* (2020). Dual-Functional Plasmonic Photothermal Biosensors for Highly Accurate Severe Acute Respiratory Syndrome Coronavirus 2 Detection. *ACS Nano*. https://doi.org/10.1021/acsnano.0c02439

[49] Moitra, P. *et al.* (2020). Selective Naked-Eye Detection of SARS-CoV-2 Mediated by N Gene Targeted Antisense Oligonucleotide Capped Plasmonic Nanoparticles. *ACS Nano*. https://doi.org/10.1021/acsnano.0c03822

[50] Kumar, V. *et al.* (2022). Development of RNA-Based Assay for Rapid Detection of SARS-CoV-2 in Clinical Samples. *Intervirology*. https://doi.org/10.1159/000522337

[51] Corman, V.M. *et al.* (2020). Detection of 2019 novel coronavirus (2019-nCoV) by real-time RT-PCR. *Eurosurveillance*.

[52] Pramanik, A. *et al.* (2021). The rapid diagnosis and effective inhibition of coronavirus using spike antibody attached gold nanoparticles. *Nanoscale Advances*. https://doi.org/10.1039/D0NA01007C

[53] Amrun, S.N. *et al.* (2020). Linear B-cell epitopes in the spike and nucleocapsid proteins as markers of SARS-CoV-2 exposure and disease severity. *EBioMedicine*. https://doi.org/10.1016/j.ebiom.2020.102911

[54] Shao, W. *et al.* (2021). Rapid Detection of SARS-CoV-2 Antigens Using High-Purity Semiconducting Single-Walled Carbon Nanotube-Based Field-Effect Transistors. *ACS Applied Materials & Interfaces*. https://doi.org/10.1021/acsami.0c22589

[55] Krsihna, B.V. *et al.* (2022). Design and Development of Graphene FET Biosensor for the Detection of SARS-CoV-2. *Silicon*. https://doi.org/10.1007/s12633-021-01372-1

Biogenic Nanomaterials: Synthesis, Characterization, Applications, and Future Remarks   Materials Research Forum LLC
Materials Research Foundations 180 (2025)   https://doi.org/21741/9781644903759

Chapter 23

Biogenic Nanomaterials-Induced Ecotoxicity

M. Ermaya[1*], H. Demir[2], C. Demir[3], A. Ozengül[4], F. Sen[4*]

[1]Van Regional Training and Research Hospital, Van, Turkiye.

[2]Van Yuzuncu Yıl University, Faculty of Science, Department of Biochemistry, Van, Turkiye.

[3]Van Yuzuncu Yil University, Vocational School of Health Care, Van, Turkiye.

[4]Sen Research Group, Biochemistry Department, Faculty of Arts and Science, Dumlupınar University, Evliya Celebi Campus, 43100, Kutahya, Turkiye

masallahermaya65@gmail.com; dpu.edu.tr

Abstract

Nanotechnology has become one of the most innovative and exciting areas of science and technology; it is extensively applied in physicochemical and biomedical fields. Nanoparticles have properties such as conductivity, large surface areas, strength, durability and reactivity. These properties influence the biological and toxicological properties of nanoparticles. Nanoparticles can easily pass through cell membranes, and engage in interaction with organs, organelles and genetic material. Nanoparticles can cause many different problems such as inflammation, genotoxicity, cytotoxicity, apoptosis and oxidative stress. The biologically toxic effects of nanoparticles in living organisms were examined using parameters such as Chromosome Abnormalities, Sister Chromatid Exchange, Micronucleus.

Keywords

Nanoparticle, Ecotoxicity, Toxic Effects, Biotechnology, Chromosome Abnormality

1. Introduction

Nanotechnology, a name we often hear today, and which conforms a new department, is developing rapidly. It does not go unnoticed that nanotechnology has spread to energy, production, health, medicine, sensor, environment, defense and many other fields [1–10]. Since nanoparticles have various properties, their usage areas are quite wide. In the medical space, nanoparticles are used in many radiological fields such as biomedical targeted technologies, computerized tomography, magnetic resonance, fluorescence and ultrasound to provide more specific analysis, molecular imaging, targeted therapy, drug and vaccine development [1]. Besides the such general and useful areas of use, it has become a requirement to scientifically examine the toxical efficacy of nanoparticles, likely which may occur on the respiratory system, nervous system, blood, gastrointestinal system and skin, due to their molecular properties [11]. Natural disasters such as forest fires, desert dust, volcanic eruptions and soil erosions, which are among the ways nanoparticles are released into nature, occur. Desert dust covers 50% of the nanoparticle particles released within the atmosphere. Again, as a result of a single volcanic eruption, heavy metals (chromium, lead, mercury, cadmium), which are known to be quite toxic, spread to the ecological system. Forest fires are similarly quite effective in the spread of nanoparticles [12,13]. Nanomaterials are antimicrobial substances similar to antibiotics also have recently prevented the growth of many pathogenic bacteria. Their reluctance to pathogenic bacteria has become an progressively considerable sanitary trouble mondial [14,15]. This problem reduces the ability of existing antibiotics to treat ailments caused by bacterial contamination [15–17]. Unluckily, this results in increased healthcare costs, hospital stays moreover annihilation proportions [18]. In accordance with, the outcome of a investigate report, it is conjectural that an medium of 300 million people will die matutinal from contagions reason by reluctance by 2050, moreover this will create an further freight of 100 trillion dollars on the national economies [17]. Also, because of the lateral influences of chemically or synthetically manufactured antibiotics, it has divulged the requirement to discover new arranged alternative infiltrators that are efficient on pathogenic bacteria, falling cost moreover have less toxic influences on human cells [18–20]. They are magnetic nanoparticles that are mostly used in biomedical applied areas. This has led to incremented exposure of humans moreover their environment to magnetic nanoparticles (MNP). Even if the aim of the application of magnetic nanoparticles is tissue, liver and spleen, other organs must also be evaluated in terms of toxicity. Therefore, there is a great need to research and analyze MNP. Potential impacts, both negative and beneficial, on human health and the ecological system are important considerations [21]. With the developing of nanotechnology, thousands of products are developed and offered on market shelves. As buyers, we purchase these products and as we use them, we begin to hear the negative effects. Generally, the test periods of these products are ignored. However, even if these nanotechnological products are not purchased with money, the nanoparticles released into the environment during their production enter our bodies either through breathing or by consuming contaminated food and beverages. Nanotechnology laboratory waste is generally transmitted to water, then to soil, and from there to our food [22]. The distribution of nanoparticles in the ecological system varies. However, one of the risks that may occur is that the proteins in the living content change their properties. Differences that occur in the living body when risks occur (such as redness, swelling, inflammation) cause effects that can damage cells and tissues. In addition, when exposed to living organisms for a long time, it will lead to the possibility of accumulation of nanoparticles in tissues and organs. This will cause some risks to arise in the coming years. Therefore, in addition to the tests that need to be performed before use, it is important to develop and implement various tests. In fact, biomaterials that will be used in living tissues come into use after undergoing certain tests. However, despite the tests performed, biomaterials containing nanoparticles have allergic, inflammatory, mutagenic and carcinogenic effects on the body moreover their effects on the immune system occur after a while [12,22]. Such situations should not be ignored, and toxicity studies in terms of whether nanoparticles have allergic reaction properties against tissues, biological compatibility and genetic effects should be subjected to longer-term tests and use should be started. Therefore, the amount of toxicity tests for the harmful effects that metal nanoparticles, which are widely used in the space of health, may cause cell poisoning on living things and genetic structure, is increasing day by day. Different nanoparticles, including metals, have been observed to have chemotherapy and genotoxic impacts in some tests on humans [23]. Thus, when the effects of nanoparticles on living things are examined more comprehensively, it will be easier to take some precautions against the risks that may arise. It is very significant to investigate the effects of nanotechnological products used in today's world. While nanotechnology is seen as the technology of the future, health should not be ignored. There are high amounts of nanoparticles in the air, and if people inhale them, the risk of cardiovascular diseases increases. Cardiac arrhythmia is the most common heart-related finding.

2. Nanoparticles

The word "nanoparticle" is applied to describe materials smaller than 100 nm in size, which are applied as building blocks in nano research [24]. Nanoparticles are small in size, contain a large amount of binding sites, and have various physicochemical and morphological properties [25]. Just as the scales and shapes of nanoparticles can be different, the physical, chemical, optical, electrical, thermal, catalytic, bioactivity moreover toxicity properties of the particles are moreover different. on account of prevent particle growth, nanoparticles can be produced at the desired scale by using separating agents. Additionally to these; nanoparticles comprise different particles may have different catalytic, magnetic, optical properties as nanoparticles occur a single particle [26,27]. Since about 40-50% of nanoparticle atoms are on the shallow, its reactivity is elevated [28]. These properties enhance the significance of nanoparticles compared to other ingredients. Metal nanoparticles can be easily synthesized. It has a wide usage area in industrial products, military field, machinery industries, consumer products and especially in the field of medicine, as it can make differences in its chemical structure [7,29]. Iron oxide, Zinc oxide, Cobalt oxide, Titanium dioxide, Silver, Platinum, Gold, Nickel, Copper and Carbon are commonly used nanoparticles [23]. Nanotechnology is a new research field that develops and produces new materials, known as nanoparticles (NP) or nanomaterials (NM), and has different applications in science and technology [30,31]. This technology designs ordinary chemicals or materials at the nanoscale and gives these materials new and unique properties. The features gained include very small size, very large surface area, ultraviolet (UV) protection, antimicrobial effect, increased power, flexibility and conductivity [32]. Nanoparticles are particles or atomic aggregates varying in scale from 1 to 100 nm, as described by the European Commission [33,34]. Based on their physicochemical properties, NPs are classified as organic (dendrimer, liposome), inorganic and carbon (fullerene, graphene, carbon nanotubes) based. Inorganic NPs are also examined in two separate subgroups: metal (aluminum, cadmium) and metal oxide (iron oxide, silicon dioxide, aluminum oxide) [35,36]. Nanomaterials attract great attention in many spaces such as medicine, pharmacy, electronics, military, agriculture, textile moreover energy because of their new physicochemical properties at nanometric size [37,38]. As nanoparticles become smaller in size, their surface area increases significantly compared to larger particles. More importantly, there is an increase in the ratio of molecules or atoms on the particle surface and therefore in their chemical reactivity. These features, which make them preferred in many areas, are also of great importance in their interactions with biological systems. Physicochemical properties of nanoparticles in addition to nanomaterials, such as scale, shape, chemical composition, physicochemical stability, crystal structure, surface area and surface energy, at large affect the toxicity levels of these nanomaterials [39]. For example, silver in nanoparticle form releases more silver ions (Ag^+) than microparticles of the same weight. These released ions enable silver to produce an antibacterial effect [40]. Because they show high antimicrobial activity, silver nanoparticles are used in many areas such as health, food, cosmetics, textiles in addition electronics [41]. Silver nanoparticles produce reactive oxygen species (ROS) that react with microbial membranes, harm their structures in addition inactivate bacteria. If the level of these reactive species exceeds the antioxidant capacity, they cause toxic effects by interacting with the lipid, protein, DNA and enzyme-like structures of the cells. For example, the oxidative destruction of polyunsaturated fatty acids, known as lipid peroxidation, is quite damaging to cells [42]. In some studies, it has been observed that 5, 10 in addition 20 nm Ag NPs are more cytotoxic in addition genotoxic than >40 nm particles [43]. Similarly, TiO2 NPs cause weakening of antioxidant defense systems in eukaryotic organisms, resulting in regional inflammation, mitochondrial damage, autophagy, apoptosis or necrosis [44]. Many studies conducted with Ag, Cu, ZnO and TiO$_2$ nanoparticles have confirmed that they can cause cytotoxic and genotoxic effects [45,46]. Due to the developments in nanotechnology moreover the increasingly common usage of nanoparticles in this technology, the exposure of both humans and all other living things to these particles has begun to increase because of the production, usage and waste of nanoparticles. Therefore, in recent years, concerns have begun to rise about the toxic effects of these particles, especially their genotoxic effects [47,48]. The genotoxic effects of nanoparticles are particularly significant on in vitro human lymphocytes, [48,49], lung (A549) [50,51], liver (HepG2) [52,53], glial (A172) [53,54] and neurons (SH- in different cell lines such as SY5Y) [55] Sprague-Dawley rat, Wistar rat [56,57], etc. in different mammal species; Chromosome abnormalities in different plant varieties such as *Allium cepa* L., *Allium sativum, Glycine max, Triticum aestivum* L., *Lens culinaris* L. and *Oryza sativa* [58,59] can be detected by sister chromatid exchange, micronucleus, *Allium* and comet test, as well as bacterial regression. It has begun to be investigated with different test methods such as mutation test-Ames test [56,60] and gamma-H2AX test [56,61]. In recent years, micronucleus (MN) and comet (single cell gel electrophoresis-SCGE) tests have been used to examine the genotoxic effects of nanoparticles. In addition, chromosomal abnormalities (KA) and sister chromatid exchange (SCE) tests are among the frequently used tests. These tests have been performed on human lymphocytes as well as on Chinese hamster ovary cells [62], HT22 cells [59], liver [63,64], blood or bone marrow cells [65] in mice, rats or rabbits, [43,66,67]. The genotoxic effects of nanoparticles on plants are widely used, especially in order to determine their effects on the ecosystem. In this area, *Allium cepa, Vicia faba*, Nicotiana tabacum and *Zea mays* are the most preferred varieties [68]. It is known that more than 70% of the bacteria that cause infection are resistant to the antimicrobial substances used. In this sense, the improvement of new moreover effective antimicrobial agents is of great significiant. Metal NPs such as copper, titanium, silver, gold and zinc each have different antimicrobial activity properties [19,69]. NPs exhibit their antimicrobial activities through the induction of oxidative stress in addition the release of metal ions [70] as well as the formation of hydrogen peroxide from the surface of NPs [71]. It is claimed that AgNPs exhibit antibacterial activity by releasing silver ions in addition inhibiting respiratory enzymes, moreover produce excessive reactive oxygen species (ROS) that damage the cell membrane and inactivate cellular enzymes [72,73]. The antimicrobial mechanism of nanoparticles moreover depends on the composition of the bacterial cell wall [74]. It has been described that AgNPs increase membrane permeability while causing membrane damage by binding and accumulating to the cell membrane of bacteria [75]. The degree of permeability of the bacterial cell membrane depends on the scale in addition concentration of NPs. This is because the small particle size easily penetrates the cell membrane in addition also interacts with cell organelles such as ribosomes, helping to cause cell loss. In a study, they showed that E. coli cells affected DNA replication after treatment with AgNPs and explained that NPs had an effect on the DNA polymerase enzyme [76]. In a study conducted by a researcher named Cha and her colleagues, they explained the synthesis of copper NP (CuNP) of casein, which has effective antibacterial activity opposite pathogenic bacteria [77]. The antibacterial activity of CuNPs and inhibition zone analysis were determined using the antibiotics ofloxacin moreover kanamycin as standards. CuNPs exhibited a good inhibition zone, almost equal to the inhibition zone generated by standard antibiotics; Thus, it can be understood that CuNPs show a similar effect as standard antibiotics. As explained above, the mechanism by which kanamycin exhibits its antibacterial activity is through its ability to inhibit enzymes participate in protein synthesis. Similarly, it can be claimed that CuNPs exhibit antibacterial activity through enzyme inhibition [77]. These studies show that NPs directly or indirectly inhibit cellular enzymes and also result in antibacterial effects.

2.1　Influences of Nanoparticles on Various Human Cells

Recent researches on the genotoxic influences of nanoparticles on human lymphocytes have revealed that these particles can cause DNA damage and differences in cell kinetic parameters. Lately researches on the genotoxic influences of nanoparticles on human lymphocytes have revealed that these particles can cause DNA damage and differences in cell kinetic parameters. Many of nanoparticles at various concentrations and at various scales have been shown to cause chromosomal abnormalities, sister chromatid exchange, micronuclei and DNA damage [78]. Copper oxide nanoparticles (CuO NP) are particles used in many different consumer products, especially in medicine, engineering and technology, like antibacterial fabric production and prevention of infections, due to their conductivity and biocidal character [48,79]. Due to their frequent use, Researches have been execute on the cytotoxic and genotoxic influences of these particles in diverse species and cell lines. As a result of the application of 25, 50, 75 in addition 100 µg/mL condensation of CuO nanoparticles application to human lymphocytes and the examination of the damages by the comet test, it was confirmed that there was an enhancement in primary DNA destruction and, as a result, substantial, increases in the comet tail density in addition tail length [54]. In a study of various CuO nanoparticles (powder, spherical, rod and needle) were applied to lymphocytes, DNA chain breaks and with oxidative destruction were encountered as a result of the comet test. It was determined that primary DNA damage increased in a dosage-addicted manner. Similarly, it was confirmed that it increased the number of micronucleated binucleate cells in the MN test and decreased cell viability in the MTT (3-(4,5-dimethylthiazol-2-yl)-2,5-diphenyltetrazolium bromide) test [23]. It has been reported that CuO nanoparticles at 0.1, 0.5, 1, 2 and 5 mM in concentrations cause cell death in human lymphocytes and this death is caused by oxidative stress caused by nanoparticles. It has been explained that these particles can suppress the immune system by causing death in lymphocytes in humans too, as they cause toxic effects in mitochondria and lysosomes, which are very important organelles for the cell [48]. A different study revealed that the genotoxic in addition cytotoxic effects of nanoparticles with oxidative destruction in human blood cells increased depending on the dose used [78]. Concentrations of 1-10 mM of Fe_2O_3 nanoparticles have been confirmed to considerable diminish cell viability in a dosage-hooked attitude [80]. After iron, copper and aluminum, the most used metal is "zinc (Zn)". It has been reported that around 10 million tons of zinc is consumed annually in the world [81]. Zinc oxide nanoparticles (ZnO NP) are secondhand in quite a variety of spaces, such as sunscreens, cosmetics, coloring, rubber, fabric dyeing, and wastewater treatment [82]. Zinc oxide nanoparticles; Have started to attract attention due to its antibacterial, antifungal, anti-inflammatory properties. These nanoparticles are used in photocatalysis, compound materials, chemical, gas, steam and moisture sensors and dye-sensitized solar cells [83]. Therefore, people are subjected to high levels of these nanoparticles in diurnal vita. The increasing consumption of ZnO nanoparticles may cause problems in terms of human being sanitary [84]. It has been confirmed that ZnO nanoparticles produce genotoxic effects even at the lowest concentrations in human peripheral blood lymphocyte cells in additon cause cytotoxicity in lymphocytes [85]. In a study in which zinc oxide (ZnO) moreover Titanium dioxide (TiO$_2$) nanoparticles were applied together, it was explained that condensation of 0, 12. 5, 25, 50, 100 and 125 µg/mL caused many chromatid-type abnormalities in peripheral lymphocytes, mainly by chromatid fracture. In a study in which rod and spherical forms were treated with human being peripheral blood mononuclear cells, it was reported that rod ZnO nanoparticles produced more reactive oxygen species than spherical. Similarly, the rod form has been shown to cause more severe DNA damage by cytokinesis block micronucleus and comet test [86]. When 50, 100, 250 and 500 ppm concentrations of ZnO nanoparticles were evaluated in terms of cytotoxicity and genotoxicity in vitro, 250 ppm concentration was cytotoxic for erythrocytes and superoxide dismutase (SOD), catalase (CAT) in addition ROTs were dose-dependent. increased [87]. In a separate study, it was investigated whether TiO$_2$ and ZnO nanoparticles influence both diverse human immune cells moreover the manufacture of exosomes, which are nanoscale vesicles that have an important feign in cell-to-cell transmission. As a result, it was determined that diverse condensations of TiO$_2$ or ZnO nanoparticles did'nt influence the viability of premier human being peripheral blood mononuclear cells (PBMC). Compare, it was watched that monocyte-derived dendritic cells (MDDC) did not respond to TiO$_2$ nanoparticles, but showed a dosage-addicted response in cell decese and caspase efficiency against ZnO nanoparticles. On the other hand, it has been announcement that TiO$_2$ or ZnO nanoparticles cannot be detected in exosomes and cannot be associated with exosome formation [88]. Silver nanoparticles (AgNP); Since they are frequently used as additives in creams, batteries, packaging packages, paints and industrial products, they have the potential to carry a genotoxic risk in terms of human health. In a study, it was explained that AgNPs exhibited genotoxic effects in human being peripheral blood mononuclear cells. These nanoparticles have been reported to increase the manufacture of reactive oxygen species in ddition the destruction of mitochondrial membranes [41]. Lymphocytes and the human being T-cell acute lymphoblastic leukemia cell line (HPB-ALL) were exposed to AgNPs for 24 hours. As a result of the MTT test used to evaluate cell proliferation, it was confirmed that AgNPs can cause cytotoxic effects in human lymphocytes [89]. Human being TK6 cells were cured with 5 nM AgNPs and silver nitrate (AgNO$_3$) to assess the occurrence of genotoxicity and oxidative stress. As a result of the micronucleus test, it was revealed that both particles were cytotoxic and genotoxic at similar concentrations (1.00 and 1.75 µg/mL), in addition both caused oxidative stress by increasing gene expression in addition reactive oxygen species in cells [90]. AgNPs of 90-180 nm in size and 25 µg/mL have been announcement to induce apoptosis moreover DNA strand breaks in human being lymphocytes [91]. In a study in which silver nanoparticles were synthesized via the bioactive fraction of Pinus roxburghii needles using a simple, cost-effective moreover environmentally comradely green chemistry technique, PNb-AgNPs did not exert a toxic effect on human breast epithelial cells (fR2) in addition human being peripheral blood lymphocytes (PBL). It exhibited a significant cytotoxic effect against A549 and prostatic minor cell carcinomas (PC-3) [44]. On the other hand, in a study in which the cytotoxicity and genotoxicity of 100 µg/mL dose of citrate coated colloidal AgNPs (30 nm) were evaluated by MTT test and Komet test, respectively, it was reported that silver nanoparticles did not cause poisonous influences on human being keratinocytes [92]. It has been observed that Cobalt-chromium nanoparticles may also have toxic effects in human cells. Human lymphocyte cells treated with cobalt showed increased free radical formation, DNA destruction, and aneuploidy occurrence [93]. It has been reported that nanoparticles activate caspase reactions and trigger cell death by increasing ROS in vitro and in vivo [94]. It has been confirmed that cobalt (II, III) oxide (Co_3O_4) nanoparticles increase cell membrane damage while decreasing cell viability. Chromosomal aberrations and mitochondrial destruction were seen at a concentration of 100 µg/mL [95]. Cobalt ferrite nanoparticles ($CoFe2O4$) enhancement the genesis of ROS at a concentration of 8 µg/mL. Similarly, necrotic cell death was found to increase in a condensation-addiction attitude [96]. Metal nanoparticles (MNPs) display unparalleled features in terms of optical, magnetic moreover electrical efficiency [97,98]. Inorganic nanoparticles; They are promising materials for applications in drug/gene to submit, cell imaging, biosensing moreover cancer therapy [95]. Therefore, it is seen that more detailed studies are needed on the genotoxic effects of these particles. It has been determined that silica-silicon dioxide (SiO_2) nanoparticles cause chromosomal abnormalities, sister chromatid alter and micronucleus genesis in human being peripheral lymphocytes [49]. In human peripheral leukocytes, all nanoparticles at a dosage of 100 µg/mL have been found to produce a important cytotoxic influence, depending on the nanoparticle concentration and scale [99]. In a separate study, it was explained that SiO_2 nanoparticles did not affect the proliferation of lymphocytes, but caused a decrease in the mitotic index value and a cytotoxic effect. It also exhibited genotoxic (DNA fracture) properties at high doses. Since the surface reactivity of SiO_2 nanoparticles can be changed by surface modification, it has a large range of implementation in medication distributing, gene therapy and molecular imaging [100]. Despite these benefits, it should be taken into account that human exposure to SiO2 nanoparticles can cause significant adverse health effects, and more detailed studies on these are required. TiO_2 nanoparticles are from an industrial point of view very important particles and have wide application areas. It is used in reducing the toxicity of dyes and pharmaceutical drugs, wastewater treatment, silkworm production, aerospace applications, and the food industry [101]. It has been reported that there is no increase in micronucleus frequency in human lymphocytes treated with TiO_2 nanoparticles, but there is an increase in DNA destruction in parallel with the increase in the application time [102]. As a result of the application of needle and spherical forms of TiO_2 nanoparticle to human peripheral lymphocytes in vitro, it was concluded that it is weakly cytotoxic, weakly genotoxic, weakly clastogenic and weakly mutagenic in terms of chromosome abnormalities, sister chromatid exchange and micronucleus test. TiO_2 nanoparticles applied to human lymphocyte cells caused a important decrease in mitochondrial dehydrogenase efficiency. DNA destruction and apoptosis increased with increasing dosage of TiO_2 nanoparticles. However, it has been reported that membrane wholeness is not influenced by nanoparticle therapy [103]. In a research using anatase-formed TiO_2 nanoparticles, short single-walled carbon nanotubes (SWCNTs), moreover short multi-walled carbon nanotubes (MWCNTs), human lymphocytes were mixed with nanomaterials of 6.25-300 mg/mL for 24, 48 and 72. hour treated. A dosage-addiction enhancement in chromosomal and chromatid-like abnormalities was seen at 48 hours [104]. Anatase, rutile, moreover a mixture of TiO_2 have been announcemented to induce DNA strand breaks moreover manufacture of reactive oxygen species in human being peripheral blood mononuclear cells [105]. There are also data to the contrary of these studies. For example, it has been announcemented that TiO_2 nanoparticles do not induce any genotoxicity in human being lung fibroblast (IMR-90) cells [86]. Because of the increasing use of TiO_2 nanoparticles in clinical applications, as in many fields, their potential risks on human being health need to be studious in detail. Tungsten (VI) oxide nanoparticles are used for many purposes in daily life, especially for electrochromic windows or smart windows, X-ray display in addition gas sensors. The genotoxic potential of tungsten ($WO3$) nanoparticles was investigated in enlightened human being lymphocytes using the micronucleus (MN) test in addition the comet test. insulated human being lymphocytes were subject to WO3 nanoparticles at varying condensations between 0-500 µM for 72 hours at 37°C. Treatment of 400 and 500 µM WO3 nanoparticles was found to cause minor increases in MN frequency in enlightened human being lymphocytes. in the same way, condensations of nanoparticles over 200 µM resulted in increased DNA destruction in lymphocytes. Due to its resistance to corrosion and water, tungsten carbide-cobalt (WC-Co) nanoparticles usaged in mining benches moreover chipless forming caused an increase in the number of micronucleated binucleate cells after 24 hours of application in the in vitro micronucleus test. has been found to occur [106].

2.2   Cytotoxic Effect Mechanism of Nanoparticles

Nanoparticles exert their cytotoxic effects by inducing apoptosis or cell cycle stopping. Nanoparticles enter cancerous cells through the nucleus moreover cause DNA fracture, which after all lie behind cell killing. In addition, the destruction of the glutathione and thioredoxin systems by the manufacture of ROS produces cytotoxic effects [107]. ROS can induce cellular destruction by interacting directly with biological molecules or indirectly by exerting oxidative stress, triggering cell killing through apoptosis or necrosis. Oxidative stress occurs as one of the centrical mechanisms of nanoparticle-originate cytotoxicity [108–110]. One study investigated the cellular answer to 1.4 nm triphenylphosphine monosulfonate titled AuNPs moreover found that cure with AuNPs was associated with incremented ROS manufacture in addition forfeit of mitochondrial potential, conclusion in necrotic cell killing [111]. The precise contraption by which AuNPs detent oxidative stress isn't fine figured out. However, it is thought to occur through the destruction of mitochondrial function as a result of high intracellular ROS. In a study, it was explained that a dosagee- addictedenhancement in intracellular ROS grades induced by 10-15 nm citrated AuNPs is related to the upregulation of caspase 3 in addition 7 and causes apoptosis through mitochondrial dysfunction [112].

2.3   Genotoxic Effect Mechanisms of Nanoparticles

While some of the studies on the genotoxic effects of nanoparticles show that these particles can cause damage to genetic material and therefore may be genotoxic, some other studies explain that they do not cause any genotoxic effects. According to literature reviews, nanoparticles can cause cytotoxic in addition genotoxic influences in cells. The coaction, of nanoparticles with genetic material and its results vary depending on whether the cell is in interphase or mitosis. Although the machinery of activity of nanoparticles is not completely figured out, two important mechanisms are mentioned. These are primary (primary) and secondary (secondary) mechanisms. Primary genotoxicity occurs by direct (direct) or indirect (indirect) action mechanisms. The direct genotoxic mechanism is the nanoparticles crossing the cell membrane and nuclear membrane either by diffusion or endocytosis, and then as a result of their physical or chemical interaction with DNA (by affecting the level of binding between DNA bases, phosphorylation, insertion in DNA or altering gene expression/regulation) occur [113]. If nanoparticles reach and interact with the nucleus and DNA during interphase, they can cause changes in replication or transcription, or mechanical or chemical modifyies in the structure of DNA. If nanoparticles reach the cell at the time of mitosis, they interact with the chromosomes mechanically or chemically and produce clastogenic (chromosome break) or aneugenic (chromosome loss on spindle fibers) effects [50]. In the indirect genotoxic mechanism;

Nanoparticles can destroy the functioning of checkpoints in the cell cycle, interact with antioxidant enzymes, generate reactive oxygen species, and even prevent the activity of proteins in the cell cycle, causing destruction in the cell cycle by interacting mechanically or chemically with nuclear proteins, not with DNA or for the purposes of direct mechanisms. Reactive oxygen species produced by nanoparticles may also cause destruction or reduction in DNA repair functions, increase in oxidative stress, damage to mitochondria or cell membranes, decrease in antioxidants and altered gene expressions. Reactive oxygen species cause both DNA destruction and cell death [50,113,114]. The secondary genotoxic mechanism occurs when nanoparticles stimulate the inflammatory response following inflammation, recruit neutrophils in addition macrophages to the region as a result of inflammation and trigger excessive ROS formation by these cells. These ROTs can eventually cause structural damage to genetic material and chromosomes. Both the physicochemical features of the nanoparticle and the exposure environment play an significiant role in the type and level of destruction triggered by the nanoparticle [49,115].

3.   Discussion and Conclusion

Although there are some conflicting results, according to the data obtained, it has been observed that there is an increase in the toxic effects in general as the dimensions of the nanoparticles decrease and the condensation and exposure time enhancement. This causes changes in cell cycle and mitotic index on the one hand, and certain abnormalities such as chromosome and chromatid breakage, fragments, adhesions, bridges, backward chromosomes and micronucleus on the other hand. If DNA repair mechanisms in the cell do not repair these abnormalities appropriately, it can result in loss of genetic material in additon mutations, which may lead to the occurrence of different diseases, especially cancer. Therefore, in order to figure out the genetic and epigenetic machineries of the destruction caused by nanoparticles, more elaborated in vitro moreover in vivo research using various particles, concentrations, application times, organisms, cells, cell lines and test systems are required. According to the findings obtained from the studies, some results show that nanoparticles are cytotoxic and genotoxic, while some other study results show that these substances do not exhibit any toxic effects. These cases also support that the toxic and genotoxic impacts of nanoparticles are still a controversial issue. Researches emphasize that nanoparticle-induced genotoxicity may occur through primary either secondary genotoxicity mechanisms [116]. Priority genotoxicity mentions to DNA destruction from direct physical coaction between particles in addition genomic DNA moreover via ROT in the poverty of inflammation. subsidiary genotoxicity mentions to DNA destruction as a conclusion of the activity of reactive oxygen species in addition reactive nitrogen species (RNT), also other subsidiary mediators (cytokines, chemokines) that take place during the particle-sourced inflammation moreover acute phase response [117]. All this information indicates that there are still debates and contradictions as regards the cytotoxic and genotoxic influences of nanoparticles, so more detailed studies with a wide variety of organisms, cells and tests are needed. The scales, shapes and compound of nanoparticles influence the physical, chemical, bioactive, optical, electrical, catalytic in addition toxicity features of the particles. For this reason, new studies are needed to better understand their synthesis, specification and predicted toxicity. Green synthesis using plant extracts; It is the most practical technique used to acquire nanoparticles with ease, economically and ambient friendly without the usage of high pressure, temperature, energy moreover toxic chemicals [7]. Due to the fact that nanoparticles have anticancer, antimicrobial, antiviral and antifungal activities, their importance in biomedical applications has increased day by day [19]. It is thought that the synthesis of nanoparticle-based new lineage medications will be realized with ongoing and newly planned researches and thus will contribute to the improvement of new treatment protocols. Thus, while protecting the toxic impcts of nanoparticles, on the other hand, it will be possible to benefit from them in the healthiest way possible. Nanoparticles, which form the basis of nanotechnology, have begun to take a place in many areas of our lives due to many physicochemical properties such as size, shape, surface area, surface charge in addition solubility, which they have newly acquired thanks to their nano dimensions. Due to their superior properties, nanoparticles are usaged in many areas such as electronics, textile, paint industry, military materials, food, biomedical and automotive sectors [118,119]. With the widespread production and use of nanoparticles, the increasing exposure of people, on the one hand, and the increase in their waste, on the other hand, pose direct or indirect health risks to both humans and all living things in the ecosystem. For all these reasons, it is of great significiant to investigate the toxic, especially genotoxic, impacts of nanoparticles. Copper oxide, zinc oxide, iron oxide, silver, cobalt-chromium, silicon, titanium and tugsten nanoparticles are widely used in medicine, engineering, pharmacy, agriculture, food, dye and textile fields due to their antimicrobial, high biological reactivity and biocidal properties. Therefore, exposure to these particles increases over time. Chromosomal abnormalities, sister chromatid exchange, micronucleus in addition comet tests are the most commonly usaged tests to determine whether exposed chemicals carry genotoxic risk. While these tests are applied to human lymphocytes in vitro, they are also applied in vivo to some plant varieties such as *A. cepa* and *V. faba*.

References

[1]   ZAFER, C. (2021). Nanotechnology, Society and National Security. *Güvenlik Bilimleri Dergisi.* https://doi.org/10.28956/gbd.845173

[2]   Göksu, H. *et al.* (2020). Oxidation of Benzyl Alcohol Compounds in the Presence of Carbon Hybrid Supported Platinum Nanoparticles (Pt@CHs) in Oxygen Atmosphere. *Scientific Reports 2020 10:1.* https://doi.org/10.1038/s41598-020-62400-5

[3]   Şavk, A. *et al.* (2019). Highly monodisperse Pd-Ni nanoparticles supported on rGO as a rapid, sensitive, reusable and selective enzyme-free glucose sensor. *Scientific Reports.* https://doi.org/10.1038/s41598-019-55746-y

[4]   Zhao, P. *et al.* (2019). A novel ultrasensitive electrochemical quercetin sensor based on MoS2 - carbon nanotube @ graphene oxide nanoribbons / HS-cyclodextrin / graphene quantum dots composite film. *Sensors and Actuators, B: Chemical.* https://doi.org/10.1016/j.snb.2019.126997

[5]   Cherif, A. *et al.* (2021). Numerical investigation of hydrogen production via autothermal reforming of steam and methane over Ni/Al2O3 and Pt/Al2O3 patterned catalytic layers. *International Journal of Hydrogen Energy.* https://doi.org/10.1016/j.ijhydene.2021.04.032

[6]   Ulus, R. *et al.* (2016). Functionalized Multi-Walled Carbon Nanotubes (f-MWCNT) as Highly Efficient and Reusable Heterogeneous Catalysts for the Synthesis of Acridinedione Derivatives. *ChemistrySelect.* https://doi.org/10.1002/SLCT.201600719

[7]   Gulbagca, F. *et al.* (2021). Green Synthesis of Palladium Nanoparticles: Preparation, Characterization, and Investigation of Antioxidant, Antimicrobial, Anticancer, and DNA Cleavage Activities. *Applied Organometallic Chemistry.* https://doi.org/10.1002/aoc.6272

[8]   Ertan, S. *et al.* (2012). Platinum nanocatalysts prepared with different surfactants for C1-C3 alcohol oxidations and their surface morphologies by AFM. *Journal of Nanoparticle Research.*

https://doi.org/10.1007/S11051-012-0922-5

[9]   Günbatar, S. *et al.* (2018). Carbon Nanotube Based Rhodium Nanoparticles as Highly Active Catalyst for Hydrolytic Dehydrogenation of Dimethylamineborane at Room Temperature. *Journal of Colloid and Interface Science.* https://doi.org/10.1016/j.jcis.2018.06.100

[10]   Lolak, N. *et al.* (2019). Composites of Palladium–Nickel Alloy Nanoparticles and Graphene Oxide for the Knoevenagel Condensation of Aldehydes with Malononitrile. *ACS Omega.* https://doi.org/10.1021/acsomega.9b00485

[11]   Medina, C. *et al.* (2007). Nanoparticles: Pharmacological and Toxicological Significance. *British Journal of Pharmacology.* https://doi.org/10.1038/sj.bjp.0707130

[12]   Buzea, C. *et al.* (2007). Nanomaterials and Nanoparticles: Sources and Toxicity. *Biointerphases.* https://doi.org/10.1116/1.2815690

[13]   Haverkamp, R.G. and Marshall, A.T. (2009). The mechanism of metal nanoparticle formation in plants: limits on accumulation. *Journal of Nanoparticle Research.* https://doi.org/10.1007/s11051-008-9533-6

[14]   AlKahtani, R.N. (2018). The implications and applications of nanotechnology in dentistry: A review. *The Saudi Dental Journal.* https://doi.org/10.1016/j.sdentj.2018.01.002

[15]   Yousaf, H. *et al.* (2020). Green synthesis of silver nanoparticles and their applications as an alternative antibacterial and antioxidant agents. *Materials Science and Engineering: C.* https://doi.org/10.1016/j.msec.2020.110901

[16]   Dodds, D.R. (2017). Antibiotic resistance: A current epilogue. *Biochemical Pharmacology.* https://doi.org/10.1016/j.bcp.2016.12.005

[17]   Lin, J. *et al.* (2015). Mechanisms of antibiotic resistance. *Frontiers in Microbiology.* https://doi.org/10.3389/fmicb.2015.00034

[18]   Kocak, Y. *et al.* (2022). Assessment of therapeutic potential of silver nanoparticles synthesized by Ferula Pseudalliacea rech. F. plant. *Inorganic Chemistry Communications.* https://doi.org/10.1016/j.inoche.2022.109417

[19]   Nesrin, K. *et al.* (2020). Biogenic silver nanoparticles synthesized from Rhododendron ponticum and their antibacterial, antibiofilm and cytotoxic activities. *Journal of Pharmaceutical and Biomedical Analysis.* https://doi.org/10.1016/J.JPBA.2019.112993

[20]   Aygun, A. *et al.* (2023). Highly Active PdPt Bimetallic Nanoparticles Synthesized By One Step Bioreduction Method: Characterizations, Anticancer, Antibacterial Activities and Evaluation of Their Catalytic Effect for Hydrogen Generation. *International Journal of Hydrogen Energy.* https://doi.org/10.1016/j.ijhydene.2021.12.144

[21]   Awaad, A. (2015). Histopathological and immunological changes induced by magnetite nanoparticles in the spleen, liver and genital tract of mice following intravaginal instillation. *The Journal of Basic & Applied Zoology.* https://doi.org/10.1016/j.jobaz.2015.03.003

[22]   Pfuhler, S. *et al.* (2013). Genotoxicity of nanomaterials: Refining strategies and tests for hazard identification. *Environmental and Molecular Mutagenesis.* https://doi.org/10.1002/em.21770

[23]   Jiang, Z. *et al.* (2019). Toxic effects of magnetic nanoparticles on normal cells and organs. *Life Sciences.* https://doi.org/10.1016/j.lfs.2019.01.056

[24]   Manivasagan, P. *et al.* (2014). Actinobacteria mediated synthesis of nanoparticles and their biological properties: A review. *Critical Reviews in Microbiology.* https://doi.org/10.3109/1040841X.2014.917069

[25]   Bogunia-Kubik, K. and Sugisaka, M. (2002). From molecular biology to nanotechnology and nanomedicine. *Biosystems.* https://doi.org/10.1016/S0303-2647(02)00010-2

[26]   He, X. and Shi, H. (2012). Size and shape effects on magnetic properties of Ni nanoparticles.

Particuology. https://doi.org/10.1016/j.partic.2011.11.011

[27]   Shin, S. *et al.* (2015). Role of Physicochemical Properties in Nanoparticle Toxicity. *Nanomaterials*. https://doi.org/10.3390/nano5031351

[28]   Sneed, B.T. *et al.* (2015). Building up strain in colloidal metal nanoparticle catalysts. *Nanoscale*. https://doi.org/10.1039/C5NR02529J

[29]   Sen, B. *et al.* (2018). Trimetallic PdRuNi nanocomposites decorated on graphene oxide: A superior catalyst for the hydrogen evolution reaction. *International Journal of Hydrogen Energy*. https://doi.org/10.1016/j.ijhydene.2018.07.122

[30]   Baranowska-Wójcik, E. *et al.* (2020). Effects of Titanium Dioxide Nanoparticles Exposure on Human Health—a Review. *Biological Trace Element Research*. https://doi.org/10.1007/s12011-019-01706-6

[31]   Yin, I.X. *et al.* (2020). The Antibacterial Mechanism of Silver Nanoparticles and Its Application in Dentistry. *International Journal of Nanomedicine*. https://doi.org/10.2147/IJN.S246764

[32]   Ahamed, A. *et al.* (2021). Too small to matter? Physicochemical transformation and toxicity of engineered nTiO2, nSiO2, nZnO, carbon nanotubes, and nAg. *Journal of Hazardous Materials*. https://doi.org/10.1016/j.jhazmat.2020.124107

[33]   Kohl, Y. *et al.* (2020). Genotoxicity of Nanomaterials: Advanced In Vitro Models and High Throughput Methods for Human Hazard Assessment—A Review. *Nanomaterials*. https://doi.org/10.3390/NANO10101911

[34]   AlQuraidi, A.O. *et al.* (2019). Phytotoxic and Genotoxic Effects of Copper Nanoparticles in Coriander (Coriandrum sativum—Apiaceae). *Plants*. https://doi.org/10.3390/plants8010019

[35]   Anu Mary Ealia, S. and Saravanakumar, M.P. (2017). A review on the classification, characterisation, synthesis of nanoparticles and their application. *IOP Conference Series: Materials Science and Engineering*. https://doi.org/10.1088/1757-899X/263/3/032019

[36]   ÇETİN UYANIKGİL, E.Ö. and SALMANOĞLU, D.S. (2020). Metalik nanopartiküllerin hedeflendirilmesi. *Ege Tıp Dergisi*. https://doi.org/10.19161/etd.698596

[37]   García-Rodríguez, A. *et al.* (2019). The Comet Assay as a Tool to Detect the Genotoxic Potential of Nanomaterials. *Nanomaterials*. https://doi.org/10.3390/nano9101385

[38]   Abrahamson, J.T. *et al.* (2013). Excess Thermopower and the Theory of Thermopower Waves. *ACS Nano*. https://doi.org/10.1021/nn402411k

[39]   Gatoo, M.A. *et al.* (2014). Physicochemical properties of nanomaterials: Implication in associated toxic manifestations. *BioMed Research International*. https://doi.org/10.1155/2014/498420

[40]   Schneider, G. (2017). Antimicrobial silver nanoparticles – regulatory situation in the European Union. *Materials Today: Proceedings*. https://doi.org/10.1016/j.matpr.2017.09.187

[41]   BEYKAYA, M. and ÇAĞLAR, A. (2016). An Investigation on Synthesis of Silver-Nanoparticles (AgNP) and their Antimicrobial effectiveness by using Herbal Extracts. *Afyon Kocatepe University Journal of Sciences and Engineering*. https://doi.org/10.5578/fmbd.34220

[42]   ALTINER, A. *et al.* (2018). Free radicals and the relationship with stress. *BALIKESIR HEALTH SCIENCES JOURNAL*. https://doi.org/10.5505/bsbd.2018.38243

[43]   Rodriguez-Garraus, A. *et al.* (2020). Genotoxicity of Silver Nanoparticles. *Nanomaterials*. https://doi.org/10.3390/nano10020251

[44]   Fresegna, A.M. *et al.* (2021). Assessment of the Influence of Crystalline Form on Cyto-Genotoxic and Inflammatory Effects Induced by TiO2 Nanoparticles on Human Bronchial and Alveolar Cells. *Nanomaterials*. https://doi.org/10.3390/nano11010253

[45]   Chang, X. *et al.* (2021). Silver nanoparticles induced cytotoxicity in HT22 cells through autophagy and apoptosis via PI3K/AKT/mTOR signaling pathway. *Ecotoxicology and Environmental*

Safety. https://doi.org/10.1016/j.ecoenv.2020.111696

[46] Agnihotri, R. *et al.* (2020). Nanometals in Dentistry: Applications and Toxicological Implications—a Systematic Review. *Biological Trace Element Research.* https://doi.org/10.1007/s12011-019-01986-y

[47] Giorgetti, L. (2019). Effects of Nanoparticles in Plants, in *Nanomaterials in Plants, Algae and Microorganisms*, Elsevier, pp. 65–87.

[48] Assadian, E. *et al.* (2018). Toxicity of Copper Oxide (CuO) Nanoparticles on Human Blood Lymphocytes. *Biological Trace Element Research.* https://doi.org/10.1007/s12011-017-1170-4

[49] SAYGILI, Y. *et al.* (2021). Metal Oksit Nanopartiküllerin Genotoksik Etkileri. *International Journal of Advances in Engineering and Pure Sciences.* https://doi.org/10.7240/jeps.875709

[50] Barillet, S. *et al.* (2010). In vitro evaluation of SiC nanoparticles impact on A549 pulmonary cells: Cyto-, genotoxicity and oxidative stress. *Toxicology Letters.* https://doi.org/10.1016/j.toxlet.2010.07.009

[51] Foldbjerg, R. *et al.* (2011). Cytotoxicity and genotoxicity of silver nanoparticles in the human lung cancer cell line, A549. *Archives of Toxicology.* https://doi.org/10.1007/s00204-010-0545-5

[52] Wang, J. *et al.* (2017). Comparative genotoxicity of silver nanoparticles in human liver HepG2 and lung epithelial A549 cells. *Journal of Applied Toxicology.* https://doi.org/10.1002/jat.3385

[53] Brandão, F. *et al.* (2020). Genotoxicity of TiO2 Nanoparticles in Four Different Human Cell Lines (A549, HEPG2, A172 and SH-SY5Y). *Nanomaterials.* https://doi.org/10.3390/nano10030412

[54] Fernández-Bertólez, N. *et al.* (2019). Assessment of oxidative damage induced by iron oxide nanoparticles on different nervous system cells. *Mutation Research/Genetic Toxicology and Environmental Mutagenesis.* https://doi.org/10.1016/j.mrgentox.2018.11.013

[55] Fernández-Bertólez, N. *et al.* (2021). Suitability of the In Vitro Cytokinesis-Block Micronucleus Test for Genotoxicity Assessment of TiO2 Nanoparticles on SH-SY5Y Cells. *International Journal of Molecular Sciences.* https://doi.org/10.3390/ijms22168558

[56] Chen, Z. *et al.* (2014). Genotoxic evaluation of titanium dioxide nanoparticles in vivo and in vitro. *Toxicology Letters.* https://doi.org/10.1016/j.toxlet.2014.02.020

[57] Meena, R. *et al.* (2015). Cytotoxic and Genotoxic Effects of Titanium Dioxide Nanoparticles in Testicular Cells of Male Wistar Rat. *Applied Biochemistry and Biotechnology.* https://doi.org/10.1007/s12010-014-1299-y

[58] Ghosh, M. *et al.* (2019). Genotoxicity of engineered nanoparticles in higher plants. *Mutation Research/Genetic Toxicology and Environmental Mutagenesis.* https://doi.org/10.1016/j.mrgentox.2019.01.002

[59] Waani, S.P.T. *et al.* (2021). TiO2 nanoparticles dose, application method and phosphorous levels influence genotoxicity in Rice (Oryza sativa L.), soil enzymatic activities and plant growth. *Ecotoxicology and Environmental Safety.* https://doi.org/10.1016/j.ecoenv.2021.111977

[60] Kim, H.R. *et al.* (2013). Appropriate <i>In Vitro</i> Methods for Genotoxicity Testing of Silver Nanoparticles. *Environmental Health and Toxicology.* https://doi.org/10.5620/eht.2013.28.e2013003

[61] Cao, Y. *et al.* (2021). Modeling better in vitro models for the prediction of nanoparticle toxicity: a review. *Toxicology Mechanisms and Methods.* https://doi.org/10.1080/15376516.2020.1828521

[62] Di Virgilio, A.L. *et al.* (2010). Comparative study of the cytotoxic and genotoxic effects of titanium oxide and aluminium oxide nanoparticles in Chinese hamster ovary (CHO-K1) cells. *Journal of Hazardous Materials.* https://doi.org/10.1016/j.jhazmat.2009.12.089

[63] Azim, S.A.A. *et al.* (2015). Amelioration of titanium dioxide nanoparticles-induced liver injury in mice: Possible role of some antioxidants. *Experimental and Toxicologic Pathology.*

https://doi.org/10.1016/j.etp.2015.02.001

[64]   Martins, A. da C. *et al.* (2017). Evaluation of distribution, redox parameters, and genotoxicity in Wistar rats co-exposed to silver and titanium dioxide nanoparticles. *Journal of Toxicology and Environmental Health, Part A.* https://doi.org/10.1080/15287394.2017.1357376

[65]   Patlolla, A.K. *et al.* (2015). Genotoxicity study of silver nanoparticles in bone marrow cells of Sprague–Dawley rats. *Food and Chemical Toxicology.* https://doi.org/10.1016/j.fct.2015.05.005

[66]   Kahraman, T. *et al.* (2021). Synthesis, Characterization, and Optimization of Green Silver Nanoparticles Using Neopestalotiopsis clavispora and Evaluation of Its Antibacterial, Antibiofilm, and Genotoxic Effects. *The EuroBiotech Journal.* https://doi.org/10.2478/ebtj-2021-0020

[67]   Golbamaki, A. *et al.* (2018). Genotoxicity induced by metal oxide nanoparticles: a weight of evidence study and effect of particle surface and electronic properties. *Nanotoxicology.* https://doi.org/10.1080/17435390.2018.1478999

[68]   Rastogi, A. *et al.* (2017). Impact of Metal and Metal Oxide Nanoparticles on Plant: A Critical Review. *Frontiers in Chemistry.* https://doi.org/10.3389/fchem.2017.00078

[69]   Malarkodi, C. *et al.* (2014). Biosynthesis and Antimicrobial Activity of Semiconductor Nanoparticles against Oral Pathogens. *Bioinorganic Chemistry and Applications.* https://doi.org/10.1155/2014/347167

[70]   Chandra, H. *et al.* (2019). Phyto-mediated synthesis of zinc oxide nanoparticles of Berberis aristata: Characterization, antioxidant activity and antibacterial activity with special reference to urinary tract pathogens. *Materials Science and Engineering: C.* https://doi.org/10.1016/j.msec.2019.04.035

[71]   Rai, M. *et al.* (2009). Silver nanoparticles as a new generation of antimicrobials. *Biotechnology Advances.* https://doi.org/10.1016/j.biotechadv.2008.09.002

[72]   Pal, S. *et al.* (2007). Does the antibacterial activity of silver nanoparticles depend on the shape of the nanoparticle? A study of the Gram-negative bacterium Escherichia coli. *Applied and environmental microbiology.* https://doi.org/10.1128/AEM.02218-06

[73]   Lazar, V. (2011). Quorum sensing in biofilms – How to destroy the bacterial citadels or their cohesion/power? *Anaerobe.* https://doi.org/10.1016/j.anaerobe.2011.03.023

[74]   Bolla, J.-M. *et al.* (2011). Strategies for bypassing the membrane barrier in multidrug resistant Gram-negative bacteria. *FEBS Letters.* https://doi.org/10.1016/j.febslet.2011.04.054

[75]   Devi, L.S. and Joshi, S.R. (2012). Antimicrobial and Synergistic Effects of Silver Nanoparticles Synthesized Using Soil Fungi of High Altitudes of Eastern Himalaya. *Mycobiology.* https://doi.org/10.5941/MYCO.2012.40.1.027

[76]   Das, R. *et al.* (2011). Preparation and Antibacterial Activity of Silver Nanoparticles. *Journal of Biomaterials and Nanobiotechnology.* https://doi.org/10.4236/jbnb.2011.24057

[77]   Cha, S.-H. *et al.* (2015). Shape-Dependent Biomimetic Inhibition of Enzyme by Nanoparticles and Their Antibacterial Activity. *ACS Nano.* https://doi.org/10.1021/acsnano.5b03247

[78]   Antonoglou, O. *et al.* (2019). Biological relevance of CuFeO2 nanoparticles: Antibacterial and anti-inflammatory activity, genotoxicity, DNA and protein interactions. *Materials Science and Engineering: C.* https://doi.org/10.1016/J.MSEC.2019.01.112

[79]   Chavez Soria, N.G. *et al.* (2019). Lipidomics reveals insights on the biological effects of copper oxide nanoparticles in a human colon carcinoma cell line. *Molecular Omics.* https://doi.org/10.1039/C8MO00162F

[80]   Assadian, E. *et al.* (2019). Toxicity of Fe 2 O 3 nanoparticles on human blood lymphocytes. *Journal of Biochemical and Molecular Toxicology.* https://doi.org/10.1002/jbt.22303

[81]   Ekman Nilsson, A. *et al.* (2017). A Review of the Carbon Footprint of Cu and Zn Production from Primary and Secondary Sources. *Minerals.* https://doi.org/10.3390/min7090168

[82]   Sun, Z. *et al.* (2019). Influences of zinc oxide nanoparticles on Allium cepa root cells and the primary cause of phytotoxicity. *Ecotoxicology.* https://doi.org/10.1007/s10646-018-2010-9

[83]   Birlik, I. and Ak Azem, N.F. (2018). Sol-jel Yöntemi ile Hazırlanmış ZnO Nanopartiküllerin Optimizasyonu. *Deu Muhendislik Fakultesi Fen ve Muhendislik.* https://doi.org/10.21205/deufmd.2018205810

[84]   Saber, M. *et al.* (2021). In vitro cytotoxicity of zinc oxide nanoparticles in mouse ovarian germ cells. *Toxicology in Vitro.* https://doi.org/10.1016/j.tiv.2020.105032

[85]   Akbaba, G.B. and Türkez, H. (2018). Investigation of the Genotoxicity of Aluminum Oxide, β-Tricalcium Phosphate, and Zinc Oxide Nanoparticles In Vitro. *International Journal of Toxicology.* https://doi.org/10.1177/1091581818775709

[86]   Bhattacharya, K. *et al.* (2009). Titanium dioxide nanoparticles induce oxidative stress and DNA-adduct formation but not DNA-breakage in human lung cells. *Particle and Fibre Toxicology.* https://doi.org/10.1186/1743-8977-6-17

[87]   Khan, M. *et al.* (2015). Comparative study of the cytotoxic and genotoxic potentials of zinc oxide and titanium dioxide nanoparticles. *Toxicology Reports.* https://doi.org/10.1016/j.toxrep.2015.02.004

[88]   Andersson-Willman, B. *et al.* (2012). Effects of subtoxic concentrations of TiO2 and ZnO nanoparticles on human lymphocytes, dendritic cells and exosome production. *Toxicology and Applied Pharmacology.* https://doi.org/10.1016/j.taap.2012.07.021

[89]   Farahani, Z. *et al.* (2020). Comparative Study of the Cytotoxic Effect of Silver Nanoparticles on Human Lymphocytes and HPB-ALL Cell Line: As an In Vitro Study. *Iranian Red Crescent Medical Journal.* https://doi.org/10.5812/ircmj.98803

[90]   Li, Y. *et al.* (2017). Differential genotoxicity mechanisms of silver nanoparticles and silver ions. *Archives of Toxicology.* https://doi.org/10.1007/s00204-016-1730-y

[91]   Ghosh, M. *et al.* (2012). In vitro and in vivo genotoxicity of silver nanoparticles. *Mutation Research/Genetic Toxicology and Environmental Mutagenesis.* https://doi.org/10.1016/j.mrgentox.2012.08.007

[92]   Lu, W. *et al.* (2010). Effect of surface coating on the toxicity of silver nanomaterials on human skin keratinocytes. *Chemical Physics Letters.* https://doi.org/10.1016/j.cplett.2010.01.027

[93]   Atli Sekeroglu, Z. (2013). From nanotechnology to nanogenotoxicology: genotoxic effect of cobalt-chromium nanoparticles. *Turkish Bulletin of Hygiene and Experimental Biology.* https://doi.org/10.5505/TurkHijyen.2013.70298

[94]   Chattopadhyay, S. *et al.* (2015). Toxicity of cobalt oxide nanoparticles to normal cells; an in vitro and in vivo study. *Chemico-Biological Interactions.* https://doi.org/10.1016/j.cbi.2014.11.016

[95]   Rajiv, S. *et al.* (2016). Comparative cytotoxicity and genotoxicity of cobalt (II, III) oxide, iron (III) oxide, silicon dioxide, and aluminum oxide nanoparticles on human lymphocytes in vitro. *Human & Experimental Toxicology.* https://doi.org/10.1177/0960327115579208

[96]   Lojk, J. *et al.* (2017). Cell stress response to two different types of polymer coated cobalt ferrite nanoparticles. *Toxicology Letters.* https://doi.org/10.1016/j.toxlet.2017.02.010

[97]   Askari, M.B. *et al.* (2021). Enhanced electrochemical performance of MnNi2O4/rGO nanocomposite as pseudocapacitor electrode material and methanol electro-oxidation catalyst. *Nanotechnology.* https://doi.org/10.1088/1361-6528/abfded

[98]   Eris, S. *et al.* (2018). Investigation of electrocatalytic activity and stability of Pt@f-VC catalyst prepared by in-situ synthesis for Methanol electrooxidation. *International Journal of Hydrogen Energy.* https://doi.org/10.1016/j.ijhydene.2017.11.063

[99]   Andreeva, E.R. *et al.* (2013). In Vitro Study of Interactions between Silicon-Containing Nanoparticles and Human Peripheral Blood Leukocytes. *Bulletin of Experimental Biology and Medicine.* https://doi.org/10.1007/s10517-013-2161-x

[100]   Lankoff, A. *et al.* (2013). Effect of surface modification of silica nanoparticles on toxicity and cellular uptake by human peripheral blood lymphocytes in vitro. *Nanotoxicology.* https://doi.org/10.3109/17435390.2011.649796

[101]   Waghmode, M.S. *et al.* (2019). Studies on the titanium dioxide nanoparticles: biosynthesis, applications and remediation. *SN Applied Sciences.* https://doi.org/10.1007/s42452-019-0337-3

[102]   Kazimirova, A. *et al.* (2019). Titanium dioxide nanoparticles tested for genotoxicity with the comet and micronucleus assays in vitro, ex vivo and in vivo. *Mutation Research/Genetic Toxicology and Environmental Mutagenesis.* https://doi.org/10.1016/j.mrgentox.2019.05.001

[103]   Ghosh, M. *et al.* (2013). Cytotoxic, genotoxic and the hemolytic effect of titanium dioxide (TiO 2) nanoparticles on human erythrocyte and lymphocyte cells in vitro. *Journal of Applied Toxicology.* https://doi.org/10.1002/jat.2863

[104]   Catalán, J. *et al.* (2012). Induction of chromosomal aberrations by carbon nanotubes and titanium dioxide nanoparticles in human lymphocytes in vitro. *Nanotoxicology.* https://doi.org/10.3109/17435390.2011.625130

[105]   Andreoli, C. *et al.* (2018). Critical issues in genotoxity assessment of TiO 2 nanoparticles by human peripheral blood mononuclear cells. *Journal of Applied Toxicology.* https://doi.org/10.1002/jat.3650

[106]   Moche, H. *et al.* (2014). Tungsten Carbide-Cobalt as a Nanoparticulate Reference Positive Control in In Vitro Genotoxicity Assays. *Toxicological Sciences.* https://doi.org/10.1093/toxsci/kft222

[107]   Menon, S. *et al.* (2018). Selenium nanoparticles: A potent chemotherapeutic agent and an elucidation of its mechanism. *Colloids and Surfaces B: Biointerfaces.* https://doi.org/10.1016/j.colsurfb.2018.06.006

[108]   Lin, W. *et al.* (2006). In vitro toxicity of silica nanoparticles in human lung cancer cells. *Toxicology and Applied Pharmacology.* https://doi.org/10.1016/j.taap.2006.10.004

[109]   Xia, T. *et al.* (2006). Comparison of the Abilities of Ambient and Manufactured Nanoparticles To Induce Cellular Toxicity According to an Oxidative Stress Paradigm. *Nano Letters.* https://doi.org/10.1021/nl061025k

[110]   Carlson, C. *et al.* (2008). Unique Cellular Interaction of Silver Nanoparticles: Size-Dependent Generation of Reactive Oxygen Species. *The Journal of Physical Chemistry B.* https://doi.org/10.1021/jp712087m

[111]   Pan, Y. *et al.* (2009). Gold Nanoparticles of Diameter 1.4 nm Trigger Necrosis by Oxidative Stress and Mitochondrial Damage. *Small.* https://doi.org/10.1002/smll.200900466

[112]   Wahab, R. *et al.* (2014). Statistical analysis of gold nanoparticle-induced oxidative stress and apoptosis in myoblast (C2C12) cells. *Colloids and Surfaces B: Biointerfaces.* https://doi.org/10.1016/j.colsurfb.2014.10.012

[113]   Barnes, C.A. *et al.* (2008). Reproducible Comet Assay of Amorphous Silica Nanoparticles Detects No Genotoxicity. *Nano Letters.* https://doi.org/10.1021/nl801661w

[114]   Wang, H. *et al.* (2013). Engineered Nanoparticles May Induce Genotoxicity. *Environmental Science & Technology.* https://doi.org/10.1021/es404527d

[115]   Magdolenova, Z. *et al.* (2014). Mechanisms of genotoxicity. A review of in vitro and in vivo studies with engineered nanoparticles. *Nanotoxicology.* https://doi.org/10.3109/17435390.2013.773464

[116]   Åkerlund, E. *et al.* (2019). Inflammation and (secondary) genotoxicity of Ni and NiO nanoparticles. *Nanotoxicology.* https://doi.org/10.1080/17435390.2019.1640908

[117]   Modrzynska, J. *et al.* (2018). Primary genotoxicity in the liver following pulmonary exposure to carbon black nanoparticles in mice. *Particle and Fibre Toxicology.* https://doi.org/10.1186/s12989-017-0238-9

[118]   Şen, F. *et al.* (2018). The dye Removal from Aqueous Solution Using Polymer Composite Films. *Applied Water Science.* https://doi.org/10.1007/s13201-018-0856-x

[119]   Hojjati-Najafabadi, A. *et al.* (2022). A Tramadol Drug Electrochemical Sensor Amplified by Biosynthesized Au Nanoparticle Using Mentha aquatic Extract and Ionic Liquid. *Topics in Catalysis.* https://doi.org/10.1007/s11244-021-01498-x

Chapter 24

Biogenic Nanomaterials from Lab to Market

S. Kaptanoğlu[1*], F. Calayir[2], A.R. Kul[3], I. Meydan[1], R. Mahious[3,4], F. Sen[4*]

[1]Department of Biochemistry, Vocational School of Health Services, Yuzuncu Yıl University, Van, 65000, Turkey

[2]Department of Biochemistry, Institute of Science, Yuzuncu Yıl University, Van, 65000, Turkiye

[3]Department of Physical Chemistry, Vocational School of Health Services, Yuzuncu Yıl University, Van, 65000, Turkiye

[3]Department of Biology, Faculty of Arts and Science, Kutahya Dumlupinar University, 43000, Kutahya, Turkiye,

[4]Sen Research Group, Department of Biochemistry, Faculty of Arts and Science, Kutahya Dumlupinar University, 43000, Kutahya, Turkiye,

semakaptanoglu@yyu.edu.tr; fatih.sen@dpu.edu.tr

Abstract

Nanotechnology involves studying materials at the nanoscale (1-100 nanometers) to enhance their properties and performance. The field has made significant advancements, from Richard Feynman's 1959 prediction to the invention of the atomic force microscope in 1981 and the development of nanomachines in 1985. IBM researchers built the first atomic-scale structure in 1991, eventually leading to the commercial use of nanotechnology in the 2000. Nanotechnology has potential applications in medicine, energy, and manufacturing, offering opportunities for more effective drugs, efficient energy sources, and stronger materials. Despite being a nascent technology, nanotechnology holds promise for revolutionizing various industries.

Keywords

Atomic Force, IBM, Nanotechnology, Nanomachines, Emerging Technology

1. Introduction

The universal and innovative nature of nanotechnology has recently attracted considerable attention. Nanotechnology has wide application areas [1–10]. Although similar commercialization of these systems has not gained significant impetus, massive government support for nanotechnology-based products has increased activation and activity in various fields, including drug development. Nevertheless, this technology perfects drug solubility, bioavailability, and toxin profile, as evidenced by numerous high-quality research papers published in colourful scientific journals and daily newspapers. Based on our decades of experience and extensive literature review with nanotechnology-based drug delivery systems, we believe that the main deficiencies in the commercialization of these nanotechnology products include the lack of public and expert support and acceptance, and expansion. Including lack of compatibility, reproducibility, and characterization. Current research examines trends associated with persistence, unregulated components, US Food and Drug Administration influence, and challenges associated with moving these products through the various clinical stages from the laboratory to clinical trials. By doing so, we are making progress on these issues [11,12].

The potential for new commercial applications of nanotechnology has the potential to significantly advance and even revolutionize many aspects of medical applications and product development. Nanotechnology has touched many aspects of medicine, including drug delivery, diagnostic imaging, clinical diagnostics, nanomedicine, and the use of nanomaterials in medical devices. This technology is already having an impact. There are many products on the market, and the number continues to grow. The trend is increasing for the successful development of nanotechnological products that are more advanced than the current ones for the diagnosis of diseases. The most active areas of product development are drug delivery and in vivo imaging. Nanotechnology is also addressing many of the pharmaceutical industry's unmet needs, such as reformulating drugs to improve bioavailability and toxicity profiles. Advances in medical nanotechnology have spanned or gone through at least three different generations or stages [13].

Some experts see nanotechnology as the next industrial revolution that will help in many fields [14,15]. Nanotechnology-based products are used in many industries. These include transportation, materials, energy, electronics, pharmaceuticals, agricultural and environmental sciences, and consumer and household goods [16]. These applications can be divided into general categories such as pharmaceuticals, food and cosmetics, agriculture and environmental health, technology and industry (Fig. 1).

Products Performing from the operation of nanotechnology include nanomaterials (similar as nanoparticles, nanocomposites, and nanotubes), nanodevices (similar as surveying inquiry microscopes and other bias with nanoscale factors), and larger It includes device manufacturing [7]. Successful investments in exploration and development of nanotechnology products have led to the emergence of further practical accoutrements with unique operations. Thus, it's important that these technologies extend the boundaries of the laboratory and contribute to working current societal challenges. Still, there are numerous challenges to overcome when bringing nanotechnology products and companies to vend [18,19].

Biogenic Nanomaterials: Synthesis, Characterization, Applications, and Future Remarks   Materials Research Forum LLC
Materials Research Foundations 180 (2025)   https://doi.org/21741/9781644903759

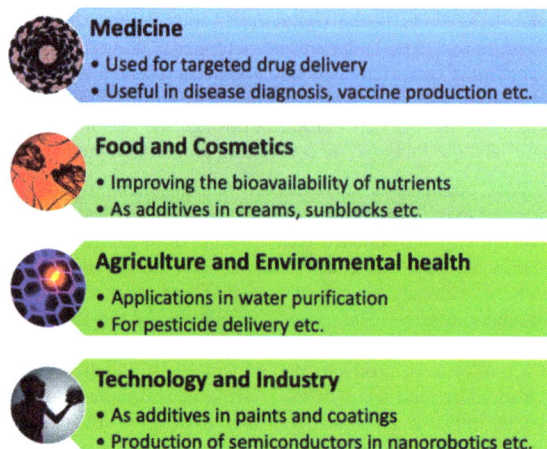

Figure 1. *Possible applications of nanomaterials (Reprinted with permission from [10], Copyright Royal Society of Chemistry).*

2.   Introduction to Nanotechnology

Nanotechnology is a technology that includes structures between 1 nm and 100 nm. In other words, the application range of these nanostructures is between 1-100 nm. To understand the technology, it would be best to compare it to another dimension before considering what it means in nanoscale dimensions [11]. To better understand the nano size, it would be appropriate to make a comparison with the thickness of a human hair. The diameter of a human hair is on average 80,000 nm. When we make a comparison, it is seen that 80,000 nm means that it fits on a hair. So, let's consider an object 1 meter high. In order to reach such a height in this object, 1 billion nm particles must be placed on top of each other [12]. Nanotechnology is a development that can affect us deeply in many areas of our lives and even open a new era. Today, this technology is defined as the new industrial revolution of our age. The word "nano" is defined as "one billionth of a physical amount". Nanotechnology is the science of understanding and controlling the behaviour of matter from 1 to 100 manometers [13]. Controlling the behaviour and structural properties of nanomaterials leads to radical innovations in many fields. It is not just the size of the nanoparticles that distinguishes nanomaterials from materials with large particles. They show different structural properties in terms of chemical reactivity, energy absorption, and biological mobility. What makes nanotechnology so interesting is that when materials are at the nanoscale, they behave differently from the macro world and even the micro world [24].

Since nature has existed, living organisms have used several natural nanomaterials. It is the most common clay mineral in nature. Nanostructured clay minerals are preferred in various manufacturing processes due to their abundance in nature. These are natural minerals with a two-dimensional nanoscale crystal lattice nanocrystalline structure. Thanks to this feature of clay minerals, it is possible to remove oils, heavy metals and organic compounds from water. Today, clay minerals are commonly encountered in advanced technical and environmental applications. Clay has a fragmented structure due to its natural nanomaterial structure. Thanks to this structural feature, it has the potential to perform tensile and absorption functions on the surface. This clay mineral feature reduces or completely destroys the effects of toxic (harmful) substances entering the body. For hundreds of years, people have used natural nanomaterial clays as therapeutic agents to treat diseases. Observations of the animal kingdom have shown that they eat and roll on natural mineral clays. It has been found that animals seek healing because of these behaviours. It is also known to be used in the treatment of many ailments such as intestinal problems, wound closure, diarrhoea and painkillers, ulcers, acne, haemorrhoids. To date, the benefits of natural mineral clays have not yet been fully elucidated. In addition, when we examined clay, which is a natural nanomaterial, we did not find anything threatening the health of living organisms. Many nano dimensions emerge when examining structures in vivo. For example, the thickness of human cell membranes is nanometers. It has also been shown that many organisms are nanoscale, such as the diameter of DNA and RNA structures, the diameter of ribosomes and mitochondria [25].

Due to technical research in the field of nanotechnology, existing structures are reduced, and more efficient structures are obtained. These highly efficient technologies result in well-crafted, durable, clean, safe and smart products. These new structures, called nanoscale materials, fall into several classes such as nanocrystals, nanoparticles, nanotubes, nanowires, and nanorods. All these structures have colorful names and are called nanoparticles. Nanoparticles are carbon-grounded (fullerenes, multi-walled carbon nanotubes, etc.), essence- grounded (gold colloids, nanoshells, nanorods, superparamagnetic iron oxide nanoparticles, etc.), semiconductor-grounded (amount blotches, etc.). Nanoparticles have a wide range of operations due to their different parcels. The functionality and operations of these accoutrements depend on the size and composition of the nanoparticles [26].

Unlike traditional production methods of nanotechnology, processes such as machining, turning, and shaping the material from the outside to the inside, the material is produced from atomic dimensions. The biggest advantage of this is the prevention of errors or defects that may occur in the material's internal structure with traditional production methods. For example, if a suitable cooling environment is not provided in a material to be produced by casting, the formation of undesirable phases in the internal structure of the material or the formation of internal stresses in the structures with rapid cooling may result. This will result in not obtaining the expected properties from the material and not exhibiting a suitable structure for the place of use. With nanotechnology, it is possible to intervene in the internal structure of the material at atomic scale, and materials with excellent properties can be produced by interfering with the arrangement of atoms in the material structure [27].

The first realization of the importance of the nanoscale dates back to a lecture given by Nobel Prize-winning physicist Richard Feynman at the American Physical Society Annual Meeting at Caltech on December 29, 1959. In a historic talk entitled "There's plenty of room for fundamentals," Feynman said that characterization of materials and devices at the first nanometer scale opens up a world of possibilities for the future. In summary, in this historical lecture, Feynman articulated the idea of manipulating and controlling things on a small scale [28]. However, in 1974, Norio Taniguchi of Tokyo University of Science first used the term nanotechnology. Fundamental Ideas of Molecular Manufacturing in the 1980, This was published by Eric Drexler in the article "Designing Proteins for Molecular Manufacturing". In his later work, K. Eric Drexler described a possible way of constructing devices and structures in the form of complex atomic properties by creating self-replicating 'compilers'. While controversial, the idea of this ubiquitous 'compiler' has become an important area of nanotechnology research today. While this universal 'compiler' view is controversial, it is an important area of application for the use of bottom-up techniques for the fabrication of nanomaterials [29]. The inventions that enabled the development of nanotechnology gained momentum with the discovery of the scanning tunnelling microscope in 1981 [30]. Thanks to this microscope, atoms can be moved on a conductive surface. In other words, this microscope can be used to change the position of atoms on a conductive surface. The realization of this process is another milestone in the history of nanotechnology. In 1985, Karl Kroto and Smalley used this microscope to discover new nanostructured carbon modifications called 'buckyballs' and 'fullerenes'. Binnig G. and Rohrer H. were awarded the Nobel Prize in Physics in 1986 for their work in this area. In the 1990, the federal governments of the United States, Europe, and Japan introduced nanoelectronics and nanomaterials. They were interested in programs in different fields of nanotechnology. In the late 1990, it was recognized that the field should be approached from many aspects of nanotechnology rather than the proliferation of different small scientific disciplines. Nanotechnology is now a technical field of government exploration and development programs, with exploration and development centers established in nearly every developed country. In 1991, as a result of the study of fullerenes, it was envisaged that the discovery of tubular structures made of carbon atoms, especially round-end graphite plates, cylinders, has great potential for applications in materials engineering in electrical engineering. As a result of the nanotechnology investment of the USA in 2000, nanotechnology research started in many countries. Development in this area is currently in full swing.

3.   The Importance of Nanotechnology

Nanotechnology should be introduced with comprehensive research outputs, followed by fundamental questions and principles regarding the risks and regulations necessary to assess the impact of nanotechnology. Finally, key areas of practice such as the environment, privacy, health, and human development should be considered, and the social and ethical implications of these practices should be carefully evaluated [31]. Given the potential hazards associated with nanotechnology, care should be taken when producing food and food-like products. Telecommunications and textiles are among the best-known and best-selling products. Consumer perceptions of opportunities in various product categories appear to be stronger than perceptions of risk. However, it has been determined that the risk perception of users is higher in products that come into contact with or enter the body [32]. Nano research shows that typically inoffensive substances can come poisonous when converted into nanoscale patches. Their small size allows them to move fluently within the body, and their large face area to volume rate makes them largely chemically and bioreactive. Adding toxicological substantiation shows that nanomaterials pose pitfalls to mortal and environmental health. Articulated types can be used for a wide variety of purposes. The possible uses and operations of these accoutrements depend on the size and composition of the nanoparticles [33].

It is important that individuals employed in the field of nanotechnology receive education. Individuals should have a working knowledge of what nanotechnology is and why it is important. For the future of nanotechnology and the eco-environment, policymakers, scientists and industry managers must recognize that they are working with a valuable resource. This is because nanotechnology has the potential to fundamentally change human life [34]. Therefore, we must carefully monitor our environment and take conscious measures to deal with it before it becomes a major risk. Nanosensors are structures that enable accurate and rapid-fire discovery and monitoring of the impact of mortal conditioning on the terrain. Nanotechnology can help clean up being pollution and use available coffers more wisely [35]. It should not be wrong to use expressions that can be included in our lives thanks to today's nanotechnology, the transmission of information to virtual screens appearing in our field of view, and the use of contact lenses in which nanotechnological computers are built. When we look at the fields of materials science and nanotechnology using advanced technology, it is seen how excited people are in the face of developments and surprises. The main source of all development today is material origin. The production of nanoparticles requires an understanding of basic physical and chemical information at the nanoscale and how to commercialize them [36]. With the introduction of new nanomaterials into our lives, many products such as computers and mobile phones are getting smaller day by day. However, this contraction does not impair the function. For example, mobile phones tend to be smaller, while larger products are often preferred because they are more practical. More or less, there are materials that are more durable and useful than ever before. When we look at nanomaterial technology today, we see that the number of applications that fundamentally change our lives is increasing. Thanks to the rapid development of electron microscopy techniques, naturally occurring nanomaterials have already been observed. Thanks to this development, today's research on nanomaterials provides control [37].

In parallel with the development of technology, the demand for nanotechnology is increasing day by day. As a result, job opportunities for graduates in the materials and nanotechnology industries are increasing day by day. Graduates of these departments can work in material production (ceramic, glass, polymers, composites), metallurgy industry, defence industry, health sector, aviation, automotive industry, consumer electronics industry, chemistry, textile industry, plastics industry, etc. all manufacturing sectors. Nanotechnology is a technology that offers great opportunities for human life and the world of science. Products produced with nanotechnology. Materials and manufacturing, nanoelectronics and computing technologies, aerospace research, medicine and health, environment and energy, biotechnology and agriculture, defense, science, etc. In recent years, many scientists around the world have been working on nanotechnology and scientific developments have been experienced in every field [38]. In nature, plants store the energy contained in nanocrystals in their chloroplasts, and nanotechnology tries to achieve this at the nanoscale. Nanotechnology is also used to store energy. Nanotechnology provides a more efficient solution for convenient storage. Energy storage is one of the biggest dilemmas of our time. Attempts are currently being made to overcome this challenge using nanotechnology. The use of nanotechnology increases the efficiency of clean energy production. Thanks to research in nanotechnology, renewable energy is replacing fossil energy. The conversion of renewable agricultural inputs and food waste into energy and useful by-products is an environmentally focused area of research that can be significantly enhanced by nanotechnology [39]. For example, solar cells can use nanomaterials to efficiently produce large amounts of energy from wind, ocean, and geothermal energy. Nanotechnology has made it possible to consume accoutrements in a way that effectively reduces the preface of adulterants from mortal conditioning into the terrain. Industrial changes using nanotechnology have enabled the production of materials that can be digested in the environment and transformed into different raw materials. Planned transitions to certain industries can help reduce environmental damage. Clean renewable energy is possible with nanotechnology [40].

Nanotechnology also brings revolutionary innovations in the field of food production. Working at the nanoscale improves food quality and reduces contamination. Different production methods have been developed using nanotechnology for each stage of food production. There are no restrictions in many areas such as packaging, biological preservation and transportation of foods. Nanoparticles provide antibacterial and antibacterial properties are still under development and are being worked on. Thanks to the developing technology, nanotechnology research continues at full speed. Nothing is more obvious than the widespread use of nanotechnology applications. The resulting nano-size gives products new and interesting properties. However, this is considered a threat to security. There are concerns that nanotechnology may pose health and environmental risks. The main reason for these concerns is that nanoproducts may cause unexpected interactions due to their very small size [31, 32].

Nanotechnology is an illustration of a new technology that can pose colorful pitfalls to the terrain and health if it isn't designed with sustainability in mind in numerous felicitations. still, we face abecedarian challenges in learning these new technologies. Green nanotechnology is presently being developed to make nanomaterials safer through rational design. Nanotechnologists don't accept a analogous obligation to address societal enterprises and are limited to addressing only environmental, health and safety issues [43]. For illustration, thanks to nanotechnology, medicines that act on blood vessels and apkins (medicine carrier nanosystems) can now only be applied to cancer cells. The objectification of nanocarriers into medicine delivery systems has led to great advances in the treatment of conditions. In the medical field, exploration on the release of anticancer medicines continues in order to reduce the poisonous goods of medicines and help multidrug resistance [44]. Cancer cells are treated with controlled medicine release and nanoscale maquillages. This operation is one of the most important operations of nanotechnology in drug. Nanotechnology is also veritably effective in probing contagious conditions. For this purpose, it offers numerous new openings to clinical microbiologists, from opinion to treatment. For illustration, exosomes are small nanoparticles ranging in size from 30 nm to 150 nm that are buried for all types of cells- to- cell communication. These are nanomaterials that have been proven to be the ultimate cell rejuvenescence in remedial and individual studies [45] .

Exosomes containing SARS-Cov proteins have been reported to accelerate negativing antibody titers when primed with a vaccine against SARS-Cov proteins and also boosted with an adenoviral vector vaccine [46]. Nanotechnology has numerous important operations in cancer opinion and treatment. moment, nanotechnology provides early opinion and effective treatment of cancer. Thanks to these early opinion and treatment options, cancer is frequently curable, making it a complaint with high cure rates and low mortality rates [47]. Cancer cell remedy is one of the most important operations of nanotechnology in drugs. Most of the studies have been completed and recovery appears to have positive issues for cases. This is a useful development in destroying cancer cells. Important exploration studies are presently being carried out in the field of health. The data attained is promising. Unnaturally, advances have been made in antitumor treatments. latterly, nano-grounded remedial strategies for atherosclerosis were developed [48].

4.   Nanotechnology from Laboratory to Industry

Before the results of scientific exploration can be applied to a new product, process, or service, they must go through a series of strategic conduct. An essential demand is that the invention first provides a result to an being problem or responds to a request or special need. In addition to invention, we also need invention. The conditioning that experimenters need to take over to increase the marketability of their scientific exploration can be epitomized as follows (Fig. 2). The blue line shows the commercialization timeline, including the decision to move to an external company or form a new company. Green arrows indicate external support that accelerates progress.

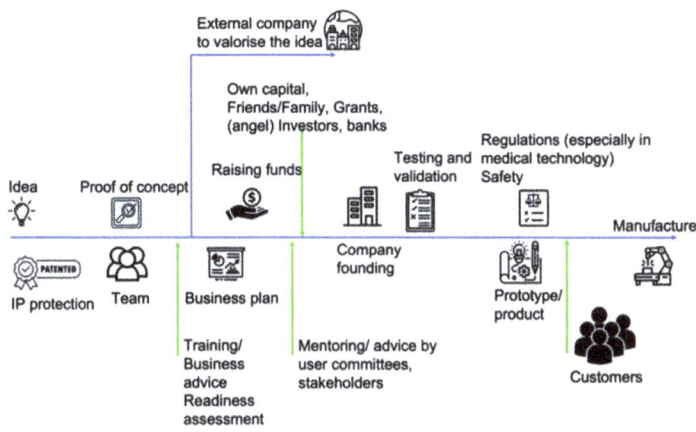

Figure 2. *A roadmap for commercialization of nanotechnology products (Reprinted with permission from [10], Copyright Royal Society of Chemistry).*

Incorporating innovative approaches into your exploration plans increases your product's chances of research success. Two possible approaches to fostering innovative ideas are exploration-based invention and data-driven invention [21]. Research-driven inventions utilize the scientific or technological talent of a laboratory as the seed of an innovative idea. Experimenters use these seeds to create exploration plans, but many ideas avoid the complexity of scientific exploration. This approach will significantly increase innovation and value creation in the private sector. This will enable the exploration group to work with relevant parties to develop slice edge technology. Data-driven invention fully captures the results of requirements analysis and also takes into account the internal capabilities of the exploration department [49].

Innovative approaches produce new perspectives in the laboratory and find new uses for being specialized chops. This also leads to the development of new products according to request requirements. Advance technologies, especially those using nanotechnology, aim to produce value. This technology creates value when it significantly improves performance or significantly reduces the cost of working problems. Still, from a business model perspective, nanotechnology invention poses significant challenges in delivering products to guests. Go-to-request strategies are important because they can be affected by a variety of factors (similar as resource limitations). Exploration agencies and organizations with strong R and D capabilities and track records can benefit from high-performance products without having to directly manufacture and sell products and services to mass consumers. This can have a significant impact on a company's success, as product development and support pose significant challenges. This issue can be resolved through a license agreement. In similar cases, a fairly binding written agreement exists between the parties stating that the intellectual property owner (the licensor) allows another party (the assignee) to use the intellectual property. The most common types of intellectual property include imprints, patents, and trademarks. Innovative approaches create new perspectives in the laboratory and discover new uses for specialized chops. This will also lead to the development of new products in response to inquiries. Advanced technologies, especially those using nanotechnology, aim to increase added value. This technology creates value by significantly improving performance or significantly reducing the cost of labor issues. However, from a business model perspective, nanotechnology inventions pose significant challenges in delivering products to guests. The go-to-request strategy is important because it can be affected by a variety of factors (such as resource limitations). The effectiveness of transferring nanotechnology inventions from the laboratory to industry depends on the efficiency of the technology transfer process. Countries that invest in perfecting nanotechnology transfer programs and practices will achieve better nanotechnology outcomes. This is clearly seen in the United States where the National Nanotechnology Initiative (NNI) was developed. This is through collaboration with the private sector and institutions involved in nanotechnology research, development and commercialization [50].

The NNI includes institutions such as the Nanomanufacturing and Small Business Innovation Research (SBIR) Program and the NNI's National Nanotechnology Coordination Office (NNCO), which is responsible for developing newly developed nanotechnology into market-ready products. It is. Spending on nanotechnology research is increasing in Asia, and concerted efforts are being made to communicate exploration results to the public [51].

Chancing the right time for a new company to enter the request can be delicate. This is true whether the company starts from scrape or operates through a cooperation. But the alternate business model can also give significant support in colourful ways. The Clover Leaf Model can be used to estimate whether a new technology is ready for commercialization and request entry. This model is so named because it includes four crucial criteria that can be compared to the leaves of a four-splint clover, making it ideal for marketable technology. These criteria, shown in Fig. 3, are technology readiness, request readiness, commercialization readiness, and operation readiness [52].

Figure 3. Cloverleaf framework for assessing the market readiness of nanotechnology innovations (Reprinted with permission from [10], Copyright Royal Society of Chemistry).

Finding a way to fund a startup can be a difficult task. This challenge can be alleviated if business innovators know who to communicate with and when. To take advantage of the maximum support options, you must first register your company. This registration process has different costs for each country. In some cases, you may need to calculate all other costs incurred by your business to determine total labor hours before reaching the implied investor. These fixed and variable operating costs typically include exploration and development costs, operating costs, product costs, corporate equipment costs (as well as ministry costs), garcon costs, and marketing costs. These costs vary depending on the business model used. Government subventions are anon-dilutive source of backing and have come a popular way to fund wisdom-grounded enterprise. This is particularly profitable because, unlike taking out a loan or chancing an investor, subventions don't bear prepayment of lenders or release of commercial capital. Scientists can make the utmost of these backing openings by learning about the types of backing available. The public sector (private and source governments) and the private sector (foundations) can be examined across major sources of funding. Each education has different operation conditions and different benefits. So, choose the stylish education for you. These backing options are competitive, so startup systems need to be innovative.

5.   The Challenge of Moving Technology from the Lab to Industry

Although the commercialization of nanotechnology is still in its immaturity, the relinquishment rate of this technology has not increased, largely due to the large quantum of government backing flowing into nanotechnology. Commercialization is the process of turning a new technology into a successful marketable adventure, involving a variety of experts with specialized, business, and profitable backgrounds to transfigure the new technology into a useful product or service. We live in a period when abomination and fear of new technology is at an each-time low the Age of No Contestation. thus, the general station is to drink new technologies similar as nanotechnology [42, 43].

Still, specialized know-style or copping new technology products alone doesn't guarantee deals success. This technology needs to be extensively accepted by the assiduity. Assiduity should only join this trouble if it's confident of its own success. One of the biggest hurdles (though not the biggest) is the gap between assiduity and academia in his medium-sized TRL (Technology Readinnes Level), also known as Death Valley. Science strives for low TRL, but if the technology isn't completely understood, at some point it becomes less intriguing for academics [54]. Still, the assiduity is generally threat-antipathetic and prefers to work in areas where short-term earnings can be achieved. For this reason, they prefer to work in areas with high TRL. This creates a gap where scientists are no longer interested, and assiduity mates remain apathetic. There are numerous impulses to close this gap. But collaboration between scientists and assiduity mates is really demanded. Once an incipiency is established, new authors also face numerous challenges. This means that numerous startups cannot survive in the early stages. Because nanotechnology operations are different, different diligence can be also affected. thus, the challenges of nanotechnology are veritably different. Common groups of these challenges include specialized, natural or environmental, profitable, nonsupervisory, etc. Specialized challenges include those related to the association's structure (both association and structure) and the specialized knowledge of the platoon. The institute's organizational structure doesn't encourage networking between nanotechnology experimenters and investors, but rather fosters entrepreneurship by fostering precious exchanges between experimenters and entrepreneurs in the nanotechnology assiduity. It promotes It doesn't promote or suppress culture. This limits the possibility of forming interdisciplinary brigades to develop specialized chops in nanotechnology. Lack of specialized knowledge can lead to significant difficulties in commercialization, as this gap can lead to industrially unsustainable inventions. Specific challenges associated with the successful development of commercially viable nanotechnology materials include the inability to sustain advantages over existing products, the inability to integrate technology, and difficulties in reproducibility and batch control. Significant changes in product manufacturing on an artificial scale [55]. When designing and developing nanotechnological products, limitations encountered in industrial production must be taken into account to ensure scalability. Similarly, the repeatability of validated manufacturing processes should be included in product development protocols. A solid understanding of nanotechnology policy, along with appropriate academic education and risk management training, can also help alleviate problems caused by a lack of knowledge on the subject.

Environmental and biological difficulties may also impede the transfer of nanotechnology from the laboratory to commercial practice [11] .Nanomaterials contain substances that are potentially harmful to natural systems and can enter the landscape at various stages of their life cycle. Three generally applicable emigrations scripts are;

✓ Emigrations from the manufacture of colorful nanotechnology products or products with nano characteristics.

✓ Released during use.

✓ Release after destruction [56].

Nanomaterials can beget numerous changes when present in the terrain. These include chemical changes (e.g. photodegradation), physical changes (e.g. aggregation), and biologically intermediated changes (e.g. the relationship between these changes and the transport of nanomaterials within ecosystems), including relations that eventually determine the fate and ecotoxicity of the nanomaterial. Implicit natural and ecological impacts of nanotechnology innovations need to be determined using in vitro and in vivo models in aquatic and terrestrial ecosystems. The manufacturing process by which nanomaterial performance is achieved must also be considered. Materials released during use, recycling, and disposal of nanodevices must be minimized. The purpose of modeling is to predict current developments [57].

Although these simulations are useful, more effective, and reliable logic tools and styles need to be developed to fully characterize and quantify nanomaterials. There is a need to develop tools to identify, coat, and fractionate nanomaterials in natural environments and complex environmental matrices. Nanotechnology decisions play an important role in profitable development. However, the difficulty of making a profit may prevent the transfer of inventions from the laboratory to industry. These generally include limited investments in viable R and D conditions, applicable mechanisms, laboratory facilities, and viable mechanisms to secure similar investments. The limits of the conditioning required to manipulate nanotechnology outcomes are also illustrated by the dynamics of socially beneficial inventions. Although many people believe that the rapid development of nanotechnology will bring great benefits, some argue that the development of nanotechnology should continue or be stopped. This group's response to nanotechnology is based on the belief that nanotechnology increases inequalities in social welfare and reinforces the power imbalances caused by inequality. They argue that this leads to inequalities in nanotechnology and therefore differences in access to nanotechnology between low- middle- and high-income countries [48, 49].

Ethical analysis is primarily concerned with inequalities based on public performance of where knowledge is developed. They must share in these processes [60].

Nanotechnology and the World This conditioning aim to promote broader access to nanotechnology and global invention and are extremely important to bring this exertion to a close. This is the final set of challenges that have important implications for moving nanotechnology from the laboratory to commercial practice. These relate to the lack of clear non-regulatory rules for nanotechnology and nanotechnology-based products. Regulatory challenges include insufficient programs to support the development and operation of nanotechnology companies and insufficient governments to attract nanotechnology companies. Similarly, there are no technology transfer protocols or formal approval conditions to facilitate the transfer of inventions from the laboratory to marketable products [61].

Therefore, new IP processes and protocols are needed to simplify the path from laboratory to commercial use and save time and cost. The special, natural, ecological, profitable and uncontrolled challenges of nanotechnology need to be urgently addressed. Programs that manage all aspects of nanotechnology discovery and subsequent commercialization must balance benefits and challenges. To overcome these challenges, efforts must be made to help the harmful substances of nanotechnology while increasing awareness of its benefits to society [62]. Therefore, it is important that academia, government, industry, and partners are involved in the decision-making process to address, minimize, and mitigate the challenges associated with the commercialization of nanotechnology.

6. Nanotechnological Advances

The growing global nanotechnology request is estimated to exceed USD 125 billion by 2024. Thus, commercializing exploration results through the conflation and operation of nanotechnology is salutary and has great eventuality to profit society through its colourful operations. As a result, governments and the private sector around the world are increasing their investments in nanotechnology. In the EU alone, roughly€ 896 million was invested in nanotechnology exploration between 2007 and 2011. Global investment in nanotechnology is estimated at roughly USD 2.5 billion, with China and the United States investing up to USD 2 billion. Although these two countries are considered nanotechnology titans, the United States remains the world leader in government investment in nanotechnology [21]. Adding global backing has told the number of scientific publications on nanotechnology. As the number of publications increased, so did the number of patented technologies.

The 25 countries with the most patents for nanotechnology development, only Europe (14), Asia (8), North America (2), and Oceania (1). Patents can be used as technological indicators to indicate research and development activities that generate commercial benefits [49]. However, translating these nanotechnology advances into commercialized final products remains a major challenge for the scientific community. There are differences from country to country regarding the bureaucracy of patent procedures. There are also differences in the scope of patent grants. In some cultures, it may be more common to keep innovations secret rather than patent them. There are also differences when it comes to patent applications. Some companies have many small patents, while others have multiple, more complex patents.

Biogenic Nanomaterials: Synthesis, Characterization, Applications, and Future Remarks   Materials Research Forum LLC
Materials Research Foundations 180 (2025)   https://doi.org/21741/9781644903759

Clothing that absorbs sweat and keeps the body dry thanks to special polymers has already been developed [63]. Japanese Kanebo Spinning Co., Ltd. has produced a polyester yarn that has 30 times the moisture absorption capacity of regular polyester. The yarn is mainly used in the manufacture of underwear. The PES yarn consists of a total of 20 layers with a total layer thickness of approximately 50 nm and is manufactured by Toray Industries, Inc. of Japan. Ultra's nano-fine nylon yarn exhibits excellent hygroscopic properties [64]. The state of nanotechnology in the last decade has supported the rapid development of the textile technology sector. Nanotechnological developments are expected in the textile industry in the next 25 years. Military clothing will be at the forefront of these developments. The Massachusetts Institute of Technology (MIT) is using nanotechnology to develop a "super uniform" for 21st century soldiers. These uniforms have certain properties such as artificial muscle building and energy storage (morph fabrics) as well as color-changing and phase-changing materials that aid in camouflage and streamlining properties that aid tearing. The fibers in this fabric contract or expand depending on the ambient temperature and airflow. Organizations equipped with nanosensors will transmit the body signals of the soldiers to the medical centre and increase the reaction speed by reporting the health information and location of the injured soldiers to the centre through the integrated communication and circulation equipment of the organization. Uniforms made with nanotechnology will be 80% lighter than currently used (the weight of paper but lighter and more flexible) and will adapt to biological or chemical hazards in the environment at a molecular level. By doing this, you lose transparency. These uniforms adapt to the temperature, light and air quality of the environment. You will easily notice the changes. A special fiber developed with nano-coating will be detectable in the dark, allowing soldiers to choose each other from kilometers away and distinguish enemies even in dark environments [65]. Improved antimicrobial properties thanks to silver threads and fabrics, color change sensitive to biological pests such as bird flu, *E. coli*, self-cleaning textile applications, jackets with integrated iPod control, shirts with electrocardiogram recording, Plasma application chemicals that can be contaminated in very low amounts are still under development. is in the process. Nanocoatings and nanoporous functional coatings for medical and sanitary applications in textile materials are also considered advanced research topics in nanotechnology. Research into the use of gold nanoparticles as a dye in luxury fashion fabrics is being conducted at Victoria University in New Zealand. Studies have shown that the proportion of gold solution used for dyeing affects particle size and therefore color. The permanence of the color depends on the nanoparticle size, which is 1/1000 of the fiber, which enables the development of new fashion fabrics and fibrous materials with an incorruptible golden color [66]. Japanese Teijin Fibers Co., Ltd. continues to work on the production of glossy polyester. The polyester substrate is coated with approximately 60 layers of polyester and nylon with different refractive indices. A thin layer such as 69 nm refracts the light, creating a 'mystical' reflection depending on the viewer's point of view and the angle of incidence of the light on the fabric [64].

In order to increase the strength and stiffness values of the fabric, carabiner molecules are being researched to strengthen the fibers of the fabric at the molecular level. Carbines are linear allotropes formed by hybridization of carbon, and their structure consists of carbon atoms forming single and triple bonds, respectively. The fact that the carabiners are in long molecular chains and have high elasticity made it possible to use the carabiners for fiber reinforcement [65]. In the near future, textile products will be equipped with functions such as computers, street computers, music players, mobile phones and internet connection [57, 58].

The properties that textiles gain thanks to nanotechnology are due to the nano-scale materials (nanofibers, nanotubes, nanocomposites, etc.) in their structures and the technology used to create nano-scale surfaces. The properties that these materials and techniques can bring to textile products are as follows: They maximize the performance properties of the fabric such as waterproof, stain-proof and anti-crease [69]. By using multi-walled carbon nanotubes, it is possible to produce multi-ply yarn with high twist strength in the yarn twisting process. Carbon nanotubes are tougher than other natural and synthetic fibers, and their tensile strength is comparable to that of spider webs. It is used in electronic textile applications such as sensors, electronic interconnects, electromagnetic wave blockers, antennas and batteries that store electricity [60, 62, 63, 64, 65].

Clay nanoparticles have excellent electrical, thermal and chemical resistance. It also has the ability to block UV rays. Therefore, fibers containing clay nanoparticles exhibit flame retardant, UV resistant and corrosion resistant properties [75,76]. Nanoclays, montmorillonite and some modified nanoclays. They are used as absorbent material for anionic, cationic and neutral dyes. To improve dye absorption, absorbent materials are physically added to the polymer matrix to form composites. Due to the dye absorption capacity of nanoclay absorbers, fiber surfaces made with this composite have properties such as very good dyeability, color fastness, low dyeing cost and less wastewater treatment problems [71]. Besides clay nanoparticles, metal oxide nanoparticles are also used for fiber applications. Metal oxides such as TiO_2, Al_2O_3, ZnO and MgO have photocatalytic effect [67, 68], electrical conductivity, ultraviolet absorption effect and photooxidation effect [79]. Nylon fibers containing ZnO provide UV protection while also reducing static on the fibers. Composite fibers containing TiO_2/MgO nanoparticles provide self-sterilization. Due to their photocatalytic functions, they have begun to replace the use of activated carbon in the manufacture of protective clothing [80]. A phase change material (PCM) embedded in a fiber material is enclosed within spheres (microcapsules) of just a few micrometers, called PCM microcapsules. PCMs are used for thermoregulation, i.e. to absorb heat and release it on demand. First, FDMs placed in clothing actively balance the thermal energy emitted or absorbed by the body, forming an insulating layer between the external environment and the human body [80,81]. Formal memory materials (SMMs) are used to improve insulating and protective properties against extreme hot or cold environmental conditions. Chromium materials are also chameleon-like fibers. In other words, they have the ability to change color in response to different environmental conditions [80]. Nanotechnology is used to increase the effectiveness of currently applied finishing processes. The finish is emulsified in nano size (using nanomicelles, nanosols or nanocapsules) for a smoother application. As a result of this process, such as crease resistance, water repellency, stain resistance, oil repellency, high dimensional stability, hydrophilicity, hydrophobicity, flame retardancy, UV resistance, deodorant, high conductivity, biodegradability. You can add features to your fabric. Nano-thin coatings increase the camouflage ability of surfaces by changing their optical properties [65,80].

Manufacturing technology for nanoscale surfaces. Surface modification with plasma, sol-gel technology and microencapsulation applications. It is used in many textile finishing processes to impart functional properties such as water repellency, stain repellency, wettability and flame retardancy to fabrics. Application areas of sol-gel technology in textiles. Water repellency, oil repellency, stain resistance, controlled odor release, biocatalytic properties, biocompatibility properties, electrical conductivity, color fastness etc. Also, sol-gel technology in the textile industry. Production of photochromic, electrochromic and thermochromic fibers, improved strength, modified barrier properties, fibers. It can also be used to make superhydrophobic (self-cleaning) fabrics, make antimicrobial fabrics, and improve filtration, adsorption, permeability, anti-wrinkle, UV protection and flame-retardant properties. Microencapsulation applications in textile. Perfumes, cosmetics (such as moisturizers), insect repellents, flame retardants, vitamin and drug applications, antibacterial applications, dyeing and phase change material applications [80] . The use of nanotechnology in the Turkish textile industry is quite common, especially in the finishing phase. Companies that produce products for important brands prepare collections with nano-enriched processes [65].

Han et al. (2011), developed self-sensing patches and demonstrated that multifunctional and smart CNT-reinforced cementitious materials and structures have great potential for traffic monitoring applications such as vehicle detection, weight and speed measurement. These materials are used in skyscrapers, highways, bridges, airports and aprons, high-speed rail links, dams, nuclear power plants, etc. is used. Compared to conventional cementitious materials, CNTs have been observed to have positive economic, social and environmental effects [81]. Saafi (2009), developed a wireless embedded sensor for in situ damage detection in concrete structures using a cement-based SWNT reinforcement material whose resistance changes with pressure [81]. Veedu (2010), highlighted the potential for in situ structural health monitoring and traffic monitoring applications in bridges using multifunctional and intelligent CNT-reinforced cementitious materials. The construction industry ranks eighth among the ten most important applications of nanotechnology [82]. Bartos (2005) points out that the construction industry is a potential main consumer of nanostructured materials, although many materials have not yet completed the experimental stage [83].

The addition of nanocomposites to polymers provides flexibility, durability, heat/moisture stability, barrier properties, light and flame resistance, strong mechanical and thermal performance, and high gas barrier properties [84]. Nanoparticles in plastics or films form an important barrier with their antimicrobial properties that prevent the passage of oxygen, carbon dioxide and moisture into food. The use of nanoparticles also makes the material lighter, more tear-resistant and able to withstand high temperatures. Applications of nanocomposites are films, coatings, plastic bottles, containers, etc. may form. It is especially used in meats, cheeses, fish, bakery products, carbonated beverages, fruit juices and beer [85]. The applications of nanocomposite films in food are mostly combined with active/intelligent packaging (antimicrobial films) and edible film/coating Technologies [86]. By incorporating nanoparticles into biodegradable plastics, new materials with different properties have been developed [87]. In other words, biopolymer-based packaging is defined as packaging in which raw materials obtained from agriculture and fisheries are used as raw materials. Biopolymers, which are important in the production of food packaging materials, can be divided into three groups. According to the raw material source used, they can be divided into three main groups as polysaccharides, oil films and protein films [88]. The application of nanotechnology to these polymers opens up new possibilities not only for improved properties but also for lower cost efficiency [87]. Biodegradation of these materials is the process by which carbonaceous compounds are broken down in the presence of enzymes secreted by organisms. The rapid degradation process has three requirements. These are temperature, humidity and microbial species [79, 80]. Smart packaging is an emerging field of packaging science and technology that offers exciting opportunities to improve the food safety, quality and convenience of seafood. Smart packaging uses the communication capabilities of packaging to facilitate decision making [91]. Nanocomposites formed by integrating several nanometals or metal oxides into polymers exhibit antimicrobial properties. It uses the antibacterial properties of nanoparticles. These substances slow down microbial growth in foods and ensure a long shelf life [92]. In addition, nanosensors embedded in plastic can be used to monitor food status [85] . Research continues on nanosensors that change color to show the presence of pathogens in foods[81, 82, 83].

Gur et al. (2022), tried to describe the structure of zinc oxide nanoparticles (ZnO NPs) attained from the factory *Thymbra Spicata L.* in different ways using a green conflation system. Some parcels of zinc oxide nanoparticles have been studied during exploration using scanning electron microscopy (SEM), energy dissipative X-ray spectroscopy (EDX), Fuller transfigure infrared spectroscopy (FTIR), X-ray diffraction (XRD), and tested using measured ultraviolet light. They set up that the size of the Zn nanoparticles was 6.5 –7.5 nm. They also delved the antibacterial goods of zinc oxide nanoparticles attained by green conflation against colorful pathogens. According to the results attained, zinc oxide nanoparticles have a positive effect on *Bacillus subtilis* ATCC 6633 with a periphery of 16.3 mm, against *E. coli* ATCC 25952, against *Pseudomonas aeruginosa* ATCC 27853, with a periphery of 10.2 mm, and against *Candida sp.* A region with a periphery of 10.2 mm was formed. Still, they set up that the revolutionary quenching exertion (DPPH) of the nanoparticles Ts-ZnO NPs (79.67) was significantly better than that of the positive control BHA. They also observed that the defensive effect of ZnO NPs against DNA damage increased with attention. The inhibition effect on DNA damage was highest at a concentration of 100 mg/L. Consistent with the overall findings, zinc oxide nanoparticles synthesized through a green method can be applied in various applications. This was confirmed to be feasible [95].

The ZnO Rr (*Rheum ribes*) NPs synthesized by Meydan et al. were validated using a reliable characterization system. The data attained from the study showed that ZnO NPs Rr had a veritably strong inhibitory effect on lipid peroxidation. The lipid peroxidation inhibitory exertion of ZnO NPs Rr was calculated to be 89.1028 at the loftiest attention of 250 µg/mL. They observed that ZnO-NP/Rr inhibited DNA damage by 92.1240 at the loftiest attention of 100 µg/mL. Since *R. ribes* has been used as an antidiabetic medicinal factory, the antidiabetic effect of ZnO NPs Rr formed by ZnO was significant. In this study, we observed that ZnO NPs Rr formed zones ranging from 8 ±3.0 to 21 ±4.5 for Gram-positive and Gram-negative microorganisms. They set up that ZnO nanoparticles have antibacterial good [96].

Meydan et al. involved nanoparticle conflation using *Arum italicum* and silver essence. They also delved the antibacterial exertion, DNA damage forestallment, and DPPH revolutionary quenching exertion of Ag NP/ Ai nanoparticles. Factory- tableware relations are anatomized by X-ray diffraction (XRD), ultraviolet-visible spectrophotometry (UV- Vis), surveying electron microscopy and energy dispersive X-ray (SEM- EDX), Fourier transfigure infrared spectroscopy (FTIR). Established. Ag and Ag NP/ Ai clusters from *Arum italicum* were set up to have antibacterial exertion against the pathogens *Bacillus subtilis*, *Bacillus cereus*, *Enterococcus faecalis*, *Staphylococcus aureus*, *Pseudomonas aeruginosa*, and *Escherichia coli*. *Pseudomonas aeruginosa* showed a stronger effect than antibiotics. Ag- NP/ Ai is believed to have defensive and preventative goods against DNA damage. The antioxidant effect of Ag- NP/ Ai was significant when comparing the DPPH revolutionary quenching exertion with the positive controls BHA and BHT [97] .

This study conducted by Altındag and Meydan (2021), delved the salutary goods of gallic acid on cisplatin- convinced testicular and epididymal toxin. Manly rats were divided into four groups control, cisplatin (8 mg/ kg single cure), gallic acid (50 mg/ kg), and cisplatin gallic acid (n = 7). Testes were examined morphometrically using stereological styles. Testicular apoptosis, DNA damage, oxidative stress parameters, and serum testosterone were also measured. The epididymis was estimated histopathologically. As a result, significant diminishments in spermatogonia, Leydig cell number, Sertoli cell number, testis volume, germinal epithelial height, Bcl- 2 immunopositive cell number, CAT, GSH, SOD enzyme conditioning, and serum testosterone situations were observed. Ta. A significant increase in the number of immunopositive caspase- 3, Bax, and 8- OHdG cells, as well as MDA situations, was observed in the cisplatin and control groups. Still, gallic acid significantly restored these parameters [98].

A study by Meydan and Seçkin (2021), set up that shops with significant medicine product eventuality were explosively favoured for nanoparticle product. Scanning electron microscopy and energy dispersion X-ray analysis (SEM/ SEM- EDX), Fourier transfigure infrared spectroscopy (FTIR), *Helichrysum arenarium* factory excerpt and ZnO ultraviolet-visible (UV-Vis) diffraction and immersion (XRD) technology. The antioxidant capacity structure of Zn NPs Ha was determined by DPPH assay. We've delved the antibacterial goods of zinc nanoparticles against six different pathogens (*Bacillus cereus* ATCC 10876, *Escherichia coli* ATCC 25952, *Bacillus subtilis* ATCC 6633, *Staphylococcus aureus* ATCC 29213, *Pseudomonas aeruginosa* ATCC 90028). As a result of the study, an inhibitory effect on some pathogenic microorganisms was observed. The content of antioxidants was also set up to be significant [99].

Nuran et al. (2022), zinc is one of the most popular rudiments due to its nanostructure and parcels. It's known to be involved in numerous places, including DNA protection, membrane stabilization, and protein, ribosome, and carbohydrate product. The end of this study is to probe several biochemical parameters directly related to mortal health as a result of zinc-grounded characterization of *Platonus orientalis* factory excerpts. Scanning electron microscopy was used for his SEM and SEM/ EDX images of the synthesized his ZnO NPs. Characterization was performed using UV-visible spectrophotometry and Fourier transfigure infrared spectroscopy (FTIR). DPPH (2,2- diphenyl-1-picrylhydrazyl) analysis shows that zinc nanostructures have significant antioxidant capacity. The antibacterial exertion of synthetic essence nanoparticles against colorful pathogens in humans was delved using fragment prolixity system. ZnO- NP/ Po was shown to have antibacterial parcels against *Bacillus subtilis* ATCC 6633, *Escherichia coli* ATCC 25952, and *Pseudomonas aeruginosa* ATCC 27853. According to the exploration results, the effectiveness of ZnO NP/ Po can be attributed to this material. It sheds light on the world of wisdom, especially drug [96].

7.   Conclusion

The developments in nanotechnology in the industrial field are realized by the commercialization of laboratory research and its transformation into production. This process is fraught with some difficulties and obstacles. Above all, nanomaterials must be produced on a large scale and produce consistently reproducible results. In addition, safety and regulatory issues need to be considered, as the environmental effects and possible human health effects of nanoscale particles are still being investigated.

When we look at the use of nanotechnology in industry, we see important developments in many areas. Thanks to nanotechnology in electronics, we can make devices that are smaller, faster and more energy efficient.

In medicine, the use of nanomaterials holds great promise in many areas such as drug delivery systems, imaging technologies, and innovative therapeutics in cancer treatment.

The use of nanomaterials in agriculture opens up new possibilities in terms of plant nutrition, pest control and productivity. In addition, the use of nanotechnology in areas such as solar cells, fuel cells and energy storage systems in the energy sector can develop more efficient and sustainable energy sources. Collaboration and knowledge sharing are essential for the industrial-scale application of nanotechnology. Enterprise similar as academic exploration, assiduity hookups, and government- funded programs accelerate the restatement of laboratory inventions into artificial operations. In addition, it is important to establish standards, develop security protocols and improve regulations in the nanotechnology sector. In this way, nanotechnology can become economically reliable and sustainable. Nanotechnology is revolutionizing many fields from laboratory to industry.

The ability to precisely control material properties in nanoscale structures offers innovative solutions in many areas. However, challenges such as mass production, safety and regulation are factors limiting the industrial adoption of nanotechnology. Therefore, it is important to engage in advanced research, collaborative research and regulatory frameworks. With these studies, the nanotechnological inventions obtained in the laboratory will reach larger masses and have great repercussions in the industrial field. Therefore, the positive effects of nanotechnology are very diverse. However, the products revealed by research are not progressing at the right pace. There are still many problems. This problem can be overcome with the proliferation of investment brokers and new ideas. The number of investment interposers is anticipated to increase in the coming times, especially considering the benefits of nanotechnology.

References

[1]   Zhao, P. *et al.* (2019). A novel ultrasensitive electrochemical quercetin sensor based on MoS2 - carbon nanotube @ graphene oxide nanoribbons / HS-cyclodextrin / graphene quantum dots composite film. *Sensors and Actuators, B: Chemical.* https://doi.org/10.1016/j.snb.2019.126997

[2]   Cherif, A. *et al.* (2021). Numerical investigation of hydrogen production via autothermal reforming of steam and methane over Ni/Al2O3 and Pt/Al2O3 patterned catalytic layers. *International Journal of Hydrogen Energy.* https://doi.org/10.1016/j.ijhydene.2021.04.032

[3]   Ulus, R. *et al.* (2016). Functionalized Multi-Walled Carbon Nanotubes (f-MWCNT) as Highly Efficient and Reusable Heterogeneous Catalysts for the Synthesis of Acridinedione Derivatives. *ChemistrySelect.* https://doi.org/10.1002/SLCT.201600719

[4]   Gulbagca, F. *et al.* (2021). Green Synthesis of Palladium Nanoparticles: Preparation, Characterization, and Investigation of Antioxidant, Antimicrobial, Anticancer, and DNA Cleavage Activities. *Applied Organometallic Chemistry.* https://doi.org/10.1002/aoc.6272

[5]   Ertan, S. *et al.* (2012). Platinum nanocatalysts prepared with different surfactants for C1-C3 alcohol oxidations and their surface morphologies by AFM. *Journal of Nanoparticle Research.* https://doi.org/10.1007/S11051-012-0922-5

[6]   Askari, M.B. *et al.* (2021). Enhanced electrochemical performance of MnNi2O4/rGO nanocomposite as pseudocapacitor electrode material and methanol electro-oxidation catalyst. *Nanotechnology.* https://doi.org/10.1088/1361-6528/abfded

[7]   Lolak, N. *et al.* (2019). Composites of Palladium–Nickel Alloy Nanoparticles and Graphene Oxide for the Knoevenagel Condensation of Aldehydes with Malononitrile. *ACS Omega.* https://doi.org/10.1021/acsomega.9b00485

[8]   Aygun, A. *et al.* (2023). Highly Active PdPt Bimetallic Nanoparticles Synthesized By One Step Bioreduction Method: Characterizations, Anticancer, Antibacterial Activities and Evaluation of Their Catalytic Effect for Hydrogen Generation. *International Journal of Hydrogen Energy.*

https://doi.org/10.1016/j.ijhydene.2021.12.144

[9]   Nagraik, R. *et al.* (2021). Amalgamation of biosensors and nanotechnology in disease diagnosis: Mini-review. *Sensors International.* https://doi.org/10.1016/J.SINTL.2021.100089

[10]   Şen, B. *et al.* (2018). High-performance graphite-supported ruthenium nanocatalyst for hydrogen evolution reaction. *Journal of Molecular Liquids.* https://doi.org/10.1016/J.MOLLIQ.2018.07.117

[11]   Kaur, I.P. *et al.* (2014). Issues and concerns in nanotech product development and its commercialization. *Journal of Controlled Release.* https://doi.org/10.1016/j.jconrel.2014.06.005

[12]   Eris, S. *et al.* (2018). Investigation of electrocatalytic activity and stability of Pt@f-VC catalyst prepared by in-situ synthesis for Methanol electrooxidation. *International Journal of Hydrogen Energy.* https://doi.org/10.1016/j.ijhydene.2017.11.063

[13]   Hobson, D.W. (2009). Commercialization of nanotechnology. *WIREs Nanomedicine and Nanobiotechnology.* https://doi.org/10.1002/wnan.28

[14]   Haleema *et al.* (2022). Recycling of nanomaterials by solvent evaporation and extraction techniques, in *Nanomaterials Recycling,* Elsevier, pp. 209–222.

[15]   Kosmowski, K. (2021). Fourth industrial revolution and new challenges in post-pandemic world.

[16]   Nasrollahzadeh, M. *et al.* (2019). Applications of Nanotechnology in Daily Life. *Interface Science and Technology.* https://doi.org/10.1016/B978-0-12-813586-0.00004-3

[17]   Nie, L. *et al.* (2021). Quantum monitoring of cellular metabolic activities in single mitochondria. *Science Advances.* https://doi.org/10.1126/sciadv.abf0573

[18]   Hälg, D. *et al.* (2021). Membrane-based scanning force microscopy. *Physical Review Applied.*

[19]   Munawar, A. *et al.* (2019). Nanosensors for diagnosis with optical, electric and mechanical transducers. *RSC Advances.* https://doi.org/10.1039/C8RA10144B

[20]   Rambaran, T.F. (2020). Nanopolyphenols: a review of their encapsulation and anti-diabetic effects. *SN Applied Sciences.* https://doi.org/10.1007/s42452-020-3110-8

[21]   Rambaran, T. and Schirhagl, R. (2022). Nanotechnology from lab to industry–a look at current trends. *Nanoscale advances.*

[22]   Tolochko, N.K. (2009). History of nanotechnology. *Encyclopedia of Life Support Systems (EOLSS).*

[23]   Appenzeller, T. (1991). The man who dared to think small. *Science.* https://doi.org/10.1126/SCIENCE.254.5036.1300

[24]   Baig, N. *et al.* (2021). Nanomaterials: A Review Of Synthesis Methods, Properties, Recent Progress, And Challenges. *Materials Advances.* https://doi.org/10.1039/D0MA00807A

[25]   Yang, Y. *et al.* (2023). Progress and future prospects of hemostatic materials based on nanostructured clay minerals. *Biomaterials Science.* https://doi.org/10.1039/D3BM01326J

[26]   Haverkamp, R.G. *et al.* (2007). Pick your carats: nanoparticles of gold–silver–copper alloy produced in vivo. *Journal of Nanoparticle Research.* https://doi.org/10.1007/s11051-006-9198-y.

[27]   Reverberi, A.P. *et al.* (2019). Nanotechnology in machining processes: recent advances. *Procedia CIRP.* https://doi.org/10.1016/j.procir.2019.02.002

[28]   Feynman, R. (2011). What is science? *Resonance.*

[29]   Maclurcan, D. *et al.* (2004). Medical Nanotechnology and Developing Nations. *Oz Nano.*

[30]   Luther, W. (2004). INDUSTRIAL APPLICATION OF NANOMATERIALS: CHANCES AND RISKS TECHNOLOGICAL ANALYSIS). VDI TECHNOLOGIEZENTRUM, GERMANY.

[31]   Mao, X. *et al.* (2016). Engineered Nanoparticles as Potential Food Contaminants and Their Toxicity to Caco-2 Cells. *Journal of Food Science.* https://doi.org/10.1111/1750-3841.13387

[32]   (2022). A Review on Nanotechnology: Applications in Food Industry, Future Opportunities,

Challenges and Potential Risks. *Journal of Nanotechnology and Nanomaterials*. https://doi.org/10.33696/Nanotechnol.3.029

[33]   Gupta, R. and Xie, H. (2018). Nanoparticles in Daily Life: Applications, Toxicity and Regulations. *Journal of environmental pathology, toxicology and oncology* . https://doi.org/10.1615/JENVIRONPATHOLTOXICOLONCOL.2018026009

[34]   Duncan, T. V. (2011). Applications of nanotechnology in food packaging and food safety: Barrier materials, antimicrobials and sensors. *Journal of Colloid and Interface Science*. https://doi.org/10.1016/j.jcis.2011.07.017

[35]   Ahmadi, M.H. *et al.* (2019). Renewable energy harvesting with the application of nanotechnology: A review. *International Journal of Energy Research*. https://doi.org/10.1002/er.4282

[36]   Charinpanitkul, T. *et al.* (2008). Review of Recent Research on Nanoparticle Production in Thailand. *Advanced Powder Technology*. https://doi.org/10.1016/S0921-8831(08)60911-5

[37]   Güzeloğlu, E. (2015). Nano innovation with smart products: The youth's nanotech awareness, perception of benefits/risks Akıllı ürünleriyle nano yeniliği: Gençlerin nanoteknoloji farkındalığı, fayda/risk algıları. *Journal of Human Sciences*.

[38]   Gunasekera, U.A. *et al.* (2009). Imaging applications of nanotechnology in cancer. *Targeted Oncology*.

[39]   ADEOLA, A.O. *et al.* (2019). Scientific applications and prospects of nanomaterials: A multidisciplinary review. *African Journal of Biotechnology*. https://doi.org/10.5897/AJB2019.16812

[40]   Babatunde, D.E. *et al.* (2020). Environmental and Societal Impact of Nanotechnology. *IEEE Access*. https://doi.org/10.1109/ACCESS.2019.2961513

[41]   Elmarzugi, N.A. *et al.* (2014). Awareness of Libyan students and academic staff members of nanotechnology. *Journal of Applied Pharmaceutical Science*.

[42]   Nesrin, K. *et al.* (2020). Biogenic silver nanoparticles synthesized from Rhododendron ponticum and their antibacterial, antibiofilm and cytotoxic activities. *Journal of Pharmaceutical and Biomedical Analysis*. https://doi.org/10.1016/J.JPBA.2019.112993

[43]   Johansson, M. and Boholm, Å. (2017). M., Johansson, A., Boholm, Scientists' Understandings of Risk of Nanomaterials: Disciplinary Culture Through the Ethnographic Lens. Nano Ethics. 11(3):229-42. (2017). *NanoEthics*.

[44]   Schiener, M. *et al.* (2014). Nanomedicine-based strategies for treatment of atherosclerosis. *Trends in molecular medicine*.

[45]   Patil, A.A. and Rhee, W.J. (2019). Exosomes: biogenesis, composition, functions, and their role in pre-metastatic niche formation. *Biotechnology and Bioprocess Engineering*.

[46]   Kuate, S. *et al.* (2007). Exosomal vaccines containing the S protein of the SARS coronavirus induce high levels of neutralizing antibodies. *Virology*. https://doi.org/10.1016/j.virol.2006.12.011

[47]   Chehelgerdi, M. *et al.* (2023). Progressing nanotechnology to improve targeted cancer treatment: overcoming hurdles in its clinical implementation. *Molecular Cancer*. https://doi.org/10.1186/S12943-023-01865-0/TABLES/15

[48]   Sell, M. *et al.* (2023). Application of Nanoparticles in Cancer Treatment: A Concise Review. *Nanomaterials*. https://doi.org/10.3390/nano13212887

[49]   Williamson, P.J. (2016). Building and Leveraging Dynamic Capabilities: Insights from Accelerated Innovation in China. *Global Strategy Journal*. https://doi.org/10.1002/gsj.1124

[50]   Roco, M.C. (2007). National nanotechnology initiative: past, present, future. Handbook on nanoscience, engineering and technology. *Taylor and Francis*.

[51]   Gao, Y. *et al.* (2016). China and the United States—Global partners, competitors and collaborators in nanotechnology development. *Nanomedicine: Nanotechnology, Biology and Medicine*.

https://doi.org/10.1016/j.nano.2015.09.007

[52]   Heslop, L.A. *et al.* (2001). Development of a technology readiness assessment measure: The cloverleaf model of technology transfer. *The Journal of Technology Transfer.*

[53]   de Sousa Victor, R. *et al.* (2020). A Review on Chitosan's Uses as Biomaterial: Tissue Engineering, Drug Delivery Systems and Cancer Treatment. *Materials.* https://doi.org/10.3390/ma13214995

[54]   Kampers, L.F.C. *et al.* (2021). From Innovation to Application: Bridging the Valley of Death in Industrial Biotechnology. *Trends in Biotechnology.* https://doi.org/10.1016/j.tibtech.2021.04.010

[55]   Sun, J. *et al.* (2016). A Camera-Based Target Detection and Positioning UAV System for Search and Rescue (SAR) Purposes. *Sensors.* https://doi.org/10.3390/s16111778

[56]   Lu, Z.-R. and Sakuma, S. (2016). *Nanomaterials in pharmacology,* Springer.

[57]   Williams, R.J. *et al.* (2019). Models for assessing engineered nanomaterial fate and behaviour in the aquatic environment. *Current Opinion in Environmental Sustainability.* https://doi.org/10.1016/j.cosust.2018.11.002

[58]   Miller, G. and Scrinis, G. (2011). Nanotechnology and the extension and transformation of inequity. *Nanotechnology and the Challenges of Equity, Equality and Development.*

[59]   Schroeder, D. *et al.* (2016). Responsible, Inclusive Innovation and the Nano-Divide. *NanoEthics.* https://doi.org/10.1007/s11569-016-0265-2

[60]   Rambaran, T.F. (2022). A patent review of polyphenol nano-formulations and their commercialization. *Trends in Food Science & Technology.* https://doi.org/10.1016/j.tifs.2022.01.011

[61]   Iavicoli, I. *et al.* (2014). Opportunities and challenges of nanotechnology in the green economy. *Environmental Health.* https://doi.org/10.1186/1476-069X-13-78

[62]   Rambaran, T. and Schirhagl, R. (2022). Nanotechnology from lab to industry – a look at current trends. *Nanoscale Advances.* https://doi.org/10.1039/D2NA00439A

[63]   Özdoğan, E. *et al.* (2006). Nanoteknoloji ve tekstil uygulamalari. *Tekstil ve Konfeksiyon.*

[64]   Kut, D. and Güneşoğlu, C. (2005). Nanoteknoloji ve tekstil sektöründeki uygulamaları. *Tekstil&Teknik, Şubat.*

[65]   Türkant, B. and Akalın, M. (2006). Sanayi ve Moda için Nano Teknolojiler ve Akıllı Tekstiller. *İtkib AR&GE ve Mevzuat Şubesi Uluslar arası Konferans Raporu. Londra.*

[66]   Üreyen, M.E. (2006). Nanoteknoloji ve Türk tekstil ve hazır giyim sektörleri. *Bilim ve.*

[67]   Kozanoğlu, G.S. (2006). Elektrospinning yöntemiyle nanolif üretim teknolojisi.

[68]   Göksu, H. *et al.* (2020). Oxidation of Benzyl Alcohol Compounds in the Presence of Carbon Hybrid Supported Platinum Nanoparticles (Pt@CHs) in Oxygen Atmosphere. *Scientific Reports 2020 10:1.* https://doi.org/10.1038/s41598-020-62400-5

[69]   Göcek, İ. *et al.* (2007). Tekstil Endüstrisinde Nanoteknoloji Uygulamaları, Nonwoven. *Technical Textiles Techonology.*

[70]   Abrahamson, J.T. *et al.* (2013). Excess Thermopower and the Theory of Thermopower Waves. *ACS Nano.* https://doi.org/10.1021/nn402411k

[71]   Erkan, G. *et al.* (2005). Tekstil sektöründe nano-teknoloji uygulamaları. *Tekstil Teknolojileri ve Tekstil Makinaları Kongresi, Gaziantep.*

[72]   Günbatar, S. *et al.* (2018). Carbon-nanotube-based rhodium nanoparticles as highly-active catalyst for hydrolytic dehydrogenation of dimethylamineborane at room temperature. *Journal of Colloid and Interface Science.* https://doi.org/10.1016/J.JCIS.2018.06.100

[73]   Ayranci, R. *et al.* (2017). Carbon Based Nanomaterials for High Performance Optoelectrochemical Systems. *ChemistrySelect.* https://doi.org/10.1002/slct.201601632

[74]   Sen, F. *et al.* (2012). Observation of Oscillatory Surface Reactions of Riboflavin, Trolox, and Singlet Oxygen Using Single Carbon Nanotube Fluorescence Spectroscopy. *ACS Nano.* https://doi.org/10.1021/nn303716n

[75]   Cireli, A. *et al.* (2006). Tekstilde ileri teknolojiler. *Tekstil ve Mühendis.*

[76]   Alin, J. *et al.* (2015). Effect of the Solvent on the Size of Clay Nanoparticles in Solution as Determined Using an Ultraviolet–Visible (UV-Vis) Spectroscopy Methodology. *Applied Spectroscopy.* https://doi.org/10.1366/14-07704

[77]   Hojjati-Najafabadi, A. *et al.* (2023). Bacillus thuringiensis Based Ruthenium/Nickel Co-Doped Zinc as a Green Nanocatalyst: Enhanced Photocatalytic Activity, Mechanism, and Efficient H2Production from Sodium Borohydride Methanolysis. *Industrial and Engineering Chemistry Research.* https://doi.org/10.1021/ACS.IECR.2C03833

[78]   Wu, Y. *et al.* (2023). Hydrogen generation from methanolysis of sodium borohydride using waste coffee oil modified zinc oxide nanoparticles and their photocatalytic activities. *International Journal of Hydrogen Energy.* https://doi.org/10.1016/J.IJHYDENE.2022.04.177

[79]   Hojjati-Najafabadi, A. *et al.* (2022). A Tramadol Drug Electrochemical Sensor Amplified by Biosynthesized Au Nanoparticle Using Mentha aquatic Extract and Ionic Liquid. *Topics in Catalysis.* https://doi.org/10.1007/s11244-021-01498-x

[80]   Han, B. *et al.* (2011). Multifunctional and Smart Carbon Nanotube Reinforced Cement-Based MaterialsMultifunctional and smart carbon nanotube reinforced cement-based materials, in *Nanotechnology in Civil Infrastructure: a paradigm shift*, Springer, pp. 1–47.

[81]   Saafi, M. (2009). Wireless and embedded carbon nanotube networks for damage detection in concrete structures. *Nanotechnology.*

[82]   Veedu, V.P. (2010). Multifunctional cementitious nanocomposite material and methods of making the same.

[83]   Bartos, P.J.M. (2009). Nanotechnology in construction: a roadmap for development, in *Nanotechnology in Construction 3: Proceedings of the NICOM3*, Springer, pp. 15–26.

[84]   Chaudhry, Q. (2009). Nanotechnology for food applications: current status and consumer safety concerns. *AAAS Annual Meeting: Nanofood for Healthier Living.*

[85]   Joseph, T. and Morrison, M. (2006). *Nanotechnology in agriculture and food: a nanoforum report*, Nanoforum. org.

[86]   Dursun, S. *et al.* (2010). Dogal biyopolimer bazli (biyobozunur) nanokompozit filmler ve su ürünlerindeki uygulamalari. *Journal of FisheriesSciences. com.*

[87]   Sorrentino, A. *et al.* (2007). Potential perspectives of bio-nanocomposites for food packaging applications. *Trends in food science & technology.*

[88]   Donkor, L. *et al.* (2023). Bio-based and sustainable food packaging systems: relevance, challenges, and prospects. *Applied Food Research.* https://doi.org/10.1016/j.afres.2023.100356

[89]   Mahalik, N.P. and Nambiar, A.N. (2010). Trends in food packaging and manufacturing systems and technology. *Trends in food science & technology.*

[90]   Şen, B. *et al.* (2017). Bimetallic PdRu/graphene oxide based Catalysts for one-pot three-component synthesis of 2-amino-4H-chromene derivatives. *Nano-Structures & Nano-Objects.* https://doi.org/10.1016/J.NANOSO.2017.08.013

[91]   Dursun, S. (2009). The use of edible protein films in seafood. *Journal of FisheriesSciences.com.* https://doi.org/10.3153/jfscom.2009040

[92]   SAKA, E. and TERZİ GÜLEL, G. (2015). Gıda Endüstrisinde Nanoteknoloji Uygulamaları. *Etlik Veteriner Mikrobiyoloji Dergisi.* https://doi.org/10.35864/evmd.513387

[93]   Şavk, A. *et al.* (2019). Highly monodisperse Pd-Ni nanoparticles supported on rGO as a rapid,

sensitive, reusable and selective enzyme-free glucose sensor. *Scientific Reports.*
https://doi.org/10.1038/s41598-019-55746-y

[94]   Korkmaz, N. *et al.* (2020). Biogenic nano silver: Synthesis, characterization, antibacterial, antibiofilms, and enzymatic activity. *Advanced Powder Technology.*
https://doi.org/10.1016/J.APT.2020.05.020

[95]   KAMÇI, H. *et al.* (2022). Biogenic Synthesis of Zinc Oxide Nanoparticles Using Saponaria officinalis L., Characterisation and Antibacterial Activities. *European Journal of Science and Technology.* https://doi.org/10.31590/ejosat.1075292

[96]   Bazancir, N. and Meydan, I. (2022). Characterization of Zn nanoparticles of Platonus orientalis plant, investigation of DPPH radical extinquishing and antimicrobial activity. *Eastern Journal of Medicine.*

[97]   Meydan, I. *et al.* (2022). Arum italicum mediated silver nanoparticles: Synthesis and investigation of some biochemical parameters. *Environmental Research.*
https://doi.org/10.1016/J.ENVRES.2021.112347

[98]   Altındağ, F. and Meydan, İ. (2021). Evaluation of protective effects of gallic acid on cisplatin-induced testicular and epididymal damage. *Andrologia.* https://doi.org/10.1111/and.14189

[99]   Meydan, İ. and Seçkin, H. (2021). Green synthesis, characterization, antimicrobial and antioxidant activities of zinc oxide nanoparticles using Helichrysum arenarium extract. *International Journal of Agriculture Environment and Food Sciences.* https://doi.org/10.31015/JAEFS.2021.1.5

Biogenic Nanomaterials: Synthesis, Characterization, Applications, and Future Remarks   Materials Research Forum LLC
Materials Research Foundations 180 (2025)   https://doi.org/21741/9781644903759

Chapter 25

Conclusion, and Future Prospective of Biogenic Nanomaterials

E. Okumus1*, F. Sen[2]*

[1]Department of Food Engineering, Van Yuzuncu Yil University, Van, Turkiye.

[2]Sen Research Group, Department of Biochemistry, Faculty of Arts and Science, Kutahya Dumlupinar University, 43000, Kutahya, Turkiye,

emineokumus@yyu.edu.tr; fatih.sen@dpu.edu.tr

Abstract

Nanomaterials/nanoparticles can be synthesized using physical, chemical and biological methods. Biological synthesis of nanomaterials involves the use of plant-based materials (fruit, flower, leaf, seed, stem), molecules (phenolics, flavonoids, proteins, vitamins, organic acid, polysaccharides, terpenoids), microorganisms (bacteria, algae, fungi), enzymes (cholesterol esterase, glycerol kinase, lipase) and polymers (DNA). Biogenic nanomaterials (BNMs) have received increasing attention in recent years due to their many advantageous usage possibilities such as non-toxicity, high stability, improved solubility, therapeutic efficacy, antimicrobial effects and sustainable properties. Today, biogenic nano-research has a wide range of applications offering numerous advantages for daily use in various fields such as chemistry, medicine, pharmacology, agricultural applications, food, packaging and wastewater management.

In this section, the situation assessment was made and inferences were made by briefly summarizing the BNMs, whose synthesis, use, effects and applications were explained in detail in the other sections. In addition, a roadmap for the future was created by mentioning the most common problems and bottlenecks in applications related to BNMs.

Keywords

Biogenic Nanomaterials, Conclusions and Implications, Future Prospective

1.   Conclusions and Implications

Research in nanomaterials has implications for many biomedical applications, including diagnostics, drug delivery systems, biosensing, and bioimaging. The synthesis of environmentally friendly, inexpensive and highly efficient BNMs, which can be used for therapeutic purposes in pharmacology and has negligible side effects, is very important. Adjusting the optimum dose of drugs, increasing their effectiveness and reducing their existing side effects are the main targets in the use of BNMs. Controlled drug release to remotely control tumour cells and cure diseased cells is now possible with BNMs. Similarly, changes in vital organs and hormonal changes can be easily detected by biomarkers in the early diagnosis and treatment of diseases.

Although developments in pharmacology continue at full speed, a notable ration of deaths globally are caused by cancer. Compared with conventional therapy, nanotechnology is effective in controlling drug release, selectively targeting release and increasing the efficacy of immunotherapy. In this way, it is possible to prolong the survival time and increase the quality of life in cancer patients without the use of toxic components with BNMs. Biogenic nanoparticles have been developed that have a high degree of immunogenicity and can carry therapeutic drugs to enhance antitumor immunity with low side effects. Thanks to their chemotherapeutic drug delivery and tumour cell specific targets, BNMs can promote apoptotic cell death and prevent tumour cell proliferation without damaging healthy cells. However, the use of BNMs in treatments and vaccines has some limitations.

Because of the widespread emergence and proliferation of antibiotic-resistant bacteria, the therapeutic benefits of antibiotics are diminishing. In addition, the continued evolution of multidrug-resistant pathogens necessitates the development of innovative antimicrobial agents for the detection and treatment of bacterial infections. Effective inhibition of drug-resistant bacterial growth by nano-antibiotics shows promise for the use of BNMs. However, the disadvantage of BNMs is their toxicity potential as they are dangerous at high concentrations and can cause various health problems and unpredictable changes in blood cells and various ecological problems. Therefore, understanding both the molecular toxicity processes of BNMs and their interaction with the body is essential for use in clinical applications.

Membrane technology in water filtration is the most common and most effective method in both residential and commercial applications. Water pollution caused by various organic and inorganic pollutants has become an important problem all over the world due to global warming. Environmental and biogenic applications of nanoparticles are promising in wastewater treatment. BNMs can be used in sewage systems, treatment plants, membrane bioreactors and water treatment devices to reduce or eliminate hazardous contaminations in water resources. BNMs can be produced sustainably, relatively inexpensively, energy-efficiently, and are ecologically safe due to their biologically renewable nature. Thanks to its high adsorption capacity, reactivity and sensitivity, it plays an important role in decontamination protocols for drinking and industrial wastewater. Compared with traditional purification methods, the most important advantages of nanomaterials include their ability to integrate various properties. Large surface areas also allow for increased efficiency with membranes that provide both particle retention and removal of pollutants.

Among the nanotechnology applications in the food industry are smart packaging, extending the average shelf life of food products, increasing food durability and bioavailability, and reducing nutrient losses. BNMs synthesized from various microorganisms in effective food preservation have critical importance in the controlled realization of the entire process from production to refining, packaging and storage. Despite its numerous benefits, there are some uncertainties regarding the use of BNMs in food packaging, particularly their accumulation in the human body and other risks to the environment. For example, smart BNMs used in food packaging may cause health complications. Considering the infrastructure available in widely used packaging processes, the synthesis and processing of BNMs is not cost-effective. In addition, developments in biogenic packaging are affected by food regulations, policies and the global energy resources.

Therefore, more research should be done to increase the success of biogenic packaging materials and to expand their use in various fields.

The rapid increase in the world population necessitates a significant increase in agricultural production in order to meet the food needs. The use of BNMs instead of synthetic agricultural chemicals, which are highly toxic to human health and the environment, can efficaciously protect plants against phytopathogens and contribute to sustainability by following an environmentalist approach. Thanks to this practice for sustainable agriculture, costs can be reduced and crop yields can be increased. In addition to these mentioned advantages, one of the biggest benefits of biogenic fertilizers for the plant is the gradual release of nutrients during the growth period of the plant. The slow release of nutrients improves crop quality and yield, while contributing to crop nutritional health, food safety and environmental quality.

Fossil fuels require alternative methods in long-term energy production because of their limited amount and non-renewable nature. Therefore, it is crucial to research and develop green technologies that are not only renewable but also cost-effective. Nanoparticles are critical in photocatalytic applications for energy technologies, thanks to their large surface area and unique optical and catalytic properties. Nanoparticles have increasing applications in improving the performance of batteries, fuel cells and solar cells, drilling fluids in the oil and gas industry, enhanced oil recovery and well excitations. These applications allow the use of nanoparticles with their high reactivity due to their small size and large volume / surface area ratio. Despite these advantages, polydispersity in biogenic synthesis is an important parameter that limits nanoparticle size, morphology, distribution and controlled use. Therefore, studies of BNMs should focus on methods that can result in monodisperse synthesis of nanoparticles.

Because the use of BNMs technology is quiet comparatively new, the potential of many natural materials such as many biopolymers, microorganisms, or synthesis from waste materials has not been extensively investigated. In addition to synthesis, *in vivo* studies should be increased to determine the antioxidant effects of BNMs and to examine their therapeutic effects on DNA damage. Although the synthesis of BNMs involves natural resources, has a broad therapeutic effect, and is low cost, there are a number of important domain-based and general issues and technical challenges that need to be addressed by researchers to improve their use. Listed below are the most common problems and future prospects in the synthesis and application of these nanomaterials.

- Purification and characterization of bioactive compounds used as reducing and capping agents is as yet only successful on a laboratory scale. For industrial-grade synthesis, several factors such as technology transfer, optimization of variable process parameters, and BNMs quality need to be re-evaluated.

- Controlling the physicochemical properties of BNMs such as homogeneity of size and shape, equal stability levels, analysis and visualization is the main problem.

- In synthesizing BNMs, there are uncertainty, lack of logical strategy, and legal deficiencies for safe clinical practice.

- Data from *in vivo* animal studies are very limited. New research is needed to simulate and predict the *in vivo* performance of the nanoparticles to be used in humans.

- Inaccurate information or incomplete characterization of synthesized nano-drug products can lead to failures in subsequent releases, so a detailed evaluation of the manufacturing processes is essential before anticipating any regulatory approval and commercialization.

- Studies on the use of BNMs in drugs and medical interventions to be applied to people in pregnancy, old age and childhood are insufficient.

- The biggest challenge in the use of BNMs in the treatment of cancer is their co-delivery of tumour antigens and adjuvants to the secondary lymphoid system.

- Increasing the protective response of BNMs against cancer cells, designing new antigens and reducing their toxicity are among the challenges in cancer immunotherapy.

- The slow biosynthesis process, low purity and complex microbial metabolism of BNMs cause serious limitations in practical applications.

- There is a need for large-scale and successful clinical studies in the synthesis of plant-derived nanoparticles that can be used in patient-specific cancer treatments.

- It is important to use animal models that closely mimic the heterogeneity and anatomical histology of human tumours in obtaining accurate results from studies.

- Factors affecting the controlled release of biogenic nano-drugs, especially in cancer treatment, should be well evaluated and the pathways they follow after release should be determined in detail.

- Physiologically appropriate research methods should be designed that can mimic complex human physiology and eliminate the need for animals in *in vivo* studies.

- Clinical success of BNMs is closely related to overcoming chemical and biological barriers such as toxicity, biodistribution, pharmacokinetics, pharmacodynamics, as well as reliable industrial scale production and reproducibility.

- Size control, stability, aggregation and sedimentation studies are still ongoing for commercial vehicles of BNMs in wastewater treatment.

- Toxicity and biosafety studies of BNMs' potential applications in wastewater treatment and water treatment are needed.

- Residues attached to the surface as a result of BNMs adsorbing pollutants from waterways may accumulate in a long time and may pose a risk by being released into the environment.

- There is a need for detailed studies that increase the adsorption capacity and selectivity of biogenic nano catalysts and nanomaterials, and facilitate their sustainability and recyclability.

- Although BNMs are simple to manufacture and environmentally friendly, the effects of reaction parameters and stability conditions must be optimized because they affect their pollutant removal performance.

- In order for the synthesized BNMs to be used in water treatment systems, high purity isolation is required. However, it is important to make cost calculations in the purification process.

- Realistic approaches should be followed by considering environmental conditions and variability levels in evaluating the performance of nano structures designed at laboratory scale.

- Various national and international regulations and laws that determine the use of biogenic nanostructures in water treatment should be established in order to prevent health problems.

- Biogenic nano-fertilizers increase nutrient release and uptake, but their effects on soil microbiota and ecosystem functioning are not yet known.

- Overuse or misuse of BNMs can have negative consequences on the food chain.

- There is a need for extensive research on nanoparticle accumulation and persistence in soil, soil fertility and biodiversity after use of biogenic nano fertilizers.

- Considering the absorption rate and accumulation status of BNMs by plants, their potential effects on agro-ecosystems and food chain should be comprehensively evaluated and long-term toxicity tests should be performed. According to the test results, the usage status, application frequency and application dose of biogenic nano fertilizers should be determined and the results obtained should be protected by legal regulations.

- Before commercialization and widespread use of biogenic nano-fertilizers, their interactions with the soil, their possible effects on plant physiology and human health should be properly investigated.

- The use of biogenic nano fertilizers may vary depending on many factors such as the type of plant grown, soil structure, climatic characteristics (precipitation, altitude, etc.), size, concentration, composition and chemical properties of nanomaterials. For this reason, it is necessary to determine the nutrient release process in greenhouse studies where different plant species are used in order to determine the synthesis source, usage rates, application period and application method of BNMs.

- The effects of BNMs will also change the plant's exposure to nanoparticles at different growth and development stages, depending on varying absorption rates. Considering this situation, the necessary conditions should be optimized.

- More research needs to be done to create databases and gather information in previously unexplored areas before BNMs can be widely applied in sustainable agriculture.

- Research on the scalability and cost-effectivity of production is important for using BNMs in plant disease control.

- It is important to elucidate the antimicrobial mechanisms underlying the action of BNMs against phytopathogens.

- A multidisciplinary approach including metabolomics, genomics, proteomics and transcriptomics should be used when evaluating the impact of BNMs. Ideas and collaborations of scientists and researchers from universities and industry are needed to realize agricultural innovation and knowledge transfer.

- Long-term studies should be conducted to determine the effects of BNMs on plant ecosystems as well as human and animal health.

- Complications in the structure of BNMs, different shapes, sizes, reactivity and electrical charges make it difficult to define the biological safety of nano pesticides. Therefore, it becomes difficult to identify and predict important aspects of genotoxicity and cytotoxicity.

- Although various techniques are used to improve pesticide pollution in the soil, the potential for use in this area is quite high by adapting nanoparticles to the bioremediation process.

- Inadequate understanding of synthesis mechanisms in many biosynthesis processes, especially microbial synthesis, makes it difficult to use biogenic agents appropriately for nanoparticle synthesis. This makes the synthesis methods unpredictable even for the production of the same material, resulting in a waste of time and resources in synthesis.

- Proven theorems and laws should be established in order to develop predictable synthesis methods. This is possible with the further development of accurate nano-analysis and detection methods to study the kinetics of reactions.

- BNMs synthesis is still mostly limited to laboratory scale. This situation causes changes from batch to batch in synthesis, especially in size and shape, and large quantities of homogeneous nanoparticles cannot be produced. Controlling the flow rate, viscosity, surface tension, mixing time and location of the reactants and selecting the appropriate solvents hinder the continuous synthesis process. Optimization of the necessary parameters and economic feasibility reports are required for the production and application of nanoparticles on an industrial scale.

- There is no data yet on the toxicity of BNMs. Therefore, there is still no mechanism for the proper storage and handling of nanomaterials without any limitations.

- Data on the direct and indirect release of BNMs to the environment are not sufficiently determined.

- The lack of studies on the selection of plant species in plant use for biosynthesis and the determination of plant species-related toxicity and bioactive components is an important problem. However, the reaction and optimum stoichiometric aspects such as plant-derived extracts and precursor concentration regarding the reproducibility of the synthesized nanoparticles have not been extensively investigated.

- Synthesis parameters such as pH, solvent used, extraction method, temperature, reaction time on the size, shape and morphological control of BNMs have not been extensively studied.

- There are many studies on plant-derived nanoparticles in the synthesis of BNMs. However, plant variations prevent the use of raw materials in global production. The fact that the same plant species does not grow in a different geography, and even if it does, does not have the same bioactive properties, it also affects the synthesis process and synthesis products.

- The natural complexity of plants is a barrier to the industrial production of plant-mediated BNMs. The phytochemicals in the plant composition to be used in the synthesis are greatly affected by external factors such as abiotic environmental factors, cultivar and mutagenesis. This negatively affects the homogeneity and reproducibility of nanomaterial synthesis and causes variations between syntheses.

- Another disadvantage in using the plant source for the synthesis of BNMs is the diversity of phytochemicals in the plant system. This can be resolved by collaborating with molecular science techniques such as genetic engineering to maximize the most dominant phytochemicals in the target plant. In this way, it can be ensured that the phytochemical composition and content in the plant are preserved and the synthesis takes place in a controlled manner.

- Priority should be given to studies aimed at increasing the low efficiency and nanoparticle conversion rate in synthesis using metal ions.

- BNMs are insufficient in removing toxic metal mixtures from the structure. New synthesis methods should be tried to solve this problem in applications.

- In the synthesis of plant-derived BNMs, plant extracts must be stored until needed. This situation causes an increase in energy use and occupying space in cold storage rooms. In addition, changes in the chemical structure of the stored plant should also be considered.

- The reaction time should be fast as it affects both the production productivity and the cost in the synthesis process. However, a rapid extraction procedure is equally unsuitable. The optimal reaction time for the source and coating agent used in the synthesis should be determined for each synthesis parameter.

- BNMs must be checked for their chemical properties and thermal behavior for a long time before they can be commercialized. Clumping of nanoparticles can lead to changes in their potential properties, leading to undesirable cell-nano interactions. Therefore, controlling the nanoparticle cluster is important in terms of increasing bioavailability and safety.

- Another important point is the conversion and reusability of BNMs. In this regard, studies should be carried out on the selection of resource and coating agents that reduce possible releases to the environment and can be recycled.

- Future study should focus on improving consumer acceptance through guidance and information sharing on the toxicological and ecotoxicological properties of nanoparticles, the development of the appropriate regulatory framework, and safety requirements.

- Future discoveries and improvements of biogenic synthesis, particularly in the health and environmental field, should be geared towards translating laboratory-based work to industrial scale.

- BNM, which can be used as 'nano drugs' in the medical field in the near future, is expected to be a potential solution to the current energy crisis in energy driven devices.

- It is foreseen that nanomaterials produced by biogenic synthesis will be widely applied in other important fields such as tissue engineering, nutrition and ornamental industries.

- Future research for the implementation and industrial scale production of BNMs is likely to benefit from machine learning. In this way, machine learning algorithms enable result prediction, specification of nanoparticles and experiment planning.

- In the future, it can be expected that researchers will be able to design stimuli-sensitive BNMs with the ability to perform multimodally, thanks to the development of materials, pharmaceutical and biomedical sciences, and the combination of nano-engineering and smart chemistry.

- New studies should focus less on investigating the primary phytochemicals or organic groups used in the synthesis of BNMs, and more attention should be paid to examining the role of the organic group in causing deviations in characterization and increasing efficacy for their therapeutic applications.

- Future research into the scale-up application of BNMs should focus on understanding the synthesis mechanisms, identifying more biogenic substances that can be used in synthesis, and optimizing the factors influencing the synthesis process.

- BNMs are highly stable and active when they can be monodispersed by varying process parameters such as temperature, pH, mixing ratio and incubation time. In order to benefit from these advantageous features, priority should be given to the optimization of the applied process parameters.

- Future studies are expected to focus on the synthesis of biogenic nanomaterials with a low cost and biocompatible improved pharmacokinetic model and therapeutic efficacy to prevent drug-resistant infections and control cancer.

- Effective collaborative work between stakeholders from industrial/academic R&D, experts in the health system, regulatory agencies and society is required to meet current challenges.

- New research planned should focus not only on the detection of specific cancer cells/bacteria and their metabolites, but also on the delivery of antibiotics and other drugs to the site of infection.

- In the future, more extensive research should be conducted primarily in acute and chronic toxicity tests, as well as the development of effective BNMs in preclinical and clinical trials in cancer theragnostic.

- The unsuitability of BNMs for large-scale processes makes it difficult for them to compete with traditional treatment methods. However, in the near future, safe and abundant nano-engineering materials offer great innovation potential, especially for decentralized treatment systems and highly degradable pollutants.

- Future studies should focus on explaining the behaviour of membranes used in water treatment in natural wastewater systems with different microorganisms and various organic and inorganic pollutants.

- The application of nanobiotechnology in the agricultural sector is important in terms of ensuring sustainability. Future research will focus on the development of R&D laboratories to advance nanobiotechnology applications and move to industrial scale production.

- It is expected that the planned new researches will focus on the long-term effects of nanoparticles on soil bacteria and in the agricultural process, as well as on the interactions between use doses and xenobiotics.

Considering all the problems, difficulties and limitations mentioned above, future-oriented BNMs synthesis and applications have significant potential for use in the aforementioned areas and in many new sectors and are promising.